From Galaxies to Turbines

From Galaxies to Turbines

Science, Technology and the Parsons Family

W Garrett Scaife

CRC Press is an imprint of the
Taylor & Francis Group, an **informa** business

CRC Press
Taylor & Francis Group
6000 Broken Sound Parkway NW, Suite 300
Boca Raton, FL 33487-2742

First issued in paperback 2019

ISBN-13: 978-0-7503-0582-2 (hbk)
ISBN-13: 978-0-367-40005-7 (pbk)

Library of Congress Cataloging-in-Publication Data

Catalog record is available from the Library of Congress

Visit the Taylor & Francis Web site at
http://www.taylorandfrancis.com

and the CRC Press Web site at
http://www.crcpress.com

Contents

Preface

My interest in the achievements of the Parsons family dates back to the 1960s when I began teaching in the Engineering School at Trinity College Dublin. Each day as I went into the stylish Museum building to reach the lecture theatres, I passed a number of glass cases in the entrance hall. One of these housed 'Steam turbo-generator N° 5, 1885', presented by Gerald G Stoney BA, BAI. Time passed, and as 1984, the centenary year of Charles Parsons' invention, approached I became increasingly interested to know exactly how this had been conceived. Also, among other items which the Engineering School possessed was a model of the 3 ft reflecting telescope built by Charles' father William, the third Earl of Rosse at Birr Castle, and this alerted me to the importance of Birr in the story. While preparing for the International Parsons Steam Turbine Conference, which was held in Dublin, several questions came to mind. Why was the turbo-generator invented in 1884, and not 1834 or 1934 for example? Why was it Charles Parsons that accomplished this? How did it happen that its development took place in Newcastle upon Tyne and not in Belfast or Dublin? The centenary was marked in 1984 by a conference at Trinity College Dublin and by an Exhibition at the Heaton Works of C A Parsons and Co in Newcastle upon Tyne, but the lack of any nationwide celebration in Great Britain of this most significant of engineering achievements only seemed to confirm the correctness of the title which Joe Clarke chose for his 1984 booklet, 'An almost forgotten great man, Charles Parsons'. As I worked with the artefacts and archival material, an awareness of the personalities of the two great men, William and Charles Parsons, began to emerge. My hope is that I can convey something of that to the reader. In the nature of the story, technical details have an essential place, and I have tried to describe them in a way that will satisfy the reader who is not an expert. At the same time this is not an attempt to write a complete history of the steam turbine.

In this effort I have become heavily indebted to two descendants of William Parsons who have each helped me in a special way. William

Brendan Parsons, Lord Rosse, has cared for the valuable collection of family papers at Birr Castle and has just overseen the successful restoration of the great 6 ft telescope, the instrument which revealed for the first time the spiral structure of certain nebulae. His cousin, Norman Clere Parsons, is a great-grandson of the astronomer and has spent his working life with Reyrolle Parsons, eventually sitting as Chairman on its Board. He has been a living link for me with Charles Parsons and the industry which he founded. Both have given me great support and encouragement.

There are three principal collections of documents from which I have drawn: the first is the Rosse Archive held by the Birr Scientific and Heritage Foundation, which has been catalogued by A P W Malcomson of the Public Records Office of Northern Ireland; the second is the Parsons Archive held by the Science Museum Library in London and which was catalogued by Janet Fyffe and Robert Sharp; and the third is the collection held by the Tyne and Wear Archive Service where my guide was R Potts. I am grateful to Lord Rosse, the Trustees of the Science Museum, London, and to the Tyne and Wear Archives Service for permission to publish material from their collections. Also, an important group of letters from Charles Parsons to his elder brother is held in the Library of St John's College Cambridge, and I am grateful to the Master and Fellows of St John's College for their permission to refer to this material. A revealing series of documents chronicling dealings with an early licensee of Charles Parsons was kindly made available to me by Andrew Strang of GEC Alstom at Rugby. The project could never have been completed without ready access to the Copyright Library at Trinity College Dublin which seems to contain most of the engineering books and periodicals which are relevant to the period, and I must express my gratitude to the many staff members who have helped me over the years. Finally, I am indebted to Mr J F Clarke whose publications have identified so much of the history of industry around Newcastle and the North East Coast of England, and who has drawn my attention to new pieces of information. Patrick Moore's book 'The Astronomy of Birr Castle' tells of the work of William and his son Laurence, and helps to set it in context. A great deal has been written about Charles Parsons and his inventions. The source books are 'The Parsons Turbine' by Alex Richardson, 'The Development of the Parsons Steam Turbine' by R H Parsons and Rollo Appleyard's biography 'Charles Parsons—his life and work', and I acknowledge permission from the publishers to make use of these. An eloquent and enduring testimony to the high regard in which Charles Parsons is held by scientists and engineers alike is the annual series of Charles Parsons Commemorative Lectures which has been organized by the Royal Society since 1936. The lectures have been presented in turn at the Institution of Civil Engineers, of Mechanical Engineers, of Electrical Engineers, of Marine Engineers, of Naval Architects, at the Institute of Physics, the North East Coast Institution of Engineers and Shipbuilders and the Institute of Engineers and Shipbuilders in Scotland. Taken together they form a valuable source of information and comment.

Museums have a special importance for projects like this, and as a child I spent endless hours in the Science Museum in London. More recently I have been indebted to staff both there and at the Discovery Museum in Newcastle upon Tyne for permission to examine the many artefacts in their care. I am also grateful to Professor J D Jackson for access to Osborne Reynolds' turbine and pumps, and for sight of a letter written by Reynolds' daughter. The recent restoration of the giant telescope by the Birr Scientific and Heritage Foundation has brought to life an enterprise which had gradually become just a distant memory. I had valuable discussions about the workings of the giant telescope with the late Professor Patrick Wayman of Dunsink Observatory, and the consulting engineer Michael Tubridy who directed the restoration. In the task of analysing and depicting the prototype steam turbo-generator with the aid of the 'N° 5 Turbo-Generator', I have enjoyed the help of two colleagues, C G Lyons and A Reid.

As the details of the story gradually unfolded I have been fortunate to have been able to discuss them with my brothers Desmond and Brendan. Finally, a long drawn-out venture such as this has been can make heavy demands on others, and I want to acknowledge this by dedicating this work to my wife Eileen and our children.

W Garrett Scaife
September 1999

Notes on Units and Abbreviations

1 lb = 0.454 kg

12 in (inches) = 1 ft (foot), 39.4 in = 1 m

1 knot is 1 nautical mile per hour = 1.15 mph = 0.524 m/s

rpm, revolutions per minute

1 psi (pound per square inch), also 1 lb/in^2 = 0.068 atm (atmospheres) = 690 Pascal

1 psig (psi gauge), a pressure measured above atmospheric

1 psia (psi absolute), a pressure measured above an absolute vacuum

1 in Hg (1 inch of mercury in barometer) = 0.49 psia

1 in wg (1 inch water gauge) = 0.036 psia

1 Btu (British thermal unit) = 1.055 kJ (kiloJoule), unit of energy

1 HP h (horsepower hour), a unit of energy = 0.746 kW h (kiloWatt hour)

1 HP (horse power) = 0.746 kW, unit of work, energy per unit time

IHP (indicated horse power), quantity inferred from measurements of pressure inside a steam engine cylinder

BHP (brake horse power), sometimes SHP, shaft horse power, quantity of power delivered to output shaft of engine. Typically 0.85 ihp = bhp

1 MW, megawatt = 1,341 HP

Currency, £1 = 20 shillings = 240 d (pennies)

Conventions:

Quotations in italics. *( )* brackets in quoted text, but ( ) implies a modification of the text made by this author for clarification.

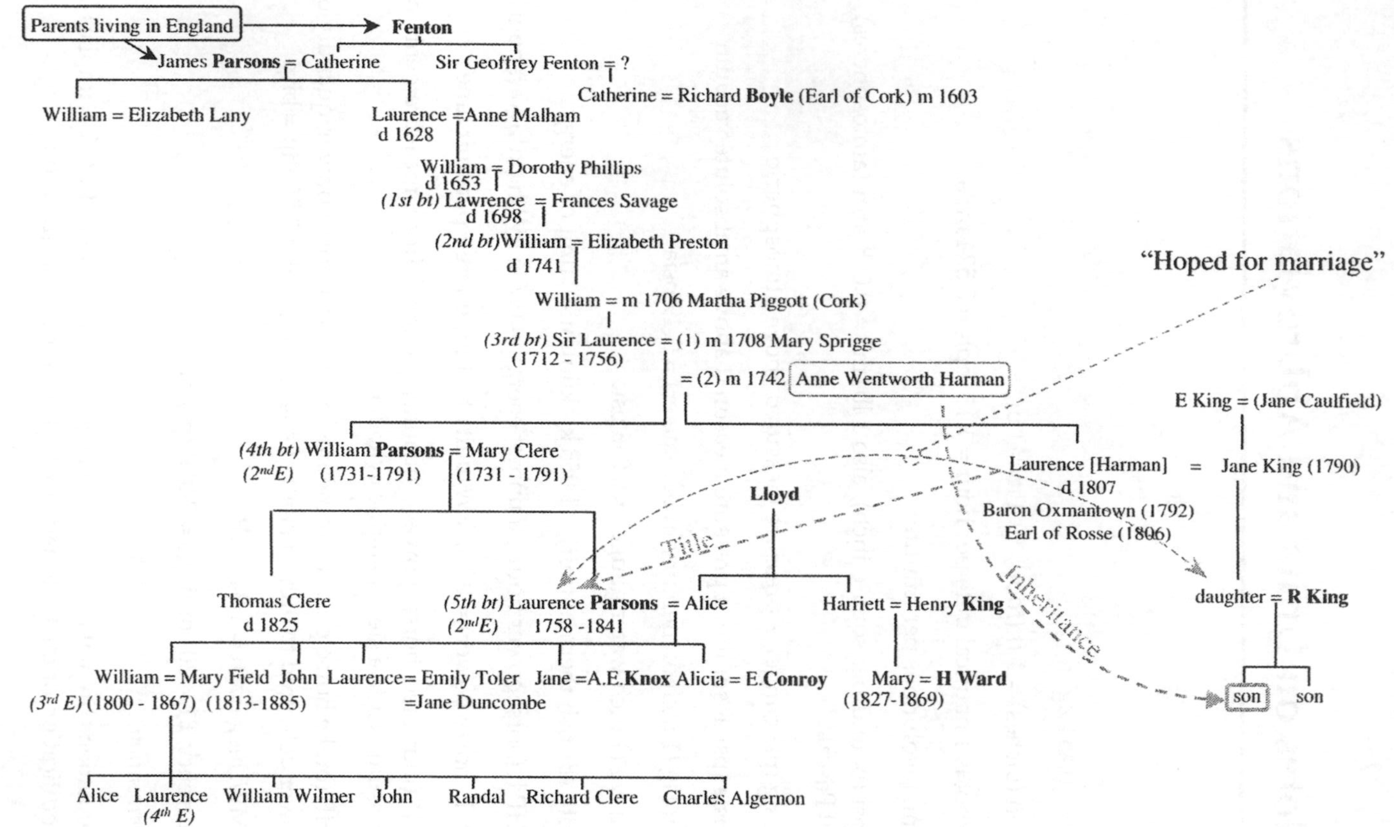

Parents living in England
Fenton
James Parsons = Catherine
Sir Geoffrey Fenton = ?
Catherine = Richard Boyle (Earl of Cork) m 1603
William = Elizabeth Lany
Laurence = Anne Malham
d 1628
William = Dorothy Phillips
d 1653
(1st bt) Lawrence = Frances Savage
d 1698
(2nd bt) William = Elizabeth Preston
d 1741
William = m 1706 Martha Piggott (Cork)
(3rd bt) Sir Laurence = (1) m 1708 Mary Sprigge
(1712 - 1756)
= (2) m 1742 Anne Wentworth Harman
"Hoped for marriage"
E King = (Jane Caulfield)
(4th bt) William Parsons = Mary Clere
(2nd E) (1731-1791) (1731 - 1791)
Laurence [Harman] = Jane King (1790)
d 1807
Baron Oxmantown (1792)
Earl of Rosse (1806)
Lloyd
Title
Inheritance
Thomas Clere
d 1825
(5th bt) Laurence Parsons = Alice
(2nd E) 1758 -1841
Harriett = Henry King
daughter = R King
William = Mary Field
(3rd E) (1800 - 1867) (1813-1885)
John
Laurence = Emily Toler
=Jane Duncombe
Jane = A.E. Knox
Alicia = E. Conroy
Mary = H Ward
(1827-1869)
son
son
Alice
Laurence
(4th E)
William Wilmer
John
Randal
Richard Clere
Charles Algernon

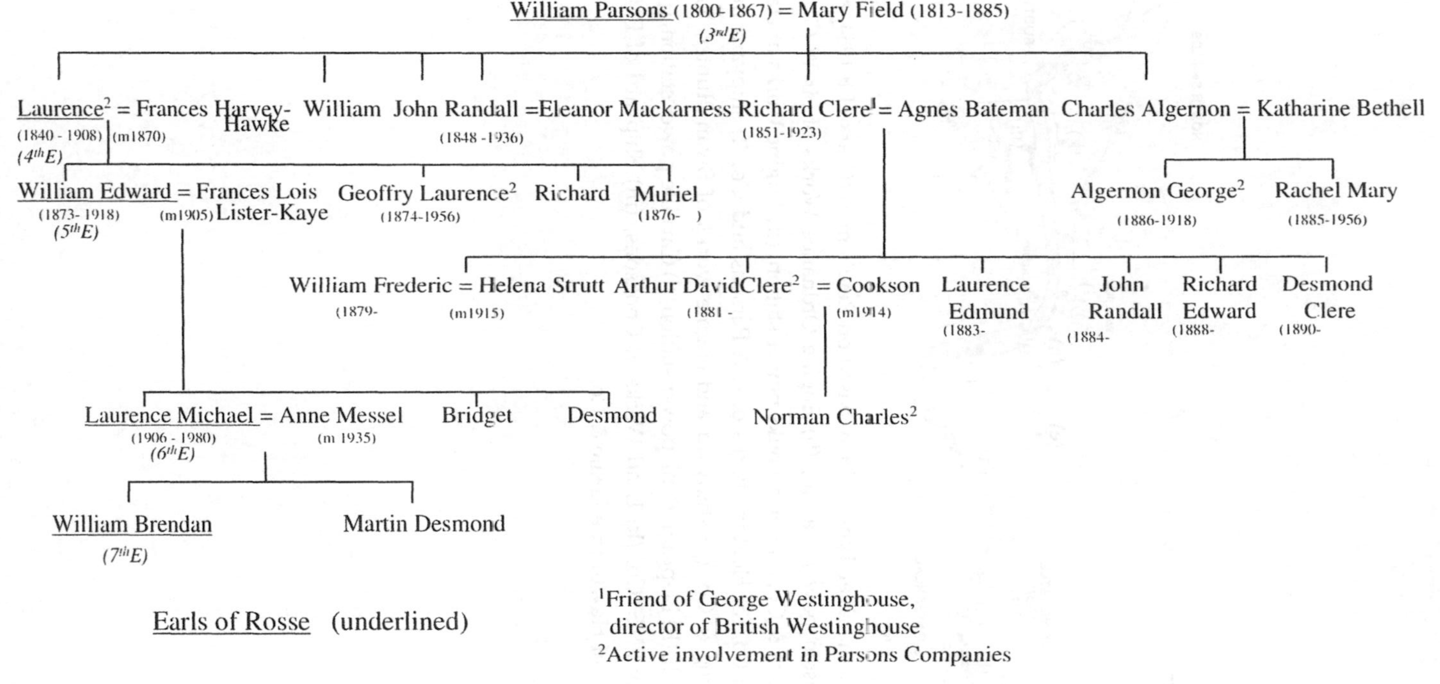
William Parsons (1800-1867) = Mary Field (1813-1885)
(3rd E)
Laurence[2] = Frances Harvey-Hawke
(1840 - 1908) (m1870)
(4th E)
William
John Randall = Eleanor Mackarness
(1848 -1936)
Richard Clere[1] = Agnes Bateman
(1851-1923)
Charles Algernon = Katharine Bethell
William Edward = Frances Lois Lister-Kaye
(1873- 1918) (m1905)
(5th E)
Geoffry Laurence[2]
(1874-1956)
Richard
Muriel
(1876-)
Algernon George[2]
(1886-1918)
Rachel Mary
(1885-1956)
William Frederic = Helena Strutt
(1879- (m1915)
Arthur David Clere[2] = Cookson
(1881 - (m1914)
Laurence Edmund
(1883-
John Randall
(1884-
Richard Edward
(1888-
Desmond Clere
(1890-
Laurence Michael = Anne Messel
(1906 - 1980) (m 1935)
(6th E)
Bridget
Desmond
Norman Charles[2]
William Brendan
(7th E)
Martin Desmond
Earls of Rosse (underlined)
[1]Friend of George Westinghouse, director of British Westinghouse
[2]Active involvement in Parsons Companies

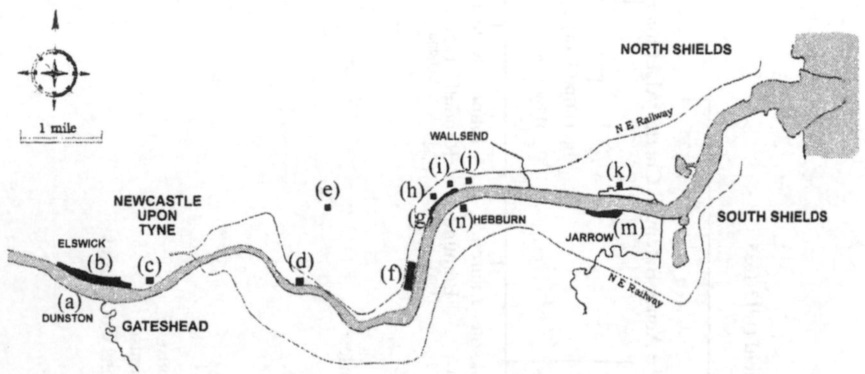

A map of the river Tyne in 1916, based on Reid's map (Clarke J F 1984, and Reid). (a) Dunston power station, (b) Engine Ordnance Works of Sir W G Armsrong Whitworth & Co, (c) Forth Banks power station, (d) Engine Works of Hawthorne, Leslie & Co, (e) Heaton Works of C A Parsons and Co, (f) shipyard of Sir W G Armstrong & Co, (g) shipyard and engine works of Swan Hunter, Richardson Westgarth, (h) Neptune Bank power station, (i) Carville power station, (j) Turbinia Works of PMSTCo, (k) Lead Works of Cooksons, (m) Shipyard of Palmers, (n) Shipyard of Hawthorne Leslie & Co.

Chapter 1

The Inventions Exhibition in London 1885

After the great 1851 Exhibition had ended and that extraordinary building of iron and glass, the Crystal Palace, had been removed to another site away from Hyde Park in London, a permanent exhibition centre and educational complex was developed nearby. It included the present Royal Albert Hall, and extended southwards down to what is now the campus of the Imperial College of Science, Technology and Medicine. It was served by an underground railway system which was still using steam locomotives, and had a station at South Kensington. By 1885 exhibitions were regular events, thus the Inventions Exhibition of that year had been preceded by one devoted to smoke abatement in 1881, to fisheries in 1883 and to health and education in 1884. These ventures provided Victorians with a chance to examine at first hand the latest developments in technology that were so transforming their daily life. The example of royal patronage shown by Prince Albert's interest in the 1851 event was continued for the 1885 exhibition with Queen Victoria now as patron and the Prince of Wales as president of the organizing committee.

As in previous years there were displays by foreign countries, from the United States of America, Austria–Hungary, Italy and Japan among others (Anon 1885a). Gardens had been laid out, and in front of the Japanese Pavilion there was a pond crossed by a Japanese style bridge (Anon 1985d). Commodore Perry's arrival in 1853 with his ironclad warships had opened up Japan to the United States and the European nations, and there had begun a period of intense curiosity about all things Japanese in Britain. In their turn the Japanese were hungry to absorb western science and technology. Describing the Japanese exhibits the official exhibition

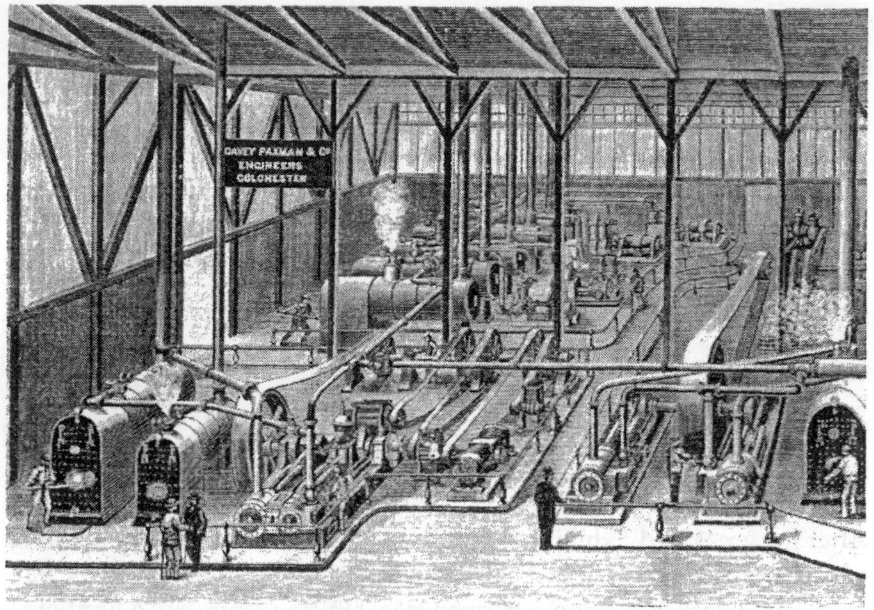

Figure 1. The Electric Lighting Machinery Shed at the International Inventions Exhibition London 1885, showing some of the boilers, steam engines and belt driven generators. Courtesy of Science Museum Library.

catalogue mentions the work of artists and craftsmen, and hints at what the future would bring. It drew attention to a firearm, '*the exquisite work of an army officer*', to complicated telegraph instruments and to some magnetic apparatus inspired by Professor Ewing. Ewing was one of those British citizens who had, in 1878, been invited to teach at the Imperial College of Engineering in Tokyo. But the most eye-catching feature of the exhibition was undoubtedly the use of electricity for lighting.

Experiments with electricity for illumination, using battery-powered arc lamps, had gone on for many years. The incredibly bright and intense light had been seen in the streets and galleries of many European cities. Dynamo electric generators that could be driven by steam engines, had been perfected by Gramme and Siemens in the 1960s, and henceforward the use of electricity for lighting could become widespread. The major breakthrough came when Swan in Newcastle and Edison in New Jersey independently found a way of 'subdividing the light', that is reducing the output of lamps down to a level that could be tolerated in relatively small rooms. This was achieved by the creation of filament lamps in 1879, but it was Swan's patented process for extruding the first cellulose artificial fibres in 1884 that made electricity a real competitor to gas. Gas burners of the time were still of the bats-wing type, and were inclined to smoke and smell as well as giving off a very considerable amount of heat. Electric lamps were free from all these disadvantages but were of course expensive; one lamp cost as much as a quarter of a ton of coal. The Inventions Exhibition offered an ideal opportunity to demonstrate these new developments, and the pathways and the exteriors of the pavilions were all lit by hundreds of bulbs. There was as yet no public source of electricity, and so a large shed was built to house the dynamos, boilers and steam engines. The firm of Davey Paxman provided 18 individual coal-fired boilers to raise steam for a total of 2300 horsepower (1.75 MW) of steam plant. Numerous large, slow speed, engines powered the dynamos by long belt drives. In another location there was a smaller installation with Siemens dynamos coupled directly to several Matthews high speed three cylinder engines. It is not difficult to imagine why 'smoke abatement' merited an exhibition to itself! In all, many thousands of lamps were used. '*The outlines and architectural features of buildings were traced out by strings of lamps which created a fairy like effect.*' The writer in the May issue of *Engineering* remarked that (Anon 1885d)

> *they were sufficient to illuminate the paths, the ripples on the ponds and the foliage on the trees with a mellow radiance which softened all harshness and hardness of the scene, and left the imagination free to wander under the inspiration of the music in the plains of Arcadia or the forest of Arden.*

Engines, dynamos and fittings of every sort and many of great ingenuity were on display. There were two quite different machines, both invented by the Hon. C A Parsons, being shown in action. Messers Kitson and Company of Leeds had his high speed, 900 rpm, epicycloidal rotary engine running directly coupled to a Brush dynamo, to illuminate the North Court with 750 lamps. This won for its inventor a gold medal from the panel of judges. The other machine, Parsons' Motor, the world's first turbo-generator, was also on display as an exhibit of the firm of Clarke Chapman, of which Charles Parsons was a junior partner. It was used to light a bar and the grillroom. The contrast that this latter development offered to all the other engines on display was obvious to observers. The slow turning steam engines made vibrations that travelled a long distance through the ground, and the pulsations in their speed, despite massive flywheels, could be seen in the flicker of the lamps that were supplied by them. By contrast the turbine turned at 12 000 revolutions per minute and its output voltage was rock steady. The total absence of vibration gave the observer the illusion that it was not moving at all. The writer in *Engineering* suggested that it would

> *constitute by far the most noticeable feature in the electric lighting department of the Exhibition, and would even rank among the foremost novelties of the entire collection in the ground. Rotary engines and steam turbines have been the dream of the inventor ever since the difficulty of converting the reciprocating motion of a piston into the rotary motion of a shaft had to be solved. Hundreds of schemes have been tried and each in turn has failed, until it has become an accepted belief that a rotary engine is an impossibility. But the electric light has created a sound demand for high speed engines, and has evoked at least two of the rotary type.*

That writer was more perceptive in his assessment than the panel of judges, for they awarded the invention a Silver Medal only.

The steam turbine developed rapidly and can truly be said to underpin our present day technology-based society. Virtually all electric power which is derived from thermal power plant is generated by machines that are still recognizably descendants of that first turbine. At the turn of the century the development of enormous battleships and fast trans-Atlantic liners was transformed by the application of turbines to marine propulsion, which was also pioneered by Charles Parsons. For nearly 50 years he was to play a central role in the continued development of steam turbines. It was a remarkable engineering triumph, and came at a time when British industry, which a century earlier had launched the Industrial Revolution, was now clearly showing evidence of its relative decline. Before we explore the development of the steam turbine it is worth while looking back to see what had prepared the way for this young Irishman's great success. In the process we may shed some light both on the development of inventive genius, and on the historical process of industrialization, and even discern some influences that are still at work today.

Chapter 2

An Elizabethan Settler Family

2.1 The Parsons family settle in Ireland

The fortunes of the Parsons family became linked to Ireland in the late 16th century during the reign of Queen Elizabeth when two brothers, William and Laurence, left Norfolk, to participate in the Tudor settlement (Anon 1826).[1] Although the English political presence in Ireland dated back to the 12th century, the post-Reformation period witnessed a new phase of conquest and dispossession in which religious allegiance was destined to play a major role. The process shared some of the features of the contemporary colonization of North America, and the Crown expected the venture to be self-financing. It intended that some of the cost of controlling the neighbouring island would be met by offering land to Undertakers who would plant it with loyal, English, tenants. Opportunities for personal gain attracted enterprising adventurers, men of ability like Richard Boyle, first Earl of Cork, who was a first cousin of the Parsons brothers (Cranny 1982). Although unable to graduate from Cambridge and having only a brief experience of the law, once in Ireland Boyle was spectacularly successful in laying his hands on the property of settlers and Church alike. Indeed he was reputed to have the largest personal income of any of the king's subjects. His youngest son Robert was the famous scientist.

The Parsons brothers were administrators rather than soldiers. William, the elder of the two, was born in 1570, and in 1588 *'made shift to raise up about £40, and with this as his whole fortune, transported himself to Ireland, where he found employment as assistant to his uncle Sir Geoffry Fenton, surveyor general, and eventually in 1602 succeeded to his office'*, (D.N.B.a). As Commissioner of Plantations and Surveyor General of Ireland he came to

[1] The family tree of the Parsons family in Ireland may help the reader to follow this account.

oversee the dispossession of landowners and their replacement by colonists loyal to the Crown. In the process he enriched himself, and *'as a single minded featherer of his own nest he was perhaps even more likely than the average public servant of his time to treat departmental records as if they were private property'* (Andrews 1985). Himself a Puritan, he was even said to have used his position as Lord Justice between 1640 and 1643 to stimulate rebellion so as to excuse further confiscations. Lecky, the 19th century historian, wrote that *'he was one of the most unprincipled and rapacious of the land jobbers who had, during the last generation, been the curse of Ireland'* (Lecky 1972).

His brother Laurence, on the other hand, had been trained as a lawyer and seemed to have been of a rather different temperament. He held several posts; eventually as Attorney General for Munster he was knighted. In 1620 he became, with William, joint Receiver General of Crown Lands. In that year he acquired over 1000 acres of land around the town of Birr in the centre of Ireland that had formerly been the property of the O'Carrolls. Birr is in County Offaly which had been one of the first to be 'planted' in the reign of queen Elizabeth, and was later known by the English name of King's County until 1922. The official policy under James I was less draconian than what was to follow later. In his instructions to *'Sir Oliver St John our deputy in Ireland'* in 1619, he noted that *'Many of the Natives are civil men, have built good houses and bawns, and some of them strong castles . . . to these we are pleased to grant them their houses and castles again.'* And *'The places where Undertakers should be planted We leave to the discretion of you our deputy and Commissioners . . . it would sooner civilise the people and keep them from private meetings to have the Undertakers mixed among them'* (Anon 1826). Laurence Parsons discharged his responsibilities as an Undertaker participating in this Plantation conscientiously. The town of Parsonstown, as Birr became known, was developed under his autocratic rule. A Huguenot family was encouraged to establish a glass factory nearby which supplied Dublin with glass. He carried out considerable work on the castle that had once been owned by the O'Carrolls and was close to the small Camcor river. It was after his death in 1628 that the castle came under attack.

The 17th century in Ireland was a period of devastating conflict. In 1641 the Catholic families rose in revolt. The resulting turmoil was aggravated by the Civil War that broke out in England in the same year, and which greatly heightened the religious aspect of the conflict. After a siege of the castle which lasted 14 months Laurence's son, William, was forced to capitulate in 1643 and he was driven out of Parsonstown. Evidence survives that shows that William and his mother who lived there had until then been regarded by Catholic families as good neighbours despite the acrimony of the war. Eventually his son, Sir Laurence Parsons, was able to return, but he too was caught up in fighting later in the century when the castle was held against his will for the Catholic king James II in opposition to the forces of William of Orange. He was captured and actually sentenced to death, but was fortunate to be rescued by the Williamite forces, who in their turn garrisoned the castle. The castle survived a siege by the Jacobite army led by the youthful Duke of Berwick in 1690, which was the one and only occasion when that soldier was thus frustrated.

The profound social upheavals that accompanied the Puritan revolution in England were paralleled by a lively scientific awakening. The Royal Society came into being in 1660. Like the Academie des Sciences in Paris, it provided a forum for discussion of scientific discoveries among its members. Important among these were the properties of and the physical principles that govern the behaviour of the atmosphere and of gases in general. Otto van Guericke of Magdeburg had perfected a pump that could create a partial vacuum in 1654. With it he had given a vivid public demonstration of the 'weight' of the atmosphere. He constructed two heavy metal hemispheres, and placing them face to face, he used his pump to extract the air between them. A team of horses was harnessed to each, but they were unable to pull them apart against the weight of the atmosphere. The Frenchman Denis Papin had explored the possibility of using the piston and cylinder arrangement that was used in the vacuum pump to do mechanical work. He did this by exploding a small charge of gunpowder beneath the piston. He also experimented with high-pressure steam on a small scale. Captain Savery, with the Duke of Worcester, had successfully demonstrated that the vacuum created when steam is condensed could be used to raise water from a considerable depth. However none of these applications yielded usable machines.

It was left to a Dissenter, the Anabaptist craftsman Thomas Newcomen, to put these scientific discoveries to practical use. By 1713 he had perfected his 'atmospheric' engine for pumping out deep mines. This successfully addressed the limitations of current technology. For example boilers could not yet withstand any significant pressure, but this was not important because the water was boiled at atmospheric pressure. The function of the steam was merely to drive out air from beneath the piston. The technology that had been perfected by more than a century of gun making was used in casting and boring the giant cylinder. This was the component that did require strength so that it could resist the crushing force of the atmosphere when the steam was condensed with a jet of water to create a vacuum inside the cylinder. The resulting force exerted on the

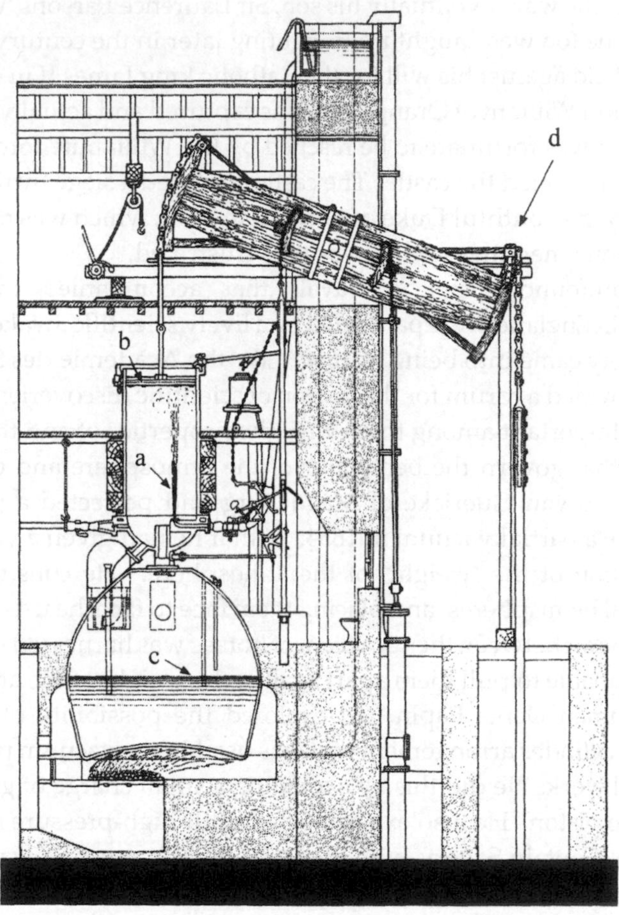

Figure 2. Atmospheric beam engine: a, water jet creating a vacuum inside the cylinder; b, piston driven down by weight of the atmosphere; c, water is evaporated in the boiler to displace air from under the piston; d, rods driving pumps below the surface (Strandh 1989). Courtesy of Nordbok.

piston by the atmosphere worked the machinery. Erecting a structure for the engine was a task for a builder, someone familiar with wooden beams and masonry walls. Many of the elements of the pumping mechanism had already been developed in machines that were powered directly by horses. Newcomen's engine was quickly taken up by mine owners for it allowed them to pump out their workings without the need to rely on animal power, and it greatly extended the depth of seams that could be successfully exploited. In the case of coal mines, the cost of fuel, for what were very inefficient engines, was relatively unimportant.

And so it was that the Industrial Revolution was ushered in by generations of men who shared something of a common world-view. On the one hand were the landowners who had not hesitated to execute their sovereign Charles I, or at a later date to replace his successor James II by someone of their own choice. On the other hand it was implemented by craftsmen many of whom had demonstrated in their religious institutions an independence of mind and a willingness to distance themselves from the 'establishment'. Their appreciation of the value of basic literacy stemmed from a wish to read the bible for themselves in the vernacular. Much of England's pre-eminence in this first phase of the industrial revolution can be attributed, among other things, to the fact that these features of society proved to be especially suited to the rapid exploitation of the early technological advances. But equally it can be argued that it was these same characteristics that in later years contributed to her failure to maintain that initial lead.

2.2 The eighteenth century and the Act of Union

The family line of William, the elder of the two brothers, which had had no connection with Birr, eventually died out in 1764 with the death of Richard Parsons, who by then had acquired the title of Earl of Rosse. The titles conferred on this branch of the family subsequently passed to a descendent of the younger brother, who was also named Laurence (1743–1807). The story is a little difficult to follow because his father had been twice married. The eldest son William (1731–1791) was the fourth baronet, while Laurence was the son of his second wife Anne Harman. She was heir to a reputed income of £8 000 per year and to £50 000 in cash. Indeed Laurence took the surname of his mother in 1790. The titles associated with the descendants of the elder of the two brothers were now revived and so Laurence Harman became Baron Oxmantown in 1792 and first Earl of Rosse of this 'second creation' in 1806. He too died without male heirs, and the titles passed to another Laurence (1758–1841), the fifth baronet and the son of his step brother William. It was supposed that this Laurence would marry his half cousin, a Harman and so also inherit the fortune of Anne Harman, but in fact the wealth went to a great grandson, a child of the Hon. Robert King who was descended in the female line (Malcomson 1982a). Though deprived of the expected wealth, Laurence was later to enjoy the advantage of political influence exercised by the Harman family on his behalf. This rather convoluted story is a reminder that wealth and power in Ireland, as elsewhere at that time, derived from ownership of land and the rentals which this could furnish.

The 18th century was a time of relative tranquillity for the descendants of the younger brother Laurence, who were among the victors of the *'Glorious Revolution'*. Ireland as a whole enjoyed increasing prosperity. Stringent penal laws and the almost total expropriation of land ownership had coerced the Catholic population and there was little fear of rebellion. Ireland's favourable situation for sailing ships, which were making the crossing of the Atlantic, gave it a special advantage as trade based on agricultural exports to the slave plantations in the West Indies began to develop. Linen manufacture and other industries, at a time before the evolution of large factories, were to be found widely distributed throughout the island. The introduction of the potato from South America had provided an excellent source of nourishment in most years, and the population in Ireland began to grow quite as fast as in the neighbouring island (Cullen 1987). From the 1740s on, the present layout of Parsonstown (Birr) began to emerge with a considerable amount of building in the Georgian style encouraged by Sir William (1731–1791), the fourth baronet, who also made improvements to the castle.

Many of the new land owning families had by this time lived for well over a century in Ireland and were coming to think of themselves as Irish. In the Parliament that sat in Dublin, pressure built up among the colonists for greater economic independence from London. The outbreak of a war of secession by the American colonies in 1773 was an object lesson to the land owners. But it also caused alarm, and when the French joined in the struggle, the fear of a possible invasion of Ireland became very real. Volunteer groups, a kind of militia, were formed by landlords and their tenants. These had several roles, not just as a protection against invasion, but also as a means for dealing with agrarian unrest for example. More importantly they helped to exert political pressure. Their existence helped to secure for the Irish Parliament considerable concessions, in respect of trade especially, from the English government. When a meeting of delegates from Volunteer groups was held in Parsonstown, in September 1781, among other things it was resolved *'That Ireland was an independent kingdom, and could only be bound by the laws enacted by the King, Lords and Commons of Ireland'* (Anon 1826). It was all part of what has been called colonial nationalism.

As the century drew towards a close various steps were taken to appease the increasingly prosperous Catholic population, by easing access to land ownership and entrance to the legal profession for instance. In March 1782 with Sir William Parsons in the chair, a second meeting of Volunteers was held in Parsonstown. It was resolved

> *That we have reason to expect that the liberal spirit of Parliament to the Roman Catholics of this kingdom, by emancipating them from restraints, which we are happy to think are no longer necessary, will be attended by the most beneficial consequences to this country*

(Anon 1826). But unlike Norman and Old English immigrants of earlier periods, the Plantation families were slower to assimilate with the native Irish population because of the religious divide that separated them. The French Revolution and the wars that accompanied it quickly transformed the scene. A Protestant reaction against Catholics set in. Catholics in the south-east, and Catholics and Dissenters in the north-east, eventually rose in rebellion in 1798. The Crown forces moved quickly and with great ferocity to crush the revolt. Soon after this the government of William Pitt introduced an Act of Union, to suppress the Irish Parliament and to centralize Government of Ireland with that of England, Scotland and Wales under the Westminster Parliament. The 1798 rebellion did not much affect King's County or Parsonstown; nonetheless sectarian tensions were heightened. In 1809 the construction of a military barracks to house 2000 troops was begun at Crinkle just a mile from Parsonstown.

Laurence (1758–1841), son of Sir William, graduated from Trinity College Dublin and represented Dublin University in the Irish Parliament from 1782 to 1790 (Malcomson 1982b, Atkinson 1961). As a student he had been admitted to the TCD Historical Society, and in 1781 became its auditor. The 'Hist', as it is known, was active in seeking to enable students who lived in rooms in the College to be registered as freeholders, and so entitled to vote at parliamentary elections in the city of Dublin. When a vacancy occurred in the representation of the University in 1782, Parsons put himself before the electorate which comprised the Provost of the College, 22 Fellows and 70 Scholars. He was successful, but at the following election in 1790, he was defeated by the son of Provost Hely-Hutchinson. Parsons petitioned the Irish House of Commons, alleging irregularities by the Provost, who was also the returning officer. By a narrow majority the members of the investigating committee favoured Parsons, but the Provost, using his *three* votes, turned the decision against him. From 1791 until 1800 he represented King's County, in which Parsonstown was situated.

He was active as a member of parliament, making a total of 54 speeches. While he shared the views of Henry Flood, he was not a member of any party. He was an informed and vocal critic of government actions and deficiencies. In 'Sketches of Irish Political Characters of the Present Day' (1799) it was said of him that

> *His voice is strong, distinct, and deep; and his language is simple, flowing, and correct; his action is ungraceful, but frequently forcible; his reasoning is close, compact and argumentative; though his manner is stiff and awkward, his matter is always good, solid and weighty.*

His concern that Ireland should not be subservient to the government in London impressed Wolfe Tone, the revolutionary leader who had been a fellow student and who wrote

> *I was exceedingly assisted by Sir Laurence Parsons, whom I look upon as one of the very, very few honest men in the Irish House of Commons. It was he who first turned my attention to this great question, but I very soon ran ahead of my master.*

While Parsons did not share Tone's revolutionary views, like many members of parliament he did believe that Catholic land holders should be entitled to vote in parliamentary elections. In this matter religion was not the issue, but rather property. In 1793 he wrote a pamphlet in which he argued that property acquired by industry and ability was '*the great prime mover everywhere*'. Political power in his view was rightly associated with property. He had no time for democracy, still less as the consequences of the French Revolution began to unfold, for Protestants faced a dilemma. On a personal level many were irked by the social divide caused by religion, on the other hand the fact remained that their near monopoly of power and its accompanying wealth depended on religious segregation. So during the debates that preceded the Act of 1793 to enfranchise Catholic freeholders of 40 shillings, Parsons took a cautious line. He proposed that the right to vote should be given, but only to relatively wealthy freeholders, and a valuation of £20 per year was suggested. He went further however and urged that Catholics should be free to enter parliament (D.N.B.b). Neither of his proposals was adopted.

Laurence Parsons had served as a subaltern in the Parsonstown Volunteers as a young man. While retaining that independent turn of mind that had brought his ancestors to Ireland, he had lost much of their ruthlessness. He was reluctant to flog his soldiers for breaches of discipline, an attitude that was to result in a complaint being made against him. It caused him to resign his command of the King's County regiment of militia at the time of the 1798 rebellion. The Act of Union was only accepted by the Irish Parliament in 1801 after the application of much pressure and bribery; Laurence Parsons was one of those who vigorously opposed it.

He was not one of the many absentee landlords who spent the income of their Irish estates in high living in London. He expressed his feelings in verse in a poem entitled 'Absentees', which runs, in part (BirrF/20),

Whence is this rage to leave our native soil,
and bear to other lands the tenants toil?
The wealthy mud our own golden harvest yields,
and flocks, and herds of ever verdant fields.
Where breathes so sweet the zephyr from the west?
Where is the glebe so green in livery drest?

He was active in steering the evolution of 'Parsonstown' during a period of rapid development near the end of the 18th century, so that even today it is outstanding among provincial towns for the good taste of its layout and architecture. The population grew, and by 1837, including the troops at the Crinkle barracks, it had reached 9500 of which one-fifth were Protestants. Despite the troubles of 1798 Laurence himself expressed no religious animosity. He distanced himself from those citizens who publicly denounced Catholic neighbours. Indeed when a Catholic church was built in Parsonstown in 1817, he gave not only the site, but a quarry for stone and £100 towards its construction. His son William, Lord Oxmantown, laid the foundation stone in his presence. This was before the achievement of Catholic emancipation in 1829 which allowed Catholics to contest parliamentary elections and it demonstrates how far the family had moved from the bitterness of the 16th century struggles. Nevertheless religious tensions did exist as an episode which happened in 1820 demonstrated. The end of the Napoleonic wars had brought the boom in agriculture to an abrupt end. The effects of the resulting depression were further exacerbated by the rapid growth in population. Agrarian unrest had existed for many years in a semi-organized form under banners such as that of the 'Ribbonmen' in neighbouring counties. While such activity tended to relate chiefly to the availability of land for the swelling rural population, landowners had on occasion been murdered. The already anxious townspeople of Birr were put into a high state of alarm by a mischievous letter writer who warned of an impending attack by a large band of armed men on Birr Castle. It proved eventually to have been a fiction, but not before troops had been summoned from the town of Templemore some distance away. The affair was dubbed the 'siege of Birr'.

2.3 The second Earl of Rosse, a reluctant politician

The Act of Union abruptly interrupted Laurence's political activities. However after the Union he was regularly elected to represent King's County in the London parliament until 1806. The electorate for these contests was small of course, just a few thousand, and was comprised largely of the tenants of the large landowners (Meehan 1983). Parsons now for the first time actively sought office, though not with as much push as his friends believed to be appropriate. It seems as if his uncle Laurence Harman was responsible for securing an appointment for him in the Irish administration. In 1804 he was made one of the Lords of the Treasury in Ireland. When his uncle died in 1807 he succeeded to the title as the second Earl of Rosse. He was successful in being elected to the House of Lords as a representative peer for Ireland in 1809 (after the Act of Union the Irish peers did not have the automatic right to sit in the House of Lords). At the time his fellow Irishman, Sir Arthur Wellesley, later Duke of Wellington, commenting on Rosse's bid to be elected wrote

> *These seats ought to be given to the persons resident in Ireland who have the largest influence and greatest property and are in themselves of the most respectable character; and on this ground Lord Rosse is the man of all others who ought to be appointed* (Malcomson 1982b).

Between 1809 and 1822 he served as joint Postmaster General for Ireland. It was a period of considerable expansion and development in the postal service, quite large numbers were being employed and there were many opportunities for theft and corruption. Lord Rosse took a keen interest in checking these and in raising efficiency. In this regard he continued with his war on waste that he had waged against the Dublin administration in the Irish Parliament in the years before the Union. He struggled with a troublesome chief executive and an unwieldy organization: for example he was just one of the two, *joint*, Postmasters. It was during his period of office that the stylish General Post Office building was constructed in Dublin (McDowell 1964).

His parliamentary duties brought him frequently to London. He had married Alice, daughter of John Lloyd, a neighbouring landowner, relatively late in life, in 1797 at the age of 39. His children were therefore young at the time. There were five in all, three boys and two girls. William the eldest boy was born in York in 1800 (Malcomson 1982b). Letters written to his wife during his absences on Parliamentary business in London contain tender enquiries about their health. He was a cultured man who showed an interest in Irish antiquities, and he had been a founding member of the Royal Irish Academy when it was established in 1785 (Parsons L 1795). He took a very individualistic approach to his children's upbringing, arranging for them to be educated at home with the help of tutors, rather than send them to public schools. Both he and their mother took a hand themselves in this, and it is said that the boys' grounding in mathematics and science originated chiefly with their mother. The boys lived in London with their father for a period in 1812 when parliamentary business made this necessary. They became fluent in French, in which language they conversed with people, possibly refugees, who lived with them. John, the second son, lived for a while with the family of Baron Strogonoff (BirrE/23).

This approach was highly successful. Having himself enjoyed a university education he sent his sons to Trinity College in Dublin in 1818, though most of their study was done at home as was permitted by the regulations of that era. Coming up to their examinations Laurence read Homer with the two boys, pointing up important points of Greek grammar. In 1821, in a move to reduce their isolation from other young people they were transferred to Magdalen College at the University of Oxford, graduating in mathematics the following year. While William was interested in scientific matters, his younger brother John began to study law, following in the footsteps of his uncle. John had endured poor health since early childhood, but when he died of rheumatic fever in 1828 his father was distraught. Some thoughts on his son's death that he penned to relieve his grief reveal the close bond that existed between father and son (BirrE/23):

> *everywhere I go, every object I see, every word that is spoken, every book that I read, in some way or other reminds me of him, and time and reflection instead of diminishing my regret, exalt him still more in my opinion.*

He was consoled by his friend the writer Maria Edgworth (BirrD/21)

> *How happy it is for you to have left such a son as Lord Oxmantown. Consider how few parents—how few of the rich and the noble have such a blessing—an eldest son distinguished among men of literature and science instead of among gamblers, boxers etc...*

His father dedicated a book to his memory. It was entitled 'An argument to prove the truth of Christian Revelation', (Parsons 1834). As he wrote, '*During the long period of deep affliction for so great a loss, I studied the subject of this argument.*' In this, although he '*accepted the main features of the Christian Revelation as given in the bible, he subjected them to a searching examination for scientific truth,'* (Malcomson 1982b). He was one of the earliest writers to attempt to reconcile the picture of the universe that was then being revealed by the latest discoveries of science with the contents of the Bible. From the text it is clear that he was as well read in science as in classic texts. He quoted from scientists like Humbolt, Herschel and Laplace. To give some examples,

> *We have just seen that the fossil remains of plants and animals prove that even in the northern parts of the earth the temperature before the flood was as high as it is now between the tropics* (Humbolt) and *La Place has proved from the elliptic form of the planet Jupiter that like the earth, it was formed originally of a fluid of varying density, decreasing from the centre to the surface* (Laplace 1802). Again *By the aid of telescopes we discern inequalities in the moon's surface, which can be no other than mountains and valleys like those now on the earth. From the lengths of the shadows of many of the more conspicuous mountains, their heights have been calculated, the highest being about an English mile and three quarters high* (Herschel 1833a).

Turning to the connection between science and religious beliefs he wrote

> *But there are several persons who admit the existence of a supreme being and that he is the creator of the world, and that we are his creatures, but who are of the opinion, that from the time the world and its inhabitants were made, it has been solely governed by the eternal laws which were then attached to it, and that he never interfered further with his work.*

At the time when he was writing this Laurence Parsons was already 70 years of age, and his eldest son was embarked upon his program of telescope construction. There must have been much discussion between father and son on such matters. It is interesting to speculate what it was that introduced this interest in astronomy. Perhaps he had seen William Herschel's great telescope at Slough on his journeys to and from parliamentary sittings in London, or it may have been an acquaintance with John Brinkley who had been appointed to the Andrews chair of astronomy at Trinity College Dublin in 1790. Whatever the truth, it was in Laurence the second Earl of Rosse that an interest in matters scientific was first displayed.

He had already shown himself to have a practical bent. At the turn of the century he continued the work of turning Birr Castle into a comfortable and stylish residence. While his contemporaries built grand new mansions in the Palladian style, he concentrated on improving the old castle. Sketches in a surviving notebook of his formed the basis of the modifications that were made to the structure at the time. The exterior was given a Gothic appearance with a new front entrance. He worked closely with his architect John Johnson who also designed the new Protestant church. He lived for a period in a rather extravagant style in the expectation of an inheritance that, as has been observed earlier, did not in the end come to him. As Governor of King's County, and as a magistrate, Lord Rosse was closely involved in the local community.

While he was helping to finance both the building of a new Church of Ireland church and a Roman Catholic church, local groups were protesting against concessions to Catholics. One particular episode illustrates his dilemma (Cooke 1875). A new curate by the name of Michael Crotty who had been educated in France was appointed to the Catholic parish in 1821. When he arrived he began to attack the local parish committee that had been established to raise funds to complete the construction of the new church. He claimed that funds were being misappropriated. A row broke out between Crotty and his own parish priest. Crotty won an overwhelming proportion of the parishioners to his side. In the end troops had to be called on to keep the peace. Crotty and several others were indicted for riotous behaviour before Lord Rosse. His son Lord Oxmantown was also involved as a foreman of the grand jury. Crotty was eventually deprived of his priestly faculties and even excommunicated. In 1832 Michael Crotty was joined by his cousin William who was also a priest and the so called 'schism of Birr' was in full swing. Predictably Lord Rosse came down on the side of authority, supporting the bishop and parish priest. But whichever way he reacted he stood to offend some Catholics. This was of significance because just then Daniel O'Connell was successfully mobilizing Catholics, some of whom now had votes for the first time, to contest elections to parliamentary seats, which in King's County had so far been held by members of the Parsons family or friends of theirs (Crotty 1847). Eventually the cousins left Birr: Michael Crotty emigrated to England and became a minister of the Church of England, while William joined the Presbyterian ministry in the west of Ireland.

In a number of letters written around this time the second Earl showed that he well understood the dilemma posed by the mixing of religion and politics. In one he wrote

> *I think that I never had any illiberal prejudice against them* [Roman Catholics] (BirrD/20)

But in another he wrote (Malcomson 1982b)

> *Neither can I agree with some who think it immaterial to England whether the principal people here are Catholic or Protestant . . . For though the maxim, divide and govern, has been often reprobated, it is nevertheless true that the division facilitates the governing of the country.*

His last speech in the House of Lords was made in 1833. He had given silent support to the government's plans for Catholic emancipation but when it came to the Act for the Reform of Parliament, he refused his support on principle, even declining a proffer of the Order of St Patrick. There is an irony in this, because what was refused as a bribe was later to be won by his son for scientific excellence.

At the time of the marriage of his son in 1836 he moved with his wife to Brighton where he died in 1841. In his long life he had sought to guide his actions by his strongly held beliefs without fear or favour, in his local community as well as on a national stage. He eschewed bigotry, but had to negotiate a difficult sectarian terrain. He was a man of considerable intellectual ability, and showed himself to be capable of effective management.

Chapter 3

Interest in Astronomy

3.1 William Parsons and astronomy

When the title of the Earl of Rosse passed to his father in 1807, William, who was his eldest son, became Lord Oxmantown. He had entered Trinity College Dublin in 1818, but with his brother John transferred to Magdalen College Oxford, in 1821. He graduated with first class honours in mathematics in the following year. In the same year he was elected to represent King's County in what was now the parliament of the United Kingdom in London (Meehan 1983). As a member of parliament Parsons travelled regularly between Parsonstown and London. It was something of an undertaking, because in those days before the railways were built the coach journey to Dublin alone took 13 hours. If he were to aspire to a political career, to become a holder of office and not just a member of parliament, he would have had to forsake his estates in Ireland. Indeed many members of the Protestant ascendancy had chosen this option, but he attended parliament out of a sense of duty, to watch the interests of his fellow countrymen. Like his father he lacked a strong ambition for power and he also felt a strong attachment to Ireland. This may have caused him to seek an activity that could be pursued at home.

An interest in astronomy was developing in Ireland as can be seen in the fact that observatories had been established by Dublin University at Dunsink in 1783, and by Archbishop Robinson in Armagh in 1790. At times both father and son would have been in London together on parliamentary business and no doubt the resultant social contacts led William to become a member of the Astronomical Society in 1824, just four years after it had come into being. It was a step that was to have great consequences for him, putting him in touch with amateur and professional astronomers. The

Astronomical Society was but one of a number of institutions that were established early in the century to cater for specialist scientific interests, often by members of the Royal Society who were dissatisfied with the stagnant state of that body (Dreyer 1987). Others included the Royal Institution (1799), the College of Surgeons (1800), the Horticultural Society (1804) and the Geological Society (1807). In the 17th century, science in England and France had set off on an almost equal footing, but England lacked the sort of central state involvement that had led to the establishment of institutions like the École Polytechnique with its links to the specialized École des Ponts et Chaussées, the École des Mines and the École du Génie Maritime which provided opportunities for employment of professional scientists.

Writing later, the astronomer Sir John Herschel said of this time (Dreyer and Turner 1987)

> *The end of the eighteenth and the beginning of the nineteenth century were remarkable for the small amount of scientific movement going on in this country, especially in its more exact departments. Mathematics were at the last gasp and Astronomy nearly so... The chilling torpor of routine had begun to spread over all those branches of science which so wanted the excitement of experimental research.*

Since the earliest times the celestial bodies, the sun, the moon and the stars, have awed and preoccupied mankind, as the construction of the Stone Age monuments like the burial tumuli at Knowth in Ireland and the pyramids of Egypt with their celestial orientation features remind us, and still the study of the universe retains an almost mystical fascination for modern man. Long ago it was the unusual regularity of the heavenly bodies as they traversed the sky that so strongly impressed, contrasting as it does with most of the behaviour of the natural world about us that is so often unpredictable. The scientific developments of the Renaissance changed this. Immediately the first telescopes came to hand, fine details of the heavens were probed. In 1610 for instance, Galileo observed the moons of Jupiter for the first time, and so put our own moon into a quite new perspective. Before long, painstaking observation revealed that some stars moved by minute amounts relative to others in the same part of the sky. A whole new area for scientific study opened up. But in order to understand why astronomy loomed so large in the scheme of things during the 18th century, it is necessary to say something about the great question of the time, something that was comparable in terms of the interest it generated, to the space race. It was the determination of longitude. The location of any point on the surface of the globe can be uniquely defined in terms of the latitude, or distance from the equator, and the longitude. Longitude is defined by an imaginary circle or meridian that passes through a given location and both of the poles. Its value is assigned in terms of the relative angle between this meridian and some arbitrary meridian such as that through Paris or Greenwich for example.

This problem of determining position accurately began to be of great practical significance during the 18th century. There was a great surge in maritime travel as improvements in ship design allowed European nations to trade over long distances. Safety at sea depends on being able to establish accurate positions during a voyage. Moreover as new lands were discovered, maps were needed. Latitude is readily determined because at noonday, local time, the sun rises to its highest point in the sky, and this height for a given time of year is a measure of the latitude. But in order to determine the longitude we need to know the time difference between noon locally, and noon at a reference meridian. With the aid of a radio signal this is a simple matter. Governments invested money to find a solution. In England, an act of Parliament was passed in 1714 to establish a Board of Longitude. This gave the board authority to award a sum of up to £20 000 for a practical way of establishing longitude accurately (Sobel 1996). There were two options. One was to carry a very accurate clock that always tells the time at, say, Greenwich. The development of such a chronometer made extreme demands on the engineering skills of instrument makers. The life's work of one such, the genius John Harrison, was crowned in 1762 when his chronometer watch was successfully tested in an Atlantic crossing. However such time pieces were costly and it was many years before they became available at a reasonable price.

An alternative approach was to note the local time at which some specific celestial event occurred, and compare this with the time when the same event would occur when observed at, say, Greenwich. Such an event could be the arrival of the moon at a specific point among the stars. Careful measurement had shown that the motion of the moon across the celestial sphere is not uniform, and it required the assistance of the Swiss mathematician Leonhard Euler to establish suitable mathematical equations to describe its course. With the aid of these, predictions could be made, and tables were computed by Tobias Mayer of Nurenberg for use at any location on the earth's surface. These allowed the longitude to be inferred by noting the local time when the moon reached a specific celestial position on a given day. This second approach was successfully tested at sea in 1757, but it made great demands on the skill of the navigators using it. Clearly the needs of both navigators and astronomers alike demanded the development of accurate time pieces and of precision instruments for the measurement of angles. By the end of the 18th century the design of astronomical instruments and telescopes especially, was developing rapidly. But the funding for astronomy which was provided by governments at the turn of the century was motivated primarily by a wish to improve the technology of navigation, rather than a desire to further scientific understanding.

In 1773 William Herschel (1738–1822), a German musician, who had settled in England, began designing and constructing his own astronomical instruments as a hobby. From the start his motivation was scientific. His discovery of a new planet Uranus in 1781 won him the patronage of King George III. From his observatory near Slough, Herschel systematically searched the northern sky. Most stars are so distant that even with high magnification they appear as points of light. Simply increasing the magnification of a telescope, by changing the strength of the eye piece, makes the images fainter. However if the light gathering power of the telescope is greatly increased by using a larger diameter mirror or lens, then faint objects can be discerned that before had been invisible to the naked eye. Indeed Herschel observed that some stellar objects did actually have an observable size. With his instruments he could only say that these appeared to be *'cloud like'* or *'nebulous'*. The question which this posed was—are these clusters of very many stars so close together that they cannot be separated, or are they indeed tenuous clouds of matter? More perfect telescopes than he possessed were needed to answer this question. As a by-product of his systematic surveys of the heavens, Herschel had also been led to conclude that the stars were not distributed randomly throughout space, but lay in a stratum or layer. In all he had catalogued a total of 2500 nebulae and clusters, and he hazarded the opinion that the nebulae could be grouped according to the stage that had been reached in an evolutionary process which led from clouds towards fully formed stars. To unravel details of the structure of the universe was therefore emerging as a great scientific challenge.

In these circumstances, an instrument with sufficient light grasp to allow the use of high magnifications promised to provide answers to many questions. Telescope makers had essentially two options. One was to use a large glass lens, the so-called refractor. The alternative was to use a mirror, the reflector. If a large aperture was used to improve the light grasp, the refractor required a large mass of optically perfect glass that would, for a long time yet to come, prove difficult to obtain. It also presented difficulties of mounting, being located at the front end of a long tube. On the other hand it lent itself readily to precise determinations of position of celestial objects of the kind sought for use in establishing the location of places on the surface of the earth. The reflector could be mounted with its mirror close to the ground on a universal joint. Because a means of silvering glass or ceramic surfaces had not yet been developed, the reflector required the use of metal mirrors. Newton and Herschel had employed mirrors of speculum metal which is a form of bronze for their telescopes. This alloy has been used from ancient times. Depending on the ratio of tin to copper, it could be employed as a mirror or forged into sharp weapons or it could be made tough enough for use in gun making. When used for mirrors those alloys that took the best polish and which were most resistant to tarnishing proved to be the most brittle and easily fractured. The strains locked in during solidification were such that a slight impact or the application of modest heat could be sufficient to cause a casting to fly to pieces. It is especially difficult to work with in large sizes. In 1787 Herschel received a Royal grant of £4000 to allow him to cast a giant 48 inch diameter mirror, but it had not been satisfactory and was eventually abandoned. When William Parsons decided in 1826 that he would become actively involved in astronomy, he chose to follow the example of Herschel in a number of respects. He decided to concentrate on scientific aspects of the subject and he too chose a reflector design. Like Herschel he intended to construct his own instruments, acquiring practical expertise in the process. William Herschel had recently died and nothing was available in print to guide Parsons in his practical work. Indeed some of the more successful instrument makers had taken pains to keep their methods secret (King 1980). In consequence a systematic program of experimentation was necessary.

3.2 Telescope building

We have no real explanation of where this wealthy young nobleman acquired the skills and experience necessary to carry out such a venture. Nevertheless he set himself two objectives (Parsons W 1830),

> (1) *to ascertain whether it was practicable to remove any of the defects known to exist in the large reflecting telescopes hitherto constructed.*
>
> (2) *to simplify the process necessary for the manufacture of good reflecting telescopes of ordinary dimensions so that the art may no longer be a mystery,*

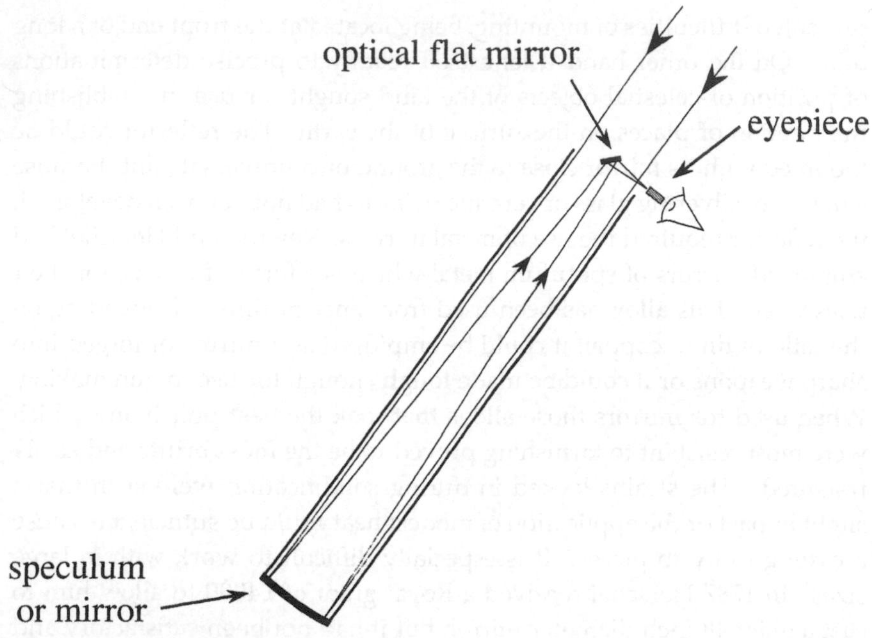

Figure 3. Reflector telescope arranged as a Newtonian.

known to but a few individuals, and not to be acquired, but after many years of laborious experiment.

His progress was reported in three papers that appeared in Brewster's, Edinburgh, *Journal of Science* in 1828 and 1830 (Parsons W 1828a, b, 1830). No doubt based on this work, he was elected to membership of the Royal Society in 1831.

He began experiments with a mirror of a couple of inches diameter that he had purchased. He quickly moved on to cast and shape his own mirrors, beginning with one that was six inches in diameter. Spherical mirrors do not produce a sharp focus. Rays from a distant object that strike the outer edges of the mirror converge to a point nearer to the surface of the mirror than rays that lie close to its central axis (figure 4). In an attempt to mitigate this aberration he made a mirror consisting of two parts, an inner disc and an outer annulus. The combination was supported on a backing plate, then ground and polished as a single spherical surface. But it was so attached to the backing plate that a small axial displacement could be given to one component relative to the other. In this way the spread of the converging rays from the annulus could be reduced to one half. While

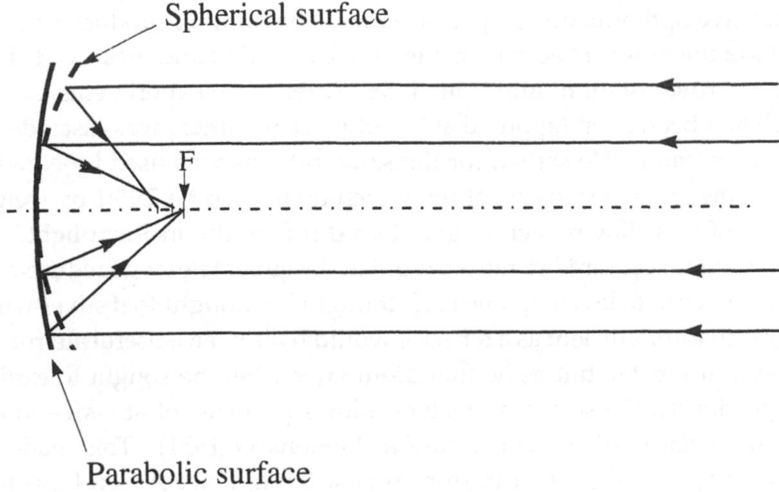

Figure 4. Rays converge to a single focus for a paraboidal surface, but not for a spherical surface.

the adjustment did work, he reported that '*The speculum was polished by hand with the utmost care more than fifteen times, but without any considerable improvement*' (Parsons W 1830). Though it had called for a high degree of workshop skill to complete, it may be considered to have been merely an apprenticeship task.

He then began a painstaking study of castings made from different alloys to establish the ratio of tin to copper that gave the best optical performance in a mirror. Conflicting advice was available in the literature of the time. Better reflecting power, greater freedom from tarnishing and better suitability for grinding and shaping were to be found in those bronze alloys that were most brittle and fragile. Ease of working varied with composition, and this ranged from 18 atoms of copper to one of tin when the metal was used for cannon, to four of copper to one of tin when employed for specula. To produce discs of such speculum alloy, Parsons built himself a small foundry. He favoured peat or turf as a fuel. This was in plentiful supply around Birr which is surrounded by bogs. It is capable of reaching the temperatures of around 1100 °C which are required for melting the copper which is necessary to form ingots of bronze alloy. The alloy itself has a melting point that is appreciably lower, around 750 °C. He found from experience that tin tended to be driven off if the surface of the molten metal was left open to the atmosphere for long periods. He carried out many experiments to establish the proportions for his alloy that would give optimum optical performance. The samples produced and the results of the tests carried out on them were recorded and preserved. They were to prove useful to him years later. After these experiments, arsenic which had been often favoured as an additive by others was discarded as having no value. He settled for the same ratio as was used by Newton, four atoms of copper to one of tin, which he took to be 2.15:1 by weight. Mirrors of this alloy reflect roughly two thirds of the incident light. The two reflections required in the Newtonian design of telescope reduced this fraction overall to less than one half. It might be thought that silver which is so much more efficient as a reflector would have been a useful alternative to speculum metal, but as he found out later when he sought to explore this possibility, the softness of silver allows particles of abrasive to bed into the surface rather than cutting it (Parsons W 1851). This made the task of shaping solid silver mirrors almost impossible. Around this time a number of other amateurs were making similar experiments, among them was the Scotsman James Nasmyth (1808–1890), who was eight years younger than William Parsons, and was casting mirrors of eight or ten inches diameter. He had spent three years working with the great tool maker Henry Maudslay, a Londoner who was familiar with foundries and machine tools of all sorts, and was establishing a reputation for himself as an engineer.

In developing a workshop at Birr, William Parsons was able to deploy some of his considerable wealth. The last 50 years had seen great advances in the building of machine tools in Britain and he seems to have had a good awareness of these developments. His journeys through England to attend sessions of parliament in the days before railway travel would have offered plenty of opportunity to visit industrial centres *en route*. It was a practice that he followed till the end of his life. Indeed a story was told of an owner of a workshop, who was showing his guest, William Parsons, around his premises, being so impressed with his visitor's grasp of the technicalities that he enquired of him whether he would be interested in working for him! (Anon 1867a).

Traditionally mirrors are ground by moving a metal blank back and forth across the abrasive surface of a convex tool of a similar diameter. The operator moves slowly around the tool to ensure that each part of the mirror passes in turn over each part of the abrasive surface. This soon forms both surfaces to the shape of part of a sphere. The reason for this is that the only shape that will allow two surfaces to slide over one another while keeping in contact at all points is a sphere. (Two planes of course can be thought of as surfaces of spheres of infinite radius.) Parsons decided to build himself a grinding and polishing machine by modifying a large power lathe which was driven by belts and shafting from a small steam engine of 2 to 3 horse power. The engine was actually made under his direction at Birr. In a letter to John in the summer of 1827, William asked his brother to purchase for him a supply of bismuth and quicksilver while he was in London. He also wrote *'I have just constructed a steam engine on a new principle. It is a rotary... I had it at work for the first time... Have you heard of any new book on steam engines?'* (BirrE/22). At a later date he wrote of the engine that *'from the distance of manufactories, it has been necessary to execute all the turning and casting work in my own laboratory'* (Parsons W 1840). This was certainly a remarkable achievement. A steam engine had already made its appearance in Birr for industrial use in a distillery, but it should be remembered that up to that time condensing 'beam engines' of massive proportions were the norm for steam power. The development of 'high pressure' engines that could work without the need for a vacuum, and that were suitable for road and rail locomotion, was only in its early stages. In 1807 Henry Maudslay had patented his small 'table' engine and this was used to drive machinery in workshops (Petree 1963). Unfortunately no details of the Birr engine survive but it is clear that to have designed and built it at this point in time was a remarkable accomplishment. The steam engine fascinated many scientific workers at that period but few sought to design and make one for their own use (Mayr 1971). This interest in steam engines was later communicated to his two younger sons.

Parsons chose to develop machinery powered by an engine, not just to eliminate the manual labour involved in hand grinding, which was considerable, but because it allowed the cause of any problems to be tracked down systematically. In his account of the development of a grinding and polishing machine he wrote (Parsons W 1830),

> *The results obtained by machinery are very nearly uniform. Where a uniform combination of motions produces a defect, that defect will uniformly recur and may therefore, with great facility, be traced to its source and corrected... the performance of the machine has improved faster than I could have anticipated.*

His recognition that the performance of machinery is repeatable, in a way that human actions are not, is clear evidence of his scientific approach to mechanical engineering. The first machine was developed to cope with discs of up to 36 in diameter, but years later a second larger machine was built to handle mirrors twice this size.

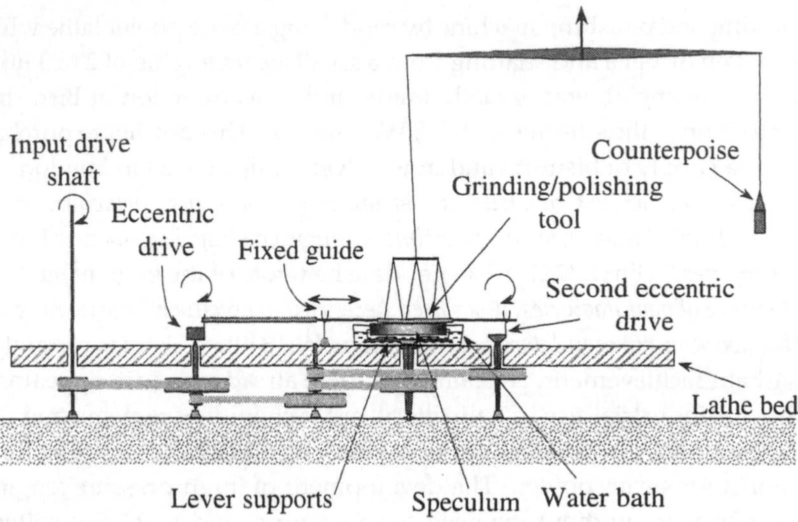

Figure 5. Machine for grinding and polishing a 36 in speculum, driven by a 2 HP steam engine.

His grinding and polishing machine comprised a succession of vertical spindles arranged along the bed of the lathe, as seen in figure 5. They were connected in a train with belts and pulleys. The speed of rotation fell successively until the last spindle. The penultimate spindle carried the speculum blank which turned in a horizontal plane. The second spindle also had an eccentric attached to it which was connected to a linkage. As the eccentric rotated, the linkage caused the abrasive tool to reciprocate to and fro diametrically across the mirror. The tool itself was suspended from a counterweighted lever that was mounted above the lathe so that the pressure exerted on the speculum by the tool could be adjusted exactly as desired. This was an important innovation, which Herschel had lacked. The tool was made of a lead and tin alloy. It was turned to a spherical shape using the same lathe, using a cutting tool that was constrained to move in an arc by a light wooden frame with the appropriate radius. For grinding, emery powder of different grades of coarseness was introduced through a central hole in the tool. As the grinding approached completion the speed was reduced. Polishing was accomplished by using another similar tool. This was covered with a thin layer of pitch, to which was applied a paste of red iron oxide powder. The ratio of reciprocating strokes to one revolution of the mirror was the subject of careful study for a given size of mirror. The lathe itself was about 18 ft long, with the bed 3–4 ft above the ground.

It was located in an out-building with a ceiling 11 ft high, that adjoins the eastern side of the castle.

A great many experiments were made on small samples to establish the factors important to the production of sound castings. As mirror sizes increased their weight rose roughly as the cube of the diameter, so he sought to achieve stiffness and strength while minimizing weight. He was aware that Herschel's 48 in diameter mirror, which though weighing half a ton was only a mere inch and three quarters thick at its centre, lacked rigidity. He designed a 15 in speculum with a rim and with ridges arranged in a hexagonal pattern across the rear face. Sometime before this he had experimented with painting on glass, which had required it to be brought to a red heat. This had taught him the need for annealing if he was to avoid cracking of the workpiece during cooling. To accomplish this he used a separate annealing vessel in which to cool his castings. His first venture fractured on cooling. The same fate befell his second, third and fourth attempts. In all five attempts were required, each time he resorted to using a poorer quality of alloy until he obtained a disc that did not shatter. Even then the optical performance was disappointing, though as he remarked *'much higher than Sir William Herschel's 24 inch'*.

For a time he abandoned attempts to cast a large mirror in one piece, and instead sought again to construct a composite speculum. This time the reflecting surface was to be made from six strips of bronze roughly 2.5 in wide by 0.25 in thick. These were soldered to a 15 in diameter disc shaped brass base designed with a rim and ridges at the rear. For the base he prepared his own supply of brass, an alloy of copper and zinc that is much easier to cast than bronze. He had already carried out experiments to determine the exact alloying ratio needed for the brass to ensure that it would have the same coefficient of thermal expansion as the bronze. The assembly was then ground and polished to a focal length of 12 ft.

At the start of experiments on casting, use had been made of an iron ladle that was purchased in London. This was found to be porous and to leak badly. Attempts at stopping up the crevices were only temporarily successful. In the end Parsons decided to cast his own ladles. The secret for avoiding leaks, he discovered by experimentation, was to arrange the wooden pattern for making the sand mould so that the mouth faced upwards. This was the reverse of usual practice, but it allowed any gases in the metal to escape easily and avoided the troublesome porosity in the walls.

Meanwhile he had partially completed a wooden stand for mounting the telescope outdoors. Initially this was a straightforward copy of the

Figure 6. Photograph of the telescope with a 3 ft diameter reflector mounted in a Ramage style altazimuthal stand at Birr. Note the counterweights which have been added for the tube and the viewing gallery. © Birr Scientific and Heritage Foundation by courtesy of Lord Rosse.

design due to the Scottish astronomer Ramage, with no refinements. It was a so-called altazimuthal design. The tube can be turned to any point on the compass (azimuthal adjustment), or raised to any angle (altitude adjustment). To maintain a heavenly body in view, both adjustments must be changed continuously. With modern computers this is relatively straightforward. When adjustment is by hand and the telescope is heavy, the task is very demanding. He was now ready to progress with a composite mirror of 24 in diameter. All this was accomplished by 1830, in a period of just four years. Although as yet there had been no serious attempt at observation with his instrument, it was a remarkable achievement. This was especially true given that it had required the assembly and training of a team of skilled workmen in a town of relatively small population and that was far removed from large centres of industry. Writing in 1830 after his first attempts at creating a compound speculum, Parsons wrote (Parsons 1830)

> *All my workmen were trained in my own laboratory without the assistance of any professional person, and none of them had seen any process in the mechanic arts; and I was not myself then acquainted with the precautions necessary to insure the production of an alloy of zinc and copper in the due proportions.*

His hopes for the future were summed up,

> *The examination of the heavens commenced by the late Sir William Herschel and prosecuted by him with such success, still continues. New facts are recorded: and there can be little doubt that discoveries will multiply in proportion as the telescope may be improved. It is perhaps not too much to expect that the time is not too far distant when data will be collected sufficient to afford us some insight into the construction of the material universe.*

3.3 Marriage to a wealthy heiress and the building of a 3 ft reflector

William Parsons was not unaccustomed to public life when in 1831 he was appointed Lord Lieutenant for King's County. In the following year while the family were away, there was a quite serious fire that caused extensive damage to the central portion of the castle. In the same year he was appointed Colonel of the Militia of King's County. He did not stand for election to parliament in 1835. In a circular he explained that he was withdrawing from the election due to *'defects of the present franchise'* and *'would not solicit support from the enemies of the Union and of the Established Church,'* (BirrE/29). Daniel O'Connell's campaigns for the Repeal of the Union, and for an end to the payment of tithes to the Established Church, had begun to influence local politics, and this decision may also have been influenced by the controversy about the Birr schism that was mentioned earlier. It did at least have the effect of allowing him more time to devote to his personal affairs. In any event he returned to Yorkshire, the county of his birth, to wed Mary Field (1813–1885). She was the eldest daughter of John Wilmer Field, a wealthy land owner with estates around Heaton Hall near Bradford. Her mother had died soon after the birth of her sister Delia, and her father remarried. The girls were brought up to London by their father where he took a house each year for the Season. No doubt it was there that she was introduced to Lord Oxmantown. The marriage took place in July 1836 at Heaton Hall. She was much younger than her husband, being just 23 years old. They were a well matched pair. He may have perceived in his intended bride some of those features that had been displayed by his own mother. Certainly the pair proved to be loving parents. The health of Lord Rosse had been failing; he was then 78 years of age, and he moved to Brighton. The newly married couple were invited to occupy the residence at Birr Castle. Mary Field quickly showed her mettle when she refused to move in until the repairs and redecoration necessitated by the fire were all completed to her satisfaction. Mary inherited the estates of Heaton and Shipley that were valued at £88 000 as well as a cash settlement of £8700 from her father (BirrG/60).

Within less than a decade from the date of his marriage, Parsons would be the possessor of a telescope that was far and away the largest in existence. It was so enormous that it was quickly dubbed the Leviathan of Parsonstown. But when he set to work on the task of building it, the full extent of the difficulties to be faced was unknown to him, as were the solutions that he eventually found for overcoming them.

The fire and its aftermath, as well of course as the wedding, must have slowed up the programme of research. While the repairs were being made to the castle, a forge, foundry and workshop were built behind the building that housed the grinding and polishing machine. A fuel store to contain turf was also constructed (Appleyard 1933). Lessons learned during earlier experiments caused Parsons to adopt a more meticulous approach when he resumed his work on telescope manufacture again. Despite the difficulties that he had encountered thus far he set about the construction of a composite 36 in diameter mirror. This time the brass backing was made as an assembly of pie shaped segments fabricated to create a stiff but light structure (figure 7). While melting zinc and copper to provide the brass, he noticed that, if the alloy was kept molten for a period of time, the alloying ratios would change as zinc was driven off. A remedy for the problem was finally found. Using a 2 in layer of charcoal dust on the surface of the molten metal, the loss of zinc was kept *'to less*

Figure 7. Fabricated brass segment for the base of a compound 36 in speculum (Parsons W 1840).

Figure 8. Base for the composite speculum, bolted together; view of underside (Parsons W 1840).

than 1/180 th'. He had repeated his earlier measurements of relative thermal expansion of brass and bronze, but this time with greater precision, and it was this information no doubt that alerted him that it was the loss of zinc which caused his difficulties. The first attempt to assemble the composite structure used iron bolts and solder, but the assembly broke during the grinding of the top face (figure 8). Eventually the joins were made by pouring a stream of molten brass onto selected spots, taking care to cover most of the surface with sand. In the few locations where this failed to produce a secure joint he bored a hole at the junction and poured molten brass into this. It was an early and novel form of welding and Parsons wrote '*this process may, in engineering prove useful as a means of connecting large masses of metal.*' After repeated experiments on a small scale, he eventually created a satisfactory base for a 36 in mirror.

His intention had been that the surface for this mirror should be made up of 16 squares of bronze roughly 9 in by 9 in (Parsons W 1840). These were to be soldered to the machined brass base using tin, with resin as a flux. It took eight hours in the oven for the assembly to reach the melting temperature of tin. He discovered that the flux must be added only at the last minute when all the components had reached the melting temperature of tin, lest the resin be degraded. Moreover if the flux was applied in the solid state, the cooling effect caused by its melting was sufficient to cause strains in the bronze pieces which would fracture them. In an attempt to obtain suitable, sound pieces of bronze he tried several approaches, and eventually cast an ingot of square cross section that was then sliced by a circular saw. The saw that he used was a soft iron disc that was fed with emery and was operated under water. In effect it was a very early application of abrasive cutting. But while these plates remained intact after assembly, their hardness varied across their surface and this made accurate grinding and polishing impossible.

He refused to accept defeat. By now he had had a lot of experience of casting bronze discs, and reflecting on what he observed, he concluded that the cause of previous failures was the stresses that set in during the process of solidification which progressed from the edges to the centre. He was aware that in Dublin the engineer Robert Mallet, a graduate of Trinity College, whose father ran a foundry, had made a scientific study of the crystalline structure that was revealed when iron castings were sectioned. Mallet was able to use these observations to identify sites of potential weakness in castings, and he noted that the long axis of the crystals tends to be parallel to the direction of heat flow that occurs in the process of solidification (Mallet 1856). Parsons therefore set out to construct a mould that would chill the metal from the base progressively to the top surface, while maintaining a uniform temperature in the radial directions. He first attempted to use a solid iron mould with the base turned to a suitable convex profile and with a jet of water or cold air playing from below. Next he substituted sand for iron around the rim, so reducing the radial heat flow. This gave sound castings, but air bubbles were trapped at the bottom surface, creating flaws in the mirror surface. The final step to a trouble free process came when he had an inspiration. He constructed the base of the mould from strips of iron, set on edge and packed tightly together. The surface was ground to give a smooth finish, and the slight gaps between adjacent strips were just sufficient to allow gases to escape without allowing any loss of molten metal. This time there was no pock marking of the surface and he obtained perfectly sound flat plates for his 36 in composite mirror. Regrettably Parsons did not record exactly how he came to hit on the critical step despite its central importance for all that followed. The plates were allowed to cool for three to four days in

Figure 9. Water-colour of the foundry at the rear of the workshops building, painted by Henrietta Crompton. © Birr Scientific and Heritage Foundation by courtesy of Lord Rosse.

an annealing oven. But having assembled them, he abandoned for the moment any further work on the composite construction.

He immediately returned to the preparation of a large single disc. This time the iron strips for the base were wound in a spiral. They were then ground to a convex shape as before. An annealing oven had to be prepared of sufficient size to house the casting and to allow it to cool without developing built in stresses that could cause later fracture. Arrangements were made to draw the speculum into the oven as soon as the metal had solidified and yet was hot enough not to be brittle. His courage in persevering in the face of so many difficulties was rewarded with immediate success. In 1839 he succeeded at last in casting a perfect 36 in disc in one piece (Parsons L 1968). It weighed one and a quarter tons, had a thickness of 3.75 in and required two crucibles of metal to complete the pour. If we compare this with his composite design that had a total weight only one quarter of this, we can recognize the attraction that the compound design possessed. In fact just such a ribbed structure is used for the largest modern telescope mirrors, even though they are made of a ceramic rather than metal. Nevertheless for the manufacture of plain discs the technique was a great success. Casting large blanks for mirrors of good quality alloy had till then presented insuperable difficulties. He quickly published a detailed account of his methods for others to follow (Parsons W 1840), and indeed many of his contemporaries like Lassell, De la Rue, Draper and Grubb adopted his methods (King 1980).

The ideal shape for a mirror is that of a parabola. With this, parallel rays converge to a single focus no matter how far they are spaced from the centre line of the mirror (figure 4). However, as has been noted, the natural shape to emerge from grinding is that of a sphere. To get some idea of dimensions, consider a 6 ft diameter speculum shaped to have a focal length of 54 ft. This requires the centre to be hollowed out to a depth of 0.5 in. To create a parabolic surface or 'figure' as it is called, extra metal has to be removed. The amount is very small. The rim height must be lowered by no more than 0.0001 in. It can be seen that grinding and polishing requires the highest precision to achieve the best results. Discussing Barton's suggestion of using a diamond tipped tool to machine the surface, Parsons wrote (Parsons W 1840)

> *When we recollect the extreme accuracy required* (for diamond turning)... *an error of figure amounting to but a small fraction of a hair's breadth would destroy the action of the speculum, it is scarcely to be expected that any process can succeed in practice, which has not like grinding, a decided tendency to correct its own defects, and to produce a result in which errors may be said to be infinitely small in comparison with the errors of any of the previous steps from which it is derived.*

Figure 10 is a plan view of the grinding and polishing machine. It can be seen that the extent of the to and fro movement of the tool is controlled by the throw of the first eccentric. The line of action of this movement is controlled by the setting of the second eccentric. The relationship of the number of strokes to the speed of rotation of the mirror under the tool was fixed by the diameter of the pulleys attached to the various spindles. Very extensive tests were carried out on mirrors covering a range of diameters to establish suitable settings for a particular operation. The lathe that had been adapted for grinding was also used to machine the moulds used for castings, and the shape of the tools used for grinding and polishing were also machined on this lathe. These tools were shaped to a portion of a sphere with a radius that is almost twice the focal length of the speculum.

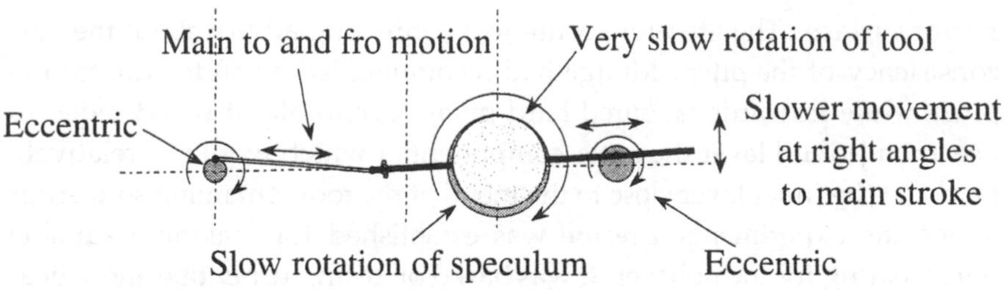

Figure 10. A plan view of the linkage of the grinding and polishing machine, showing the adjustments available.

For modest diameters, this was done using an attachment for the lathe which carried the cutting tool at the end of a long arm of suitable radius. For radii that were too great for this, gauges were made up by hand to guide the steps in the machining process. Initially it had been thought that a parabolic figure was unobtainable when using a machine, but in fact a way was found to approach this objective. A great deal of effort was devoted to achieve perfection, and he wrote

> *An immense number of experiments, where the results were carefully registered, eventually established an empirical formula, which affords at present very practical results.*

The grinding and polishing processes were subjected to the same scientific study as the casting methods. An important step was the sub-division of the surface of the grinding tool into small squares by channels cut in its surface. These prevented local pile-ups of debris and abrasive which could upset the shape. The polishing tool was given a coating of pitch of 30 to 40 thousandths of an inch thick. Some moistened iron oxide paste was spread over the surface. The purpose of the polishing operation was to remove the scratches left by the particles of grinding abrasive. Close observation of the polishing process revealed that failure was often associated with local accumulations of powder. Up to this it had been usual to sub-divide the surface of the pitch into small squares. As in the case of the grinding tool, any excess debris tended to be moved laterally into a channel without any build-up developing. But as the size of mirrors increased it was found that this was insufficient. The remedy adopted was to subdivide the underlying metal surface of the polishing tool with circular and radial channels. These channels facilitated whatever lateral movement of the pitch was necessary to accommodate any tendency for a build up of polishing material. The soft pitch layer quickly assumed the figure of the mirror surface. The literature contained conflicting advice about the best consistency of the pitch. Mudge had recommended a soft texture for the pitch, while Edwards favoured hard pitch. Eventually it was decided to use a compound layer the outer component of which would be relatively hard and the inner layer close to the metal of the tool remaining soft. After systematic experiments a recipe was established for making a suitable outer coating for the polisher. It was based on resin, turpentine and wheat flour, boiled to expel moisture and to achieve the correct hardness. This was harder than pitch and also contained fewer impurities in the form of particles, and yet still allowed the particles of polish to embed successfully. It was rolled out as a 30 thousandth of an inch thick layer and then applied as a second coating over the pitch. Moistened iron oxide paste was applied as a polishing agent. Refinements were introduced. It had been observed that temperature was of great importance; as far as possible it should be kept constant and close to the eventual operating value. Arrangements were therefore made to partially submerge the speculum in a water bath maintained at 55 °F during both grinding and polishing (see figure 5). The fact that the polisher was counterweighted allowed a very precise adjustment of the contact pressure between the surfaces. In the case of the 36 in mirror this amounted to no more than one hundredth of a pound per square inch.

The removal of the very slight extra amount of metal at the outer radii needed to create a parabolic figure was achieved at the polishing stage by altering the adjustments of the machine so that it tended to produce a surface with a greater radius of curvature at the edges. A suitable adjustment was found which gave an acceptable approximation to a parabolic shape. Obviously from what has been said it was quite impossible to judge the progress towards a true parabolic surface by mechanical measurements alone. The only adequate test was to view an object, and judge the quality of the image. Ultimately the object had to be stellar, but it was possible to make interim tests by viewing the face of a pocket watch mounted at a distance. This could be done because the mirror, while sitting on its horizontal bed, could be drawn from the workshop to the base of an adjacent tower which was contiguous with it. Each floor in the tower had a trap door cut in it so that there was a clear view of the tip of a flag-pole mounted at the top of the tower (it can be seen in figure 9). For mirrors up to 36 in diameter and 27 ft focal length, the image of a watch dial fixed to the flag-pole could be viewed by using a temporary eyepiece placed at a suitable position previously determined by calculation. Parsons followed Mudge in using discs and diaphragms

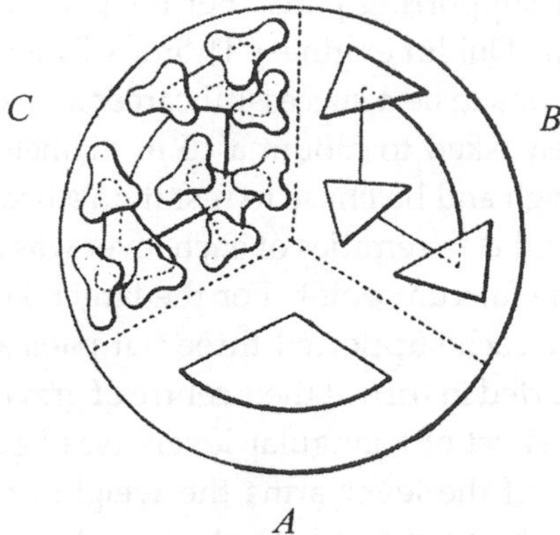

Figure 11. Lever support for speculum, viewed from below, *A* is one of three primary levers, *B* is a set of three secondary levers and *C* a set of nine tertiary supports (Parsons W 1861).

to explore the quality of the figure at different radii. From his records of earlier experiments carried out on his polishing machine, he was able to make a suitable choice of settings so that the polishing machine would correct any deficiencies that were encountered. It was not until 1859 that the French scientist Leon Foucault described his knife edge test that made it possible to assess accurately the optical quality of any region of the surface, not just circular or annular areas (Ashbrook 1984b).

To complete the telescope, a flat mirror was required to divert the converging rays of light from the main speculum along a line at right angles to the tube axis, to a point where the observer could be located. This was made of the same alloy, in the shape of an ellipse 3 in by 2 in. For this a 3 in diameter polisher was employed. It was practicable to create a plane surface because it had been found that short strokes of the polishing machine tended to create a concave surface, while long strokes yielded a convex surface, so that by a judicious choice the desired result could be achieved.

Having taken so much care to figure the main speculum as a parabola, it was important to avoid any flexure caused by the mounting. Despite its great size and weight it was much less stiff than the smaller mirrors and required much ingenuity if it was to be supported without causing the surface to change. In modern instruments like the 4.2 m William Herschel reflector telescope that was supplied in 1985 by the firm of Grubb Parsons for the La Palma observatory, this task is accomplished by the use of 64 pneumatic supporting pads. For his part Parsons borrowed an idea employed by the Dublin engineer Thomas Grubb who was just then embarking on what was to be a successful career as a telescope builder. In 1835 Grubb had been asked to mount a 15 in diameter speculum for the observatory at Armagh and he chose to rest it on a bed of levers. The idea was that the thrust at the extremities of each lever was transferred by arms of equal length to the fulcrum point. For the larger 36 in mirror, triangles were used and these each supported three flat pads at their apices. The triangles were supported in turn at their centre of gravity by pivots located at the tips of a further set of triangular levers (see figure 11). By suitable choice of the length of the lever arms the weight of each region of the speculum could thus be supported evenly over the entire surface. As the diameter of the mirror was increased a further layer of triangular levers could be added, thus yielding nine, 27 or 81 supports. When machining the 36 in speculum Parsons mounted it on its bed of levers immediately after it was removed from the annealing oven, and it remained there during all subsequent activities of grinding and polishing. To avoid vibrations the levers were designed to be massive. Another idea was taken from the observatory at Armagh, where the Rev. Dr T Romney Robinson (1792–1882) had become director in 1823 (he is not to be confused with the Archbishop Robinson who had established the observatory at Armagh (Bennett 1990)). He had designed a box of sheet metal to cover the 15 in speculum when it was not in use. This carried a charge of quick lime that counteracted the tendency of the mirror to tarnish in moist air. To ensure a snug fit of the box, the circumference of the mirror was first carefully machined. Over the years Robinson became a close friend of William Parsons and of his eldest son, and was a frequent visitor at Birr Castle.

Meanwhile the wooden Ramage style altazimuthal stand had been completed. It was made more manageable by counterweighting both the telescope tube and the moving platform that gave the observer access to the Newtonian focus (figure 6). These changes considerably reduced the physical effort required from the attendants who helped the observer to keep a heavenly object continuously in view. The massive mirror assembly was located at the base of the 26 foot long tube, and rested on two wheels. It was carried around in a circle by the wooden superstructure that pivoted about a central support. This pivot could be levered so as to take most of the weight off the wheels on which the structure rested. Of course the instrument was out in the open, so a small observatory building was also constructed nearby. Records and instruments could be kept there and it also provided some warmth at night time in the intervals between observing.

There were now two 36 in mirrors, the one a composite arrangement and the other a single piece design. The telescope was ready for use in September 1839 and it was turned towards a number of 'test objects' that had been listed by John the son of Sir William Herschel. It was thought, mistakenly, that the nebula listed in Messier's catalogue as M27 was resolvable into individual stars, in other words that the cloudlike appearance was due only to a multitude of close packed stars. Magnifications of 100 to 600 were employed, but Parsons remarked

I am anxious to guard myself from being supposed to consider it certain that they are actually resolvable . . . nothing but the concurring opinion of several observers could (establish) it as an astronomical fact.

The maximum magnification usable for viewing the moon he found was 300. Details were observed that were not visible to other observers until many years later (Moore 1971). In October 1840 Romney Robinson visited Birr in the company of Sir James South the noted English astronomer. They were able to make observations with both mirrors. In many respects the two performed equally well, but with very bright stars the diffraction effects caused by the joins between the multiplicity of mirrors in the compound design were clearly evident. Robinson judged this to be '*the most powerful telescope that has ever been constructed.*' Both men were amazed to hear of Lord Oxmantown's plans for a 72 in diameter instrument. Writing in 1840 he said (Parsons W 1840)

I think that a speculum of six foot aperture could be made . . . , but added that "*the construction of such an instrument would be a serious task, and I should be sorry to attempt it till, after additional experience in observing, and further opportunities of comparing the two three foot specula already finished, I felt more competent to do justice to the undertaking.*

Meantime he planned

to continue the experiments already in progress, and to arrange the details of the mechanism necessary to render so large a tube conveniently manageable. Everything then, having been previously determined with care, subsequent alterations would not be required; tedious experiments would not now be necessary, either in constructing the speculum, or in the less interesting but necessary task of acquiring a practical knowledge of the mechanic arts; and an instrument even of the gigantic dimensions I have proposed might, I think be commenced and completed within one year.

With the exception of the time scale, this proved to be an accurate forecast.

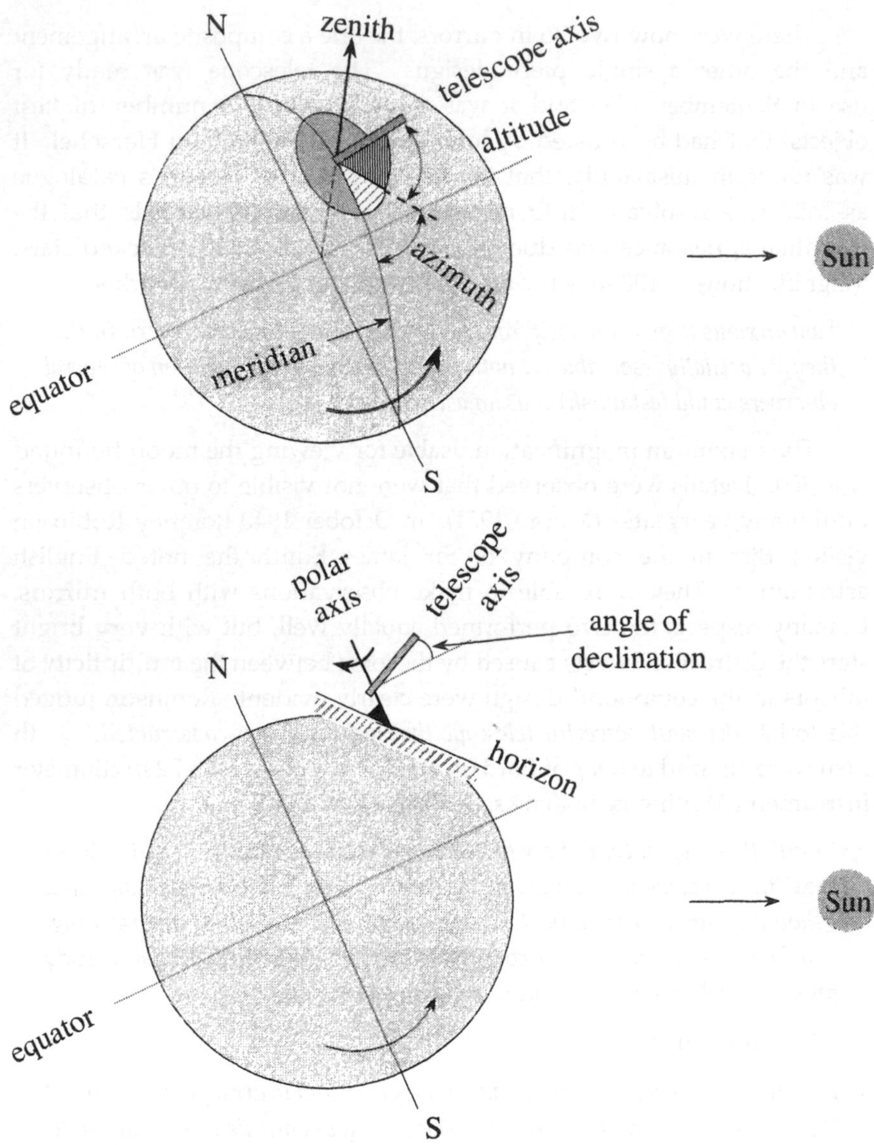

Figure 12. Schematic illustration of telescope mounts, altazimuthal (top) and equatorial (bottom).

At an early stage Parsons had contemplated whether to chose a refractor design or a reflector, and had come down on the side of the reflector. There was a further choice between an altazimuthal mounting and a so-called equatorial mounting. With the latter the telescope is attached to an axis that is parallel to the earth's axis. If the tube is pointed to some object in the heavens it will remain trained on it provided only that the mounting rotates uniformly about this axis at a rate of once every 24 hours. As a practical task this is much easier to achieve than the two simultaneous adjustments that are required for the altazimuthal mount. The great advantage of the equatorial mount only became fully evident with the invention of photography. The longer a photographic plate is exposed, the stronger is the image that is captured on film. Consequently lengthy exposures of a faint celestial object can make up for small light gathering capacity in a telescope, that is for small lenses or mirrors. But pending major developments in photographic technique, reliance had to be placed on the acuity of vision of the naked eye. In this situation a mirror or lens of large aperture was the only method of revealing faint objects. Some nebulae are large and stretch over regions of the sky that are comparable in size with the moon, but their vastly more feeble light meant that nothing like the same detail could be made out, and so some imprecision was acceptable in exchange for a brighter image. Given that nebulae were to be his chief concern, and that photography of stellar objects was not to become a reality until long after 1850, William Parsons' choice of a very large reflector design was entirely appropriate, even if its useful life was certain to be limited.

It should be remembered too that refractors mounted equatorially presented their own problems. In 1845 Robinson reported that the French Government on the recommendation of Mon. Arago proposed to construct an achromatic refractor with an aperture of 1 m (Robinson 1845). Such a lens was expected to weigh 400 lb. *'Supposing homogeneous discs of glass can be obtained and wrought of that magnitude, there remain other difficulties.'* It would be difficult to mount since *'it is known that a very modest pressure produces in glass a double refraction'*. Even with relatively quite small lenses, much engineering effort was required before a sufficiently stiff mounting was achieved for them to be useful for long duration photographic exposures. Parsons' resort to the relatively cumbersome altazimuthal design was therefore not such a grave disadvantage as it might first appear. Neither was precision in measuring the orientation of the telescope of critical importance for the program of work that he had in hand, because it was the shape and structure of the nebula which mattered, not their exact location.

Chapter 4

Engineering of a Giant Telescope

4.1 Scientific engineer

It was not until 1839 that Mary Parsons gave birth to her first child, a girl Alice, at the age of 26 years. As for all married women of the time, whether rich or poor, the next 15 years were to be dominated by child bearing. It was to be a period of much tragedy. In the years just prior to this, the death rate in England and Ireland had declined perceptibly and as a consequence the population was increasing sharply. All the same, infectious diseases were no respecters of wealth. Four infants born in 1842, 1843, 1845 and 1850 were to die even before they could be christened. Then her eldest child, Alice, was to die of rheumatic fever at the age of eight years. That left only boys, and two of them, William and John, were just eleven years of age when they too died in 1855 and 1857. Eventually only four boys grew to manhood, Laurence, Randal, Clere and Charles (Davison 1989). A not dissimilar experience was the lot of Mary Ward (1827–1869), who was a first cousin of William Parsons. Though 27 years his junior, she was a constant visitor at the Castle and became a good friend of Mary Parsons. She too had 11 pregnancies but gave birth to only eight live infants (Harry 1988). In one sense both mothers were fortunate in that they at least did survive the hazards of childbirth which carried off so many women at the time. Despite her commitment to rearing her family, Mary Ward became an accomplished artist and illustrator and was the author of several popular books on telescopes and microscopes. Like Mary Parsons she was known for her independent personality.

In 1841 Laurence died at Brighton at the good age of 83 years. His widow continued to live in Brighton where her grand-children would later visit her. William, now Earl of Rosse, began to receive recognition

Figure 13. Photograph of Mary Countess Rosse, probably taken by the third Earl of Rosse in the early 1850's. © Birr Scientific and Heritage Foundation by courtesy of Lord Rosse.

for his scientific work. The following year he was awarded the degree of Ll.D by Cambridge University. When the British Association for the Advancement of Science held its meeting in Cork in 1843 he was invited to preside. It was at this meeting that James Prescott Joule reported on his early experiments to show the equivalence of heat and work (Cardwell 1975). He had to address the Chemical Section because the Physics Section was full up. He had three supporters, Lord Rosse, Professor Apjohn of Dublin University and an engineer, Eaton Hodgkinson. The Association had come into being in 1831 at a meeting in York (Macleod and Collins 1981). The founders had been influenced by reports of similar ventures in Germany that had been publicised by Brewster and Babbage. Like the Royal Astronomical Society it was in part a reaction to the exclusive and unrepresentative character of bodies like the Royal Society. Its objectives were stated to be '*to give a stronger impulse and a more systematic direction to scientific enquiry; to promote the intercourse of those who cultivate science; to obtain more general attention for the objects of science*'. Further it was asserted that '*the Association contemplates no invasion of the ground occupied by other Institutions*'. It met once each year in different cities, always outside London. William Parsons was an active participant and made five written contributions to meetings between 1843 and 1859 (S.P.(1)). Such involvement gave him a convenient access to the scientific community as well as providing a vehicle for making his work more widely known outside that community.

Big houses like Birr Castle were normally the centre of hospitality for visitors coming from near and far, and this made demands on the mistress of the house who was regularly pregnant during this period. In later years Randal Parsons wrote a private account of his childhood at Birr Castle in which he listed the Irish relatives, the English relatives, the neighbours, the clerics and the scientists who all enjoyed the hospitality of Mary Rosse (Parsons R). Many like Romney Robinson, James South and Edward Sabine[1] were associated with the Earl's scientific interests. As Lord Rosse's reputation grew such guests steadily increased in number.

The programme of systematic experimentation from 1826 to 1839 had produced solutions to the main technological problems in constructing giant reflecting telescopes, but still Parsons had not attempted to engage in a programme of conventional observations. Instead he set about constructing an instrument with a mirror of 72 in diameter and a focal length of 54 ft, twice that of the 36 in. To put this in perspective it may be noted that John Herschel had finished his work with his 20 in speculum in 1838 when he returned home from South Africa (Ashbrook 1984). In county Sligo in Ireland, E J Cooper had installed at Markree what was the largest refractor in existence at the time. In 1831 he had purchased an object glass of 13 in diameter from Cauchoix of Paris, and the Dublin telescope maker, Thomas Grubb, supplied an equatorial mounting for him (McKenna-Lawlor 1988).

In this phase of his work the obstacles to be faced by Parsons were not so much technological as logistical. The much more ambitious scale of the enterprise required careful planning and close supervision to bring it to a successful conclusion. The responsibility all fell on one man's shoulders. In later times the construction of giant telescopes was a task given to teams of workers, supervised by committees of experts. But in this venture there was neither committee to steer the project, nor sub-contractor to take on sections of the task. It has already been noted that William Parsons had gathered and himself trained the team that was necessary for the manufacture and experimentation which was conducted at Birr Castle. Regrettably little is known of the personal history of these individuals. One exception is William Coghlan who was recruited in 1841. He was from a neighbouring village and was a locksmith by trade. At Birr he learned iron and brass founding as well as metal turning and grinding. He helped in

[1] E Sabine FRS was born in Dublin of English parents. He was the General Secretary of the BA for 20 years and later was President of the Royal Society.

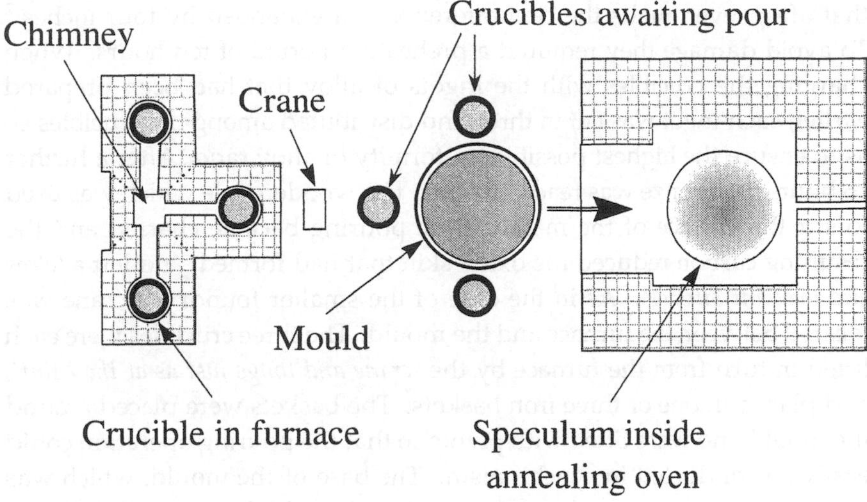

Figure 14. Plan view of the second foundry and oven, located in the moat, based on Anon (1844).

the casting of the 72 in mirrors as well as in the erection of the telescope mounting. He retired as mechanical assistant at the age of 79 years, and died one year later on 7 March 1896 (Parsons L 1896).

4.2 A 6 ft diameter reflecting telescope

The first step was to construct a new and larger furnace complex further from the Castle (Woods 1845, Robinson 1845, Parsons W 1861). A moat had recently been excavated around the rear of the Castle, and it was here that it was built. Everything was in the open air, which must have been an advantage for the workmen involved in handling molten metal, but it also carried a penalty from the uncertainty about the weather. The quantity of metal required would be much greater than for the previous designs. In order to prepare the four tons of bronze needed for a single casting, three cast iron crucibles had to be heated simultaneously. Each was seated in its own furnace four feet in diameter by six feet deep, with all three furnaces sharing a common chimney. The crucibles were manufactured in London by Dewars who followed the specific instructions that Parsons himself had evolved in the light of his own earlier experiences. Each weighed a half ton and measured two and a half feet deep by two feet in diameter internally. The duty that they were required to undergo was severe; it was observed that after seven melts the circumference had expanded by four inches.[2] To avoid damage they required a preheating period of ten hours. When charging the crucibles with the ingots of alloy that had been prepared earlier, each ingot was cut in three and distributed among the crucibles so as to ensure the highest possible uniformity of alloy ratio. After a further 16 hours the bronze was ready to pour. The wooden pole which was used to stir the surface of the metal before pouring became charred and the resulting carbon reduced the oxide skin that had formed, and left a *'clear and brilliant surface'*. As in the case of the smaller foundry, a crane was located between the furnace and the mould. The three crucibles were each lifted in turn from the furnace by the *'crane and tongs just as at the Mint'*, and placed in one of three iron baskets. The baskets were placed around the mould and were fitted with pivots so that the pouring operation could easily be conducted in synchronism. The base of the mould, which was formed from coiled iron strips four inches deep, had an overall diameter of seven feet, and its sand walls had a depth of five and a half inches. The mould was carefully levelled and preheated to ensure that it was quite dry. The pouring took only seconds. The ripples and waves on the molten surface died away slowly and within 20 minutes the whole had solidified.

The annealing oven was placed at the other side of the mould, and was directly opposite the melting furnaces. This was convenient for drawing the solidified casting straight into the oven before its temperature could drop appreciably. As soon as the casting had been separated from its base, a hoop attached to a chain was placed around it. The chain passed through the oven and was wound in by a capstan on the far side. In this way the hot disc was slid in over two tracks. This oven is the only part of the complex that has survived to the present time. It was constructed of brick with walls four foot thick at the sides and two foot thick at the rear. It has an interior space of eight foot by ten. The floor on which the casting rested is located over four arches that required to be strengthened by bolts, because it was important to ensure that the shape of the casting, which weighed four tons, was not distorted during the six week long annealing period. A preparatory period of several weeks was needed to raise the interior of the oven to a dull red heat of around 900 °F. Once the casting was safely inside, it was packed around by charcoal and all openings into the oven were closed off. Great quantities of turf were used as fuel. It was estimated that for the complete operation 2000 cubic feet were required and a large store

[2] The tendency for cast iron to 'grow' when exposed to elevated temperatures later troubled his son Charles.

was constructed to hold it. Supplies had to be laid up in advance because turf must be harvested in the summer and dried. He had to make several attempts, five in total before two usable specula were obtained. The first successful cast was made in the evening of 13 April 1842. It was described by T R Robinson, the visitor from Armagh,

> *The sublime beauty can never be forgotten by those who were so fortunate as to be present. Above, the sky, crowded with stars and illuminated by a most brilliant moon, seemed to look down auspiciously on their work. Below the furnaces poured out huge columns of nearly monochromatic yellow flame, and the ignited crucibles during their passage through the air were fountains of red light, producing on the towers of the castle and the foliage of the trees, such accidents of colour and shade as might almost transport fancy to the planets of a contrasted double star. Nor was the perfect order and arrangement of everything less striking; each possible contingency had been foreseen, each detail carefully rehearsed; and the workmen executed their orders with a silent and unerring obedience worthy of the calm and provident self-possession in which they were given.*

As has been mentioned, the casting was placed on an array of flat supports. These were located in such a way that each carried an equal part of the weight of the mirror, and were supported at the tips of nine triangular levers. The latter were in turn supported by another layer of three triangles that sat on three supporting columns. Each column carried a universal joint and was fixed to a sturdy cast iron carriage running on four wheels. Three supports were chosen because, not unlike a stool with three legs, it would remain steady on any surface. Any larger number of support points would present difficulties of alignment if each were to bear an equal load. The upper surface of the casting was of course flat. The operation of withdrawing the cold disc from the oven and turning it so that the concave surface faced upwards ready for grinding and polishing required great care. It was important to avoid subjecting the casting to strains that could distort or fracture it, for it was now extremely brittle. A pit four feet deep was made near the oven and the casting was slid into it on planks of wood until it was vertical. Cloth wrapped iron bars and wooden planks were bolted to it as protection while it was hoisted onto its array of supports and levers. Once on its carriage it remained there during grinding and polishing and eventual mounting in the telescope tube. When in use the mirror face, and its carriage, could be tilted until it was almost vertical. So that the mirror did not slide off its supporting levers two iron ridges attached to the carriage which were arranged to

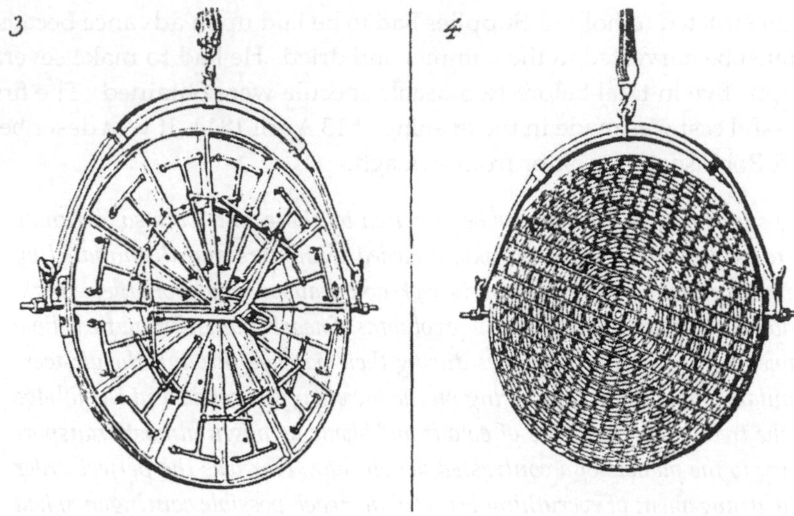

Figure 15. Grinding tool for 6 ft speculum. On the left is a view of the upper face showing the load sharing levers and shackle, and on the right can be seen the working surface machined with circular and rectangular channels (Parsons W 1861).

support the rim of the mirror. Also, in this position the not inconsiderable weight of the secondary levers was carried by wires. These were attached to a series of weighted, bell crank, levers at one side of the carriage (figure 19). With these precautions it was hoped to avoid putting stresses on the mirror that would spoil its parabolic shape. These arrangements added an extra one and a half tons to the weight of the mirror assembly.

A second grinding and polishing machine was built especially for the larger 72 in specula and located in the same building as the earlier machine. Grinding and polishing of large mirrors even today is an art as much as a scientific procedure. The grinding process alone could take up to two months to complete, and must then be followed by a polishing process. The machine that Parsons had perfected for the purpose adopted an arrangement where the polishing or grinding tool was arranged above the mirror, which was the reverse of the usual practice. It was of course just as important that these tools should not lose their shape during storage or use. Therefore they too were hung from a shackle that supported the weight through an arrangement of levers similar to that used for the speculum itself. Also, to allow access for coating the surface of the polisher with pitch the tool could be supported in a gimbled ring (figure

15). As before, the grinder or polisher was counter-weighted so that its entire weight did not press on the surface being ground. The design of the machine was continually evolving. Since the grinding or polishing tools were only partly counterpoised, they were dragged around slowly by the frictional forces of the mirror as it was rotated beneath them. Although the original machine used leather belts and pulleys such as were commonplace in workshops for distributing power from a central source, later versions employed chains. A travelling crane arrangement was built above the machine to facilitate the moving of tools and the lifting of the mirror and its carriage onto a trolley for transportation to the telescope. Reflecting on the evolution of the grinding and polishing machine Parsons wrote

> *Some are surprised that machinery so rude should be employed, and successfully, in a mechanical operation where the utmost precision is required, a precision almost fabulous.*

The first casting was made with slightly less tin than intended. It fractured while being ground. Meanwhile work was completed on '*a powerful lathe for turning the grinding tools, with a slide rest moving in the proper curve*'. It may be remarked that James Nasmyth, who was also an amateur astronomer, became a friend of Parsons, no doubt because both were members of the Royal Astronomical Society. Nasmyth was famous as an engineer for his invention of the steam hammer, but he also contributed to the development of the slide rest for use with lathes. This new lathe was used to turn the iron base of the mould to a suitable convexity.

For the second casting the proportions of the metal mix were corrected, but due to fissures that appeared in the concave surface it required two months of continuous grinding before it was ready for polishing. This was possible because the machinery was powered by a steam engine, and it could work night and day if necessary. The fissures were caused because the base of the mould had only been turned to shape. Its surface had not been ground and so it was rough enough to resist the contraction of the bronze as it solidified. For the next attempt a tool made from sandstone was assembled from many pieces and used to grind the surface of the base of the mould to a smooth finish. But once again there was a failure, and the casting fractured while still in the annealing oven. This was attributed to the existence of temperature gradients in the oven caused by the fact that the thickness of the end wall was less than that of the others. A fourth casting proved to be porous, and it too fractured. It was thought that the tallow coating that was applied to the iron base of the mould to prevent rusting after it had been machined was not completely burned off before the metal was poured. Finally the fifth attempt was completely successful. Grinding and polishing took only one month to accomplish.

For the 6 ft speculum the grinding tool, like the polisher, was made of iron and itself weighed one ton. Therefore the forces required to move it to and fro were quite substantial and great care was needed lest it should make unintended contact with the speculum causing it to shatter. The mould for the casting divided the surface of the tool into squares 2 in wide, and afterwards a series of concentric circular grooves was cut in the face (figure 15). The surface of the tools was turned to the desired shape. Gauges were constructed of brass and used to check the surface profile with an accuracy of one thousandth of an inch. The counterpoise weight used when grinding was adjusted progressively, to reduce the net force from 784 lb at the start of operation down to 224 lb at the end of the procedure.

It is clear that despite all the earlier work on the 36 in mirrors great persistence and determination were required to achieve success. From the time the fires were lit in the annealing oven until the cold speculum was removed for machining, up to three months could elapse. Since metal could not be kept for long periods in a molten state careful planning was needed so that all the processes dovetailed together. Parsons had the courage to persevere despite three failures because his past experience had made him confident of his ability to master the difficulties. Even while the manufacture of the mirrors was still in progress a start was made on the mounting of the telescope.

The size of the mirror was so great that the wooden structure that was used for the 36 in was quite inadequate and instead it was decided to build a structure out of cut stone. The decision was no doubt influenced by the experience in 1839 of the 'Great Wind', a devastating storm which raged all over Ireland and which had toppled the wooden Ramage style mounting from its base. He was now committed to an altogether greater commitment of resources. The work of actually designing the telescope tube and the universal joint upon which it was to be supported, together with the working out of the details of the masonry, of the access platforms and of the method to be used for precision positioning of the tube, must have taken a long period to work out. Strangely, little was written about this aspect of the work. There is a wooden model of the telescope, now in the Science Museum in London, which was originally loaned for an exhibition in London in 1879. Because it contains an extra viewing gallery that was never built, it seems probable that this model predates the construction of the instrument, and may have been used to work out details of dimensions and layout (Tubridy 1998). Correspondence in 1843 with the Astromomer

Royal, Sir G B Airy (1801–1892), makes reference to 'Lord Rosse's model'. As a result of the decision taken by the seventh Earl of Rosse in 1996 to restore the 72 in telescope to its original condition, there is now a much better understanding of the engineering aspects involved in its original design and construction.

In February 1843 work was started on two massive stone walls that were built on a north–south alignment, some 250 yards in front of the Castle in a north westerly direction (Robinson T 1845). Occupying such a prominent position in front of the castle, the masonry on the western wall was given architectural treatment, with gothic arches and crenellations. It was just another item in the considerable quantity of stonework built around this period as improvements and decorations, partly to give local employment. The walls were 70 ft long and 50 ft high. A massive stone plinth was placed between them and it carried the cast iron universal joint to which was fixed the mirror and its 46 ft long tube. Recent strength tests have demonstrated that the plinth is indeed quite rigid enough for this purpose. This arrangement allowed for access to a full range of altitudes, but azimuthal movement on the other hand was restricted by the walls, so that a celestial object could be followed for a period of no more than 30 minutes or so on either side of the meridian. Given the purpose for which the telescope was designed this was an acceptable limitation. An equatorial mount, in which one axis is arranged parallel to the earth's axis, requires that only one co-ordinate needs to be adjusted in order to follow the diurnal motion of a star. A relatively small, 18 in instrument had already been successfully installed in such a mount at Birr, but the great size and weight of the 72 in instrument ruled out such an approach for it (Parsons L 1868).

The main mass of the speculum together with its carriage was directly supported by the cast iron universal joint (figure 17). The tube of the telescope was made of 1 in thick wooden staves held in place by iron hoops and strengthened internally by iron diaphragms. It was 46 ft long and 7 ft in diameter, swelling to 8 ft in the middle. When its construction was completed a hole had to be made in the gable end of the building in which it had been manufactured to facilitate its transportation to the telescope mounting. A cube shaped wooden structure, the 'mirror box', with sides 8 ft long served to join the tube to the universal joint. It also housed the mirror on its carriage and protected it from the elements. In order to make it possible to adjust the elevation of the telescope tube a chain was attached to the top of the tube. This ran over a pulley mounted high up to the north and then down to a windlass on the ground below. When

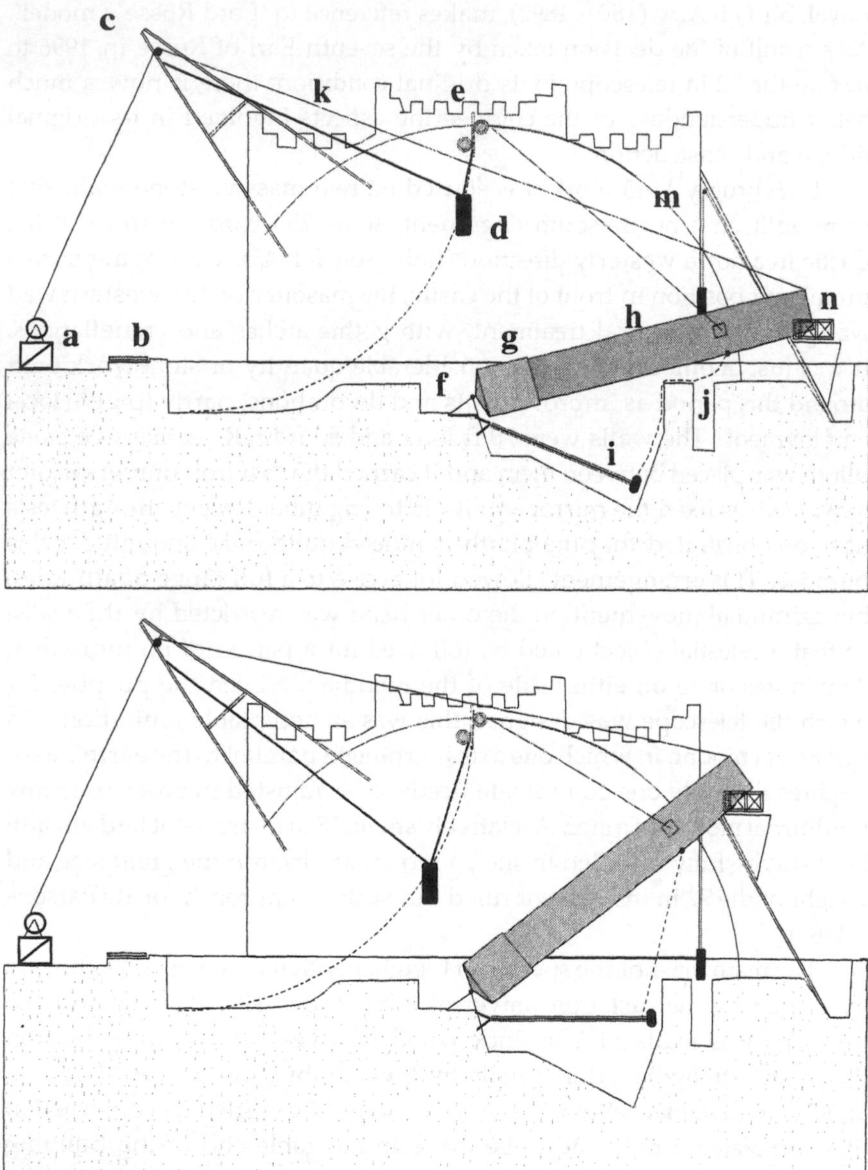

Figure 16. (Top) Arrangements for elevating the giant telescope: a, elevating winch; b, turntable; c, wooden 'horns' ; d, counterweight for tube; e, swivelling pulleys; f, universal joint; g, mirror box; h, tube; i, counterpoise lever; j, counterpoise chain; k, elevating chain; m, segment of a meridian circle; n, lower observing platform. (Bottom) Telescope tube in an elevated position.

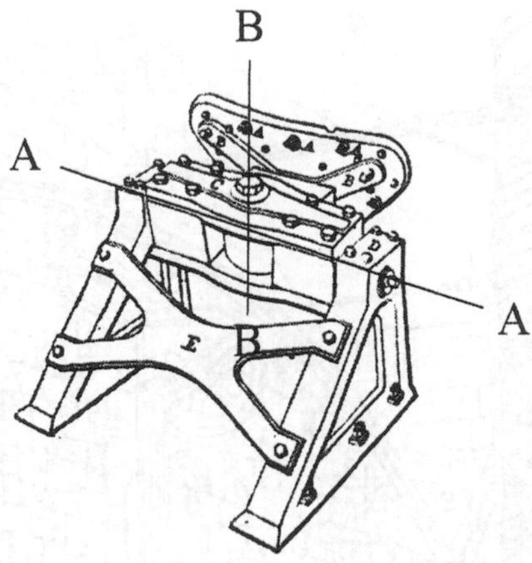

Figure 17. Detail of the universal joint to support the speculum. A–A is the horizontal axis, while B–B allows east–west motion (Parsons W 1861).

the tube was 'at rest', roughly at an angle of 15 degrees to the horizontal, this gave poor leverage, which meant that the tension in the chain must be very great. In order to ensure that the effort required at the windlass would be manageable, the idea of applying counterweights, first used for the wooden Ramage style mount, was again employed (see figure 16).

An elaborate arrangement of counterweights and chains was evolved. Two extra chains, attached at points two thirds of the distance along the tube, ran over a pair of swivelling pulleys mounted on the inside faces of each wall (figure 18). The other ends of these chains were attached to a wooden 'horn' mounted high up at the northern end of the walls. Massive cast iron counterweights were hung near the midpoint of each chain. Each was said to weigh some two tons, though it is more likely that this was near the combined weight of the two (Wayman 1995). As the tube was raised closer to the vertical, the force required to support it became smaller. In part this was allowed for because the counterweights were constrained by the chain to move in an arc. A further refinement was added. An extra chain was hung from the front of the tube into a pit that was located to the south. As the tube became more erect this chain raised heavy iron levers from the pit, one after the other. The forces exerted by these levers helped to balance those due to the main counterweights. In this way they also ensured that the tube could still be controlled after it had passed the vertical and was pointing in a northerly direction. While calculations were made for the design, the details were modified several times in later years

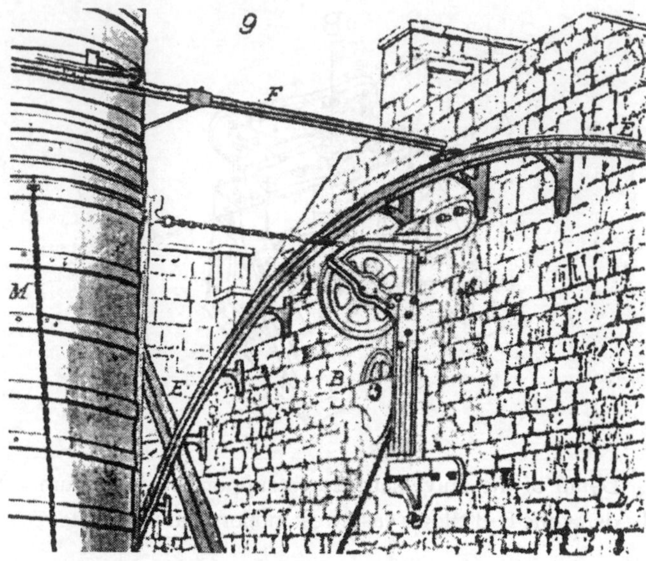

Figure 18. View of east wall, showing the meridian circle and swivelling pulleys to carry the counterweight chains (Parsons W 1861).

in the light of experience. In total the weight of iron used was stated to have been around 150 tons, but again this seems to be an overestimate. This work was largely completed by the early months of 1844 (Robinson 1845). The writer of the obituary for Lord Rosse, published by the Institution of Civil Engineers (Anon 1867b), wrote that

> *all this massive work was executed in Lord Rosse's laboratory at Birr Castle, and* (that) *he had so thoroughly considered the whole, that in no instance was it found necessary to deviate from the drawings which had been prepared.*

The elevation angle of the telescope tube was judged with the aid of a circular arc 42 ft in radius that was fastened to the east wall (see figure 18). The circle was made up of machined 5 ft segments carefully set to lie precisely in the plane of the meridian (Robinson 1845).

> *Sir James South took charge of this delicate operation, and performed it with such precision that when a transit instrument was adjusted by this line it gave the passage of Polaris to a small fraction of a second.*

A horizontal bar carried a wheel that rested on this circle. The tube could be moved from side to side by a worm drive attached to this bar. This drive could be operated by an attendant below, with fine adjustment by the

Figure 19. Speculum (mirror) on its carriage viewed from the west, showing the cover to exclude moisture when not in use, and the bell crank levers to counterweight the triangular supports (Parsons W 1861).

observer. A fixture at the universal joint supplied with a spirit level and a circle of 18 in radius facilitated approximate azimuthal adjustment. Also a pointer of 6 ft in length was attached to the horizontal axis of the universal joint to assist in setting the altitude of the tube's orientation. The telescope did not have the usual 'finder' telescope, with a wide field of view, to assist the operator when searching for a specific object. Instead a special low power eyepiece was used that gave a field of view of 31 minutes of arc with a magnification of 216 times (Robinson 1848). The angle subtended by a full moon is near enough equal to 31 minutes. A holder permitted either of two eyepieces to be slid into place as required with a minimum of difficulty. Sir James South was on record as saying that from uncovering the speculum he could locate a selected star in eight minutes, with the aid of the team of assistants.

The observer had to be positioned at the mouth of the tube where a small plane mirror turned the light rays through 90 degrees to reach the eyepiece. In August 1844 work began on the construction of a wooden gallery upon which he could stand. It had to move up and down along inclined guides that were fixed to the southern edges of the walls in order to accommodate different angles of elevation (figure 23). On it was mounted a carriage running on rails that could be traversed horizontally to allow for changes of the azimuthal angle. Counterweights were fitted so that the necessary adjustments to the height of the gallery could be made by

Figure 20. View looking south east of the 3 ft and 6 ft telescopes and the little observatory building, with Birr Castle in the background. The 6 ft speculum is being moved into position at the northern end of the telescope. From a water colour by Henrietta Crompton. © Birr Scientific and Heritage Foundation by courtesy of Lord Rosse.

one attendant. The wooden structures were subjected to careful testing for strength before they were put in place. Up to 12 persons could be carried on the observing platform, and such was the attraction of this novel instrument that on occasion such numbers had to be accommodated. The access gates were designed to open inwards and were provided with safety catches. Shields were placed to avoid injury to attendants below in the event of an accident such as a dropped eyepiece (Robinson 1848). Such attention was not always paid to safety on large telescope installations of the time. In fact there appears to have been no record of a serious accident during the working life of the instrument. To make changes in elevation, the main windlass was operated by assistants who could be up to 100 feet distant from the observer. Once the elevation of the tube had reached a certain angle it was necessary for the observer to move to a second gallery. This was arranged to extend outwards from the west wall across the space between the walls as far as required. A third similar gallery allowed for observations closer to the celestial north pole, but a planned fourth gallery was never built, figure 23a. The massiveness of the whole construction meant that it was not easy to manipulate, but on the other hand there was the advantage that even in strong winds the orientation of the tube remained rock steady.

4.3 The Observatory of Birr Castle

The fame of the telescopes at Birr was spreading rapidly. The issue of 9 September of the newly launched *Illustrated London News* of 1843 carried an article, no doubt the work of someone local, with pictures of Birr Castle, the workshop building and the 36 in telescope mount 'on the lawn' (Anon 1843). It is rather adulatory in tone but it does give a detailed and accurate account of the steps in construction from the casting to the grinding and the polishing of the giant mirror, with an account of the future arrangements planned for supporting the tube. The author concludes

> *Lord Rosse's pleasure grounds are most elegantly and tastefully laid out. A large lake has been lately added to the other beauties of the place, and has given his lordship an opportunity of trying his skill as an engineer... As the bed of the river was low near where the lake was intended to be, an aqueduct was cut communicating with the river higher up its source, and when it was brought to the required situation, a tunnel was sunk under the original bed of the river, and thus one stream runs over the other, both supplied by the same source. It would be an injustice to the Countess of Rosse were this short notice of the demesne concluded without acknowledging the debt which the people of Parsonstown owe to her. She has with the most exquisite taste improved and made delightful the grounds of the castle, and freely opens them for their accommodation... She has raised the tone of its society; but she has done what reflects much more credit on her mind; she has taken the most lively interest in the poor, and is constantly improving and changing to afford them work. The lake was commenced solely to give them employment, and, since then, hundreds have been daily hired to do what but for beneficence might well remain undone. The consequence of this conduct is, she is universally esteemed and looked up to, and that her town is almost free from the discontent and distress that is so rife in other places.*

By August 1844 a first polish had been given to the mirror. When it was ready, the mirror on its carriage was lifted in the workshop onto a sturdy four wheeled cart. It appears that it was first used, in a rough, temporary mounting on a slope in front of the castle (Kilgour 1844). Those who looked at its images were amazed. One witness could not understand why there was so much excitement until he was told that until then the two stars in view had always been taken to be one object. It was then moved a quarter of a mile or so further by a score of men to a point to the north (rear) of the telescope. Here the carriage was lowered from the cart onto an inclined railway track 60 feet in length which could be placed in position to allow the mirror to be run into place. The telescope tube had already been elevated to a vertical position and the carriage and the three supports for the bed of levers were firmly secured to the surface of the universal joint inside the wooden cube. The rails were then removed and the mirror was ready for use. A short length of track and a turntable were provided so that as one speculum was being brought out for installation, the other could be removed and returned to the workshops on the cart. To prevent the surface of the mirror becoming tarnished by moisture and so losing its reflecting power, it was kept covered by a box containing quick lime (figure 19). Nevertheless it was necessary to bring the mirror back to the workshops at intervals for re-polishing, and so this procedure had to be perfected. As Parsons remarks in his paper to the Royal Society (Parsons W 1861)

> *If the vivid polish of a speculum employed in the open air was as enduring as that of glass, the difficulty of the process and its uncertainty without continual practice would have no great objection to it; but when on the contrary, it is necessary to repeat the process at intervals perhaps so long that minute details are not fresh in the memory, the task becomes the labour of SISYPHUS.*

The procedure for grinding and polishing required a team with great skill and experience. This became very apparent many years later when a 48 in diameter metal speculum, made by Grubb in Dublin, was sent to Melbourne in Australia in 1868 (Glass 1997). This was made very much along the lines followed by Parsons, though it differed in having an equatorial mounting. It was unsuccessful largely because the necessary know-how for grinding and polishing was not available locally. In a letter written in 1843, J W H Herschel had drawn attention to the fact that '*Steinheil of Munich can gild by electroplating*.' But such an approach to preserving the surface seems never to have been used.

Obviously the best conditions for viewing were found in winter months, during cloudless nights when incidentally, frost was most likely. The observer was exposed to the elements on a platform that was cantilevered out from the wall, poised at a height of up to 50 ft above the ground. Romney Robinson in his account of 1845 noted that (Robinson 1845) *almost always the thermometer was at 22 or 20 deg* (Fahrenheit) *when they ceased working; and on one occasion it was as low as 17 deg, the lowest he remembered in Ireland.*

It is not surprising that contemporary paintings of the astronomers show them to be wearing heavy coats and stove pipe hats. As will now be evident, each step in the project was preceded by painstaking experiments and trials to establish, on a scientific basis, the best solution to problems. Some idea of the motivation that drove Parsons to such exertions over such an extended period may be gained from his Presidential Address to the Cork meeting of the British Association in 1843.

The love of truth; the pleasure which the mind feels in overcoming difficulties; the satisfaction in contributing to the store of knowledge; the engrossing nature of the pursuit so exalted as that of diving into the wonders of the creation; all these are powerful incentives to exertion; and under their influence great works have been undertaken in the cause of science, and carried through to a successful termination.

In February 1845 Dr Romney Robinson and Sir James South were invited to make a return visit to assess the new telescope. However the weather was cloudy. On the 15th of the month there was a chance to observe a magnificent double star in the constellation Castor, and Messier's M67 which displayed a mass of faint stars shining very brightly. When the weather failed to clear in March the mirror was returned to the workshops for repolishing. Although Parsons was not satisfied with the results he acceded to his guests' pleading and it was taken out again. During the first weeks of March the moon, double stars and nebulae were all examined. The potential of the new instrument was clear. The telescope was formally opened by Dean Peacock of the Church of Ireland. To demonstrate its size he walked along the inside of the tube wearing his top hat and with an umbrella unfurled over his head.

In April William Parsons was able to apply himself in earnest to observing with the new instrument. He chose a nebula, Messier 51 in the constellation Canes Venatici. William Herschel had described it as '*a bright round nebula, surrounded by a halo or glory at a distance from it, and accompanied by a companion.*' His son John had observed it himself with his 28 in diameter reflector, and his drawing of it was printed in a paper published in 1833 (Herschel 1833b). This showed that a portion of the encircling halo was split, giving the appearance of an ellipse lying on a circle. It suggested to him an analogy with the Milky Way, of a '*brother system*' to

Figure 21. (a) The drawing of M51, the Whirlpool nebula, which 'was handed around the section at the Cambridge meeting' of the British Association in 1845. It has been turned into a negative for comparison with a modern photograph shown in figure 21(b). © Birr Scientific and Heritage Foundation by courtesy of Lord Rosse.

our own. Although daguerreotype photographic plates were acquired at Birr in 1842, just three years after the publication of details of photographic processes, the art of photography was as yet too undeveloped for use in astronomy. And so Parsons had to make his own drawing in order to communicate what he saw to other people. This was no easy task although the nebula was a relatively large object: expressed as a fraction of the moon's diameter it was roughly 40% by 20%. He was perched high up, in the open air, shielding his eyes from lights that would reduce their sensitivity to faint images. He did not use a micrometer to measure relative distances in the image, but returned to the task on successive nights to refine the details in his drawing. With the greatly enhanced light gathering power of his instrument he could see much more detail than before: in particular he noted arms that seemed to spiral outwards from the centre. Comparison with modern photographs of the nebula shows that his sketch gives a very true likeness (figure 21). The British Association for the Advancement of Science was meeting in Cambridge in June of 1845, and the drawing was passed around among those in attendance (Hoskin 1982).

Figure 21. (b) Photograph of M51 taken with the Mount Wilson telescope, by courtesy of the Science Museum.

John Herschel who gave the presidential address praised Lord Rosse's reflector as *'an achievement of such magnitude... that I want words to express my admiration for it'*. Coming from such an expert this was praise indeed. The appearance of the spiralling arms suggested some form of motion, and the nebula was soon dubbed the 'Whirlpool' (Robinson 1848). Few if any other telescopes of the time could reveal such details and many were incredulous, suggesting that it was merely an instrumental artefact. It was not until 50 years later that Isaac Roberts put the matter beyond dispute with his photographs (Macpherson 1926). Galileo's observation of Jupiter's satellites marked the beginning of an era in which for the first time new celestial objects were added to our inventory of the sky. Parsons' observation of spiral nebulae added a new category which still remains today a subject of intense research.

Almost as soon as the giant telescope was completed, the first of many editions appeared of a booklet by Thomas Woods MD entitled 'The Monster Telescopes erected by the Earl of Rosse, Parsonstown, with an account of the manufacture of the specula, and a full description of all the machinery connected with these instruments'. As the article in the *Illustrated London News* which has been quoted earlier makes clear, Lord Rosse had pinned his colours to the mast in a most public manner by disclosing his plans well in advance. Yet he had no way of knowing *in advance* that there existed such relatively large objects that were so faint that only an instrument such as his was capable of making them visible to the naked eye. Despite this his relentless programme of work was executed with a conviction that in the end the effort would prove to have been worthwhile. He must have experienced the greatest pleasure when his efforts were so quickly rewarded. It has been observed that the greatest achievement of the giant telescope, the discovery of the spiral nebulae, was made within just one month of its completion (Hoskin 1982).

4.4 Public responsibilities in Ireland

In the year 1845 Lord Rosse resumed his parliamentary career when he was chosen as one of the Irish representative peers in the House of Lords, so following in his father's footsteps. He was created a knight of Saint Patrick in the same year, earning a distinction on his merit as a scientist, which had been declined by his father when it had been offered as a political inducement. As a land owner he opposed the repeal of the Corn Laws in 1846, but his involvement in Parliamentary business was more usually as a committee member and expert advisor. A number of other public appointments followed. He was named as a member of the Board of Visitors for the Greenwich Observatory. When an act of Parliament in 1845 arranged for Visitations of the Roman Catholic College of Maynooth near Dublin, his name was added to the Board of Visitors (D.N.B.c). He was already very friendly with its President Rev. C W Russell who was a frequent visitor to Birr Castle. Indeed in that year Russell published in the Dublin Review a lengthy account of the 'monster telescopes erected by the Earl of Rosse'. This was but one of many popular accounts of the construction of the telescopes that appeared in print around this time. Parsons was also friendly with the physicist Professor Nicholas Callan, the priest who taught science to the seminarians at Maynooth. Callan was the inventor of the induction coil, sometimes known as the Rumkoff coil after the individual who exploited it (McLaughlin 1965). The seminarians were frequently required to submit to the harsh discipline of experiencing the shocks from his coils as a way of judging the strength of their output! As well as producing voltages of tens of thousands of volts with his induction coils, he constructed enormous batteries that powered electromagnets of huge lifting power. In some respects Callan resembled Parsons. Except for his charitable contributions during the famine, his professorial salary and his life's work were devoted to research into electric phenomena. A student yarn about the two men was current at Maynooth around this time. It ran like this. Callan called to Birr to see the telescopes, but for some reason was not admitted. Later the Earl visited Maynooth to see Callan's giant coil. When notified, Callan sent word to the gateman presenting his respects, but suggesting that if the noble lord would care to return to Birr he could look through his giant telescope to view the coils!

But now once more the work at Birr was slowed almost to a halt. The year 1845 saw the outbreak of yet one more episode of potato blight, which as it turned out, was to prove the most devastating ever. The previous 50–100 years had witnessed a number of technological advances, in agriculture and transportation especially, that had reduced the death rate in the British Isles. For centuries before this, both death rate and birth rate, though high, had been generally balanced. Now the relatively modest reduction of the death rate without a corresponding fall in birth rate had caused an increase in population that had been especially rapid since the turn of the century. In England the Industrial Revolution was well under way by the middle of the 19th century. Because of this there was enough work to sustain much of the surplus population in the new industrial centres, admittedly often at the price of appalling living conditions. In Ireland, despite hopes of emulating this English experience, the necessary prerequisites for industrial growth, such as indigenous supplies of coal and iron, were lacking. Nevertheless the population grew, in some measure at least because the potato proved to be an excellent and plentiful source of food. Tenant farmers could benefit from the ready supply of labourers who required minimal space in which to grew their potato crop and build their inadequate dwellings. It was these labourers who would prove to be the most vulnerable. Their miserable circumstances were very visible to English visitors who travelled by road, and they had already been vividly portrayed for readers in publications like the *Illustrated London News*.

William Parsons sought remedies among which was the founding of the Parsonstown Agricultural Society in 1841, largely financed with his help. In his annual speech to the Society in 1843 he gave voice to his alarm at the rapid increase in population and warned that (Parsons W 1843)

> *a year of scarcity would at length come, and with it a visitation of the most awful famine, such as the history of the world affords very many examples of, a famine followed by pestilence, when the utmost exertion of the landlords*

of Ireland, of the government, and of the legislature, aided by the unbounded generosity of the people of England,... would be totally inadequate to avert the most fearful calamities.

His prophecy was borne out all too soon. The widespread outbreak of a potato virus disease in 1845 caused severe privation. The new season's potatoes were rotten and so there were no seed potatoes for the coming year. Matters were made worse by the English aversion to allowing the state to intervene. The Imperial Parliament and the civil service placed their reliance on the forces that governed the affairs of the marketplace, the so called 'laissez faire' doctrine. This meant that no effective redistribution of food took place; indeed corn continued to be exported from Ireland during the period of the famine. As people became weakened by hunger, and with unusually bleak winter weather, infectious diseases took hold. These spread so that even affluent members of the community were struck down, among them doctors and even members of the gentry who sought to assist the destitute. It is not impossible that some of the infant deaths in the family of Mary Countess Rosse were due to contagion during such epidemics. The cost of any relief, whether in the newly built workhouses or by soup kitchens, had to be met from local resources, that is by landlords. In western counties where distress was greatest the landlords themselves were often already impoverished, as the boom days for agriculture that characterized the period of the Napoleonic Wars were replaced by a period of low agricultural prices.

In these circumstances the Parsons family faced a crisis. William as Lord Lieutenant of King's County and Colonel of the local militia had a key part to play in finding remedies for local problems. With his wife he turned his full attention to the task, and the greater part of his income in succeeding years was spent on relief. Already various ornamental schemes were being pursued in the castle domain with a view to offering employment. For many who would otherwise have remained destitute this employment meant that they were able to buy food. The Office of Public Works had recently been formed, and its engineering staff were now given the task of organizing relief projects aimed at alleviating the worst distress. Among the parliamentary papers for 1847 is a report by Mr Fitzgerald written to the chief engineer Mr Mulvany, in response to his request for information about the situation in King's County (Parl P 1847),

I have yours of the 25th instant and hasten to give you a general view of the state of this part of the country... The people are generally very quiet; indeed they are not able to act otherwise; and with the exception of a few cases

of sheep and cow stealing, there has not been any violation of the laws. With respect to future crops, from what I can learn, the quantity of wheat sown is less than the average of the former years, in consequence of the small (five acres) farmers being unable to sow it . . . indeed they seemed so paralysed in mind and body that they are unwilling to look beyond their present wants, which are extreme, and which are very scantily supplied by Relief Works, whereon they are nearly all employed. Their appearance and colour of their faces is sickening to behold. Perhaps there is no part of Ireland where there is so much land under potato crop as here owing to the facilities for getting bog stuff which when mixed with other manure produced good crops on limestone soil . . . I do not know any landlord with the exception of Lord Ross (sic) *and another small proprietor who have taken any measures to enable poor tenants to sow their lands. I have heard of but one case of death from destitution in this barony . . . but people are dying fast from want of food and clothing, and fever brought on by bad and insufficient food. They are wasting and dying by inches. If some proper measures are not taken, I fear the summer will be awful.*

Birr and its neighbourhood was by no means the worst hit locality at this time, and for this the Parsons family could take some credit. In previous years Laurence Parsons, second Earl of Rosse, had acted as a magistrate, and William Parsons, then Lord Oxmantown, had served as foreman of the Grand Jury, at a time when agrarian unrest and civil disturbances over matters like the Crotty schism had led to legal proceedings. Much of the local unpopularity that seems to have accrued to William Parsons during that period was erased by the public appreciation for the work of the Earl and Countess of Rosse during the famine years. Randal, who was born in 1848, recalled that during his childhood (Parsons R)

there were times of great unrest, murders and robberies of arms. My father used to go out to the telescope to observe with pistols in his pockets. The laurels near the telescopes were kept cut down to a foot or two in height, so as not to afford cover for an evilly disposed person to be concealed. Yet I do not think there was any real danger, as the family was very popular.

Romney Robinson the astronomer at Armagh observatory had been giving reports at regular intervals to the Royal Irish Academy on the work of Lord Rosse. In his report of the spiral M51 nebula in 1848 he began,

He has often been asked why this instrument (the 72 in) *had given no further results. Those who put the question had but a faint idea of the overwhelming*

pressure which the last three years exerted here on all who were resolved to discharge the duties which men owe their country. Lord Rosse is not a person to seek knowledge or enjoyment in the heavens, when he ought to be employed on earth! And he devoted all his energy to relieve the present misery and provide for the future.

Generally, throughout most of history, the wealth of a land owner depended every bit as much on having sufficient workers available to him to work his land, as it did on the acreage of that land. Now, perhaps for the first time, there was an excess of workers, and society was unequipped to deal with the problems that this created. As a legislator, William Parsons, like his father, sought to understand the influences that bore on Irish society. In 1847 he was provoked by a leading article in the *Times* of London to write to its editor. His letters were not printed, so he published them as a pamphlet (Parsons W 1847). He speaks of himself as an Irish landed proprietor, with a knowledge also of England, and of Yorkshire in particular. In defending landlords he points to the different situation in Ireland compared to England.

It is clearly the interest of a landed proprietor not to be negligent; and where we fancy that people are acting palpably against their own interest, it is not unreasonable to suspect there is something in the back-ground we do not understand.

He argued that in the face of a rapidly increasing population, the ratio of capital to population had not remained stable in Ireland in recent years as it had in England, but had fallen drastically. The consequent fall in wages had meant that

. *it is very easy therefore, to see why under these circumstances the Irish labourer, who is industrious abroad, should be the reverse at home...*

All this has been greatly aggravated by failure of the potato crop.

There remains therefore, but a choice between two things either to throw more capital into the district or take away the people; but to remove the people would be the work of time, therefore capital must be thrown in. ... there is no third course,... still a variety of projects have been started, all resolving themselves ultimately into a local tax, ... such projects appear to me like the schemes of the busy gentlemen who are incessantly at work planning machines to produce perpetual motion, by new combinations of wheels levers and springs: but the machine will not move. Projectors of this class overlook the fundamental laws of political science, as the others of mechanical science.

In fact of course as things turned out during the following century and more, unpopular as such a view would prove to be, emigration did indeed become a prominent feature of Irish life, whether under British rule or under a native government. Touching on the current disaster he says

Whatever cavil there may be as to the quarter from which the State is to obtain the capital which will be required, there is one thing certain, that it is only by a very large expenditure that all the worst horrors of famine can be averted, or even mitigated, and that too, provided that large supplies can be obtained... Although no doubt landlords will usually be willing to contribute... still a very large portion of the expense will have to be borne by the state.

He considers next why population has grown despite the absence of any increase in resources, and why emigration has not removed the excess numbers before this. He concludes that the cause is the subdivision of land. Each new occupier means a new household with a new family to feed. The practise in England and Ireland in this respect was totally different. An English tenant would not contemplate sub-dividing his farm. In Ireland this was not so, for a variety of reasons. There were differences in the law that encouraged both landlord and tenant to turn a blind eye to the practice. Even if landlords had recourse to the law, the cost of enforcement was too high, and the chance of success too small to make it worth while. And of course there was a political dimension. As he puts it

the notion was prevalent that numbers would eventually overpower the owners of the soil.

It is frequently inferred that enforcement of laws that could be contemplated in England would create excessive civil strife in Ireland. Despite the Act of Union, he argued, Ireland was treated differently from England by the legislature. The members of the Irish Executive were poorly acquainted with the country and its peculiar problems.

The management of Irish affairs has, therefore, usually rested with a few clerks and lawyers; a more inefficient executive there could not well be.

At a later date Lord Rosse found himself in a strange position. In 1852 he was chairing a House of Lords Select Committee of Inquiry into the *'Operation of the Acts relating to the Drainage of Lands in Ireland, as administered by the Board of Works'* (Parl 1852). The Board of Public Works had been established in 1831. Gradually it was given various duties: development of the Shannon navigation, management of fisheries and in 1842 the administration of the Drainage Act. Land drainage was seen as one way of reclaiming land for the increased population. Under the act a scheme would be drawn up, and if the consent of two thirds of the landed proprietors was forthcoming, the board's engineers would supervise its execution. Government funds were available but the proprietors had also to meet part of the cost. Once the scheme was commenced the board had very strong authority to modify details as events dictated. The commissioner with responsibility for drainage and for fisheries was W T Mulvany. During the famine, the treasury official C E Trevelyan saw construction works as one means to provide relief for the destitute. By 1847 some 14 000 were employed in supervising relief schemes (Griffiths 1987). The pressure on staff was such that some work was carried out without the full consent of some proprietors, even though they would have to find some of the extra funding required. A case in point was the Brosna river in King's County, where an initial estimate of £47 329 crept up to a predicted £76,000 to finish the job. The proprietors refused to pay, and it was as a result of such resistance that this Select Committee was set up. Lord Rosse was very well equipped to appreciate the matters in contention, and yet Mulvany was made a scapegoat and was forced to resign despite receiving warm praise from Sir John Fox Burgoyne, Richard Griffith and Thomas Larcom. He subsequently left Ireland and went to the Ruhr valley where he enjoyed a highly successful second career developing coal mines in what had been until then a prosperous agricultural area.

Chapter 5

Scientific Achievement

5.1 The astronomer and the search for spiral nebulae

Little astronomical observation was done during the duration of the famine, though in 1846 a second spiral nebula was discovered. It was during a visit in the February of 1848 that Dr Romney Robinson observed Parsons' efforts to bring the large instrument back to its peak condition. The mirrors had become tarnished and were in need of polishing. As the weather was freezing, stoves were used to raise the temperature of the workshop and of the polishing tool to around 55 °F. Despite many attempts both the finish and the figure of the mirror were unsatisfactory. Eventually in order to try to trace the source of the problem Parsons turned his attention to the 3 ft mirror. Because of its shorter focal length its performance could be tested at the base of the tower, without removing it from the polishing machine or the need to install it in a telescope. After some time it was noticed that the stoves had caused the humidity of the workshop to fall, and the iron oxide paste was drying out. Once the problem was understood the solution followed quickly. The boiler that powered the engine was used to feed a jet of steam which raised the humidity of the room. Success followed immediately.

The Astronomer Royal, Sir George Biddell Airy, visited in the summer of 1848 and wrote daily to his wife. His letters give a feel for those early days during which Lord Rosse and his team were making adjustments to the instrument (Airy 1896). They also remind us that success depended not only on perfection of the telescopes, but on the weather, on clouds and on the degree of turbulence of the atmosphere. Airy's own extensive experience as an observer must have been of considerable assistance at this point.

The Castle, Parsonstown
1848, Aug 29

... After tea it was voted that the night was likely to be fine, so we all turned out. The night was uncertain: sometimes entirely clouded, sometimes partially, but objects were pretty well seen when the sky was clear: the latter part was much steadier. From the interruption by the clouds, the slowness of finding with and managing a large instrument (especially as their finding apparatus is not perfectly arranged) and the desire of looking well at an object when we had got it, we did not look at many objects. The principal were Saturn and the Annular Nebula of Lyra with the 3 feet; Saturn, a remarkable cluster of stars, and a remarkable planetary nebula, with the 6 feet. With the large telescope, the evidence of the quantity of light is prodigious. And the light of an object is seen in the field without any false colour or any spreading of stray light; and it is easy to see that the vision with a reflecting telescope may be much more perfect than with a refractor. With these large apertures, the rings around the stars are insensible. The planetary nebula looked a mass of living and intensely brilliant light: this is an object which I do not suppose can be seen at all in our ordinary telescopes. The definition of the stars near the zenith is extremely good: with a high power (as 800) they are points or nearly so—indeed I believe quite so—so that it is clear that the whole light from the great 6 feet mirror is collected into a space not bigger than the point of a needle. But in other positions of the telescope the definition is not good: and we must look today to see what is the cause of this fault. It is not a fault of the telescope, properly so called, but it is either a tilt of the mirror, or an edge-pressure upon the mirror when the telescope points down which distorts the figure, or something of that kind. So I could not see Saturn at all well, for which I was sorry, as I could have compared his appearance with what I have seen before. I shall be very pleased if we can make out what is the fault of adjustment, and so correct it to get good images everywhere. It is evident that the figuring of the mirror, the polishing, and the general arrangement is perfectly managed.

The following day they caught up a little on lost sleep and he wrote,

The Castle, Parsonstown
1848, Aug 30

Yesterday we were employed entirely about the Great Telescope beginning rather late. The principal objects had relation to the fault of definition when the telescope is pointed low (which I had remarked on the preceding night), and were, to make ourselves acquainted with the mechanism of the mirror's mounting generally, and to measure in various ways whether the mirror does actually shift its place when the telescope is set to different angles of elevation. For the latter we found that the mirror actually does tilt 1/4 of an inch when the tube points low. This of itself will not account for the fault but it indicates that the lower part is held fast in a way that may cause a strain which would produce the fault. These operations and reasonings took a good deal of time. Lord Rosse is disposed to make an alteration in the mounting for the purpose of correcting this possible strain.

The following day the weather was poor,

The Castle, Parsonstown
1848, Aug 31

The weather here is still vexatious: but not absolutely repulsive. Yesterday morning Lord Rosse arranged a new method of suspending the great mirror, so as to take its edgewise pressure in a manner that allowed the springy supports of its flat back to act. This employed the workmen all day, so that the proposed finish of polishing the new mirror could not go on. I took one Camera Lucida sketch of the instrument in the morning, dodging the heavy showers as well as I could; then as the afternoon was extremely fine, I took another with my head almost roasted by the sun. This last view is extremely pretty and characteristic, embracing parts of the mounting not shewn well in the others, and also shewing the Castle, the Observatory and the 3-feet telescope. The night promised exceedingly well: but when we actually got to the telescope it began to cloud and at length became hopeless. However I saw the fault which I had remarked on the two preceding nights was gone. There is now a slight exhibition of another fault to a much smaller extent. We shall probably be looking at the telescope today in reference to it.

The following day he described the results of the modifications made,

The Castle, Parsonstown
1848, Sept 1

Yesterday we made some alterations in the mounting of the great mirror. We found that sundry levers were loose which ought to be firm, and we conjectured with great probability the cause of this, for correction of which a change in other parts was necessary. The mirror was then found to preserve its position much more fixedly than before... At night, upon trying the telescope, we found it very faulty for stars near the zenith, where it had been free from fault before. The screws which we had driven hard were then loosened, and immediately it was made very good. Then we tried with some

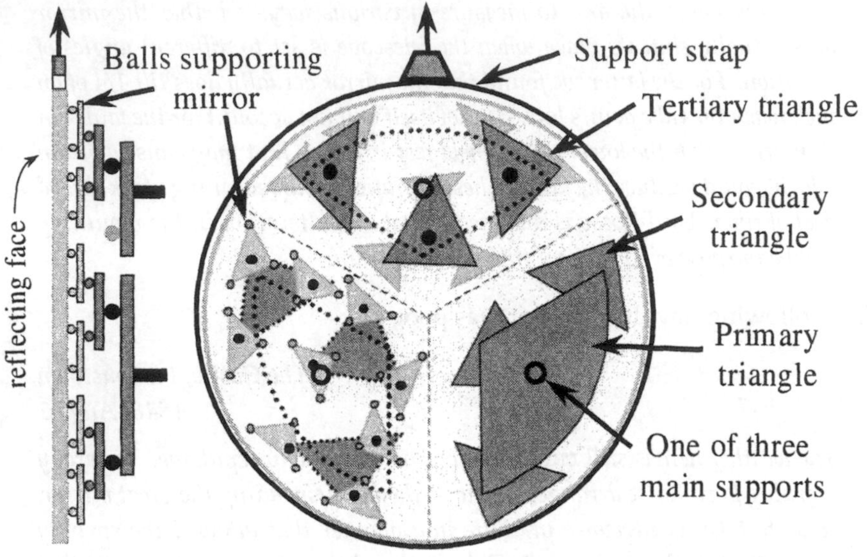

Figure 22. Final arrangements for the support of the 6 ft speculum on brass balls.

lower objects, and it was good, almost equally good there. For Saturn it was very greatly superior to what it had been before. Still it is not satisfactory to us, and at this time a strong chain is in preparation, to support the mirror edgeways instead of the posts that there were at first or the iron hoop which we had on it yesterday. Nobody would have conceived that an edgewise gripe of such a mass of metal could derange its form in this way. Last night was the finest night we had as regards clouds, though perhaps not the best for definition of objects.

Finally his last letter has a brief note,

The Castle, Parsonstown
1848 Sept 2

I cannot learn that the fault in the mirror had been noticed before, but I fancy that the observations had been very much confined to the Zenith and its neighbourhood.

When Airy came to report on the outcome of his visit to Parsonstown to a meeting of the Royal Astronomical Society he said that the mirror displayed signs of astigmatism (Airy 1849). The image of a star appeared as a line, with one orientation if the eyepiece was placed at one focus, but turned through a right angle if the eyepiece was moved by one half inch to

Figure 23. (a) The Leviathan of Parsonstown as the 6 ft telescope was popularly known. This photograph was taken by Mary Countess of Rosse sometime around 1857. It shows Lord Rosse with three unidentified ladies and two of his sons, Randal standing and Charles seated on the steps. © Birr Scientific and Heritage Foundation by courtesy of Lord Rosse.

an alternative focus. Moreover these distortions changed as the inclination of the mirror was adjusted. The cause was traced to the fact that the pads supporting the back of the mirror had become stuck in place and were stressing the surface. A remedy was found. The 27 flat plates were each replaced by triangular shaped levers carrying three brass balls located at the vertices, each ball being one and a quarter inches in diameter (see figure 22) (Parsons W 1850). The 81 balls were individually secured in place by a fine brass wire. In addition a change was made to the way in which the side thrust of the mirror was supported when it was moved from the horizontal. Friction tended to build up between the rim of the mirror and the circumferential supports that were part of the carriage. This prevented the mirror moving freely in a direction perpendicular to the surface. The size of the friction force depended on the angle of tilt and interfered with the even distribution of weight by the lever system. A new arrangement was designed with a band or sling that encircled the rim of the mirror.

Figure 23. (b) Detail from an engraving made from a photograph, showing Lord Rosse, Clere and Charles Parsons (Parsons W 1861).

The sling was supported from a pillar to the north of the mirror so that it could hang from the carriage, but yet was completely free from constraints perpendicular to the surface (see figure 22). These modifications were very successful, and it was noted that pressure from a hand on the back of the mirror, which weighed three tons, was sufficient to produce a visible distortion of the image. No contemporary instrument came anywhere near the size and mass of the 6 ft mirror and the problems encountered were quite unique. Given the great leap in scale that this instrument represented compared with contemporary practice, such teething problems were not unexpected. Moreover since the immediate objective was to make out very faint objects, and not to make precise positional measurements, these difficulties were of lesser importance.

In concluding his account the Astronomer Royal made reference to the '*enormous light of the telescope*' that was evidenced in the details hitherto unseen in nebulae listed by Sir John Herschel in his catalogue, in the great numbers of stars outside the Milky Way, and in '*the prodigious brilliancy of Saturn*'. He concurred with another observer, Sir James South, who spoke of the appearance of Jupiter as resembling a coach-lamp. It is difficult perhaps for the modern reader to appreciate the impact that was made on the observer looking into the eyepiece. Light pollution has made present-day city dwellers almost unaware of the night sky. If they have viewed images of the heavens, it is likely that these will have been in the form of photographs. The huge light gathering power of the telescope made faint stars bright, and bright stars dazzling. With a strong magnification a conventional refractor could reveal details of the moon and planets, but the act of magnification dimmed the image. The great telescope could both magnify and brighten such objects. Perhaps more astonishing was the revelation of the structure of the nebulae. Although their images as seen in the eyepiece remained faint, their size was unexpectedly large. This unique experience was shared by many privileged guests at Birr Castle who had no pretensions to being professional astronomers. Tales of what could be seen spread widely, and contributed to the high profile that Lord Rosse achieved as a scientist in contemporary society. Until very recently it has been necessary to rely on written accounts and on photographs to form an impression of what it was like to operate the giant telescope in practice. During the last 80 years the walls became covered with ivy, the wooden galleries, the chains and counterweights, the winches and pulleys were removed one by one. But in 1996 and 1997 the Birr Scientific Foundation raised the necessary funds to restore the telescope to its original state.[1] Under the supervision of the civil engineer and amateur astronomer Michael Tubridy it has been completely renovated with all its wooden galleries, chains and counterweights. Once more visitors can watch for themselves as the giant tube is moved to the desired direction. What strikes one is the excellence of the original engineering design. It might be thought that the use of chains and windlasses would result in coarse movement, but because of the length of the lever arm this is not the case.

By now William Parsons' reputation as a scientist was well established and this was recognized when he was honoured in 1848 by election as President of the Royal Society. It was a step that placed him at the centre of science policy making in Britain. In 1849 his engineering achievements that underpinned his scientific work were recognized by his election to honorary membership of the Institution of Civil Engineers. In 1851 he became one of the first recipients of the Royal Society's Gold Medal. From 1850 onwards this medal was awarded *'for the two most important contributions to the advancement of Natural Knowledge published originally in her Majesty's dominions within a period not more than 10 years and not less than one year of the award'* (Macleod 1971). When plans were being made for

[1] Brendan Parsons, the seventh Earl of Rosse, has directed the drive to win financial support for this ambitious restoration project.

the holding of a Great Exhibition in the Crystal Palace in Hyde Park in London, in 1851, Lord Rosse was appointed as one of the commissioners. His knowledge of both science and engineering qualified him to play a key role, but he was also required to underwrite some of the costs should it fail financially. The event was used as an excuse for Lord and Lady Rosse to lay on an entertainment for friends, neighbours and townspeople at Birr Castle. According to the *Illustrated London News* more than 20 000 visitors watched a fireworks display that was held in front of the castle. Mary Ward joined in the manufacture of the pyrotechnics, and Lord Rosse ignited them. Laurence, now 11 years of age, made out the programme. It was an excellent show and once more Parsonstown made the headlines. In 1853 Lord Rosse was admitted to membership of the Imperial Academy of St Petersburg, no doubt in consequence of his scientific collaboration with the astronomer Otto Struve of the Pulkovo observatory near St Petersburg. He was created a knight of the Legion of Honour by emperor Napoleon III as the Paris Exhibition drew to a close in 1855. Lord Rosse had a long association with Dublin University, being one of the commissioners who were appointed to inquire into the state of the University in 1852. In 1862 he became Chancellor of the University, in which post he served until his death. An invitation to attend the installation ceremony had been extended to his friend Sir James South, but he was not well enough to attend. He wrote (Wayman 1987)

> *My Lord,*
>
> *Deeply lamenting that bodily infirmities will prevent my being present at your installation, I cannot resist sending you a substitute. It is my 12 inch achromatic object glass. Will you do me the honour of presenting it and its appendages to the Dublin College for its use at its observatory as a mark of veneration for the University and my public and private esteem for his Lordship.*
>
> *I am, my dear Lord, with sincere respect, ever yours faithfully,*
>
> *James South*

The lens had a diameter of $11\frac{3}{4}$ inches and a focal length of 19 ft. It had been purchased by James South from Cauchoix in 1829 (Glass 1997). The firm of Troughton and Simms had mounted it for him, but South was so dissatisfied with the instrument that after losing a legal battle over payment, he smashed the mount into pieces. The lens was eventually used in an equatorial mounting which had been made by Thomas Grubb in 1853, but which had remained unsold. The telescope was purchased for the Dunsink Observatory where it remains today.

Most attention at the Birr observatory was directed at stellar objects, but inevitably items like the planet Jupiter and the moon also became the subject of interest. Two German astronomers Beer and Mädler had produced a map of the moon's surface in 1838 and members of the British Association at a meeting in Belfast appointed a committee to study the moon's surface and compare it with that of the Earth. Consisting of T R Robinson, E Sabine who was president of the Association, John Phillips a geologist and Lord Rosse himself, the committee was invited to meet at Birr in September 1852. While there, various observations and measurements were made, and a programme was drawn up for a study to be conducted by collaborators in Britain or abroad. Not a great deal came of the venture, and Birr did not feature in subsequent developments. In 1854 Parsons informed the Royal Astronomical Society, through the Astronomer Royal, that he had attempted to construct a clock mechanism that would move a photographic plate at such a rate that a time exposure photograph could be made of the moon. But he found that although the clock worked well, the plates were too insensitive to capture the sort of detail that was visible to the eye (Parsons W 1854a).

Obviously with his many public commitments in Ireland and England, the Earl of Rosse could not himself make use of the telescopes in an efficient manner, and so he began to rely on assistants to carry on his work. Romney Robinson during his visit in March of 1845, instructed an assistant *'on whose zeal and steadiness he can rely'* so that the instrument would be actively employed (Robinson 1848). After the famine William Hautenville Rambaut worked at Birr until 1849 before going to join Romney Robinson at the Armagh Observatory. Next to follow him were the two brothers Stoney, whose family had farmed an estate at Oakley Park near Parsonstown until the time shortly after the Napoleonic Wars when farming income collapsed, and they were forced to sell their property. Like the Parsons family, they had emigrated from Yorkshire, though at a date some 100 years later. The eldest, George Johnstone, was born in 1826 and was employed at Birr for four years from 1848. With the help of Lord Rosse, he was appointed to a chair of Natural Philosophy in the newly created Queen's College Galway in 1852. He was elected a Fellow of the Royal Society and became its vice- president. In later life he was appointed Secretary to the Queen's University in Ireland (Cox 1990). His brother Bindon Stoney, who graduated with distinction from the Engineering School at Trinity College Dublin in 1850, spent two years at Birr. Both men were very able and were skilled in the art of depicting their observations as drawings, and their work was acknowledged in the publications of William Parsons and his son (Parsons W 1850, Parsons L 1868). A succession of other able young men worked at Birr. Robert Ball during his time at Birr, from 1865 to 1867, as well as making astronomical observations by night also acted as tutor to the two youngest surviving boys, Clere and Charles. Through the good offices of Lord Rosse, he was appointed Professor of Applied Mathematics at the Royal College of Science in Dublin in 1867. Subsequently he had a distinguished career as a mathematician and astronomer.

The first scientific publication concerning the 6 ft telescope, and observations of nebulae made with it, appeared in the *Proceedings of the Royal Society* in 1850. Particular attention was paid to Messier 51. It is important to remember that when examining such faint objects with the naked eye, even with the most powerful instruments, the light intensity would be so low that reliance would often be placed on 'averted vision'. This requires the use of the cells at the periphery of the retina, rather than on those at the centre. Although the latter are colour sensitive and have the ability to distinguish finer detail, they require very much brighter images to function properly. The skilled observer can also improve his ability to make out faint details by causing his eye to 'scan' an image. Parsons was cautious in interpreting his observations of spiral structures (Parsons W 1850).

> *Much however as the discovery of these strange forms may be calculated to excite our curiosity, and to awaken an intense desire to learn something of the laws which give order to these wonderful systems, as yet, I think, we have no fair ground even for plausible conjecture; and as observations have accumulated the subject has become, to my mind at least, more mysterious and inapproachable. There has therefore been little temptation to indulge in speculation, and consequently there can have been but little danger of bias in seeking for the facts. When certain phenomena can only be seen with great difficulty, the eye may imperceptibly be in some degree influenced by the mind; therefore a preconceived theory may mislead, and speculations are not without danger. On the other hand, speculations may render important service by directing attention to phenomena which otherwise would escape observation, just as we are sometimes enabled to recognise a faint object with a small instrument, having had our attention called to it by an instrument of greater power.*

He also explained the difficulty of sketching while observing.

> *In sketching we necessarily employ the smallest amount of light possible, very feeble lamp light, especially where the objects or their details are of the*

last degree of faintness. To see the sketch as we proceed it is often necessary to mark it too strongly: this would be of little moment if the excess of colour was always in the same proportion, especially as different eyes form a very different estimate of the relative intensities of a nebula and its representation on paper, but it is not so; the contrast between the faint and bright nebulae and between faint and bright parts of the same nebula is very liable to be made too slight. The most important error to guard against is that of supposing that the well-marked confines of the nebula on paper really represent the boundaries of the object in space in all cases.

Some protection against subjective response to an observation was achieved by involving more than one observer and repeating the observation on many different occasions.

A much fuller account of the construction of the 6 ft telescope appeared in *Philosophical Transactions of the Royal Society* in 1861. This paper was illustrated with engravings made from photographs giving different views of the telescope and its mirror. It contained a selection of notes and some thumb nail sketches of close to 1000 of the nebulae that had been listed by John Herschel. Most objects were observed from once up to five times but others had observations in double figures. In addition to these sketches, there were 43 plates prepared by the two Stoney brothers, S Hunter and Mitchell. The paper also contains an exchange of correspondence with Otto Struve at Pulkova dated 1851 and 1852. This concerns a comparison between published star positions that had been determined by Bindon Stoney at Birr, and those made by Struve himself. A mean error of 1°57′ was found. Once this was pointed out, the source was quickly recognized. It lay in a fault in the method of calculation used. The episode points up the contrast between the large 15 in diameter Pulkova refractor that was dedicated to high precision determinations of position, and the 72 in reflector with a light grasp over 20 times greater, but which was lacking in the quality of its measurements of orientation.

As the years have passed, the memory of the giant telescope has been kept alive chiefly by virtue of the massive masonry structure and the wooden tube attached to the universal joint, everything else having vanished. One of its mirrors on its carriage was sent to London in 1914 where it is displayed in the Science Museum. Since its heyday there have been some who have written rather dismissively about the venture. Not all visitors to Birr were convinced of its superiority. Certainly its performance was critically dependent on the state of the mirror. Just like modern instruments its surface deteriorated with time. The difference is that refurbishing of a metal mirror is required quite frequently, and unlike glass mirrors with aluminium coating, this may involve grinding as well as polishing. In effect it became a new mirror each time. As Parsons wrote (Parsons W 1861)

> *... although everything has usually gone smoothly when we are in the midst of experiments and in constant practice, yet after the lapse of even one year, when we have had occasion to repolish a speculum, there have often been disappointments. The difficulty arises from the necessity of employing two strata of resinous matter, one so hard, and both so thin*

These processes were a challenge, even to a man like William Parsons who had had enormous personal experience of developing suitable techniques. Once he ceased to be closely involved it was perhaps inevitable that the quality of the mirror should deteriorate. Unfortunately, even when in later years photography began to be applied to astronomy, the instrument could not be fitted with an automatic drive of sufficient quality to allow a photographic record to be made. Therefore any judgement of its performance must rely heavily on the evidence of experienced contemporary astronomers like Romney Robinson and Sir James South, and on the drawings created by the Earl and his assistants.

If we take the classic example of a spiral nebula, M51, the 'Whirlpool', and compare the drawing with a modern photograph, as in figures 21(a) and 21(b), it is clear that agreement is remarkable. A great deal of effort was devoted to another object, the great nebula in Orion. The drawing of this which was published can be compared with other contemporary drawings and modern photographs (Ashbrook 1984). Again the agreement is good, especially given that in this case the subject is indeed a true nebula largely composed of dust, and the image displays a very great range of brightness. For the layman a useful book is K G Jones' *Messier's Nebulae and Star Clusters* (Jones 1968). This gives details of the 102 items listed by Messier, with comments made by early observers like William Herschel and Lord Rosse, as well as numerous, modern, photographs and up to date assessments. In 44 cases the drawings made by William Parsons and his astronomer assistants are reproduced. From this it is clear that their work was sound and reliable.

Many years after William Parsons' death Johnstone Stoney attended a paper read by his son Laurence, and in the discussion afterwards he said (Parsons L 1880a)

> *The test usually applied was the performance of the mirror on the star of the 8th or 9th magnitude, magnification 750. Such stars are bright in the*

great telescope. They are usually seen as balls of light, violently boiling in consequence of the atmospheric disturbance. If the night is good there will be moments now and then when the atmospheric disturbance will abruptly seem to cease for a fraction of a second, and the star is seen for an instance as the telescope really presents it. It is by opportunities of such moments that the performance of the telescope must be judged. With the best of your father's mirrors that I saw, the appearance at such opportunities was like that of a light shining through a minute needle hole in a card placed in front of a flame. I think any practical astronomer will agree with me in the opinion that the mirrors of six feet diameter that bore the test bordered very closely indeed on theoretical perfection.

In the same year, 1880, Laurence published in *Nature* a letter he had received from Otto Struve (Parsons L 1880b).

My dear Lord Rosse,

Yesterday evening a friend conveyed to me a note inserted in the Times of April 3rd under the title 'Three Giant Telescopes' in which I am told of having expressed myself in a very uncourteous manner on the optical qualities of the great reflector constructed by your late father. I beg leave to say that those expressions are altogether invented by the anonymous author of the note, or at least quite a voluntary and thoroughly wrong interpretation of what I may have said. I am sorry my name is abused in such a manner by people who probably have a design of their own in deprecating the performance of the instrument, the construction of which marks a high progress in optics and mechanics, and which in space penetrating power has not had any rival until now, though certainly with regard to definition (particularly when the mirror is considerably out of horizontal position) there are other instruments superior to it.

Pulkova April 14th.

Otto Struve

If the eyepiece of a reflecting telescope is dispensed with, and the eye is placed at the focal point of the mirror, the image will be quite free of colour distortion, but at that time the significance of colour was little appreciated. The object H 262, seen in 1848, was observed to be blue, but in general little reference is made to colour. The task of conveying the reality of a visual impression with no more than engravings and brief verbal descriptions was formidable. Even with the light gathering power of the giant mirrors, the greatest skill was required. Writing of Herschel's nebula

131, Parsons observed that '*I should not have seen the spiral arrangement had I not observed it before*' (Parsons W 1861). He added '*The details of faint nebulae with curved or spiral branches have usually been made out by degrees, not only on successive nights, but even in successive years.*' While other astronomers like Otto Struve were able to identify spiral nebulae with refracting telescopes of much smaller aperture once they had been described to them, none made new finds until photographic techniques were perfected (Ashbrook 1984). Slowly the general term of nebula was shown to cover several distinct objects. Two powerful tools were emerging, photography and spectroscopy, and these would in time settle many arguments about correct assignment, but as yet they were in their infancy. True nebulae that consist of gas illuminated by nearby stars include such as the great nebula in the constellation of Orion. There are other cases where a cloud of gas surrounds a central star, the 'planetary nebulae'. These are now thought to represent a phase in the evolution of a dying star after it has ejected its outer layers of mass. Other objects proved to be simply clusters of countless related stars, the 'globular clusters'. Given a sufficiently powerful instrument and good atmospheric viewing conditions, such objects could be 'resolved' into individual stars. Finally there were those objects of which the spiral nebulae were examples, that were at the limit of resolvability but which seemed to show evidence of dynamic structure.

The first effective step in measuring the distance of stars had been taken by Bessel in 1838 who successfully estimated the parallax of a near neighbour to our sun. But it was not until 1925 that Cepheid variable stars, which provide a reliable gauge of distance, were found in spiral nebulae, and showed them to be extremely distant; their light takes millions of years to reach us. Only then did it become clear that these spirals were in fact stellar galaxies comparable in size and composition with our own Milky Way but enormously distant. So in the case of M51, the Whirlpool, for example, light takes 37 million years to reach the Earth. In time it was realized that many of the nebulae are indeed galaxies like M51, but, because of their orientation relative to the Earth, the spiral structure is not evident. From the beginning of his reports, Parsons and his colleagues agonized about the question of whether such nebulae were in fact 'resolvable' or not. Unlike some of his contemporaries, he was always cautious on this point. In fact it was not until 1925 that the matter was settled in the affirmative. At the present time the orbiting Hubble space telescope is still revealing fresh information about the countless galaxies that make up even the apparently most empty parts of the universe. Their number is of an order equal to the number of individual stars in our own Milky Way galaxy. William Parsons' giant telescopes in their day did play an important role in revealing new details of the structure of the universe.

5.2 Bringing up the children

William Parsons first showed his interest in photography in 1842, which was just three years after the announcement of the photographic process (Davison 1989). He had purchased equipment from London but, after some experiments with the Daguerreotype process, he seems to have left the subject until 1852. In that year he wrote to his friend Fox Talbot for advice on his efforts to photograph the moon with the 36 in telescope. This was more convenient to adjust than the larger instrument, and a holder was devised for the plates that would compensate for the rotation of the earth. Despite the power of the 36 in mirror, he found the wet collodion process was too slow and he seems to have abandoned attempts to use photography. However Mary Countess Rosse now took an active interest in photography herself. In 1854 while she was pregnant with Charles Algernon, her last child, Lord Rosse sent to Fox Talbot some examples of her work. These were highly praised and she was soon embarked on a successful career as a photographer. Her work was characterized not merely by technical excellence as in the case of her stereoscopic photographs of the large telescopes, but by her artistic skill in composing human subjects to form interesting studies. The chemical processing which was required at this early period in the development of photography was all carried out in a dark room set aside at Birr Castle. It required the acquisition of quite new skills by a woman who until then had been primarily a housewife. Her husband's research into the problems associated with the successful manufacture of specula had required the development of a competence in the procedures of the chemical laboratory. Most of the facilities that she would require were therefore likely to be readily available. Before this she had already evinced a close interest in her husband's engineering activity, and an aptitude for being involved in practical work. For example she made use of the foundry when it was not making components for the telescopes. Modelling in wax, she made moulds for the decorative features required when the main gates to the estate were cast (Parsons R). And of course it was largely her fortune that provided the funds needed for the construction of the giant telescope, which were estimated to have been £30 000 (Anon 1867b). Looked at in another way one could say that the telescope was built thanks to the hard work of a great many tenant farmers in Yorkshire!

Figure 24. William third Earl of Rosse with two of his sons, Randal and John, photographed by his wife Mary Countess of Rosse sometime around 1856. © Birr Scientific and Heritage Foundation by courtesy of Lord Rosse.

In the 19th century, children in wealthy families very often saw little of their parents, and their early years were given in care to nannies, and later the boys were dispatched to boarding schools. This does not seem to have been the case with the Parsons family. Among the surviving photographs taken by Mary Rosse, there are several of the children. One is of a frowning Charles seated in a three wheeled perambulator with his brother Clere standing beside him. Both are dressed in skirts as was usual of young children then. Charles is perhaps two or three years of age. The engravings of the giant telescope that illustrate the Royal Society paper of 1861 were made from photographs and they show various adults, men and women, and two young boys, of whom one at least must be Charles (figures 23(a) and 23(b)). This suggests a certain laid back attitude on the part of the (noble) author.

One of the brothers, Randal, in later life a clergyman, wrote his recollections of childhood (Parsons R). He recalled that his brother Laurence, who was eight years older than he, was taught by a tutor. Until he was ten years of age Randal was taught Latin, English literature and religion by a governess, Miss Payne. Lessons began at 7.30 am, breakfast at 8 am, lessons until noon. The period until lunch at 2 pm was spent outdoors. Lessons resumed from 5 until teatime at 6.30 pm. The assistants who were employed by William Parsons to use the telescopes served as tutors to the boys as they grew older. Laurence graduated from Trinity College in 1864 as a non-resident student, which meant that he was not required to attend lectures. His university education would therefore have done little to dissipate his natural shyness. He seemed to have a special fondness for his youngest brother whom he calls Charlie in his diary.

A niece recalled her uncle as *'the biggest man she had ever seen, with the kindest face and manner'*, someone of whom it was impossible to be afraid. The painting that portrays William Parsons as Chancellor of Dublin University certainly conveys this impression. He was inseparable from his sons. He brought them sailing with him, and encouraged them to use the workshops, Clere and Charles especially. The girls too were encouraged to try their hand with machinery, turning wooden candlesticks on a lathe. The cousins dubbed them, father and sons collectively, 'the boys'. Late in life Charles recalled (Parsons C A 1926b) that

> *as a youngster of ten years old, when not at lessons, I spent much time making contrivances with strings, pins, wire wood, sealing wax and rubber bands as a motive power, making little cars, toy boats and a submarine. Since that time I have often thought that a boy in this way learns more than by playing with bought toys and mechanics. When working in my father's workshop, dealing with metals became more attractive, and amongst other things I made a spring trap, and I well remember the delight of catching my first rabbit with it.*

He enjoyed fishing and shooting for the rest of his life. Randal had a dislike for hunting and killing animals, and preferred to read,

> *standard books of the day or anything I could get hold of... while my brothers never opened a book of this kind unless compelled to do so.*

They were keen on rowing, riding and shooting; but Randal treasured his own patch of garden (Parsons R).

These childhood preoccupations foreshadowed adult activities. An air gun that Charles constructed still survives. It made use of small cylinders of compressed gas to propel a projectile. Later he was to use the same cylinders to propel an experimental boat that marked a first step in the exploitation of the turbine principle. Nor were all the contrivances mere toys for he demonstrated early on the ability to create things of practical utility. At the age of ten years Charles constructed a depth sounding device that he used in the lake at Birr (Parsons C A 1926b). It was made of glass tubing and, as it was lowered into the depths, the air inside was compressed so that water was able to enter through a capillary tube. The capillary was so positioned that when the device was raised to the surface this influx remained trapped, and the quantity of water was a measure of the pressure or depth to which the device had been subjected. It was a serious invention, though as Parsons jokingly remarked when mentioning it '*the idea was vastly improved by Lord Kelvin, and I think his was quite independent*'. A photograph taken some time around 1855–1860 shows a small steam boat on a carriage with some of the children standing nearby.

When their father took a house in London in the Season, the family accompanied him. Indeed it was at 13 Connaught Place that Charles Algernon, the youngest son, was born on 13 July 1854. Later, in the summer the children would visit their grandmother who continued to live at Brighton after her husband's death in 1841. Full advantage was taken of their time in London. Laurence's diary of 1865 conveys an impression of their activities (BirrJ/25). In June that year, at the age of 24 he was brought to the House of Lords, to the Royal Institution to hear a lecture on the moon by Nasmyth, the engineer and astronomer, and to the Institution of Civil Engineers to attend a 'drawing master'. The naval architect and Fellow of the Royal Society, Scott Russell, came to lunch with the family armed with drawings of some Woolwich guns which had been bored out, lined and fitted with breech blocks. Randal's attention, characteristically, was drawn to an exhibition of the Horticultural Society in South Kensington, which he thought had improved since the previous year. There was time too for entertainment, visits to exhibitions of paintings by day and to operatic performances at Covent Garden in the evening, *La Somnambula* and Gounod's *Faust*. The soldier's chorus in particular impressed the diarist. Laurence attended a ball at the Palace.

The house in Connaught Place was the scene of much social activity. Soirées were held at which distinguished scientists would attend, often armed with models to illustrate their discoveries. The children were not excluded from this company. Randal described the seemingly endless dinner parties where cousins and other guests were entertained. The girl cousins were encouraged by the Countess, 'Aunt Mary', to accompany Charles when he played the violin. In general he did not enjoy such social occasions. Mama and Papa made social calls on family friends in the company of the children. Their mother would delight in driving in the nearby Hyde Park with her boys. Her personality was dominating but benevolent. Family life gave her much satisfaction and she enjoyed children. Her devotion to her sons was no doubt rendered almost obsessive after two of the surviving boys had died at the age of 11 years, William in 1855, followed by John in 1857. She refused permission for Laurence to travel abroad with friends, even though the party included a medical doctor.

The Earl's prominent role brought him into the company of the royal family. He served with Prince Albert in helping organize the great London Exhibitions of 1851 and 1862. When he was younger, his eldest son had received an invitation to Buckingham Palace to kick a football with Edward the Prince of Wales, but the boys do not seem to have hit it off (Parsons R). Hesitating in responding to a question from the Prince about what he thought of Charles I, Laurence received a sharp slap on the face! Nevertheless relations with the Royal Family remained cordial, and in 1865 Laurence attended a Levée. *'There was much less ceremony than I was led to expect'* he wrote in his diary, *'It consisted in walking past the Prince of Wales and the Duke of Cambridge'* (BirrJ/25).

The family retained quite strong connections with England and William Parsons was actually born in York. After the Act of Union, Parliamentary business meant regular journeys by sea to England. Besides, after his marriage the estates at Heaton and Shipley required attention. At a time before steam ships were introduced for the crossing of the Irish Sea this could be a very slow journey depending on the weather and the direction of the prevailing winds. Still it is a little strange that William Parsons, whose estates were land locked, should have become a dedicated sailor. His son Randal recalled that in his later years his father suffered from bronchitis, and proximity to the sea was thought to be beneficial for his health. This may have been a factor. After the London Season the family would stay for several months of the summer with their grandmother at Brighton. When they first took up sailing, their boat was a small 20 tonner. But the following year the *Themia* was purchased. It was built of iron and displaced 150 tons. They spent much time aboard their yacht that was kept during the winter either in Dublin or at Cowes or Southampton. With it they made excursions in the English Channel, to Cherbourg and round Lands End to Ireland.

A couple of years later Parsons purchased the *Titania*, which displaced 188 tons, for £1800. She was the second yacht of that name, and had been built in 1853 by Scott Russell (1808–1882) for Sir Robert Stephenson the railway engineer (Emmerson 1977). She was sold after Stephenson's death in 1859. It had been used for quite long journeys, both on business and for pleasure (Clark and Linfoot 1983). It had a deep draft of 14 ft which was of advantage when it was used for ocean racing. Her hull had been designed for speed, on the wave-line principles advocated by Scott Russell, which ensured that the front section was slender and was given a concave shape. Indeed she was equipped with a large balloon jib and balloon topsails for racing. The crew numbered 13 sailors plus a cook and the captain. When the ship was laid up for the winter, all but the captain were laid off. Randal describes several quite lengthy trips that he evidently enjoyed very much (Parsons R). The first brought them up the Irish Sea to Oban on the west coast of Scotland, then via Stornaway on the Isle of Lewis, rounding Cape Wrath and the Pentland Firth, in a gale, to reach Wick. They continued on down the east coast calling at Aberdeen and Leith. The shorter route across Scotland via the Caledonian Canal was not available because the canal was closed for repairs. In 1862 they sailed to Antwerp up the river Scheldt. The boat was drawn by horses along canals to Rotterdam. It was here that Laurence saved the life of a girl who had fallen into the water while washing clothes. A crowd of spectators gathered to watch but did nothing to assist her. Before diving into the water, Laurence handed his umbrella to a bystander to hold. When he emerged, bystander and umbrella were nowhere to be seen! Sight seeing trips brought them to a High Mass in Antwerp Cathedral, to the Hague and to Haarlem where steam engines built by Harvey and Co. in Cornwall had been recently used to drain the lake. In Amsterdam they were able to watch diamond cutting in the workshops. By now the family were seasoned sailors. Randal remarks that they faced a severe gale and heavy seas on the journey back to Deal on the coast of Kent, but observed that the *'Titania behaved splendidly'*.

In July 1865 after a spell in London, the family travelled to Brighton, then on to Portsmouth. They crossed to Cowes and went aboard their yacht. They met the *Aline* and tried *'to see if they could do better than we had done in the Scheldt in 1862'*. They more or less matched her speed, and sought advice on possible improvements that they could make from Scott Russell her builder. Russell was a graduate of Glasgow University and was married to an Irish gentlewoman from Tipperary. He was elected a Fellow of the Royal Society in 1849 and must have known Parsons quite well. *Titania* competed in the Royal Cup race around the isle of Wight though without success. The boys were shown around the frigate *Galatea* by Captain Maguire, and the following week they visited the *Royal Sovereign.*

Another trip brought the family from Plymouth to Coruna on the north west tip of the Spanish coast. The diary that Laurence kept on this occasion mentions 'Charley' who had gone off sculling on his own. 'Mama' was with them and she took some photographs, processing them on board (BirrJ/25). Another entry records '*We were very glad to get to bed and settled that the decks were not to be scrubbed till after breakfast.*' Randal recalled that the weather was extremely hot and the journey continued slowly along the north coast. A call was made at Ferrol '*where my father wished to see the Arsenal there and where there was a colony of English workmen and engineers engaged in the Dockyard*'. Then they sailed via Santander and San Sebastian (newly rebuilt after its destruction by the English in 1813) across the Bay of Biscay to La Rochelle. Lord Rosse had obtained permission from Paris to allow them to visit the Arsenal at Rochfort. Leaving the coast at Brest they returned home.

The childhood experiences of the Parsons children were quite unusual. The fact that none of them attended a public school had one great advantage: they avoided contamination with the all pervasive contempt for active involvement in industry that characterized wealthy members of the establishment in Britain. On the other hand it did deprive them of the opportunities given to most young people for mixing and interacting with their peers on an equal footing. It is in such encounters that important skills are acquired, such as the ability to read and assess the signs of response by others to the actions of oneself. It is true that the boys did meet with cousins on occasions, but for the most part their company was made up of adults. Their father had married relatively late in life, and though he spent time with them when he could, their childhood coincided with a period when his public duties were quite demanding. Their mother always took a close interest in her boys. She took good care of them especially after their father died, when all but Laurence were teenagers.

The adults at Birr Castle who were outside the family circle would be very much aware of a class distinction that separated them. This deference is evident in contemporary writing where phrases like 'the noble Lord'... are generously sprinkled throughout. Not surprisingly this created a tendency towards an autocratic approach to servants and to those working for them, though the personality of each individual could modify such effects. Laurence's diary describes how the boys made forays in the company of individual crew members when their yacht brought them to the Isle of Wight. Writing of William Parsons, Sir Robert Ball emphasized that he at least had a talent for getting over such social barriers when dealing with the craftsmen and astronomical assistants on whom he relied so heavily. The obituaries of the eldest surviving child, Laurence who became the fourth Earl of Rosse, and of the youngest, who became Sir Charles Parsons, both refer to shyness as one of their characteristics. Because of the deaths of his two younger brothers while he was a teenager, Laurence especially was set apart. There was a gap in age of eight years between himself and Randal. Moreover from childhood Laurence, Lord Oxmantown, had his life mapped out ahead as the future Lord Rosse. Clere and Charles, the youngest, were close in age, and shared interests that later led both to become professional engineers. Laurence too had imbibed from his father an interest in science and technology and though fourteen years separated them, he remained very close to Charles in later years.

5.3 A high priest of science in Victorian Britain

During his tenure of the office of President of the Royal Society Lord Rosse tackled a number of issues. He mentioned some of these when he gave his presidential address to a meeting of the Society on 30 November 1854 (Parsons W 1854b). One problem that was inherited from his predecessor seemed simple enough: it was the date of the main meeting, the Anniversary meeting. It was fixed by custom for 30 November, the date on which the society had been founded. In recent years the peak of the London Season when most members were likely to be available for meetings, and which used to be in November, had been put back by six months. This created problems in allocating the society's medals. Because of the very wide compass of subjects embraced by the society, the president was keen for members of the council to consult with the membership when they were trying to adjudicate between the claims of two candidates in different disciplines. This was made much more difficult when so many had left London at the end of the Season. To support his argument he cited the situation in the University of Dublin. Here a new regime had been introduced that made the filling of vacant fellowships subject to public, oral, examination (McDowell and Webb 1982). This had ensured that true merit was rewarded. Besides, in the process, a beneficial consensus emerged in the University about the choice that eventually emerged. At this time the University was enjoying the substantial fruits of this procedure, but the Fellows of the Royal Society were unmoved by his arguments. He made clear his views on current provisions for higher education when he spoke of the recent appointment of a commission to effect improvements in the University of Oxford.

> *A man therefore after having very creditably passed a public school, and having taken his degree with a first class in literis humanioribus, may find that he knows no more than was known 1800 years ago. He may be ignorant of physics in its most elementary form, and may therefore be incapable of comprehending the first principles of machinery and manufactures, or of forming a just and enlarged conception of the resources of this great country.*

Another matter had its origin as far back as 1823. The council had been requested by the Lords of the Treasury to appoint a committee to inquire into a proposal of Charles Babbage to build a Difference Engine or Calculating Machine. Born in 1791 and graduating from Cambridge University in 1814, Babbage had been elected a fellow of the Royal Society in 1816 (Swade 1991). Between 1828 and 1839 he occupied the Lucasian chair of Mathematics at Cambridge. At that time there was a need for some means of creating mathematical tables that would be more reliable than the current, purely manual, method and this was becoming ever more evident as scientists and commercial users uncovered more and more errors in existing tabulations. The council's committee reported positively and the funds were provided. Progress was slow and the council sought further funds on his behalf in 1828. This request was also granted. His Difference Engine No 1 would have contained 25 000 parts and weighed several tons. A small section, roughly one seventh of the whole, was assembled. It worked successfully and today represents one of the finest surviving pieces of precision engineering of the 19th century. However Babbage ran into more trouble, losing the services, and the tools of his engineer Joseph Clement. Meanwhile his ideas had matured so that his aim was now to create a calculating machine. This would be programmed by the use of punched cards of the kind then used to programme the Jacquard loom to weave elaborate patterns. By 1834 he had received a total of £17 470, a huge sum (Anon 1872). After this Babbage had to keep the project alive using his own funds, but in 1842 he was notified that no further government funding would be provided, or as Parsons put it (Parsons W 1854b)

> *Science was weighed against gold by a new standard, and it was resolved to proceed no further. No enterprise could have had its beginning under more auspicious circumstances: the Government had taken the initiative, they*

> *had called for advice, and the adviser was the highest scientific authority in this country;- your Council,* (was) *guided by such men as Davy, Wollaston and Herschel... That the first great effort to employ the powers of calculating mechanism, in aid of human intellect, should have suffered in this great country to expire fruitless, because there is no tangible evidence of immediate profit, as a British subject I deeply regret. ... Circumstances had changed, mechanical engineering had made much progress, the tools required and the trained workmen were to be found in the workshops of the leading machinists, the founder's art was so advanced that casting had been substituted for cutting in making change wheels, even of screw cutting engines, and therefore it was very probable that persons would be found willing to undertake to complete the Difference Engine for a specific sum.*

Acting now on his own initiative Parsons consulted eminent engineers and scientists once more, and with their unanimous approval approached Lord Derby's government. The timing was politically unpropitious and the request was turned down. The writer of Babbage's obituary reflected on the whole affair and concluded that (Anon 1872)

> *there was no provision in the Government for the proper consideration of such a question—no minister, no department, no official advisers who could properly be made responsible for thoroughly investigating the matter, and for pronouncing a well weighed decision on it.*

In 1991 to mark the bicentenary of Babbage's birth, his Difference Engine No 2 was built exactly to the original drawings, using only the same level of technological precision as was available to Babbage (Swade 1991). It is now on display to the public at the Science Museum in London. It comprises some 4000 parts, stands 7 ft high, is 11 ft long and 18 in wide and works as was intended!

There are parallels in the experiences of Babbage and Parsons. Each pursued a great scientific project with determination. The sums involved were not dissimilar. While both men were able scientists each was well versed in the practicalities of industry and was aware that the scientific project on which they were engaged had the potential to produce valuable technological benefits. But it was Parsons and not Babbage who brought his project to a successful conclusion.

As President of the Royal Society, Lord Rosse also became involved in a somewhat similar venture, later dubbed the 'Great Melbourne Telescope' (Glass 1997). John F W Herschel had travelled to South Africa with his 20 ft telescope, and between 1835 and 1838 he had produced for the southern hemisphere a catalogue of nebulae somewhat as his father had done for the northern. His work had caught the imagination of scientists in Britain. Astronomy, which was making such great progress in probing the universe, showed off science very well. At the same time it was, apparently, unconnected with the grime and din that characterized the industrial technology which was generating such great wealth. Indeed it could be said that Lord Rosse himself had, through his pursuit of astronomy come to be regarded as one of the 'high priests' of science. The Dublin born Edward Sabine was currently general secretary of the British Association and had recently travelled to South Africa in connection with his work on geomagnetism. Together with Romney Robinson he promoted the idea that a large telescope should be set up permanently in the southern hemisphere. The resulting approach by the British Association for government funding was not successful. The matter was reopened by the Royal Society, but their Memorial, which was addressed to the prime minister, Lord John Russell, never reached him. The Secretary of the Treasury had asked the Astronomer Royal G B Airy for his views, and he had effectively aborted the venture. Despite this, in 1852 the Royal Society was persuaded to set up a Southern Telescope Committee. As well as professionals like Airy, Robinson and J C Adams, it included amateur astronomers in the persons of Lord Rosse, Nasmyth, Lassell, Herschel and Cooper of Markree. Of the latter Rosse, Lassell and Nasmyth all had experience of making large reflecting telescopes. The reasons which Robinson advanced to persuade the Committee to opt for a reflector remind us that the refractor at that time was still not a serious competitor for a telescope intended for viewing very faint objects.

> *... I must express my belief that the opinion expressed by several members of the Committee as to the difficulty of making a 4-feet reflector is exaggerated. I have seen so much of Lord Rosse's operations, that I feel authorised to say this, absolutely for a 3-feet reflector; with high probability for one of four* (feet). *And were the difficulty and the uncertainty even as great as those gentlemen suppose, that is the strongest reason for pursuing this our purpose. It is only by the reflecting telescope that we shall reach the remotest parts of the visible universe. There is no likelihood that an achromatic of 3 feet (the equivalent of a 4 feet reflector) will be made in the next century; if made it must be of enormously greater cost, and will be embarrassed by the evils of flexure and the polarising action of pressure. It is therefore specially desirable that the reflector shall be made as perfect as possible; and never was a better opportunity offered than now, when we shall be authorised to*

experiment under the guidance of Lord Rosse, Sir J. Herschel, Mr Lassell and Mr Nasmyth.

Romney Robinson obtained a quotation from Thomas Grubb for the cost of making and mounting a speculum of 4 ft diameter, and he urged the committee to make it the basis of their request for funding from the government. At its meeting on 5 July 1853, with Lord Rosse in the chair, the committee approved Grubb's proposal, arranged for a deputation to communicate with the government and also appointed a sub-committee *'for the purpose of superintending Mr Grubb's undertaking.'*

The outbreak of the Crimean War in 1854 delayed matters, and when the Committee next dealt with the matter it was with a site in Australia in mind, at the Melbourne observatory in the Colony of Victoria. In the meantime Foucault had succeeded in making a small speculum from silvered glass. He had spoken about it at the meeting of the British Association held in Dublin in 1857, but Robinson was cautious. It was not known what difficulties might be met in figuring a glass disc of 48 in diameter; besides temperature changes might affect glass quite differently from the more thermally conducting metal speculum. After some discussion it was decided to stay with the original decision of a metal mirror. The major step forward was the decision to adopt an equatorial mount in a Cassegrain configuration. The question of transferring the necessary 'know-how' to Australia was considered. Lord Rosse wrote

> *I think it essential that someone should accompany the telescope as a mechanical assistant—he having been the principal operator in grinding and polishing the specula. Some intelligent optical glass grinder who was sufficiently educated to read and understand all that has been written on the construction of reflecting telescopes , would probably answer.*

Robinson added

> *but perhaps it might be better if the Australian observer himself were to come over, and while the process was going on, make himself familiar with all the details. The polishing of large specula is 'skilled work' of the highest order, and I do not think it can be effectually supervised by a mere mechanic.*

An up to date quotation was provided by Grubb in December 1862, which was accepted. The mirrors were eventually cast, with not a little drama, in September 1866.

5.4 Weapons of war

Coming after a rather long period of relative peace between the European Powers, the Crimean War provided a shock for both the army and the navy, supplied as they were with outdated equipment. Reading an account of the battle of Inkerman in November 1854 which revealed how cumbersome was the army's field artillery, William Armstrong was provoked to apply his inventive mind to gun design. His engineering business at Newcastle upon Tyne had until then specialized in applications of hydraulics to dockside cranes and similar equipment, but in December he outlined for the Secretary of State for War, Lord Newcastle, a proposal to produce a gun with quite novel features. The barrel would be rifled, and it would fire elongated, lead, shot. By replacing the cast iron barrel with a composite design of a steel inner sleeve wound round tightly with an outer wrought iron coil, he was able to reduce the overall weight enormously. His designs developed rapidly and although they were not available before the war ended with the Peace of Paris in 1856, the seeds were sown for what would in time become something quite new, a very large, commercial, armament industry centred at Newcastle upon Tyne.

The Earl of Rosse for his part was taken by the problem that confronted the Navy, of attacking the Russian shore batteries using existing naval vessels. The design and construction of ships had progressed by trial and error over the centuries, until eventually the globe was circumnavigated in the 16th century by Ferdinand Magellan and yet naval architecture remained a craft, subject to unpleasant surprises like the capsizing of Henry VIII's newly launched warship, *Mary Rose*, in full sight of the land. In the 18th century the French had sought to apply scientific principles to the art, but it was only when steam replaced sail as a means of propulsion, and iron supplanted wood in the construction of ships, that this became a realistic proposition. The British Navy had been tardy in its response to the challenges. It was the French and Americans who were carrying out experiments to determine the effect of gunfire on iron plate, and the French barquentine *Gloire* would be the first iron clad war vessel when it was built in 1858 (Emmerson 1977). Parsons had obtained details of the French experiments on the strength of armour that had been carried out several years earlier, and he prepared a proposal for Sir John Fox Burgoyne at the Ordnance Office (S.P. 1926). After a successful military career Sir John Burgoyne had been appointed chairman of the Board of Public Works in Ireland when it was established in 1830, and he was therefore no stranger to Lord Rosse. The scheme was for a steam powered, iron clad vessel of

1500 tons that would be used to ram ships moored under the protection of the guns of the Russian shore batteries. He also sketched in broad outline a design for iron clad floating batteries that could be used to bombard the forts at Cronstadt. He approached a personal friend, Mr Nassau Senior who was a member of the government and was advised to raise the matter in the House of Lords, but he did not do so for fear of alerting the Russians.

> *The scientific Russians I have been in contact with have always represented the Czar as a very able man thoroughly conversant with everything relating to the matériel of war. He has a magnificent foundry with every mechanical appliance. He has Jacobi and other ingenious and able men at his elbow*

He was told that if he wrote to the Lords of the Admiralty, his letter would most likely find its way into the waste bin, for '*though fully conversant with routine business,* (they) *had not the general knowledge necessary to deal with new subjects.* In July of 1854 he wrote to the Duke of Newcastle to urge that the Admiralty should carry out its own tests on iron plate armouring. He noted that only after several vessels had actually been built with thin iron plate had any tests been carried out at all.

> *Strange to say the authorities were then satisfied and proceeded no further with experiments, instead of completing the inquiry by investigating the other case, the effect of shot on thick plate iron of sufficient substance to stop the shot altogether.*

Receiving no response to his letter, he himself ordered some iron plates for a target. He already possessed an 18 lb howitzer that he had cast at Birr four years before this. In the end ill health prevented him from making his own investigation. In December he wrote to Sir Baldwin Walker of the Admiralty to set out the basis of his calculations on the thickness of armour that would be appropriate.[2] Since the summer, there had been rumours that rifled ordnance was being developed, and from his earlier experience of casting his own howitzers, he feared that the Russians could, if they wished, quite quickly make much larger rifled guns of their own. He also drew attention to the problems of stability to which vessels with heavy iron cladding would be prone. He had evidently carried out some experiments, because his son Laurence mentioned a 'punt-like model' 3 ft 3 in long, by 10 in wide and 4 in deep, which was eventually presented to the Science

[2] Admiral Sir Baldwin Walker's beliefs that '*this country ought not to take the lead in naval improvement, but only . . . follow on a larger scale the improvements of others*', were being quoted a generation later in debates on naval policy (see Manning 1923).

Museum. In November 1855, the naval architect Scott Russell also wrote to Walker with proposals for an armoured ship constructed completely of iron.

At the mid-point of the century the scale of British commerce and manufacturing industry were the envy of the world. Werner von Siemens, the inventive German entrepreneur, recalled in his memoirs how stimulating and encouraging he and his brother had found the atmosphere during his early visits to England (Siemens 1892). His experience had encouraged him in his resolve to use the application of scientific principles to develop his own industrial enterprises. It was a theme that Lord Rosse returned to frequently. When the British Association for the Advancement of Science met in Dublin in 1857, he was again invited to preside, this time over the Mechanical Science section. In his address he defended the existence of a section devoted to engineering. He refuted the argument that men, realizing the importance of engineering in peace and war, would seek to acquire the general knowledge necessary for them to know when and where to look for help in times of difficulty.

> *In this eminently practical country, private individuals, very often relying on experience, neglect the means to render calculation effective. Experience however, is not always at hand, and is often very costly. How often do we see the ingenious mechanic working on false principles, vainly perhaps attempting to accomplish something which a little elementary knowledge would have shown to be impossible!.... Some years ago he was invited by a physician of eminence in London to visit the works of an ingenious mechanic, who was endeavouring to employ air heated by gas as a prime mover. The physician had embarked £12,000 on the project; a lady of wealth had speculated in it to the extent of £30,000; and various individuals had advanced sums altogether to a large amount. At the entrance to the premises there was a wreck of a gigantic machine of unknown construction: other machines in a dilapidated state were lying about in all directions. It appeared from the explanation of the mechanic, that these huge masses of ruined machinery had been constructed partly for the purpose of ascertaining facts to be found in every elementary treatise, and partly for the accomplishment of objects manifestly impossible. In the construction of the engine itself there was a striking display of great ingenuity constantly engaged in a struggle with the laws of nature. It was perfectly evident that the whole was fated to end in disappointment; still the mechanic and his patrons, undismayed by repeated failures, and heedless of repeated warnings, which where there is no science, were without force, struggled on till the project came to an end*

from exhaustion. Some of the parties were ruined, while all lost the capital they had embarked on the speculation.

He then turned to the importance of engineering science in the service of the state. He described how when sailing through Portsmouth harbour some years back, he was given a running commentary on various ships that had had to be laid up there because of defects in design. He was by then an experienced sailor himself, but no doubt from modesty he made his point by quoting from Captain Washington who held the post of Hydrographer to the Admiralty, who had said

During the late war our ships were copies, and not always very successful ones, of foreign models... our ships were inferior to those of France and Spain in speed, stability, and readiness in manoeuvring.

He attributed their inferiority to the fact that while in France and Spain, and other continental countries, *'the aid of science had been called in'* and *'... the only English treatise at all of a scientific character was published by Mungo Murray, who died a working shipwright'*. Of the 42 men who were educated at the School of Naval Architecture during its brief existence from its foundation in 1811, until it was closed in 1832 for motives of economy, only five had risen to positions of responsibility. Of the rest, they *'could be seen, men familiar with differential calculus, chipping timber in the dockyards in the company of common mechanics.'* It was a damning indictment, often to be repeated by those who were concerned at the clear evidence that despite its early headstart, Britain was now falling behind its industrial competitors because of a failure to provide adequate education for those engaged in manufacturing industries.

Parsons' close connection with the British Association led to a request in 1859 that he should preside again, this time over the meetings of the Mathematics and Physics section. On this occasion his thoughts were more philosophical, rather in the vein of his father's book 'An argument to prove the truth of Christian Revelation' though without any overt theological argument (Parsons W 1859).

How wonderful from its origin has been the progress of geometrical science! Beginning perhaps 3000 years ago almost from nothing, one simple relation of magnitude suggesting another; the relations becoming gradually more complicated, more interesting, more important, till in our day it expands into a science which allows us to weigh the planets; more important still, to calculate long beforehand the course they will take acted upon by forces

> *continually varying in direction and magnitude. When we ask ourselves such questions... we become in some degree conscious of the immense moral benefits which the human race has derived, and is deriving from the gradual progress of knowledge. The discoveries, however, in physical science are often immediately applicable to practice, giving man new powers, enabling him better to supply his many wants.*

He then looked at the staggering technological advances that had occurred in his own lifetime, the harnessing of steam power, its effect on transport by sea and by rail, and the transmission of messages instantaneously over enormous distances. Making the connection between science and technology, he pointed out that

> *The mere utilitarian, has been often reminded that discoveries the most important, the most fruitful in practical results have frequently in the beginning been apparently the most barren, and therefore that the discoveries in abstract science are not without interest even for him.* He concluded, *I confess that the gradual development of scientific discovery... appears to me to serve far nobler purposes than merely to minister to the corporal wants of man. What those purposes are, to some extent I think we can clearly see, though to fathom the full depth of such an inquiry would be beyond my powers.*

Between 1845 and 1860 work continued on papers giving the details of construction of the giant telescopes, and giving an account of the observations made with them. His '*Observations on the Nebulae*' was read to the Royal Society in June 1850 (Parsons W 1850). This gave a preliminary account of the 6 ft telescope, together with some sketches of nebulae, including M51, the 'Whirlpool'. A much more detailed account of the construction of the 6 ft instrument was given in the paper read in June 1861, '*On the construction of specula of six-feet aperture; and a selection from the observations of nebulae made with them*' (Parsons W 1861). This was illustrated with engravings made from photographs, and it also gave comprehensive details of the machinery devised to manufacture the specula. Lord Rosse's motivation in writing his papers was to facilitate those who might wish to build their own telescopes. In this he was successful for when, in 1866, Howard Grubb came to the task of making the 48 in speculum for the Melbourne telescope, he was guided very much by the contents of this paper.

Chapter 6

The Family After the Death of the 3rd Earl

6.1 Teenagers without their father

The boys, like their father and grandfather before them, would complete their education by attending university. Laurence, the eldest, graduated from Trinity College Dublin in 1864, having studied science. In May of 1867 their grandmother, Alice dowager Countess of Rosse, died at Brighton. In that year her son William himself became ill, and the family holiday was spent not aboard their yacht but travelling in Germany and Switzerland. A steamer journey brought them up the Rhine from Cologne to Mainz from where they travelled by train to Switzerland, visiting several cities before returning home. Instead of returning to Parsonstown, Lord Rosse took a house on the coast just south of Dublin. His younger cousin Mary Ward was living close by, and she gave very valuable support to Mary Rosse and her boys as his illness progressed (Harry 1988). Eventually an operation was performed to deal with a tumour that had developed on his knee, but, after suffering much pain, he died on 31 October 1867. His body was brought to Trinity College for a memorial service, and thence by train to Parsonstown where he was buried in the old church of St Brendan. Some 4000 attended his funeral. Queen Victoria, herself not long since widowed, wrote a touching letter of condolence to his widow (BirrJ/19).

Windsor Castle
Nov 13 1867

Dear Lady Rosse,

You will I trust, excuse me from intruding on your grief, but I cannot refrain from expressing to you personally my deep concern at the loss of your excellent and distinguished Husband.

He is a great public loss, but to you and your family it must be most gratifying to feel how his name will for ever will be known and gratefully remembered for the great benefits he conferred on Science.

I personally shall ever retain a most pleasing recollection of those associations when I met Lord Rosse, for whom my beloved Husband had the greatest respect.

May God support and comfort you always,

Believe me always,

dear Lady Rosse,

yours very sincerely,

Victoria R.

At the time of his death his estate was valued at £20 000 in Ireland and £60 000 in England.

As well as the expected eulogies for a respected citizen, one rather homely recollection taken from the *Bristol Times* was reprinted in the local news sheet, *the King's County Chronicle* of 4 December 1867 (Anon 1867a). Headed 'A noble working man' it ran

About ten years ago I was in Parsonstown close to which is the demesne of the noble working man, now no more... (being near) *I availed of the opportunity* (to see the giant telescope). *I saw not only the great telescope but I saw the Earl, the telescope maker himself—not in state, with his coronet and ermine robe on, but in his shirt sleeves, with his brawny arms bare. He had just quitted the vice at which he had been working and, powdered with steel filings, was washing his hands and face in a coarse ware basin placed on the block of an anvil, while a couple of smiths sledging away on a blazing bar on another, were sending a shower of sparks about his lordship which he as little regarded as though he were a 'Fire King'. This was in a spacious, rude, smithy which almost occupies one side of the court yard of the castle and in which not only were swing bridges and force pumps, and tackle for scientific instruments constructed, but common and everyday articles in the shape of agricultural gates, sub-soil ploughs etc for use of his farms. I alighted on the noble Vulcan by accident, and feared I might be thought intrusive, but it was not so; for he spoke in an easy friendly way to the gentleman with whom I was, and whom he knew by name, giving permission to see everything about the house and demesne. As he drew on his coat... the Earl looked an intelligent foreman . . . a man with his 'head (as the phrase is) screwed in the right place'.*

Though he did not live to see three score years and ten, William Parsons had witnessed giant progress in technology. The cumbersome 'atmospheric' steam engines of Thomas Newcomen and James Watt, which still powered industry in the early years of the century, had given way to more powerful, more compact, high pressure, high speed engines pioneered in the first place by Richard Trevithick. The rapidity with which these were applied to travel by land and sea is reminiscent of the revolution in air travel that the advent of the turbo-jet engine effected in the years following the second World War. At first the journey from Ireland to London that he, like his father and other Irish members of parliament, faced was by horse drawn vehicles averaging at best perhaps 8 miles per hour, and the sea journey was by sailing ship. Depending on the weather, the time for the crossing from Dublin to Holyhead in Wales averaged from 15 to 20 hours (Baines 1895). In his lifetime the journey was enormously speeded by trains and ships powered by steam engines, until eventually mail from London could reach Dublin within 12 hours.

Wheatstone's experiments with the electric telegraph in 1837 led to a gradual growth of a commercial service. In 1855 a fellow Irishman, William Brooke O'Shaughnessy, completed a network with a total length of 3756 miles in India (Bridge 1998). In just two years, he linked Ootacamund in the south with Peshawar on the northern frontier, Bombay in the west and Calcutta in the east. By the exertions of the Siemens brothers a link between Europe and India was also accomplished, and just before Parsons' death Ireland and the American continent were successfully connected by submarine cable (Von Siemens 1892). The experiments of Volta on the first batteries, and the work of Michael Faraday and Hippolyte Pixii on electromagnetic machines, had laid the basis for the development of electricity as a rival to steam power in industry. Exploitation of materials like cast iron and brass had made available a supply of cheap mass produced goods in a way that foreshadowed the impact of plastics 100 years later. Parsons' active scientific and engineering involvement ensured that he remained very well informed of what was going on around him. It was a misfortune that there was no place in British political life where such a gifted man could have influenced industrial development and could have helped to provide a more adequate scientific superstructure for the second phase which the Industrial Revolution was now beginning.

In the year following her husband's death, when her son Randal entered Trinity College, Mary Rosse took a house in Dublin. During the university vacations the family continued to gather at Birr Castle. In 1869 a tragic accident occurred there (Anon 1869). A steam road carriage had been constructed under the Earl's direction by the youngest boys. It had a flat base with the front pair of wheels being steerable (Ewing 1931). A boiler supplied steam to a vertically mounted engine. Sufficient power was fed to the rear wheels to achieve a speed of 7 miles per hour. This was at a time when, although steam trains had penetrated everywhere, the state of the roads was such that the horse remained supreme for several decades yet to come. The carriage had a bench for passengers to sit on. One day their cousin Mary Ward and her husband were passengers, with the boys' tutor, Mr Richard Biggs, who was steering. Charles and Clere were stoking the boiler. Randal was following on foot. Just beyond the gates of the castle the vehicle overturned. Mary was thrown from her seat and killed. Dr Woods, who lived close by, found that *'she had a broken neck, her jaw was greatly fractured, and she was bleeding from the ears'*. She was only 42 years old and had displayed great talent as an illustrator of scientific books, both her own and those of others. Her gifts had flourished in no small measure due to the encouraging atmosphere that was provided by her cousin at Birr. Her death must have been a severe blow to Mary Rosse.

In 1871 Randal graduated from Trinity College Dublin with a silver medal and a couple of other prizes. He was ordained in the Church of Ireland and went to a curacy at Reading in England. Richard Clere followed Randal to Trinity where he attended the Engineering School. Entering the church was one thing, but the engineering profession! His father's brother, Laurence, had written to him admitting to some prejudices against his becoming a civil engineer, but conceding that *'so many young men of good pretensions are now going into any kind of profession where they think they can make money'*.[1] Then he cited the brothers of noble Lords who were stock brokers, wine merchants or the eldest son of Sir Hiram Maxwell who was a guano merchant. Evidently his letter did not produce the intended effect, for in 1873 Clere received his diploma in Civil Engineering having already graduated BA, securing a gold medal for his performance in science subjects.

On his father's death Laurence, Lord Oxmantown, had suceeded as the fourth Earl of Rosse, and so he became responsible for the Birr estates. In 1870 he married the Hon. Frances Harvey-Hawke, the only child of

[1] I am indebted to Mr N C Parsons for a copy of this letter.

Baron Hawke. She, like her mother-in-law, was a Yorkshire heiress. Mary the dowager Countess of Rosse had a difficult relationship with Frances and only rarely returned to Birr after this. She took a house in Dublin. Clere was followed into Trinity College by his younger brother Charles in 1871. Like all entrants he studied the course in arts. He won a prize for 'proficiency in the German language', and another for 'progress in Mathematics'. In 1873 Mary Rosse returned once more to London, taking a house in Connaught Place, at number 10 close to number 13 in which Charles Parsons had been born. He was transferred from Trinity College Dublin to St John's College, Cambridge. It is not clear why this step was taken. His father and uncle had both graduated from Oxford. His mother of course owned estates in Yorkshire. Whatever the reason may have been, it certainly helped in later life for Charles to have been a member of the Oxbridge establishment. There was still no School of Engineering at Cambridge despite the recommendation of the Royal Commissioners who investigated the situation between 1850 and 1852, so instead he studied for the mathematics Tripos. Although the subject matter of the examination papers was chiefly mathematical they did contain references to thermodynamics, optics, electricity and magnetism (Anon 1878). There was a wide choice of questions but all were demanding and the help of 'coaches' was sought to prepare students. There were two examinations each day for four days in January, an ordeal in itself. If performance thus far was considered adequate students were required to face five more days of papers written in the unheated Senate House (Leedham-Green 1998). Parsons graduated as 11th 'Wrangler' in 1877. A student from Japan, Dairoku Kikuchi, was placed 19th in that year (Anon 1878). Eleventh place was not an especially remarkable performance, but it placed him in the top third of a class of 36. Little detail is known of the courses that he attended. James Stuart, a native of Scotland and graduate of Cambridge, held the chair of Mechanism and Applied Science at that time, and Parsons attended some of his lectures with half a dozen other students (Parsons 1926b). Some experimental work was also included. He attended lectures by E J Routh, who some years after this, in 1887, was to publish a paper on 'The stability of motion' that dealt in part with the dynamic behaviour of governors. Certainly Parsons showed a very firm grasp of dynamics, and different means for taming vibrating systems frequently featured in his patents in later years. He seems to have been rather less well prepared in the behaviour of fluids, unlike the ship builder J I Thornycroft who was sent to Glasgow by his father to study under Professor Rankine.

In later life Parsons never used a slide rule, and his notebooks show calculations carried out approximately, with simple fractions rather than precise decimal numbers. There is no sign of analysis or algebra. He recalled that (Parsons C A 1914)

> *After that* (schooling at Birr) *were interposed five years of pure and applied mathematics including the Cambridge Tripos; and I recall that the strain was more severe than anything I have experienced in business life. Luckily for me boat racing interfered with reading.*

The lack of engineering laboratories did not mean that Parsons ceased to be concerned with engineering projects. Ewing wrote that

> *the table in his room, with its litter of models, bore witness to his continued interest in matters which the Tripos did not touch. He was scheming then a quickly rotative epicycloidal engine to which he gave practical form during the apprenticeship which followed his college course.*

Lord Rayleigh[2], in his introduction to the collected works of Charles Parsons, quoted a contemporary of Parsons who had seen a paper toy engine in his rooms at Cambridge, the wheels of which *'simply flew around'* when Parsons blew on it. Parsons said of it that it would *'run twenty times faster than any machine'*. This in fact was no mere toy, for in the year in which he graduated, he took out his first patent 2344 for a rotary steam engine on 15 June 1877. It was the first manifestation of his genius for engineering innovation. He must have been working on his design while preparing for his final examinations.

When considering the early education of Charles Parsons, account must be taken of the probable influence of his brother, Richard Clere. The latter, after distinguishing himself during his engineering studies at Trinity College Dublin, had, as a student member of the Institution of Civil Engineers, been awarded the Miller Scholarship in 1876 for a paper that he had delivered on the theory of centrifugal pumps (Parsons R C 1877a). This was based on experiments that he had carried out as a trainee for William Anderson of the firm of Eastons and Anderson at Erith in Kent. Later Clere collaborated with Charles, and there can be no doubt that they would have had extensive engineering discussions, though no correspondence or other evidence of this has survived.

2 Robert John Strutt (1875–1947), 4th Baron Rayleigh, was a son of John William Strutt (1842–1919), 3rd Baron Rayleigh and winner of the Nobel Prize in 1904. In 1915 a nephew of Charles Parsons married a member of the Strutt family.

Given the nature of their schooling at Birr Castle, it is not surprising that both Laurence the eldest, and Charles, the youngest son, were described by those who knew them as shy. Charles was a keen oarsman, and a friend of his, R F Scott, later Master of St John's College, remembered (Appleyard 1933).

> *I first made his acquaintance as a comically shy freshman at the Boat-house of the Lady Margaret Boat Club. Parsons I believe had been educated at home and was never at a public school. It was I fancy a new experience to him to dress and undress in the presence of others, and just at first he was always the first and last at the Boat-house for the sake of privacy. But that soon wore off, and though one would not have expected it from his appearance, he developed into a very good 'oar' and rowed in our first boat. In after years he often said to me that his real friends at the University were those whose acquaintance he had made at the Boat-house, and he kept up his friendships with them throughout his life. When we met we just talked of old times, of long forgotten boat races and friends. He was always so simple and unassuming that it was difficult to realise the real greatness and eminence of the man.*

Another friend, A H Prior, who rowed with Parsons in the races of May week 1877, just before the final examinations, recorded his memories. During one race, the stroke of a rival boat lost his temper and as a result holed and sank their boat. It was Parsons who helped Prior ashore after the collision, extricating him from the entangled wreckage. Shy he may have been, but his experience of seamanship with his father had developed a firm self-confidence.

6.2 Astronomy continues at Birr

On his father's death, Laurence followed in his father's footsteps when he was elected in 1869 to the House of Lords as one of the representative peers for Ireland. He sat in the House of Lords as a Conservative until his death. After graduation Laurence had turned his attention to continuing his father's astronomic researches. He inherited a formidable collection of astronomic equipment that still had a considerable life expectation, but almost none of which had been built on his initiative. An exception was a water powered clock that he constructed in an attempt to control the adjustment in right ascension of the large telescope (Parsons L 1866). In 1869 he made a further attempt to use clockwork. But the force to be overcome when changing the east–west orientation of the tube could be of the order of 400 lb and he was unable to achieve the precision that was necessary to allow the use of the newly developed spectroscopes and cameras. He also turned his attention to the wooden altazimuthal mount that was used for the 36 in speculum. The tube had been replaced by a lattice iron structure of square cross section, but the wooden frame was in need of repair and he decided to replace it with an equatorial mounting (Parsons L 1880c). This time no attempt was made to carry out the work at Birr, and instead he turned to William Spence, who ran an engineering firm in Cork Street in Dublin. Its design was inspired by ideas of Lassell and Grubb, and was executed with the help of Samuel Geoghan, the engineer to Guiness's brewery. It was supervised by Bindon Stoney, who since 1862 had been Chief Engineer at the port of Dublin (Cox 1990). It was completed in 1875. No attempt was made to provide a dome that would give protection from the weather (figure 25). A counterweighted seat that carried the observer gave him access to the eyepiece. The carriage on which it was mounted ran along a circular track that was mounted on top of a wall surrounding the telescope. The 'fork' which carried the main lattice tube was aligned parallel to the earth's axis of rotation. The arrangement has a very modern look about it. The telescope tube was much better balanced than the larger instrument, and a clockwork drive was provided for this polar axis. The secondary mirror at the mouth of the tube was mounted on a ring that could be rotated to suit the observer no matter what the orientation might be. While the new arrangement was a very considerable improvement on the altazimuth mount, it failed to reach the standard necessary for photography or spectroscopy.

The first paper that appeared under Laurence's name, when he was still Lord Oxmantown, dealt with observations of the Great Nebula in Orion, Messier 42 (Parsons L 1868). In that year, 1867, he was elected Fellow of the Royal Society and Fellow of the Royal Astronomical Society. The paper summarized the work of many individuals at Birr over a period of 20 years. The engraving which illustrates the text was made from a drawing, and can be seen to be remarkably faithful if it is compared with other such efforts (Ashbrook 1984). An attempt was made to employ a spectroscope to examine the continuous spectra of stars as distinct from the line spectra of gaseous clouds, but it was not successful. Despite the light gathering capacity of the telescopes, the light when spread out over a continuous spectrum became too faint for the eye to discern. Time exposure photography was essential, and the telescope mountings were unsuited to this.

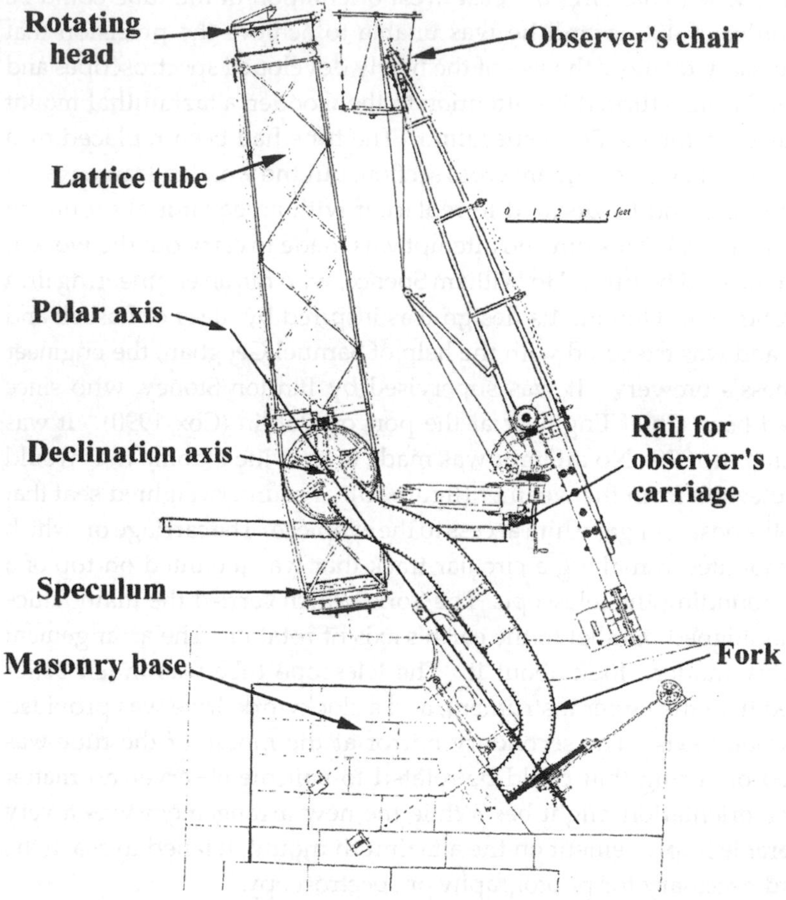

Figure 25. The equatorial mounting for the 36 in speculum completed in 1875 (Parsons L 1880c).

Laurence continued his father's example and funded from his own resources the employment of assistants, many of whom earned international reputations in their later careers. In 1868 he began a study that was to become a major interest for the rest of his life, the measurement of the temperature of the moon's surface. To measure the thermal radiation from so distant an object required that an instrument of very large aperture should be used, and he possessed two such. Unlike the sun, the temperature of the moon is relatively low, and the wavelength of its radiation is relatively long. Because of this, glass is largely impervious to its heat rays. This is the characteristic that allows glass-houses to trap solar heat. The optics of a conventional refractor telescope are therefore not suitable for such work. While it is true that each time radiation is reflected from a surface of speculum metal only 60% of the energy is passed on, this process is less dependent on wavelength than transmission through glass.

There remained the major problem of detecting and quantifying the tiny amount of energy coming from the moon, and lying chiefly in the infra-red. Laurence chose to make it heat the junction of a thermopile. With the help of a very sensitive galvanometer recently invented by William Thomson (later Lord Kelvin) he could measure the tiny current that was caused to flow by the heat of the moon when it was concentrated on it. Two thermopiles were used, connected back to back in opposition; one was illuminated by the moon, the other was turned to the nearby sky. The difference in voltage represented the contribution of the moon's radiation. When desired, the position of the thermocouples could easily be interchanged. The next problem was to convert these galvanometer deflections to an equivalent temperature. For this he used a metal vessel containing water at a known temperature, the walls of which had been blackened with soot to approximate to a perfect radiator. This was mounted in front of the telescope where it would subtend an area equivalent to that of the moon. He reported his first results in 1869 (Parsons L 1869, 1870).

The interpretation of his measurements was not straightforward. The illuminated surface of the moon is not a homogenous disc facing the earth, but rather part of a sphere. Moreover the efficiency with which its surface radiates in a given direction may not be the same as at right angles to its surface. The invisible, thermal, radiation will to some extent be absorbed on its way through the earth's atmosphere to the instrument. The extent of this will depend on how near the moon's position is to the horizon. Measurements taken when the full moon is at different altitudes helps to clarify this effect. Extra information can be gleaned by observing an eclipse of the moon by the earth. But this too brings problems. The earth's atmosphere which absorbs thermal radiation, but not visible light, extends outwards so that the earth's shadow affects heat rays before light begins to be cut off. At that time there were no spectrometers that could detail the relative intensity of thermal radiation at the different wavelengths emitted by the moon. However Parsons was able to make good use of a sheet of glass as a 'colour' sensitive filter. He found that while this only transmitted around 10% of lunar thermal radiation, it transmitted 86% of the sun's heat. In 1870, an estimate of the maximum temperature of the moon's surface of 500 °F was made.

The following year, with the help of his assistant Dr Ralph Copeland, this was revised to 230 °F. In 1873 he reported his results in the Bakerian lecture to the Royal Society (Parsons L 1873). He later revised his estimate down to 197 °F (Parsons L 1877). In the United States S P Langley at Allegheney was carrying out careful studies of the radiation from surfaces at relatively low temperatures, and had built a spectrometer that could now explore this region of the spectrum quantitatively. Turning his attention to the moon he came to an estimate of maximum surface temperature that was significantly lower than that of Lord Rosse. On the basis of observations of the eclipse of 1885, Langley concluded that the temperature never rose above the melting point of ice (Langley 1889). In the previous year, 1884, while Parsons was in America among a party that was attending a meeting of the British Association, an eclipse was carefully observed at Birr by the assistant Otto Boeddicker (Boeddicker 1883–87). Strangely there is no evidence in Laurence's diary that he met Langley during his trip to the United States, even though he called in to the Smithsonian and the Observatory in Washington (BirrM/8/1). While he makes little mention of Langley's work, the latter was certainly familiar with his contributions, even if he was rather critical of them. A more recent reassessment of Parsons' data by W M Sinton in 1958 gave an estimate of 158 °F which agrees very well with modern observations (Elliott 1998). In an obituary notice, Boeddicker wrote that *'the results of these investigations were treated with remarkable coolness by the scientific world'* and the lack of recognition *'was a life-long disappointment to him'* (Elliott 1998). It may well be that this was, in some way, a consequence of a rather retiring character, quite different to that of his father. But one trait he did share with both his father and his youngest brother, and it was *'his extraordinary tenacity of purpose... no failure could turn* (him) *away from a project he once considered feasible'*.

When Copeland left Birr in 1874 he went to Dunsink Observatory near Dublin. Eventually he became Astronomer Royal for Scotland. He was succeeded at Birr by a Dane, J L E Dreyer, who continued to observe nebulae. It was at Birr that he began work on what became the New General Catalogue of Clusters and Nebulae, or NGC (Dreyer 1888). Messier's list numbered around 100 objects; John Herschel's catalogue had 5079. The NGC had 7740 objects. After a spell at Dunsink, he became Director at the Observatory in Armagh. Dreyer's successor was Boeddicker, a German, who returned home in 1916. He had spent much of his time making a drawing of the Milky Way, but this effort was rendered obsolete by the advances that had been made in photography. As Laurence Parsons observed, *'can the pencil of the draughtsman be any longer profitably employed*

Figure 26. Painting of Laurence Parsons, fourth Earl of Rosse in the robes of the Chancellor of Dublin University; courtesy of the Provost and Fellows of Dublin University.

upon nebulae as seen through the 6-feet reflector when photography, to say the least, follows so closely on his heels?' (Moore 1971). From time to time attention was directed to the planets, Venus, Jupiter and to Mars with its newly discovered satellites (Moore 1971). As late as 1886 improvements were being made to the larger telescope, but publication of new observations had ceased by 1880.[3]

Some time after this his brother Charles started to make silvered searchlight mirrors at Newcastle. Charles obviously amused himself with their potential for harnessing solar power as the following letter illustrates (C.L.(3))

Elveston Hall
Ryton-on-Tyne
20 June 91

My dear Laurence,

I have been trying a 2 ft. mirror today (a very hot bright day) in the sun and got some rather (to me) surprising results. It melted holes in the sheet

[3] A recent publication has reviewed this work (Hockey 1998).

iron enclosed in from 2–3 seconds—melted about 2 oz of solder in a sheet iron ladle in about 30 sec[ds]*—and when a little locomotive (like the ones you have) was hung in the focus it got up steam and ran the wheels around at about 500 a minute.*

The day seems to have a good deal to do with it as I never got nearly so powerful effects—an arc carbon rod at the focus became white hot and burnt away rather rapidly.

Your affectionate brother

Charles

Such a mirror was used to construct a portable device with which to make lunar measurements more conveniently (McKenna-Lawlor 1988). The so-called lunarscope was described in 1905, but it was only shown in public shortly after Laurence's death in 1908 at a meeting of the British Association in Dublin. As the century drew to a close a number of political developments gradually brought about a radical change in the position occupied by the Earls of Rosse and other big land owners in Ireland (McDowell 1997). Their monopoly of political representation was ended in 1884 by the Reform Act which extended the franchise to small farmers and labourers. The Land Purchase Act in 1891 began the process of buying land from landlords and making it available for purchase by small farmers and in 1898 the Local Government Act made changes that removed power from landed oligarchies and transferred it to elected urban and rural councils.

Chapter 7

The Parsons Brothers Enter Manufacturing Industry

7.1 Industrial development on the North East coast

William Parsons was very knowledgeable about industrial activity, and although he had no direct involvement himself, his two younger sons began their engineering career in manufacturing industries as soon as they graduated. In 1873 Clere went as a trainee to Messrs Easton and Anderson at Erith in Kent, and four years later Charles went to Sir William Armstrong's Elswick Works in Newcastle-upon-Tyne. This was several years after their father's death, and it suggests that their mother may also have played a constructive role in making this transition. At the same time it has to be said, Charles himself must have had a very clear picture of where he wanted to go in his future career. Looking at his life as a whole he seems to have been drawn as by a magnet to a career as an engineer inventor, a path that inexorably led him to become an industrialist on a grand scale. Such a choice by a graduate was most unusual in England at that time. In his farewell lecture to students in 1853, Professor R V Dixon gave some statistics for the 75 students who had already obtained a diploma in engineering from Trinity College Dublin since the course had begun in 1841. An almost equal proportion had found employment in Ireland, 42%, as had travelled abroad either to Australia, Canada, the United States or India. Only three had found work in England and none were engaged in manufacturing. The picture did not change appreciably in the following generations.

During the lifetime of William Parsons industrial activity had progressed very rapidly. Nowhere was this more evident than on the North East coast of England which was truly one of the seed beds of the

Industrial Revolution. In the 16th century, even though mining technology was still primitive, coal had been mined in that region at locations that were close to the coast. Furthermore a market for coal as a fuel had developed in London. This came about as the forests were decimated by a surge in demand for wood in the 17th century. Increasing quantities of timber were needed to build wooden ships, for house building and for charcoal for iron making. Eventually a point was reached where an alternative to wood as a fuel for heating had to be found and coal met this need. The only way to transport coal economically was by sea, and so a shipbuilding industry grew up close to the mines. Horse drawn wagons from the many collieries used railways which criss crossed the neighbourhood of Newcastle-upon-Tyne at the end of the 18th century, giving easy access to transport by sea. It is therefore no surprise that it was in Newcastle that George Stephenson set up his first steam locomotive works in 1821, nor that numerous shipyards were already well established by then. The early adoption of Newcomen's 'atmospheric' steam engine (figure 2), as a substitute for horse driven machinery, had made it economic to mine deeper than ever before. Now mines could be exploited which had to be continuously pumped out in order to keep them dry. The efficiency of these early engines was very poor, but being sited at collieries they had a plentiful supply of cheap coal. As time passed the collieries of the north east came to dominate the supply of coal in England.

From 1838 the town centre of Newcastle was completely remodelled, largely by the efforts of a single property developer, Richard Grainger, and it had taken on much of its present form (Cochrane 1909). At about that time William Armstrong, a lawyer who was also a self-taught engineer and inventor, formed a partnership and began manufacturing hydraulically operated machinery, and dockside cranes in particular. It was an era when the manufacture of machinery was changing from 'one off' construction of an individual item by millwrights to a situation in which machines were built complete in factories using machine tools. Enterprising individuals, like Thomas Newcomen in the previous century, were finding themselves increasingly on the pay roll of factory owners such as William Armstrong. His business prospered and a river side site was developed at Elswick, a mile or so up stream and close to the newly constructed railway. By the end of 1851 he had 310 men on his payroll.

There was a ready supply of workers for, just as in Ireland, the population of England and Wales was rising rapidly. A study in 1866 noted that from 8 million in 1800, the total had already grown by two and a half times and it was estimated that by 1966 it would reach 576 million if the existing trend continued. The explanation of this growth which was equally evident in both islands, was to be found in a reduction of the death rate. Until then both death rate and birth rate were very high by modern standards, but roughly equal. As long as the birth rate remained unchanged, it required only a modest fall in the death rate to start quite rapid growth. This was brought about by agricultural and technological advances. To give but one example, the potato which was so helpful in sustaining the rapid growth in numbers was after all an innovation, imported from the Americas. Its availability was a direct consequence of the technical advances in maritime navigation that had made new continents accessible. But living conditions remained poor, and the industrial towns like Newcastle, into which the many displaced agricultural workers flocked, suffered from overcrowding. The housing in the older parts of the town with its narrow streets, and the crammed, newly built two storied 'Newcastle flats', was unsuitable, and was swept by regular epidemics of infectious diseases like cholera and typhoid. Many women and children kept themselves alive by undertaking hazardous and physically demanding work in mines and factories. And yet it was from among this population that the tough and courageous workforce of Geordies sprang, to man the mines, the shipyards and the factories that grew up along Tyneside.

An account appeared in the *Engineer* in 1863, of a sight seeing visit by members of the British Association along the Tyne (Anon 1997). It gives a vivid picture of what manufacturing industry in Victorian times entailed.

> *The visitors proceed... by steamboat, and so had an admirable opportunity afforded them of peering beneath the banks of smoke which almost hide this river and the tremendous ranges of factories and works which line its steep banks on either side. Though the picture is a very dark one, still, in its way, there is no river in the world which presents such a wonderful picture of manufacturing industry as the Tyne. Everything around—houses, workshops, the wharves and the river itself—is blackened to a blackness that that would be scarcely believed possible even in the grimy districts of the Black Country itself, while the countless chimney stacks rise into the air in all directions, pouring forth dense volumes of white smoke from the chemical alkali works, or black smoke from the foundries, which as they mix make regular stratifications in the air, almost thick enough to keep out the sunlight. No matter where the eye turns, the view is all the same—it is steam, fire and smoke in every direction for miles, with occasionally a background more hideous over all. Never did industrial labour*

assume a more unattractive aspect than it does on the Tyne, where peaceful employment looks more tremendous and dangerous to the eye than almost any horror of war. The great yards of Messrs Palmer at Jarrow are some miles down the river... tolerably isolated, so that from the water they are pretty equally divided between flame, steam and Smoke.

Mention has made before of the fact that in 1854, Armstrong had turned his attention to the design of artillery. He obtained permission from the Duke of Newcastle, Secretary of State for War, to demonstrate a revolutionary gun that he had designed. It was rifled, had breech loading and the steel barrel was enveloped in coiled iron cylinders (Cochrane 1909). It fired lead coated cylindrical shaped projectiles. Compared to contemporary cast iron guns that fired round cannon balls, it was lighter for a given calibre, more accurate and much more easily handled across rough territory. Although it arrived too late to influence the Crimean War, the design was hailed as a brilliant advance. Development was carried forward with great speed and not without some danger to the inventor who supervised some of the tests. There was little scientific basis to guide the new method of construction. Claims by others were being made on Armstrong's patents, but possible litigation was bypassed in 1859 when Armstrong agreed to make over the patent rights to the government (Emmerson 1977). In return he was knighted and appointed Government Engineer of Rifled Ordnance. A separate Elswick Ordnance Company was established, independent of Armstrong, and for a period it supplied the national requirements for the new design. However by 1862 total demand had eased, and orders were once again directed to the state owned Arsenal at Woolwich. As a result compensation was paid for the termination of the arrangement, and the old 'Engine Works' at Elswick and the new Ordnance Company were amalgamated in 1864 as Sir W G Armstrong and Company. Moreover government permission was granted to seek customers from abroad for the new guns. This was a most significant development. These orders filled the gap caused by the withdrawal of government support, and so a thriving armaments industry came into being. A Scot, Andrew Noble, who had been involved in the original trials of the Armstrong gun, left the Army at this time and became one of the original partners in the new venture. Great advances were made in gun design and manufacture based on Noble's careful scientific and technical studies that were in time to earn him election to Fellowship of the Royal Society. In 1882 an amalgamation took place with the Walker Shipyard of Charles Mitchell and Company, and steel making and shipbuilding were added to the firm's activities. By 1889 the workforce had grown to nearly 15 000 men.

In the early years of the Elswick works, there was an easy friendliness between William Armstrong and his employees. He built housing, and provided schools and a library. He gave dinners and attended social functions organized by his employees. He encouraged the Elswick Works Institute, a Mechanics Institute set up by one of the shop foremen. Such institutes had grown from the efforts of industrial craftsmen throughout Great Britain and Ireland to improve their own education, and were not part of any state initiative. The harmonious relations that existed between the paternalistic employer and the independent minded employees were finally shattered by a strike in 1871 (Fairburn). A tremendous boom that followed the ending of the Franco-Prussian war encouraged engineering workers in the region to seek some improvement in conditions, specifically a reduction in working hours from nine and a half down to nine. The demand was resisted and for the lack of adequate negotiating machinery, a strike quickly engulfed the industry. With the passage of years Armstrong, living in his magnificent residence at Cragside, had largely lost touch with his workforce, and he led his fellow employers in a determined fight. His own senior managers lacked his personal experience of close contact with many of the workers and their families that he had enjoyed in the early days of the enterprise. G W Rendel had come from a public school, and Noble had come straight from the Army, displaying military attitudes to discipline (Fairburn).

> *Noble was a typical example of the now outmoded Victorian industrialist, imperious, autocratic, emphatic in his expressions, sudden, almost volcanic in his decisions; he treated everyone in the Works alike, so that managers of departments and the latest junior office boy all hurried to carry out his orders. His passage through his Works was marked by strained attention, and no one, high or low, could expect to be exempt from the searching inspection.*

Armstrong imported over a thousand workers in an attempt to break the strike. But after four and a half months the employers had to concede defeat, leaving behind a legacy of much bitterness.

Britain's manufacturing industry was a Juggernaut which, when coupled to the rapidly expanding empire, had given her a commanding position in world trade. But like a Juggernaut it was a cumbersome machine and not all that efficient. This was especially evident in the case of shipbuilding. The great sail-powered navy of Nelson's era was challenged for a time by fast American ships, cheaply built from abundant softwood.

The outbreak of the American Civil War in 1861 ended that threat, but it was reckoned that the yield of 400 000 acres of forest were needed each year to meet the current demands of British shipbuilders for timber (Pollard and Robertson 1979). Gradually in Britain, builders were adapting to steam power, and in time iron and then steel came to replace wood. The strength of metal hulls gave them greater cargo carrying capacity, and made them better suited to screw propulsion by steam engines.

7.2 Theory and practice of shipbuilding

During the quarter of a century that followed William Parsons' address to members of the British Association for the Advancement of Science at its Dublin meeting in 1857, some real progress was made. Parsons had deplored the lack of a School of Naval Architecture and the weaknesses in British ship design. In 1860 the Institution of Naval Architects was brought into being, though, as its secretary remarked at the time, it would be many years before it could number many professionally educated naval architects among its members (Barnaby 1960). Prominent among the members of the new Institution was the friend of William Parsons, John Scott Russell (1808–1882). It was he who had, in collaboration with the great engineer Isambard Kingdom Brunel, built the giant *Great Eastern*, a vessel of 18 914 tons and 680 ft length. Immediately members agitated for the creation of a College of Naval Architecture. In 1864 their efforts were crowned with success, and the Royal School of Naval Architecture and Marine Engineering was established at South Kensington in London. It was one of a number of institutions devoted to technological education that were being established on land purchased with the proceeds of the Great Exhibition which had been held nearby in 1851. It was planned that its intake would include both fee paying and Admiralty sponsored students. One of the first graduates was William H White (1845–1913) who passed out with highest honours and was immediately employed by the Admiralty (Manning 1923)[1]. But the initiative was only partly successful. It was reminiscent of the attempt at establishing a School of Engineering at the University of Durham in 1840. This had had to be abandoned after a few years because of a lack of support from

[1] White rose rapidly in the service of the Admiralty. In 1880 he proposed that a Royal Corps of Naval Constructors should be created. Instead of a reliance on apprenticeship, he sought to institute a scheme similar to that in the French and many foreign navies. He resigned in 1883 to become designer and manager at Armstrong's new warship yard being built at Elswick. In 1885 he returned to the Admiralty as Chief Constructor.

employers; its graduates were largely ignored by industrialists (Preece 1982). Speaking in 1877 White observed that so far not one fee paying student had been sent to the School of Naval Architecture by any of the large shipbuilders (Barnaby 1960). Of the 25 fee paying students admitted to that date, nine were foreigners. The state had undoubtedly been remiss in not providing suitable educational establishments, but as much blame must be attached to the deeply entrenched attitudes of the managers and owners of manufacturing industries who consistently failed to recognize the importance of a sound theoretical training for their own engineers and designers.

In 1871, the *Captain*, a new turret ship of 4000 tons, went down in the Bay of Biscay with the loss of over 500 lives. The disaster was attributed to design deficiencies, inadequate freeboard and a lack of stability. It shocked members of the Institution, some of whom had a hand in her design. But now at last moves were being made towards the adoption of a more scientific approach to ship design such as had been pioneered by continental workers. William Froude (1810–1879), who was an Oxford graduate and the son of a clergyman, had worked for Brunel, and had been experimenting for some time with models of ships. He measured the resistance encountered as models of different shapes were towed at different speeds, using techniques pioneered at the end of the previous century by Beaufoy in London and by Chapman in Sweden. The problem of scaling up the measurements made on models for application to full sized vessels is complex, which was why most shipbuilders were so sceptical about such an approach. Scott Russell had himself tested 120 models, ranging from 24 in in length to 12 ft, and from 30 to 60 ft. Speaking in 1870 he acknowledged that '*the most agreeable period of my life was that romantic period of about two years in which I was mainly occupied with the amusement of making experiments*' which he described as today's '*organised play for adults*'. And yet the '*results on a very small scale and a very large scale were... far removed from each other*'. Despite such scepticism from eminent men, Froude had made sufficient progress to persuade the Admiralty, in 1871, to make funding available for him to build a proper experimental tank in which accurate measurements on scaled models could be carried out (Crichton 1989).

The resistance to motion of a hull of a given shape depends partly on the friction forces that act on the wetted surfaces, and partly on the energy that is absorbed by the creation of the waves which accompany the progress of the vessel. The way in which each of these components varies with speed is different, and it needed the inspired experimental skills of Froude to discover a way to deal with the problem. He made careful measurements of the force required to tow planks at different speeds through the water. Using these data, he found that the friction forces experienced by a model could be related to those experienced in the full size vessel, if the wetted surface areas of the two were known, as well as their respective velocities through the water. The second component of resistance, which is due to wave making, could then be established from the model tests by subtracting the calculated drag due to friction from the measured force required to tow the model. Froude next established the scaling law that allowed this, wave, resistance to be worked out for the full scale vessel.[2] This could then be added to the calculated, friction, drag to give the total resistance experienced by the full size vessel at any speed.

Now that results of tests on models could be connected with tests on full scale ships, it was worthwhile to validate tests on models by carrying out towing tests on a full sized vessel. For as long as wind had been the source of power for ship propulsion, little progress could be expected from experiments on models, but the advent of the steam engine changed all that. Shipbuilders needed to know what size of engine they should install. When the results for towing a full scale vessel, the *Greyhound*, were analysed in 1874, it became clear that there was a considerable difference between the power that was delivered by the engines and what was needed to drive the ship through the water (Barnaby 1960). It was this fact that focused attention on the design of screw propellers.

The application of screw propellers to ship propulsion began early in the 19th century. In 1829 the *Civetta*, a 33 ton steam ship from Trieste, was fitted with a screw. An English patent was granted in 1836 to Francis Pettit Smith (Bourne 1867). Admiralty trials began in 1838. Obviously paddle wheels were vulnerable to gun fire, and screws were an attractive alternative for warship designers. But until W J M Rankine in 1868 proposed a suitable idealized model, design was a matter of trial and error (Rankine 1881). In his model, it is imagined that the screw is replaced by a disc that creates a jump in the pressure of the water flowing through it.[3] This jump in pressure when multiplied by the disc area gives the thrust that is generated by the propeller. Ideally the jump in pressure comes in two equal parts, a suction in front of the disc, and a rise in pressure following the disc.

[2] In fact the law had been published in 1853 by Reech of the École Génie Maritime, but it was Froude who showed how to apply it in practice (Moor 1984).

[3] The reader may find it helpful to consult the simple account of the factors which determine the behaviour of fluids interacting with solid structures in appendix 1.

7.3 Clere's apprenticeship and education for industry

In 1875 while Clere Parsons was training with Easton and Andersons at their works at Erith on the Thames below London, he was set the task by Mr Anderson of investigating the performance of centrifugal pumps and if possible improving it. He published a paper based on this work for which he was awarded the Miller Scholarship of the Institution of Civil Engineers (Parsons R C 1877a). With Hesketh, a fellow student member of the Institution, he made tests using three different impellers, as well as comparing the effect of using a casing of a spiral shape, with one of a plain cylindrical shape. He used a dynamometer to compare the power input with the power actually delivered in raising the water pressure. The resulting figures for efficiency allowed him to rate the various designs objectively. In his discussion of the results, he took Professor Rankine's book on 'Applied Mechanics' as a starting point, but he also cited a paper in French by Combes and another in German by Rittinger. Centrifugal pumps and ventilating fans have much in common. Whereas the latter had been in use for many years in collieries, the high speeds that are needed for centrifugal water pumps required steam engine designs that were only just beginning to emerge at this period. His paper does not break much new ground but it shows a firm grasp of the theory of hydrodynamics applied to pumps. What is more unusual is the mathematical analysis that is developed and used to assess the experimental results. It contrasts sharply with later published work of his brother Charles which is quite devoid of mathematical symbolism. It was a respectable effort for a 24 year old.

The Parsons boys had all been given a taste for ships and sailing, but it was Clere who, while an apprentice with Anderson and Easton, was the first to make a contribution to marine engineering. Clere's engineering curiosity led him to explore the behaviour of the screw propeller that was now increasingly being chosen for the newer ships. He began by making a model screw of 3 in diameter which was unusual in being shrouded by a cylindrical tube, and which was mounted in a water trough of circular plan. The propeller was driven and he took measurements, of the power consumed, and of the velocity of the water. The preliminary measurements, though rather crude, gave promising results, and so with the permission of William Anderson, the firm's steamer *Louise* was used for full scale experiments. A modified propeller of 3 ft 6 in diameter was fitted with shrouding, and with a set of fixed, guide blades placed behind it and attached to the shrouding (see figure 27). The latter were

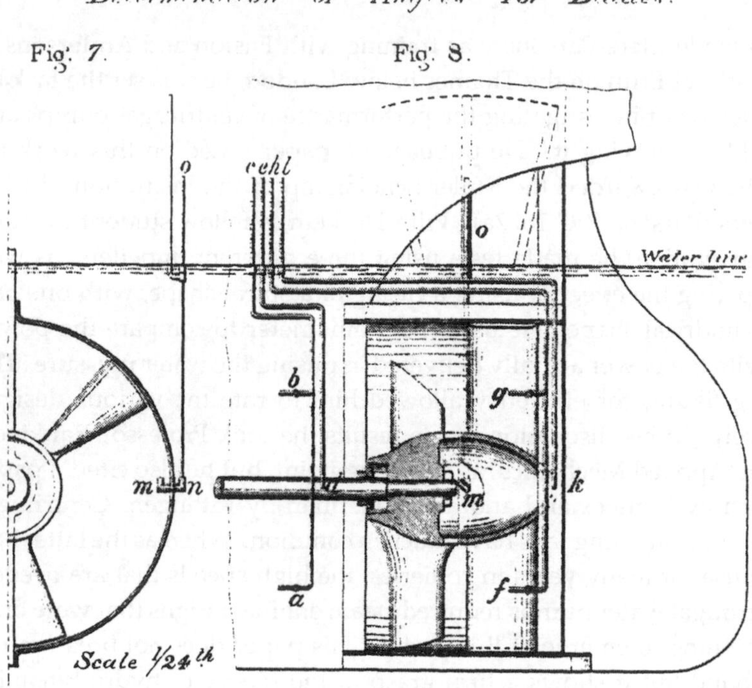

Figure 27. Clere's experiments with propellers. A front, half view, of a propeller can be seen on the left, while a section on the right shows the propeller surrounded by a cylindrical shroud with guide blades aft of it. The pressure sensing tubes *a–o* monitor flow directions (Parsons R C 1879).

designed to harness the energy in the vortex that was discharged behind the propeller. The angle of the fixed blades was carefully calculated so that it would correspond to the direction of the water reaching them (figure 28). The pressure probes seen in figure 27 could be rotated until the column of water reached a maximum height. This happened when the probe pointed directly at the oncoming flow. The correctness of his calculations was fully verified. The results that were reported in a paper read at a meeting of the Institution of Mechanical Engineers held at Manchester showed that some improvement in efficiency had been achieved, but pointed to the need for more work (Parsons R C 1879). It is clear that the author was well informed on the subject, having previously discussed some points with William Froude and other experts. He was aware of

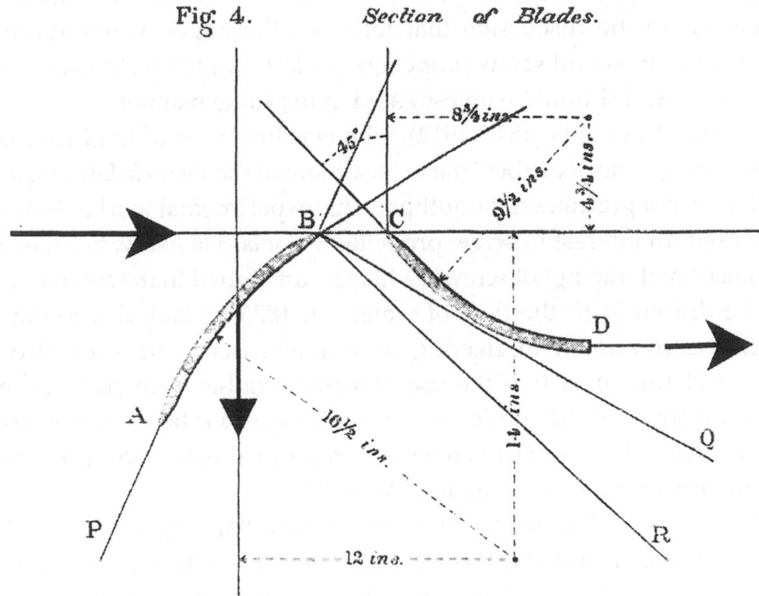

Figure 28. The propeller blade AB moves from top to bottom, and water flows from left to right. After passing through the propeller, water travels parallel to BR. The fixed blade CD deflects this flow so that it leaves D in a horizontal direction (Parsons R C 1879).

the appreciable power loss due to the surface friction of the blades and attempted to calculate this using Froude's data. As can be seen in figure 28, he also had ideas for the shape that the propeller blades should be given.

At the same time as he was writing his paper for the Institution of Civil Engineers, Clere applied for two patents (Parsons R C 1877b). One on 'Pumps' was given 'provisional' cover only, but in view of Charles' later work it has some interest. It envisaged a 'fan' (propeller), mounted with its axis vertical. Above it, blades were fixed to extract the rotary component of motion and so increase the pressure rise. For greater pressure rises he suggested adding further pairs of blades, which amounted to pressure compounding, a course of action described by Osborne Reynolds in his recent patent (Reynolds 1875). The second patent, on 'Propelling Vessels' was sealed in December 1877, and covers the application of the same idea to marine propulsion, as was later to be described in his paper to the Institution of Mechanical Engineers. In fact the drawings are almost identical in the paper and the patent specification. As William Anderson pointed out in the discussion that followed the paper, water turbines, centrifugal pumps and screw propellers *'formed varieties of the same species of machinery'* and should be investigated in the same manner.

Osborne Reynolds (1842–1912), who was Professor of Engineering at Owens College, shortly to become a constituent of the Victoria University of Manchester, congratulated the author on his experimental work. Reynolds had himself an interest in screw propelled vessels. He had written on the phenomenon of 'racing' of screws, which he attributed to the tendency for air to be drawn in to the flow of water. In 1875 he had also examined problems encountered in steering of steamers that were propelled by screws, and the effect that the use of screws, rather than paddles, had on their response to the rudder. For his investigations he used one model that was 2 ft 6 in long and driven by a spring, and another that was 5 ft 6 in long and driven by a steam engine (Allen 1970).

Both the Chief Constructor of the Navy, Barnaby, and the Chief Engineer of the Admiralty, Wright, had been to see the *Louise* in action. Following the death of William Froude in 1879, the matter was then passed to his son for examination at the government hydraulic laboratory that had been established in 1871 at Torquay. The shipbuilder J I Thornycroft (1843–1928) tried out the idea later (Thornycroft 1883).

When his apprenticeship at Erith was over Clere went north to Kitsons' Airedale Foundry at Leeds in Yorkshire, where he became a partner, remaining until 1887. This firm made railway engines and had been founded in 1837 by James Kitson. Meanwhile Clere's brother Charles had entered the Elswick works of Sir W G Armstrong at Newcastle-upon-Tyne as an apprentice in 1877. He was a premium apprentice because he paid £500 for the privilege of receiving his practical training at Elswick. Other apprentices, having only elementary schooling, continued their education for the 'engineering profession' by attending evening classes, provided of course that overtime working did not make this impossible. What distinguished the British industrial scene from that in many of its continental neighbours at this time was the inadequacy of the educational underpinning for those entering industry. Attention had been drawn to this by several witnesses who gave evidence to the Parliamentary Select Committee set up in 1867 to *'inquire into the instruction in theoretical and applied science for the industrial classes'* (Parl P 1868). In its summary report it states that *the foremen are almost without exception, persons who have been selected from the class of workmen, by reason of their superior natural aptitude, steadiness and industry. Their education and that of the workmen during the school age, has been received in elementary schools; and owing to the defective character of the instruction in some of these schools, and to the early age at which the children go to work, it is rarely sufficient to enable them to take advantage of scientific instruction at a later period.*

Later it reports that

In Scotland, where the superior primary instruction of the artizans removes one of the obstacles to their acquiring scientific instruction, the Watt Institution of Edinburgh and the Andersonian University of Glasgow have rendered good service... they can boast amongst their scholars such names as those of Nasmyth, James Young and many others.

Some insight into the causes that had led to this situation can be obtained by comparing the experience in England with that of its neighbouring island, Ireland (Scaife 1997). It had been possible to legislate for the establishment of a national system of primary education in Ireland in 1831, largely because the administration of education there was colonial in style, and the behaviour of centralized authority was not unlike that to be seen in European states at the time. By contrast in England funding for primary education by the state was delayed until 1870. The delay was in part due to difficulties in obtaining agreement between different religious groups, but it was also another manifestation of an English unwillingness to grant control to the state. This same lack of state initiative was responsible for the failure to establish Technical Institutes, independent of the traditional universities, such had grown up in other European countries in the wake of Napoleonic reforms in the early years of the century (Green 1985).

In a confidential report prepared in 1880 by W H White the attitudes displayed by the Admiralty to technical education are described. He had written the report to make the case for the formation of the Royal Corps of Naval Constructors (Manning 1923).

... The Constructive Staff at the Admiralty and in the Dockyards has been recruited almost exclusively from the ranks of those persons who have entered the service as 'workmen' or 'apprentices'. This statement holds good for the apprentices who receive the higher education at the Royal Naval College. When their period of apprenticeship expires they are established

as 'shipwrights', and on completing their course at (the Royal Naval College at) *Greenwich are appointed supernumerary draughtsmen in the Dockyards... From this point on the ex-students are on exactly the same footing as other Dockyard draughtsmen as regards promotion. Moreover, the character of the examinations, at which they have to compete, although quite high enough for the positions competed for, are not such as give scope for any display of the advantages of the special training received at the College.*

We can obtain an idea of the attitudes of industrialists to the sort of education that was considered appropriate for men engaged in manufacturing from some of the papers read at the North East Coast Institution of Engineers and Shipbuilders. One such 'on works organisation', was read in the year 1899 (Westgarth 1899). In the discussion afterwards, speakers highlighted the inadequacy of the training of apprentices. Little effort was made to teach them the appropriate skills, and one speaker asked

would it not pay from the point of view of the employer to subsidise a few of their best artificers to train the lads into a true use of their tools ... There was too much subdivision of work. When a lad was out of his time he was proficient at one thing only.

Nor was much effort made to ensure that they continued to attend evening school. They were of course still children when they began work. Even the 'premium' apprentice who paid for the privilege of joining the work force

was, in many works, from the foreman downwards, looked upon with considerable prejudice, and placed in an altogether false position. And unless the manager took them under his immediate supervision and organised their training, they were treated with indifference and allowed to go their own sweet way, which, as a rule, was far from satisfactory.

The following year the training of young engineers was debated (Borrowman 1900). Mr W C Borrowman drew attention to the inadequacies of British educational structures when compared to American and Continental practice. He also went on to detail the current situation in the Japanese Imperial dockyards. The engineering students at Tokyo University who were destined for the naval service, spent three years studying science and mathematics, followed by three more years of study of technical specialisms. The Tokyo Engineering School had been established with the help of English, Scottish and Irish staff. The Professor of Mechanical Engineering and Naval Architecture until he died in 1908, was C D West, a graduate of Trinity College Dublin (Latorre and 1989). Students acquired the necessary practical experience during summer vacations, in the workshops of the dockyards. However, it is clear, from the printed account of the discussion that followed Borrowman's paper, that few of his hearers shared the author's conviction of the need for change. The management of the shop floor was entrusted to foremen who had entered the workforce as apprentices with very modest schooling. Although managers and owners would have had better schooling, it would have had a 'classical' rather than scientific content and they, too, would have acquired their 'third level' education as premium apprentices, and so they were likely to favour the only form of education that they were familiar with. Among them were individuals of great natural ability. Nevertheless men trained in the sciences could see clearly a connection between the gap in industrial performance that was opening up between England and its European rivals, and the weakness of an education system that was so dependent on apprenticeship. The captains of industry on the North East Coast for the most part were themselves products of the system and showed themselves to be singularly unaware of the problem.

One of the most influential industrialists in the region was Lord Armstrong, and he set out his views in the periodical *Nineteenth Century* in an article 'A vague cry for technical education' (Armstrong 1888a). It was provoked by a meeting organized in the Mansion House to consider the establishment of Polytechnical Institutes for London. He was critical of existing primary schools, which he felt should concentrate more on developing manual dexterity. Colleges of physical science, he went on to say,

> *are apt I think, to be too scholastic for popular requirements.* He continued *when colleges can be established by joint subscription and private munificence, they are worthy of approval... but where the state and local governing bodies have to furnish money for education in relation to national industry, they must look to attaining the required results at the least possible expense, and I am inclined to look upon colleges as luxuries in education rather than necessities.*

He was replied to by Professor Lyon Playfair who observed that labour could be divided into two categories, labour of *quantity* and labour of *quality* (Playfair 1888). Already the importance of the latter was evident in industries in which technical innovation was displacing workers. He noted

Lord Armstrong belongs to the distinguished body of engineers who have advanced so much the industry of this country. They are generally indifferent to technical education because they feel, with just pride, that their offices and workshops have been the schools for engineers and have produced admirable men.

But he gave an example,

While Coventry and Spitalfields were losing their silk industries, the town of Crefeld in Germany was spending £215,000 on its lower schools and £42,500 on a special weaving school. It has doubled its population and quadrupled trade.

Further,

When it is seen that all the Nations of Europe as well as the United States are vying with each other to promote technical training, and that they are spending vast sums from national resources in order to get ahead of one another in the race, Lord Armstrong may regret he has tried to check our modest efforts to effect by private energy, results which have elsewhere been accomplished by strong governments.

Lord Armstrong was unconvinced (Armstrong 1888b). He observed that of the 13 000 workforce at Elswick, not to mention the many other local industries, only 350 on average attended the local Mechanics Institute. He reminded Playfair that

the men who carry on the great industries of nations so far from being rarely in the category of self made men, predominate in it.

But it was not just industrialists who held such views. Although Osborne Reynolds in his inaugural lecture in 1868 made powerful pleas for suitably educated engineers for industry, he also said (Allen 1970)

We have as yet no national system of education; why should we? Because other nations have found it an answer is no reason why we should. The strength of the English character does not lie in state organisation. Our railways and other great enterprises have been the work of individual energies, brought out by the force of circumstances.

An impression of the environment in which workers of the time passed their days can be gleaned from the reminiscences of E K Clark. He was a relative of the Kitson family which owned the Airedale Foundry in which both Clere and Charles Parsons worked for a time. Remembering his apprentice days at that period he observed (Clark 1938).

> *The shops themselves were far from attractive, in an atmosphere where smoke from furnaces and even boilers, grit from smiths' fires, dust from machines, were accepted as the outward and visible signs of prosperity by both employers and employed, there was little encouragement in a movement for their elimination. The foolish slogan—'Where there's muck there's brass' still drew a clap at local engineers' dinners, and when some years later I painted pillars and girders in a dark shop white, it was received with the cynical comment, 'better to use a colour that won't show the dirt'.* The discipline of those days treated many offences, but not that of spitting. Its instrument was the fine. *For smoking, larking, neglecting work, throwing Time Board over gate, reading, bad work, card playing, leaving work before time-bell, fighting, one shilling. For breaking machine, gambling, climbing gates, drunk, half a crown.*

7.4 Charles' apprenticeship

Charles Parsons came to his practical training in 1877 having already had much experience in his father's well equipped workshops. As a premium apprentice, and no doubt as the son of an earl, he had some privileges. He was allowed at his own expense to work on a high speed steam engine of a quite novel design. To appreciate the timeliness of this project it is important to recall that in 1866 Charles Wheatstone and William Siemens each claimed to have invented the self-excited dynamo. This had matured with the appearance of Gramme's ring dynamo in 1871 and Hefner von Alteneck's drum armature in 1872. Electrical dynamos had now become a practical source of large quantities of power. Ideally they required to be driven at speeds of around 1000 rpm which was much faster than the speed of contemporary steam engines. The availability of dynamos offered a more practical way of supplying electric arc lighting for large scale installations than chemical batteries. Such installations now began to appear all over Europe and in the United States and Canada before 1880. Public interest in electric lighting was greatly enhanced by the work on incandescent electric lamps carried out by Edison in New York and by Joseph Swan in Newcastle (Chirnside 1979). Unlike arc lights that had an enormous luminous output, incandescent lamps with their more modest output were suited to the many applications in which candles or gas light had been used up to this time. Swan began his attempts to make

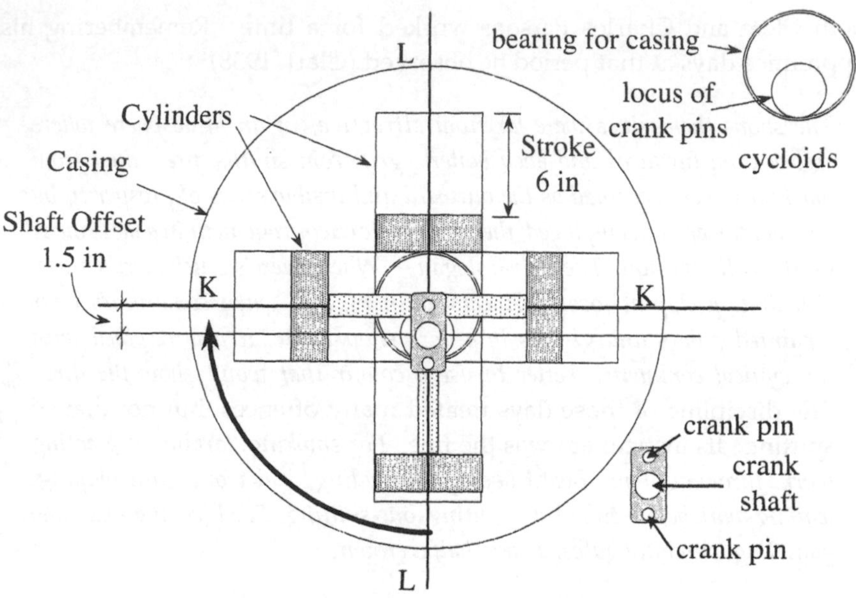

Figure 29. Schematic showing the principle of Charles Parsons' so-called epicycloidal, rotating cylinder, steam engine. K–K is a pair of pistons, with L–L another pair located behind them; each pair is coupled by a rigid connecting rod which engages with the crank shaft. The casing rotates with the cylinder block.

suitable carbon conductors for incandescent lamps as far back as 1848, but it was 1879 before he was able to demonstrate, at a public meeting of the Literary and Philosophical Society in Newcastle upon Tyne, a practical lamp that used a thin carbon rod. He had taken a very important step when constructing this lamp. He heated the carbon element to its working temperature while using the newly invented Sprengel vacuum pump to exhaust the lamp bulb continuously, a procedure now called out-gassing. In 1883 he patented a practical process for making the necessary fine carbon filaments. This involved squirting a fine jet of cellulose solution from a nozzle and was incidentally the first example of a man-made fibre. These events were a few years into the future but, since much of the activity was taking place in Newcastle, Parsons would have been very much alive to their importance.

The idea of a high speed steam engine was something that had occupied his thoughts for a long time. Almost 50 years before, his father, as a young man, had not only employed a steam engine at Birr, but had built it himself in his own laboratory. Indeed Charles had helped to construct the steam engine used in the road vehicle that was involved in the tragic accident at Birr. Since he had been a student at Cambridge, he had been considering ways of applying an idea that he says he took from a Siemens meter mechanism. He wrote (Parsons C A 1927a)

> *I had a little engine (Epicycloidal) which I patented—I had made a model in paper, sealing wax and steel wire when I was an undergraduate at Cambridge. The idea was taken from Siemens.*

When one attempts to visualize the operation of his engine from the quite detailed drawings that have appeared in print, it comes as no surprise that he used models to assist his reasoning. He had already taken out his first patent, number 2344, in the summer just before he began his apprenticeship in 1877 (Parsons C A 1877). It represented his first attempt, and sought to create a product that would meet an emerging demand in the market place and at the same time offer the prospect of creating a business of his own.

His design had two advantages. The first was that it could minimize the vibration that is associated with the conventional design of reciprocating engine. In such engines this arises from the fact that the force exerted by the pressure of gas or steam acting on the piston must be transferred by means of a connecting rod to the output shaft to cause it to rotate. The connecting rod has a bearing for the crank pin at one end. As the crank rotates the connecting rod swings from side to side, and this creates a force that acts at right angles to the axis of piston movement and which is difficult to neutralize. Rubber mountings are used in automobile engines to isolate such forces from the body of the vehicle, but this technique is not practical for large steam engines. The second advantage of his design was that its geometry permitted high speeds, and met what was a growing demand for such engines. Existing engines ran at maximum speeds of a few hundred rpm and required the use of long leather belts to connect the flywheels of the engines with the small diameter pulleys on the dynamos in order to step up the rate of rotation, as can be seen in figure 1.

The basis of the Siemens mechanism is the following. When a disc rolls along a flat surface, a point fixed on its circumference traces out a curved path in space that is described as a cycloid. If that surface is now bent into a cylindrical shape, the curve is called an epicycloid.[4] Somewhat surprisingly, when the cylinder is reduced until its diameter is just double that of the disc, then the path of any point on the edge of the disc is no longer curved, but is a quite straight line. Also the disc rotates twice every

[4] Strictly speaking, if the disc rotates *inside* a cylinder, it traces a hypo-cycloid. However, the term used by Parsons was epicycloid.

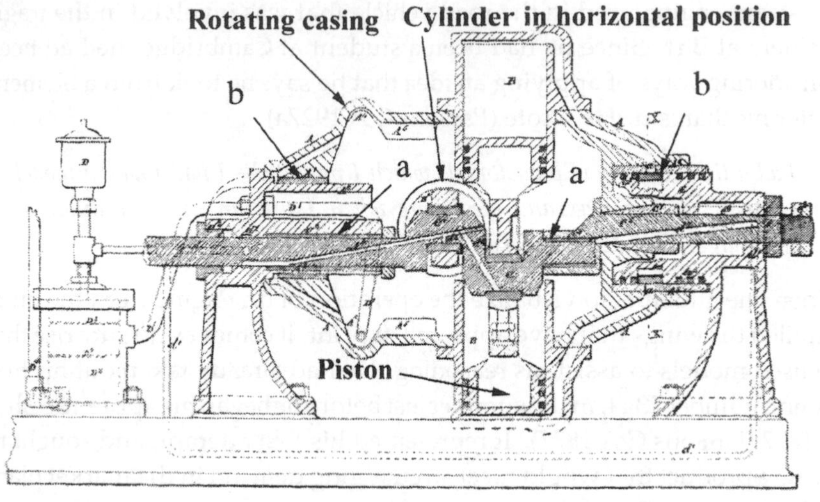

Figure 30. Parsons' patent 4266 of 1878, which shows his steam engine with rotating casing and cylinder block, a a are the bearings for the crankshaft, and b b are the trunnion bearings for the cylinder block. The holes bored in the crank shaft to supply oil to the bearings can be clearly seen. Oil and water which escape into the casing are conducted by a pipe shaped as a scoop to the separator vessel on the left. Power take off is at the left hand end of shaft.

time it completes a circuit of the enclosing cylinder. Parsons was 23 years of age in the summer of 1877 when his ideas had matured sufficiently for him to submit the specification for a patent to cover this application of cycloidal motion to an engine. When he began work at Elswick he was allowed to continue to develop the design, provided he paid for any work carried out in the Works. In 1878 a second patent was taken out in his name.

What patent 2344 describes is a quite novel engine in which the whole cylinder block rotates. Since Watt had adapted Newcomen's beam engine to enable it to transform the reciprocating motion of the piston in its cylinder into rotary motion of its output shaft, a large variety of geometrical arrangements had been tried. An early design allowed the cylinder to rock or 'oscillate' to and fro on trunnions fixed to the cylinder casing. Another, Brotherhood's engine, patented in 1871, used three cylinders arranged on a circle around a common crankshaft and spaced at angles of 120 degrees. Other complex geometries were tried around this time including Tower's 'spherical' engine (Kennedy 1898). But if we leave aside the latter, which used discs instead of pistons and spheroidal chambers in place of cylinders, it was quite unique for the whole cylinder block to rotate. Pairs of cylinders were arranged so as to face each other at the opposite ends of a diameter. These housed a pair of pistons K–K and L–L as in figure 29, that were joined together by a rigid connecting rod moving to and fro along the axis of the cylinders, but with of course no side to side motion. At the centre of the rod was a bearing through which the crank pin passed. The whole cylinder block rotated around a single large diameter, hollow, trunnion bearing. The output crankshaft was also supported by a journal bearing that was placed at the other end of the engine. However its centre line was offset from that of the cylinder block by a distance equal to one quarter of the piston stroke. The throw of the crank pins was made equal to one half of the 6 in stroke. The effect of these features taken together was that, when viewed by an observer rotating with the cylinders, the crank pin appeared to travel to and fro in a straight line along the cylinder axis, with none of the side to side motion associated with conventional piston engines. This removed one source of vibration, and it also had a second benefit. The speed of rotation of the output shaft was twice that of the cylinder block.

In the second patent, number 4266, dated 24 October 1878, arrangements were described for the rotating cylinder block to be supported by *two* hollow trunnion bearings, one at each end, instead of being overhung from a single bearing. Also the crankshaft was given two journals (see figure 30). Obviously he had already run a prototype because he noted that these changes appreciably improved the steadiness of operation. As before, the crankshaft passed through the interior of one of the trunnions to provide an output drive point. Besides this, arrangements for supplying steam to the rotating cylinders were improved by an ingenious design of rotary valve. Patent number 4266 also breaks new ground in another most important respect. Up to this time, delivering lubricating oil to the big end, or crankshaft bearings of steam engines was difficult. One approach was to use small reservoirs that delivered oil under the influence of gravity. These reservoirs, which were attached to the moving connecting rod, had to be topped up by hand by an attendant. This was costly in labour and practical only for slow speed engines. An alternative in which the end of the connecting rod was caused to splash in an oil-filled sump was also employed. But this was uncertain in performance, especially at higher speeds. In Parsons' design, for the first time oil was supplied to the bearings under continuous pressure. The problem of accessing the rotating, crank, bearings was solved by drilling passages in the crank shaft itself to connect the big end, crank, bearings with the stationary, journal, bearings. These received their supply of oil under pressure from a pump driven from the output shaft. This is the method used in most modern internal combustion engines, and the same idea was patented years later by Albert Charles Paine for use in Bellis engines in 1890 and 1892 (Dickinson 1939). Obviously in those days the procedures of the Patent Office for searching previous patents for priority were far from perfect. Lubricating oil, and water from the steam that was exhausted from the cylinders, collected inside the outer casing that surrounded the cylinder block, and they were flung by centrifugal force to its inner surface as it rotated. A stationary pipe, suitably placed, scooped up this mixture and fed it to a stand pipe in which the oil and water were separated, before the oil was fed back to the bearings. The gravity feed shown in figure 30 was replaced by a pump in the final design. Further modifications relating, among other matters, to a suitable design of valve gear for regulating the admission of steam to the rotating cylinders, were covered in a third patent, 4797 of 1882. This was taken out jointly in his own name and that of his brother Clere as well as John Hawthorne Kitson. The patents for the rotary engine were the first of many. It is clear that he had a pretty good idea of the current state of the technology of the steam engine, as well as of the possible opportunities for marketing his special variant. He was intent not just on protecting a clever idea, but he wished to defend what he hoped would be a commercial product. This is evident from the fact that he made no effort to protect a feature of the patent 4266 of 1878, for giving a supply of oil under pressure to the crank bearing of a reciprocating engine. As events turned out this idea was of more lasting importance than that of an epicyclic engine.

Armstrongs were not enthusiastic to manufacture the engine, but at the suggestion of William Cross, who was at the time an under-manager in the Ordnance Works, the prototype was put to work driving machinery in the millwrights shop.[5] Charles was approached by Clere who was then working with Easton and Anderson at Erith in Kent. As we have seen, they had a need for a high speed engine to drive pumps, and so Charles sent the prototype, which was his personal property, to Erith. It proved very successful, and one or two more were built by Easton and Anderson. This caused problems because a director of Armstrongs, Percy Westmacott, was on bad terms with Sir William Anderson, and he was angered that the machine had been sent without his knowledge (Parsons C A 1927a). It may have played a part in Parsons' decision to leave Armstrongs at the

[5] William was the son of Lord R A Cross (1823–1914), a Cambridge graduate who served as Home Secretary for a long period.

end of his apprenticeship. When Clere left Erith to join Kitsons at Leeds, arrangements were made to transfer manufacture to them.

7.5 Labour saving in India!

During this period Clere and Charles kept in close touch. Both joined the British Association in 1878. In the following year Charles followed Clere in joining the Institution of Civil Engineers (Parsons C A 1926b). Soon after, Charles was involved on behalf of his brother Clere in another venture that contrasts sharply with the eminently practical, and potentially commercial, projects that would characterize most of Charles' life's work. In March of 1880, 'Charles Algernon Parsons of Elswick' left a provisional specification for a hydraulically operated apparatus for working punkahs at the Office of the Commissioner of Patents on behalf of 'Richard Clere Parsons, of Calcutta, in the Indian Empire' (Parsons R C 1880). Clere found himself in India, perhaps on business for Anderson and Easton. The punkah was used in the monsoon season in India to assist cooling in conditions of high humidity. It consisted of a screen hung from the ceiling, at the lower edge of which was a horizontal rod to which was fixed a rope. A servant was employed to tug on the rope and so keep the screen swinging gently. The resulting air movement gave some measure of cooling to a person beneath. In 1878 Clere wrote a pamphlet 'On the working of punkahs as at present carried out by coolie labour and the same operation effected by machinery' (Parsons R C 1878). In it he points out the benefits of automation. When used at night to make sleep more comfortable, there was a tendency for the punkah walla himself to doze off. To some extent this could be countered by causing the servant to sit on a high stool!

> *The object of placing the coolie upon a high stool... is that in the event of his falling asleep, which is an occurrence by no means rare, he may have a severe fall, thus inspiring him with a wholesome dread of committing an offence.* Cessation of cooling *puts the sleeper in a state of violent perspiration,* and on its resumption *in the heated sleeper a feeling of chill is experienced, and this is a notorious cause of fever and diseases of the chest and lungs.*

In 1868 a clockwork arrangement for driving such fans had been tried without success, and in 1874 John Imray obtained a patent for a compressed air operated device that would replace the punkah walla. Sir W Palliser and Co displayed such a mechanism at an exhibition at South Kensington in the same year. In 1877 Clere was consulted about the matter and was granted a patent with E Palliser for a mechanism powered by compressed air. It was an improvement on this that the 1880 application described. Clere estimated that the annual wage bill for operating punkahs in various government buildings and barracks was £75 000. Provided that compressed air supplies could be supplied by a public main, Clere reckoned that this could be reduced by one third. It is true that a large hospital in Madras already used a steam engine, driving line shafting to operate punkahs in the many wards, but at the time of his publication, India was in the grip of a famine brought on by a two year drought. In a sub-continent with an enormous population, labour saving can hardly have had much priority. The idea was a non-starter as a commercial product; nevertheless Clere subsequently had a successful career as a civil engineering consultant.

There were other projects to be pursued. In 1880 with William Cross, Charles patented an arrangement for 'Communicating fluid pressure to work moveable machinery' (Parsons and Cross 1880). Essentially this made available to mobile cranes and riveting machines the same advantages that hydraulic power had given to fixed installations. The idea sprang no doubt from experience with the hydraulic systems that had been applied so successfully to dockside cranes, and which were the first products to be developed by William Armstrong. It is interesting that the two brothers each registered patents in the same year for pneumatic/hydraulic mechanisms, and is evidence that they shared their engineering expertise among themselves. By a lucky chance, a batch of letters from Charles to Laurence has survived covering the last decade of the century (C.L.). These provide written evidence of a close relationship between them, but nothing of a similar nature relating to Clere survives. When speaking in later years of the evolution of his ideas on steam turbines and screw propellers, Charles makes no reference to his brother and yet they must have talked about such things since Clere did not leave manufacturing industry for a career as a consulting engineer until after 1885 at least.[6]

7.6 A gas turbine?

More significant perhaps was the collaboration of Charles Parsons and William Cross in a program of experiments on a novel torpedo. Obviously Armstrongs, who were heavily engaged in the manufacture of the largest

[6] An article in the *Kings County Chronicle* about Laurence in 1885 describes Clere as a manager of a large engineering works.

naval guns, would have an interest in a design that could possibly compete, as a naval weapon, with the Whitehead device. Whitehead, an Englishman, had invented a practical submarine torpedo in 1866 (Gray 1975). It was a response to the very recent adoption of iron in warship construction. A great variety of torpedo designs were developed around this time: most used submerged explosives which were activated from the shore, or were delivered at the end of a boom attached to an attacking vessel. But this one was locomotive, or self-propelled, and is the antecedent of the modern weapon. Its success was due to the incorporation of a device for keeping the missile at a stable depth beneath the surface, the details of which were not patented but were kept a closely guarded secret. The missile was driven by a propeller coupled to the novel, triple cylinder, Brotherhood motor, powered by a supply of highly compressed air. At the time, the technology necessary to manufacture cylinders for air storage capable of sustaining the necessary high air pressures, was available only in England. As so often with technical developments, the Royal Navy showed little interest at first and so Whitehead set up his factory to manufacture the weapon in the north of Italy. He, like many other Englishmen who were to be found working in other European countries at that time, was a manager of an engineering works, the Stabilimento Technico Fiumano at Fiume.

Together Cross and Parsons worked on a plan to use the gases evolved by the combustion of gunpowder, 'composition' as Parsons calls it in his notes, as a source of energy to propel a torpedo (Parsons C A 1881a). Gunpowder technology would be very familiar in a factory making guns and in which Andrew Noble was a designer. Indeed the records of weapon trials at Elswick for 1881 show that a '1.875 inch rocket mortar' was fired in November of that year (Fairburn). The notebook kept by Parsons shows that he used cases of gunpowder of much this size for his torpedoes. The tests could be hazardous. Lord Rayleigh recorded that on one occasion the windows of the directors' dining room at the Elswick Works were shattered by an unplanned blast (Rayleigh 1934). Cross left Elswick to become managing director of Hawthorn's locomotive works in Newcastle, and Parsons, having completed his apprenticeship, also left in 1881. A letter accompanying his indentures, signed by Sir William Armstrong, responds to his request for a reference, *'we have the pleasure of bearing testimony to your high theoretical knowledge, your constructive abilities, and your promising business qualifications'*. During his apprenticeship he had the opportunity to watch Sir William Armstrong at close quarters. Speaking in later years he said (Parsons C A 1920)

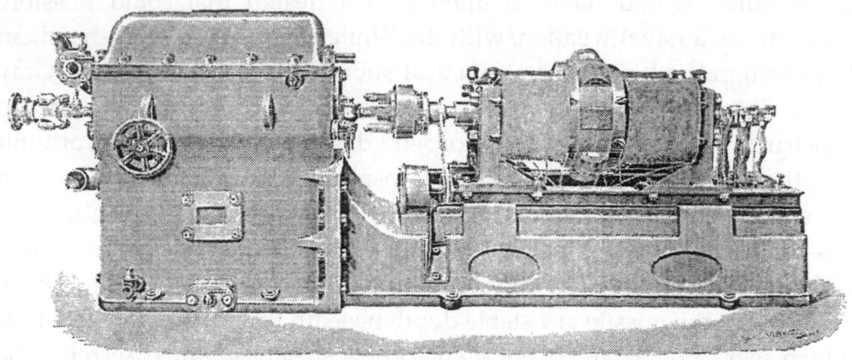

Figure 31. Parsons' high speed engine driving a Brush generator, as seen at the Inventions Exhibition (Anon 1885d). The casing of the cylinder block rotates within the fixed, outer, cover.

> *I entered Elswick Works as a premium apprentice, and served three years. During this time I learned from Sir William Armstrong and his staff... the methods of mechanical research and construction that have made the Works famous throughout the world, methods closely analogous to those which the workers in physical science have followed from the time of Faraday. In those days I was immensely impressed by Sir William Armstrong's mechanical genius, the layout of his experiments, his judicial selection of the fittest, and perhaps above all, by his extraordinary attention to details in critical cases of difficulty, while ordinary general administration he deputed to others. When I recall these facts, and that he excelled Whitworth in the battle of the guns, I picture him as the cleverest mechanical engineer I have ever known.*

These object lessons would have simply reinforced much that he had absorbed from his father. Indeed it is probable that the most significant result of his apprenticeship was not the acquisition of practical skills, but the experience of seeing at first hand the operation of a large manufacturing plant. Regrettably it was, in many respects, no model of enlightened management. The mention of Faraday reminds us that Charles' father as President of the Royal Society had dealings with Faraday, so that to Charles Faraday was no remote figure in the distant past.

Shortly after leaving Armstrongs in 1881 Charles joined his brother Clere who was now working as a director at Kitsons in Leeds (Ewing 1931). An agreement was made between Charles and James Kitson to manufacture the epicycloidal engine. About 40 were produced to power dynamos or centrifugal pumps, and some were exported (Clark 1938). Had it been his only invention it would have been a remarkable achievement. Its steam consumption was relatively heavy, something like 53 lb kW h^{-1} (Anon 1885b). It exceeded the Brotherhood three cylinder engine in efficiency but was later eclipsed by Willans' central valve high speed engine, that was patented in 1884 and 1885. Parsons was awarded a gold medal at the London Inventions Exhibition of 1885, but as the Exhibition catalogue, as well as the accounts in contemporary journals make clear, a great number of rival designs sought to exploit the market for high speed machines that was then opening up. The design was slow to establish a place for itself amongst its competitors, and it was not suited for use with a condenser. The necessary enthusiasm to push the product to commercial success was just not forthcoming. It may be that Parsons recognized deficiencies in the design that would affect its prospect of ultimate success. Whatever the reason, despite many years of commitment, he turned his back on the venture. The technical concept embodied in the engine has a continuing validity. In more recent times Volkswagen, the German car manufacturers, have experimented with an adaptation of the hypocycloidal principle for an internal combustion engine because of its freedom from vibration (Anon 1984).

Perhaps the most valuable consequence of this venture for Parsons was the fact that he became actively involved in the race to build high speed steam engines. He saw at first hand the benefits to be gained, as well as the difficulties to be overcome. The first beam engines had been slow and massive because they depended entirely on the modest pressure of the atmosphere acting against a partial vacuum and so required pistons of large diameter. As technology improved, pressures increased and metal working became more precise, and speeds of hundreds of revolutions per minute were achieved. Parsons', Brotherhood's and Willans' engines all appeared at around the same time with configurations well suited to high speed operation. When the reciprocating steam engine reached its peak at the end of the century, it would prove to have been the terminal stage of development.

As will emerge later, Charles and his father William had much in common when it came to their dedication to science and technology but they differed in one respect. When William began his researches into telescope construction he stated that his objective was to make his methods widely known, so that anyone who wished could copy them. He was lavish in his hospitality to those who sought to visit his laboratory and workshops at Birr. On the other hand Charles and Clere both hoped to

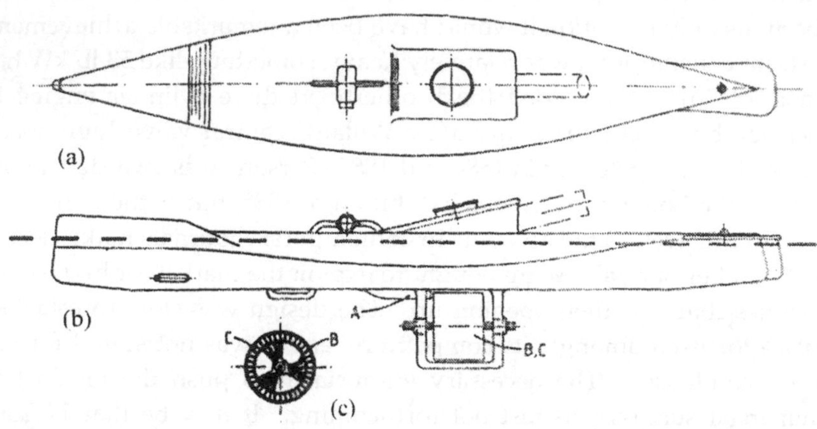

Figure 32. Model boat (0.86 m long) made of bronze sheet (kept in the Discovery Museum at Newcastle upon Tyne). It is driven by a jet of compressed air A, which impinges on the blades of a turbine B. The blades are located in the shrouding which surrounds a three bladed propeller C. (a) Plan, (b) elevation and (c) front view of propeller/turbine (Scaife 1988).

earn a living from their inventions, indeed Clere took out several more patents around this time, four in 1886 and one each in 1887 and 1889. The protection of patenting was invoked by Charles when he was only 23 years of age but Charles Parsons did not use patenting as a means of suppressing competitors. When the time came he adopted a positive policy of licensing manufacturers to use his inventions. This contrasted with the behaviour of James Watt who had sought to protect his crucial steam engine patents by vigorous use of the law against competitors, in a way that without doubt ultimately slowed the pace of technical progress.

When Charles joined James Kitson he made an agreement with him to continue the work on torpedoes, with each partner sharing equally both costs and profits. Accordingly a section of an old workshop at the Airedale Works was fitted out and a team was assembled consisting of a skilled mechanic, a clerk and a labourer/driver. Permission was granted for Parsons to carry out tests on Roundhaye lake nearby. The torpedo, or 'fish' as it was called in his notes, was roughly 6 ft in length. It had been developed so that by now it could reach speeds of up to 20 knots and had a range of 200 to 300 yards. His notebook covering the period 1882 to 1883 gives some idea of the design, though no complete drawing survives

(Parsons C A 1881a). Help in understanding his design is provided by a model boat made of bronze sheet that is preserved in the Tyne and Wear Museum (figure 32). Small cylinders of compressed air provided its motive power. These were available at the time for use in air guns of the type that Charles had made at Birr as a boy. A jet of air was directed to the perimeter of a screw propeller that was located beneath the hull of the vessel. This three bladed screw was shrouded by a double walled cylinder which contained a series of passages that traced out a helical path. By acting on these passages the jet of air turned the propeller. Some arrangement similar to this, described in the notebook as a 'spinner', drove the torpedo (Scaife 1988). In the case of the torpedo, because the whole vehicle was submerged, the spinner could be fitted at the rear of the vehicle which also carried tail fins. It is not clear when the little boat was made, but in later years Parsons spoke of spending much of his time as a ten year old *'making contrivances with strings, pins, wires, wood, sealing wax and rubber bands as motive power, making little cars, toy boats and a submarine.'* The ornamental lake that had been constructed at Birr had presented opportunities for Charles as a small boy to experiment with models, and he was to make good use of such facilities in the years ahead.

The powder that filled the rocket cases was compressed to a high pressure in a hydraulic press, no doubt with a view to improving the stability of combustion. The hot combustion gases emerged from the casing through one or more nozzles. Arrangements were made to instrument the 'fish'. Records were made of the pressure inside the casing by a device carried on board. This marked a disc of paper that rotated once in 30 seconds. Presumably the device was driven by clockwork and was sufficiently waterproofed to be carried on board the 'fish'. The pressure generated by the combustion gases was of the order of 120 psi, but could be up to 290 psi. A comparison was made of the pressure with and without a 'spinner' in place. In addition a distance–time record was kept that appears to have been made with the help of a length of fishing line attached to the missile. This rotated a drum on which a time record was made by a pendulum or similar device that marked the surface of a small diameter cylinder fixed to it. Thus he could compare speed, time and distance of each run, as well as observing combustion chamber pressure. A test was made to compare the speed achieved when cases of different diameters were used. This confirmed that the propulsive effect was simply related to the rate of burning of the charge. During the period of tests at Leeds, attention was being concentrated on optimizing the 'spinner' design.

Although the sketches in the notebook are not easy to interpret, it

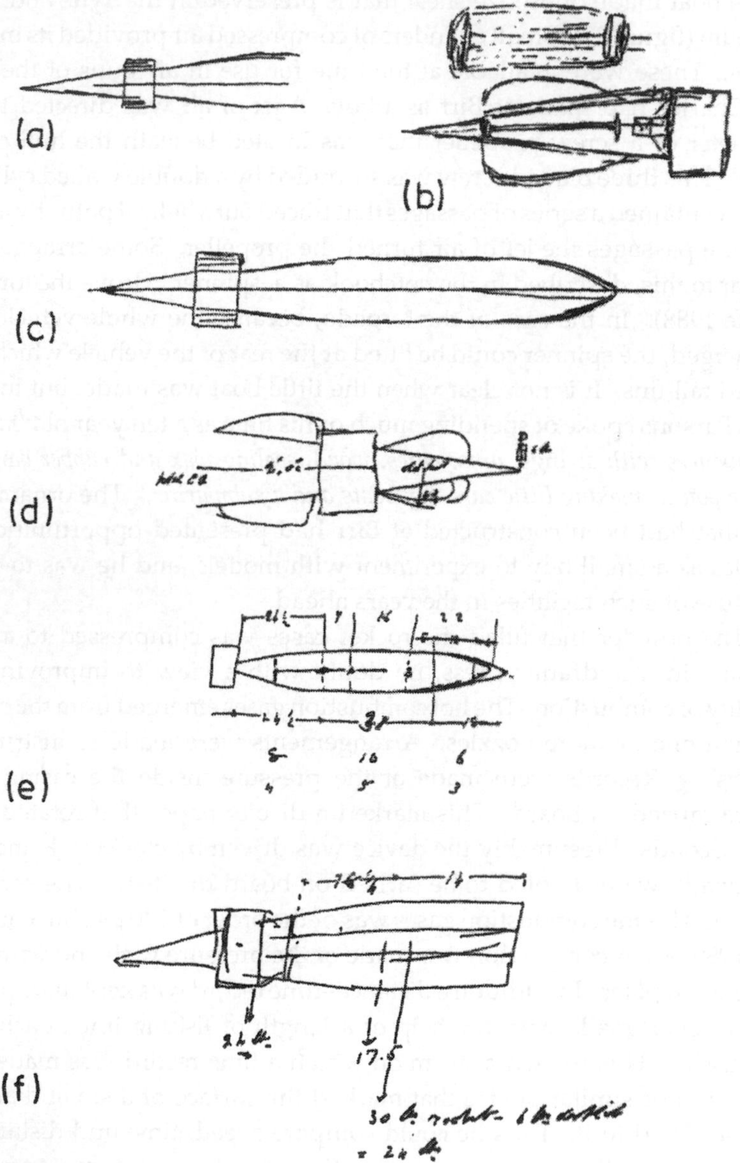

Figure 33. Sketches from Parsons' notebook showing the torpedo body (c), and the tail and spinner (a), (b), while (d), (e) and (f) relate to calculations dealing with the centre of buoyancy (Parsons C A 1881a).

seems that, like the little boat, the 'fish' was in fact driven by a gas turbine. One important difference from later steam turbines was that, in the case of the torpedo, it is probable that the gas flowed *radially* across the blades on the spinner. No measurements of temperature were attempted, but Noble in his book quotes a peak combustion temperature for gunpowder of 2200 degrees (Noble 1906). The device was soldered together, and the cooling effect of the water in which the metal structure was submerged must have been adequate in normal circumstances to avoid melting the solder. However on occasion the spinner jammed, and the cooling effect was interrupted, so that the blades on the spinner were burned. The performance was respectable, 16 to 20 knots, which compares well with 18 knots achieved by Whitehead's recently invented torpedo in 1876, and 24 knots in 1884 (Gray 1975). Up to three cases, or containers, of gunpowder were used, each containing about 2.5 lb. Using figures given by Nobel, we can estimate an energy consumption rate of 500 kW to achieve a mechanical effect equal to about 3.5 kW.

As the gunpowder charge was steadily consumed, the centre of gravity of the missile would tend to move relative to the centre of buoyancy, and so cause it to change its depth. In fact some entries in the notebook can be interpreted as references to an arrangement that was devised to maintain a level trim. Occasionally the 'fish' broke surface. On one such occasion it was noted that the '*note of the spinner... (was) much too low*'. Perhaps the spinner was seizing and the 'fish' was losing speed. In the end the trials were abandoned because the device was too unreliable. As he said toward the end of his life (Parsons C A 1927a)

> *There were unseen difficulties met with, especially the blowing up of rockets which made the apparatus unreliable and under some circumstances dangerous.*

It was only during the Second World War that a new form of cordite was developed that burned in a sufficiently stable manner to be useful in guided weapons (Gibb 1954). Had it been available to Parsons his career might have taken a very different direction. From the technical point of view the notebook shows that he had a good grasp of the factors affecting the buoyancy of the device. He also estimated the expected frictional drag encountered by the device and its dependence upon speed using Froude's measurements. However he shows little awareness of the behaviour of the gases as they flowed through the turbine spinner, or of the need to match gas velocity to the velocity of the blades upon which it acted. At the same time it is clear that Parsons' thoughts had been running along the lines

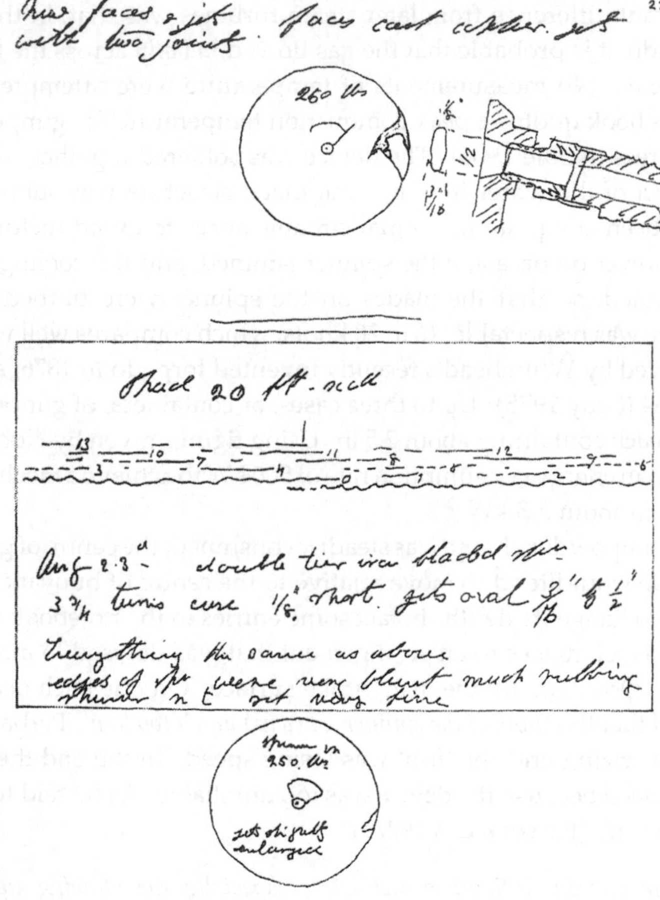

Figure 34. Page 22 from Parsons' notebook. The circular discs are pressure–time recordings. The rectangular strip carries the time distance record used to determine speed. At the top is the oval shaped mouth of a nozzle. The text reads (Parsons C A 1881a)

New faces and jets, faces cast copper, jets with two joints

Speed 20 ft sec^{d}

Aug 23rd *double tier iron bladed spinner* $3\frac{3}{4}$ *turns case 1/8″ short*

Everything the same as above

edges of the spinner were very blunt with much rubbing

spinner not set very fine

spinner on

250 lbs

jet slightly enlarged.

of a purely rotary, turbine, engine for some years. It will be recalled that while at Kitsons he was working closely with Clere who was well versed in the related problems of the design of centrifugal water pumps and screw propellers. As William Anderson had observed, turbines, pumps and screw propellers have much in common, and yet Charles' notebook shows no evidence that he shared Clere's attention to the principles of fluid flow.

Other topics drew his attention at this period. A practical internal combustion engine had been patented by the German engineer Nikolaus Otto in 1876. It was powered by town gas as its fuel. The early machines were of small power output but their high thermodynamic efficiency suggested that in time they would challenge the steam engine. One of Parsons' notebooks begins with an entry dated 20 October 1881, headed gas engine experiments' (Parsons C A 1881b). It is unusual in giving a contemporaneous overview of a project on which he was about to embark. He begins

> *The defects of all gas engines now in the market are,*
>
> *(1) Their great weight and costliness*
>
> *(2) Their want* (i.e. need) *of careful, skilled attention, and tendency to go wrong.*
>
> *The first arises from (a) the small number of explosions and consequent inefficiency of moving parts (b) the low limit of speed which is fixed by the slide ignition arrangement (which arrangement is the only one at present successfully used) To determine the best course to pursue and to obtain data, and to test new arrangements that may lead to the design of a suitable, light, powerful and cheap engine. It is proposed to carry out a series of comparatively inexpensive trials of the explosion of mixtures of* (town) *gas and air at various pressures in a closed cylinder similar to an engine cylinder of 6 in diam and 6 in long.*
>
> (indicator) *diagrams are to be taken to determine the rise and fall and duration of pressure in the various cases. This will determine the probable gain in efficiency arising from an increase of speed due to the diminution of the cooling action of the cylinder in contact with the exploded gases.*

Pressure–time records similar to those taken in the torpedo experiments were attached to the pages of his notebook. At this period, ignition in engines was achieved by means of a pilot flame and a form of slide valve. He tried applying this to a cylinder spinning at 200 rpm, which carried 16 axial passages, rather like the barrel of a revolver. Each passage in succession was filled with a gas and air mixture, and ignited in turn to create a high pressure stream of combustion gases. These gases were intended to eject a jet of water from a nozzle mounted on an oscillating U-tube. It is clear that he was scheming a quite novel device. He was aware that a successful machine would need to run at high speeds, to be light in weight and to be free of the need for continual skilled attention. He did achieve these objectives eventually, but it was with the steam turbine, not the internal combustion engine.

7.7 Settling down to married life

Obsessed as he was with technical projects, Charles Parsons still managed enough leisure time to find himself a wife. Like his mother, she was a Yorkshire woman, Katharine (1859–1933), the youngest of ten children of William Bethell, a landed gentleman from Rise. Rise is just north of Hull and about 60 miles from Leeds. She was five years younger than him and, when they met in 1882, she recalled that (Ewing 1931)

> *he had the character of being an extraordinary and weird young man socially but it was understood that he was a genius.*

Nevertheless the shy, lean and red headed young Irishman and the independent minded young Yorkshire woman were attracted to one another. Perhaps he saw something in her of his own mother. While they were courting he developed an interest in needlework. Possessing a very special manual dexterity, he soon could embroider more skilfully than his bride to be. They were married on 10 January 1883 in the Church of All Saints in the parish of Bramham, by the Rector of Rise, William Whateley. The witnesses were William Bethell, her father, Mary Rosse, his mother, his eldest brother Laurence, now Lord Rosse, and John Conroy. The father of the last named had eloped to Gretna Green with Charles' aunt Alicia in 1837; William Parsons had been sent after them in hot pursuit.[7] Obviously relations between the two families had been patched up in the meantime.

The bride was to receive an early introduction to what being married to an inventive genius would entail. She recalled that (Ewing 1931)

> *after a few days honeymoon* (we) *went into lodgings in Leeds. Charles was immensely keen about the torpedo and used to take the mechanic and me to*

[7] From documents in the Birr archive it is clear that Alicia's mother was very upset by the affair, but that her father was incapable of perpetuating the quarrel.

Roundhay Lake at 7 a.m. There they spent hours trying the torpedo while I shivered on the bank.

Before long she became ill and contracted rheumatic fever. Perhaps to make amends, he brought her on an extensive tour of the United States, that took them to New Mexico and California. Los Angeles, Chicago and Pittsburg were among the places visited. Not infrequently engineering projects caught his attention. In the following year his brother Laurence joined Robert Ball, the one time astronomer/tutor at Birr, and many other scientists to travel to a meeting of the British Association for the Advancement of Science that was held for the first time in Canada. Both kept diaries (Wayman 1986, Birr M8/1). They took the opportunity to visit scientists as well as factories as occasion offered. Laurence describes a visit to one of the first of the oil wells in Pennsylvania. At that time the accompanying gases could not be employed usefully and he describes how they were burned in a flare, day and night, scorching the vegetation round about. On one visit at least, Laurence followed in his brother's footsteps, noting that he was visiting 'Charlie's friends'.

Returning home, Charles continued experimentation with the torpedo into the late summer. However it became clear that as a family man a more settled occupation was required. Once more he sensed that the time for his inventions was not ripe, and without hesitation he moved on to other things. He ended the arrangement with James Kitson and paid his share of the expenses incurred at the Airedale Foundry. There had been considerable financial outlay involved in developing his two inventions. Admittedly he had been able to spread the cost by partnership deals, yet his various inventions had so far failed to yield the promise of a steady income. If they were to become a commercial success he would have to establish his own works and push forward by himself. Obviously it did not seem to him that he could afford to risk his inherited capital on either of the devices developed so far. It was at this point that he had a stroke of good fortune.

The firm of Clarke and Chapman at Gateshead, opposite Newcastle on the south bank of the Tyne, manufactured equipment for use on board ships. William Clarke had left Armstrongs in 1864 and set about manufacturing winches that he supplied for use in ships. He had been joined in 1874 by Henry Chapman who, though from a Quaker family, had served in the army in India.[8] They had become aware of the market that

[8] By virtue of their conscientious refusal to swear an oath, Quakers were for a long time unable to join the armed forces, attend university or start a legal practice.

was opening for the use of electricity on board ships. Electricity as a source of artificial light was of limited use until the creation of the new small, incandescent, filament lamps. These were well suited to illuminating parts of a ship that were completely cut off from natural light, and Edison's lamps were first used on board the paddle steamer *Columbia* in 1879. The Edison system of electric lighting was a feature of the International Exhibition in Paris in 1881. Joseph Swan began commercial production of his lamps in Newcastle in the same year. Of course Parsons, while working on his engine at Elswick, had already gained experience of electric machinery. He had used his epicycloidal engine to drive a directly coupled Siemens dynamo, and had used it to supply electric arc lights. He had also experimented with the carbon rods used in the arc lights. These he bored out and filled with chalk, so greatly increasing their light output. And so Parsons could be seen as a very suitable junior partner by Clarke and Chapman, and he was engaged as their chief electrical engineer. He was fortunate to have inherited considerable personal wealth, and he was able to invest £14 000, to secure a one eighth share in the business. It is difficult to put a modern value on this investment, but given that a skilled tradesman at the time earned from £1.10s to £2 a week, it must be equivalent to £1 or £2 million. The partnership document was dated 1 February 1884, and it credited him with an additional £6000 in respect of goodwill, presumably relating to torpedoes. It stipulated that the rights to any patents taken out in the course of his work would accrue to the partners.

When he moved into the partnership, he brought with him the experimental gear associated with his torpedo research; indeed to the public this seemed to be the chief reason for his joining the firm. The *Shipping World* of December 1883 carried a note

> *Clarke Chapman & Co. have taken into partnership the Hon. C.A. Parsons, son of Lord Rosse, who has invented an instrument of war of a very formidable character. New and spacious workshops are to be erected for its manufacture.*

This was followed in February 1884 by another note

> *The union of Sir William Armstrong and partners with Charles Mitchell & Co. indicated the future policy of the amalgamated firm. Ships of war are destined to occupy most of the berths of the Company, which will also furnish the vessels with guns. And we have now to announce that the Hon. C.A. Parsons has joined the firm of Clarke Chapman & Co., Victoria Works, Gateshead-on-Tyne, with the view to the construction on a large scale of torpedoes and kindred machines of war.*

But events showed that the firm had no interest in involvement with weapon development. Once more Parsons, without regret, turned his back on a product the development of which had evinced his remarkable gifts of inventiveness and experimentation. Looking back years later he said (Parsons 1920)

> *in my own business I have endeavoured to follow the same methods and principles* (as were used by my teachers) , *and have made many more failures than successes (numerically speaking) in the effort to progress as my teachers had done before me, yet the failures should, and have been soon discerned and force concentrated on the successes.*

In 1881, Parsons joined the British Association and attended the annual meeting that was held in York that year. The opinion among respected scientists and engineers was that little further improvement could be expected from the steam engine (Smith 1936). Sir William Armstrong said '*even supposing the steam engine to be improved to the utmost extent that practical considerations give us reason to hope for, we should still have to adjudge it a wasteful though a valuable servant*'. Sir Frederick Bramwell commented '*however much the British Association may today contemplate with regret even the more distant prospect of the steam engine becoming a thing of the past, I very much doubt whether those who meet here 50 years hence will then speak about that motor except in the character of a curiosity to be found in a museum*'. The following year when Sir William Siemens was President, he predicted that '*before many years engines using producer gas would take the place of the complex and dangerous boiler*'. In the year Parsons joined Clarke Chapman, Lord Rayleigh reckoned that '*the efficiency of the steam engine is so high that there was no margin for improvement*'. Certainly while the adoption of triple and even quadruple expansion in engines had improved efficiencies, they still failed to take full advantage of the degree of vacuum that could now be achieved. Attempts at improving efficiency by using steam with higher pressures and temperatures also ran into difficulties because piston lubricants suitable for such conditions were not available. Despite all this pessimism, Parsons at this point turned his attention to creating yet another steam powered engine suitable for driving electric generators, but this machine was to be unlike any that had gone before. Technology had moved on, and the time had arrived for the steam turbine to be perfected, the steam turbine whose performance would be free from the narrow restrictions that limited the steam engine, but there was no time to delay, for already there was an active competitor in the person of the Swedish member of parliament and graduate engineer Dr Carl Gustaf Patrik de Laval (1845–1913).

Chapter 8

A Practical Steam Turbine

8.1 The design of a steam turbine is conceived

When Parsons took up his position at Clarke Chapman, his wife looked for somewhere outside Gateshead to live, and they settled for Corbridge on Tyne. The choice seemed sensible: she came from a farming background and he enjoyed outdoor activities. But Corby was nearly 20 miles up river from the Victoria Works of Clarke, Chapman, Parsons and Partners, and it soon became clear that for someone like Parsons, whose working day was long and unpredictable, this involved much too much travel time, especially in the days before the general availability of motor cars. So within the year they moved house to Elvaston Hall, at Ryton on Tyne, which was less than half the distance away.

Parsons seems to have settled down to the design of a new engine almost immediately because the provisional specifications for patents 6734 and 6735, which are a remarkably precise and accurate description of the prototype, were lodged on 23 April 1884. To quote a contemporary, '*An engineer reading this specification is at once struck with the apparent practicability of the motor therein described, compared with most of its predecessors of a similar type,*' (Neilson 1912). He had been thinking about the possibility of building a steam turbine for some time, writing to Kitson Clark he said (Clark 1921)

> *... I did not entertain the commencement of any experiments to develop the steam turbine till some time after leaving Leeds, though since leaving Cambridge I discussed with several persons the subject as one that might be tackled, in view of modern developments of machinery of high precision.*

Elsewhere he said that he was impressed by the high efficiencies that were now being achieved with water turbines. It was the French in particular

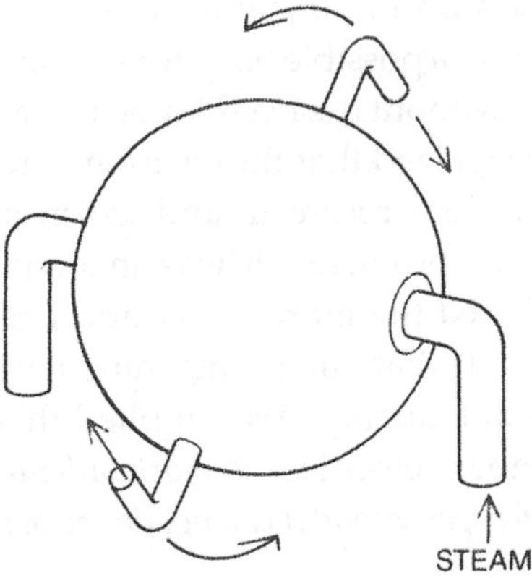

Figure 35. The principle of Hero's aeliopile, or reaction turbine, from Scaife (1985).

who had subjected the design of the water wheel and water turbine to mathematical analysis, spurred on by the offer of a 6000 franc prize by the Société d'Encouragement pour l'Industrie Nationale. The prize was won in 1833 by Benoît Fourneyron, a young graduate of the École des Mines. But it was the work of the English engineer J B Francis that Parsons quoted (Parsons C A 1888b). The latter had printed in 1855 the results of his very careful experiments on the Tremont turbine at Lowell in Massachusetts in America (Francis 1855). Parsons may just have read this publication, but it is also possible that during his visit to America in the previous year he either visited Francis or was made aware of his work. And of course, as has been noted, his brother had published two papers on closely related matters, pumps and propellers.

The idea of harnessing the power exerted by a flowing fluid on a structure was very old. Water wheels were known in Roman times. The use of a jet of steam in a completely rotary engine, the aeliopile, was described by the writer Hero of Alexandria some time between 100 B C and 100 A D (figure 35). This was a hollow sphere filled with steam that had two hollow arms on opposite sides, bent so that they pointed along opposite tangents of the sphere. The sphere was supported between two bearings that also served as supply points for steam. It was set to spin by the rush of steam ejected through the jets at the tips of the arms. The original machine was seemingly used in a temple during religious ceremonies and was for long regarded as something of a toy. But once experience had been gained in building boilers that could withstand pressures that were significantly greater than that of the surrounding atmosphere, such a device suggested itself to many inventors as a possible basis for a machine that could produce rotary motion directly, without the need for beams and linkages. Certainly engineers had long recognized that the creation of rotary motion in a shaft by causing a piston to reciprocate to and fro in a cylinder was a very devious approach to the problem. It was inadequacies in the available technology that precluded the more direct approach. In 1784 Dr Joseph Priestley told James Watt that such a machine had been constructed by Baron von Kempelen in Hungary. Watt replied that it would necessarily have to rotate at enormous velocities, '*in short without god makes it possible for things to move at 1,000 feet per second it can not do much harm* (to his business),' (Dickinson 1939).

Nevertheless by the middle of the 19th century at least one practical application had been found. In 1837 William Avery in Syracuse in New York State, among others, was using such a device to drive a circular saw for cutting timber. Quite a modest pressure is capable of giving to a jet of steam a speed that approaches the velocity of sound. To harness its kinetic energy efficiently, the jet should travel at a speed approaching this. Avery's jets were fixed to long arms 5 feet in diameter, but even so the speed of rotation had to exceed 3000 rpm (Parsons C A 1911a). Thus, as well as being very inefficient, they made a horrendous din. More recently, in 1883, C G P de Laval had found another application for such 'reaction' turbines (Jung 1973). It was to provide a direct drive for the centrifugal cream separator that he had just invented, and which required a speed of 6000 to 8000 revolutions per minute (Bergh 1882). A really practical form of gearing that could reduce such very high rotational speeds was not as yet available, so that further applications for his machine were restricted. During the 19th century a great many patents were awarded in Britain alone to inventors of turbines, more than 200 in the century before, and half of these in the 20 years before 1884 (Neilson 1912). But none of these efforts had yet produced a machine capable of producing useable amounts of power. The broad concepts were known, but no one had implemented them commercially.

All the turbines described above derive their torque from the force of *reaction* that the accelerating steam jet exerts as it leaves the nozzle, very much in the way rocket propulsion functions out in space. A second class of machine was illustrated by Giovanni Branca in his book *Le Machine* in 1629 (Dickinson 1939). This contained a drawing that showed a jet of steam blowing onto blades that were placed around the circumference of a wheel (see figure 36). It produced all of its torque, or turning power, from the

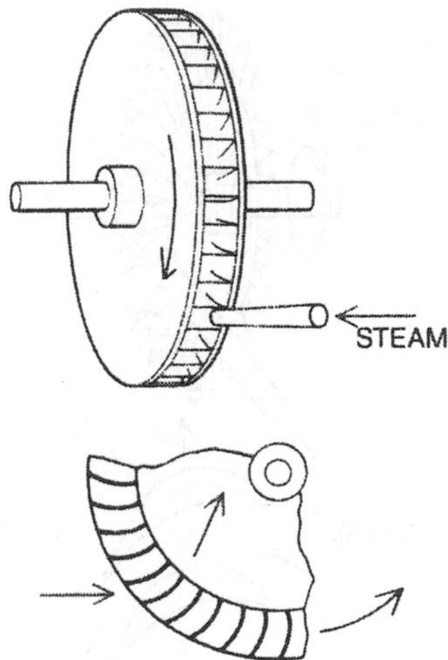

Figure 36. The principle of Giovanni Branca's impulse wheel, from Scaife (1985).

impact of the jet, its *impulse*, as the steam was deflected and slowed by the moving blading. There is also a third class of machine in which the force on the blades is partly due to a reaction effect and partly due to an impulse (figure 37). It was to such a machine that Parsons directed his attention.

When trying to appreciate what makes the task facing the pioneer of steam turbine construction so difficult, it is important to realize that a prototype must of necessity be of modest power. It will therefore have relatively small steam consumption, and consequently modest physical dimensions. Consider now the order of velocities to be handled. If steam from water, boiling at atmospheric pressure, is released into a vacuum through a simple converging conical nozzle, we shall be dealing with a steam speed of over 1400 feet per second. The consequences of these considerations are twofold. First, the necessary cross section of the steam passages in the turbine need only be quite small. Second, the laws of dynamics require that if most of the kinetic energy is to be extracted from the steam, the blading upon which the steam impinges must have a velocity comparable to that of the steam. Since the dimensions are small, the speed of rotation of the shaft carrying the blades must be very high. However once the design has been developed and applications requiring very large flows of steam have been identified, then much larger areas and smaller

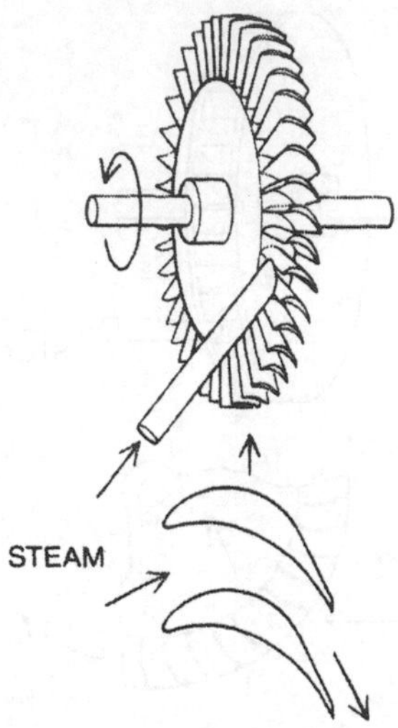

Figure 37. Flow of steam in a mixed impulse and reaction turbine, from Scaife (1985).

rotational speeds are possible. The problem becomes much easier to solve as the scale increases.

The experimental work for the development of the steam turbine was carried out at an incredible pace in Park House, an old building in the Victoria Works. Certain important facts were already clear to Parsons. The torpedo tests that had ended in the summer of 1883 had been a valuable experience and the results had been carefully analysed in his notebook (Parsons C A 1881a). Despite this, his entries at the time of the trials show no awareness of the actual magnitude of the gas velocities, or of how they related to the velocity of blades on the spinner (Scaife 1988). Yet by early in 1884 he had become fully conscious of the importance of this point. If discussions with Clere about the experiments on centrifugal pumps and screw propellers had not made the matter clear to him, then the results of J B Francis' careful experiments on water turbines, with which he was familiar, would certainly have done so.

The calculation of the velocity of steam emerging from a nozzle that is subject to a given pressure drop had been set out by W J M Rankine in 1869 in an article discussing experimental data of Weisbach (Rankine 1869a). In an approach which was rather typical of Parsons, he had arrived at a much simplified way of estimating the magnitude of velocity that would be imparted to steam by a given pressure drop. He did this by analogy with a calculation of the velocity c that can be imparted to a jet of water when fed from a supply with a head H, using the equality $c^2 = 2gH$ (where g represents the acceleration due to gravity). He also made the assumption that the temperature remained constant in the process, when in fact it would be more correct to consider that the quantity 'entropy' stayed constant. In this case the resulting difference could be ignored. But it is interesting that he balked at the somewhat abstract concept of entropy that seemed to arise purely from mathematical considerations and to possess no 'physical' reality. Indeed this was an attitude that remained with him for the rest of his life.

Water turbines generally speaking have one set of fixed passages in which the cross sectional area reduces and the water is speeded up, and this is followed by a set of passages formed by blades attached to a rotating shaft. In passing through the rotor passages the water velocity is reduced by the moving blades which, as a consequence, experience a torque that can be made to perform work. Even with very high heads of water, the ratio of pressure at the inlet of a turbine to that at the outlet is small enough for most of the energy in the flow to be harnessed in one stage. But with steam this ratio can be very much greater. Besides steam is more 'elastic' than water and, because of this, volume increases very rapidly as pressure falls. So, faced with the need to handle a range of steam pressures of several atmospheres, he decided to adopt 'pressure compounding'. He did what many inventors before him had proposed, namely arrange for the total pressure drop across the machine to occur in many small steps. The exhaust of each step or stage would serve as input to its successor. In this way the speed of the steam would always remain manageable. Moreover it would give a greater justification for treating steam as if it were an incompressible fluid like water. Relatively modest steam velocities were a distinguishing feature of all subsequent Parsons turbines.

The configuration of the machine was now becoming clearer. The shaft would have discs locked on to it with the moving blades arranged around the circumference of the discs. There would be spaces between consecutive discs. Steam would flow generally in an axial direction, that is along the shaft. A second series of rings with stationary blades projecting radially inwards would be held inside a cylindrical casing. These too would have spaces between them and would be made in two halves. When assembled the latter would form a cylinder enclosing the shaft and its blades. The steam would pass first through passages formed by the blades of a fixed ring, then through similar passages between the blades on one of the discs on the shaft. There would be an equal pressure drop across

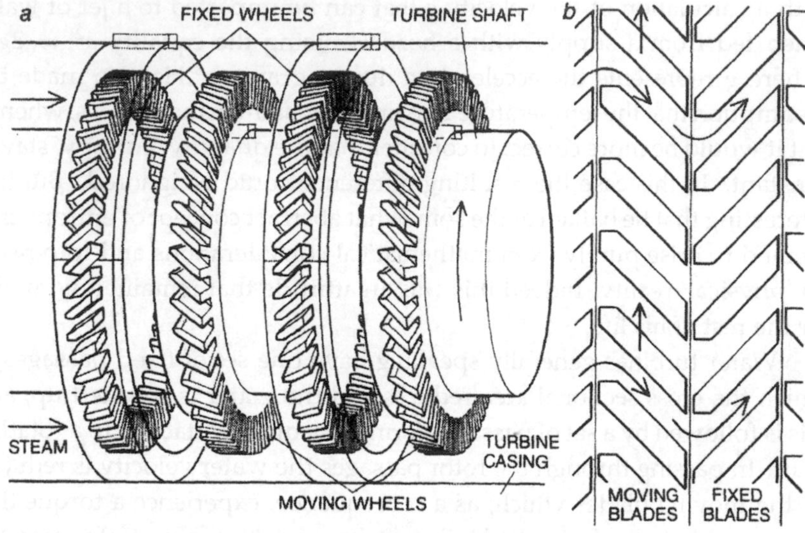

Figure 38. Steam flow through pairs of alternately fixed and moving blades. (a) Steam expands and accelerates as it traverses fixed blades from where it is directed onto the moving blades. Steam expands while crossing through the moving blades, and propels them by reaction and deflection. (b) Detail of one of the two sets of 15 pairs of fixed and moving blades. From Scaife (1985). Copyright © 1985 Scientific American Inc. All rights reserved.

every set of blades, whether fixed or moving. As well as possessing a component of velocity in the direction of the shaft, the steam would also have a component of velocity in the circumferential direction, and it was this which would create a torque on the shaft. While each of the pressure drops would tend to increase the velocity, each encounter with the moving blades would reduce the velocity. The axial component of velocity would thus remain roughly constant from inlet to exit. Of course as the steam density diminished with reducing pressure, the area between the blades would need to be increased to accommodate the increased volume of flow.

The provisional statement in patent 6735 of 1884 also envisaged a second configuration. In this the blades would be arranged in several concentric circles on one or both faces of a disc. Steam would be introduced at an inner radius and move radially outwards to the perimeter of the disc. Between the successive circles of moving blades there would be corresponding rings of fixed blades attached to the walls of the enclosing housing. The velocity which would increase as a result of the pressure drops across a row of fixed blades would be reduced as a result of the encounters with the moving blades and the latter, as before, would experience a torque that could be harnessed. Something like this seems to have featured in the design of the spinner of the torpedoes because, although the sketches in his notebook are not very revealing, there are mentions of two, and on other occasions of three 'tiers' of blades. There is no evidence to explain the fact, but this radial configuration was completely omitted from the claims of the final specification. As events developed subsequently this omission was to prove to be a life saver for the inventor, but that is a story for later.

It would be interesting to know something of Parsons' thinking at this time. Certainly he left many addresses and technical papers after him that are clearly expressed as far as they go, but he did have an apparent lack of awareness of what his hearers especially wanted to know, namely the steps in reasoning that had led him to certain conclusions. If his hearers did not suspect the importance of some underlying fact that he had grasped, then it could well remain hidden. It was a reality that he freely acknowledged. In later years when dealing with the framing of patents he was greatly helped by Dugald Clerk and Fletcher Moulton. However on an occasion when Clerk was unavailable, his assistant was pressed into service with the following instructions (Appleyard 1933).

> *When Sir Charles Parsons comes to you, you sit in this chair of mine and let him talk. For the first quarter of an hour you will not understand what he has in mind; but in about twenty minutes it will dawn on you, and you can put it together. And by the time it begins to dawn on you, put a neat statement of it before him. Say to him 'Before going any further, I want to see if I have it rightly in my head'. Do not say you do not understand a word he is saying. About four or five minutes later give him the note. He will say 'Oh yes, that is better than I have put it myself'. Then say, 'Well if there is anything you have to add, we will add it'. When you have got that, Sir Charles will start on another part of the same field, and piece by piece you will obtain the whole.*

The chief uncertainty facing Parsons was whether it would be possible to design a shaft with rotating blades that could be run safely at the enormous speeds that he knew would be necessary. In this respect his work on the high speed epicycloidal engine had been of crucial importance. It had demonstrated the importance of forced lubrication of bearings. It had also allowed him to try out a system where the oil pump was driven directly from the output shaft, and in which passages bored through the shaft gave access to the journal bearings. The lubrication of bearings was still a subject that lacked a scientific understanding. Although the Institution of Mechanical Engineers had appointed Beauchamp Tower to carry out an experimental investigation of bearing lubrication in 1878, the true explanation of the results that he obtained was only made clear in 1886 when Professor Osborne Reynolds set out his theoretical analysis in a paper to the Royal Society (Reynolds 1886). As his patent of 1878 clearly shows, Parsons' own experimental experience with the epicycloidal engine had placed him far ahead of the theorists.

But important matters had still to be resolved. The shaft of the turbine would have to rotate at incredibly high speeds; in fact the prototype ran at 18 000 revolutions per minute. The only contemporary mechanisms that even approached such a speed were de Laval's newly invented centrifuges and the spindles of some textile machines. Engineers had been familiar for some time with the behaviour of long lengths of shafting because these were used to distribute power from a central engine. A variety of machines in a workshop could be driven using leather belts running over pulley wheels mounted on a common shaft, so-called line shafting. If the shaft was driven too fast, it would whirl, that is it would tend to bow outwards, vibrating vigorously, and become damaged. The speed at which this happened is known as the critical speed, but the phenomenon was little understood at the time, even though in his note on the subject in 1869 Professor W M Rankine of Glasgow University had stated that the whirling speed was the same as the frequency at which the shaft vibrated transversely if struck, and he had set out how this was to be calculated (Rankine 1869b). We do not know whether Parsons was aware of this publication. Anyway it was not until 1895 that S Dunkerley's paper appeared in the *Philosophical Transactions of the Royal Society* with a mathematical method for predicting the critical speed for shafts with different loading configurations.

8.2 Critical experiments

Parsons made a number of shafts. One of them was 1.5 in in diameter and 2 ft long, and with bearings at each end that were 3/8 in diameter. These were run at speeds up to 40 000 rpm No doubt a friction drive from a large diameter wheel was the method of achieving such great speeds. Summarizing the results of these tests Parsons stated that (Parsons C A 1904a) *no difficulty was experienced in attaining this immense speed, provided that the bearings were designed to have a small amount of 'give' or elasticity, and after the trial of many devices to secure these conditions, it was found that elasticity combined with frictional resistance to transverse motion of the bearing bush, gave the best results, and tended to damp out vibrations in the revolving spindle.*

We can only surmise how much trial and error was required to arrive at the critical insight. The correct solution involved a bearing with heavy damping and some 'give'. Parsons' rival, C G P de Laval, was forced to make his own study of the phenomenon, and eventually he would find his own unique solution to the problem.

A speed of 40 000 rpm is of course several times greater than the critical whirling speed for shafts like the one described, if Parsons had cared to calculate it from Rankine's formula. In the patent specification, all that was said was that, if arrangements were made to control the vibrations arising from the inevitable small unbalances in the shaft, it would rotate about its actual centre of mass rather than about its geometrical centre. It is not clear whether Parsons was aware that there exists a stable region at speeds above the critical speed in which the shaft runs very steadily about its centre of mass. He may just have assumed that the inevitable tendency for out of balance masses to cause vibration had to be tamed. The problem is to reach this stable speed without encountering damage on the way. His objective in providing the shaft with a bearing that had some radial 'give' was to allow it to rotate about a point that was not the geometric centre, while at the same time it could provide frictional damping of any resulting movement. This damping drained away the energy that otherwise would have tended to build up the level of vibrations in the shaft, especially near the critical speed.

The actual method that he devised to achieve his purpose was to arrange a series of washers in such a way that the cylindrical, journal, bearing within which the shaft rotated, sat inside the washers, while the outer diameter of the washers sat within the supporting casing. Alternate washers differed in size. Members of one group were a tight fit on the bearing but smaller than the housing, the others were a loose fit on the bearing and a tight fit in the housing. The assembly was compressed by a spring so that some relative radial motion was possible, but only by overcoming friction forces. Each such bearing had one washer that acted as a fulcrum, being a tight fit both on the journal bearing and in the housing.

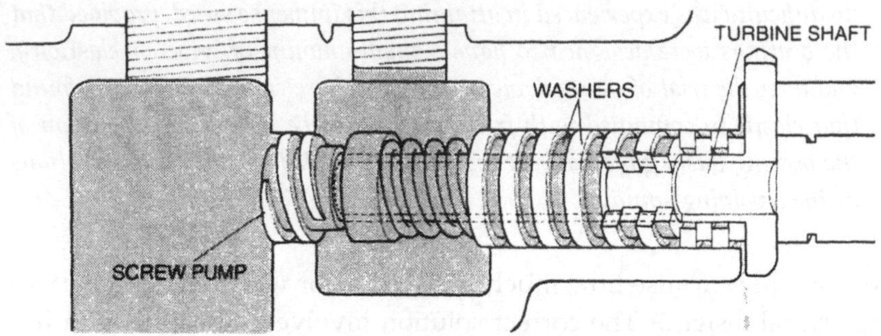

Figure 39. The end of the shaft runs in a bearing tube which is supported by a sequence of washers, each alternately a tight fit on the bearing, but loose in the housing, or loose on the bearing and a tight fit in the housing. The assembly of washers is compressed by a spring and allows lateral movement of the shaft, but the friction between the washers damps vibrations. The screw at the end of the shaft is fed by oil from a reservoir to the left, and acts as pump to supply oil under pressure to the various bearings, from Scaife (1985). Copyright © 1985 Scientific American Inc. All rights reserved.

This arrangement was completely successful. Other approaches had been tried but had had to be abandoned. He tried using just a spring to accommodate the motion of the shaft when it was turning about the centre of mass rather than the geometrical centre of the shaft, but this was abandoned because it lacked sufficient damping, and vibrations built up to a destructive level. It is important to realize that he simply had to solve this problem otherwise no further progress could have been made on other aspects of the turbine. He explained his reasoning in his reply to the discussion on his paper which was read some years later to the Institution of Mechanical Engineers in Dublin (Parsons C A 1888b)

Another point which was perhaps rather interesting from a scientific point of view, was that the body of the steel spindle itself of say 4 inches diameter, if it were running in fixed bearings at 10,000 revolutions per minute, would be very apt to 'whip' or bend; the whole spindle would bend bodily, and its mass revolving eccentrically about the fixed bearings would cause intense pressure on the bearings, causing them to heat. The washer bearing combined the resilience of the spindle itself with the frictional resistance of the washers to any motion. The same principle held good here as with the gyroscope, that when pressure was applied in one direction it produced a motion at right angles, and so brought the spindle back to its central position, involving a minimum motion of the washers.

He was able to arrive very quickly at suitable design estimates, relying on only relatively few trials. He achieved this with the aid of 'the rule of three', that is from two trials at outlying conditions he would predict the behaviour at a third, intermediate, condition (Stoney 1938). For success in such a technique a very sure grasp of the underlying physical phenomena was required. He would rely on uncanny mental processes that were not overtly mathematical and that he could rarely put into words. It was said of him that (Ewing 1931)

the intellectual discipline of Cambridge left its mark. It had given him a sound working knowledge of dynamics which experience converted into something like instinct.

It should be pointed out that a solution to the vibration problem was important for Parsons' particular design in one respect that did not affect the designs of some of his competitors. Because there was a drop in pressure across every set of blades, it was important to minimize the fraction of steam that leaked around the tips of the blades without doing any useful work. If vibrations had been allowed to build up it would have required a larger clearance around the blade tips to prevent them from rubbing against the casing. The task of keeping the clearance between the blade tips and the casing as small as possible was made even more difficult by the small scale of the prototype.

In the prototype machine running at 18 000 rpm the blades were subject to colossal centrifugal forces, up to 14 000 times that of gravity. That is to say each pound of material experienced a pull equivalent to six tons. To withstand this the blades were made in one piece as part of the discs that carried them. The blades were created by sawing slots at an angle of 45° in the rims of the one piece bronze discs or 'rings'. Those on the shaft were secured in place by keying so that they could transmit torque. They were far from the ideal shape, but at least they could safely sustain the forces to which they were subjected. Parsons made a correct decision to leave the niceties of correct blade profiles to be dealt with at a later date. It turned out that turbines are quite forgiving of deficiencies in this area. In the patent specification 6735, it is also stated that

in some cases it may be found convenient to make the blades of sheet metal and to secure them in suitable grooves or recesses in the rings—In this way the blades may be made thinner and they may be accurately formed either before or after insertion in the rings.

While this proved later to be a useful method, it does not feature among the listed claims of the final specification.

For the first machine he used 30 sets of moving blades, 15 on each side of the centre. With a mean blade diameter of 3 in this gave a blade velocity of 235 ft s^{-1}. Remembering that there were equal numbers of fixed and moving blades, the inlet pressure of say 45 psig was reduced in thirty steps of about 4.7% each before exhausting to atmospheric pressure. This gave an approximate steam velocity of between 400 and 450 ft s^{-1}, which though somewhat high was not unreasonable for a blade speed of 235 ft s^{-1}. He quickly embarked on variants of this design. By the time he constructed 'number 5' machine, the shaft speed had been reduced to 12 000 rpm, two thirds of the speed of the prototype. He achieved this by increasing to 50 the number of pressure reducing steps.[1]

It was not practical to use a piston pump to supply the lubricating oil as in his steam engine, so he attached a small propeller to one end of the turbine shaft (see figure 39). At the very high speeds used, this served as an effective oil pump for the surrounding lubricant. The pressurized oil was delivered to the journal bearings at either end, from whence it ran down under the influence of gravity to a reservoir located at a lower level. The propeller would only work properly as a pump if it was drowned with oil, and that meant that the surface of the oil had to be raised from the level in the reservoir to a point above the centre line of the shaft upon which the propeller was fitted. Parsons accomplished this by also mounting a small centrifugal fan on the shaft (see figure 40). This was of simple and robust design, a metal disc with holes drilled radially that connected with inlets located in the boss. It was used to create a suction, or sub-ambient air pressure. The suction was used to draw oil up from the reservoir into a stand pipe to a height sufficient to prime the oil pump. This fan fulfilled another requirement. The speed of contemporary steam engines was regulated by a variant of the governor used by James Watt in which weights are flung outwards with a force that increases as the speed increases. At a speed of 18 000 rpm this was clearly not a practicable arrangement. So the suction fan was also made to act on a leather bellows that was strengthened with a metal disc. The bellows was arranged to

[1] This machine was presented by Dr Gerald Stoney to the Engineering School at Trinity College Dublin in 1912. Another number 5 was reported to be in a museum in Milwaukee in the USA.

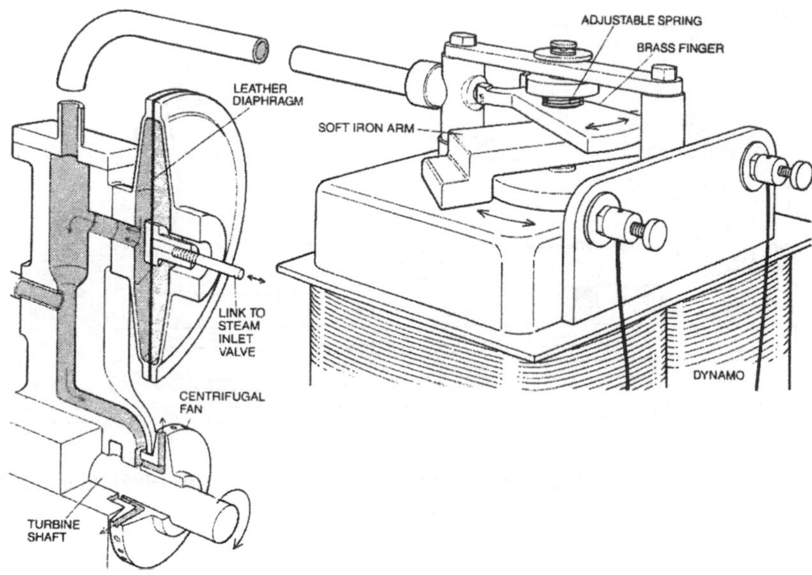

Figure 40. Details of the governor for controlling steam flow and maintaining the desired turbine speed. The centrifugal fan creates a suction which acts on the leather diaphragm and so controls the steam valve. Fine control of this servo motor is achieved by a brass finger which varies the leak of air into the rear of the bellows. This component is joined to a soft iron sensor which is spring loaded and responds to variations in the magnetic field of the dynamo. It is very sensitive to dynamo voltage, and the combination maintains speed, and voltage, constant to 1%. From Scaife (1985) Copyright © 1985 Scientific American Inc. All rights reserved.

pull on the arm of a rotary valve controlling the steam supply. The actual equilibrium speed could be set by adjusting the tension of a spring that resisted this pull, but it could also be adjusted by modifying the suction pressure acting on the bellows. This was done by arranging for a leak of air into the suction line. Increased speed meant more suction and a greater force to close the steam supply valve. By adopting a rotary design for this valve, operation was made easier because the forces on the valve due to steam pressure were balanced out.

A number of other features required attention if a practical machine were to be built. The action of the steam on the moving blades, as well as creating a torque on the shaft, also exerts a push in an axial direction. To build a thrust bearing to resist this at such high speeds would be very

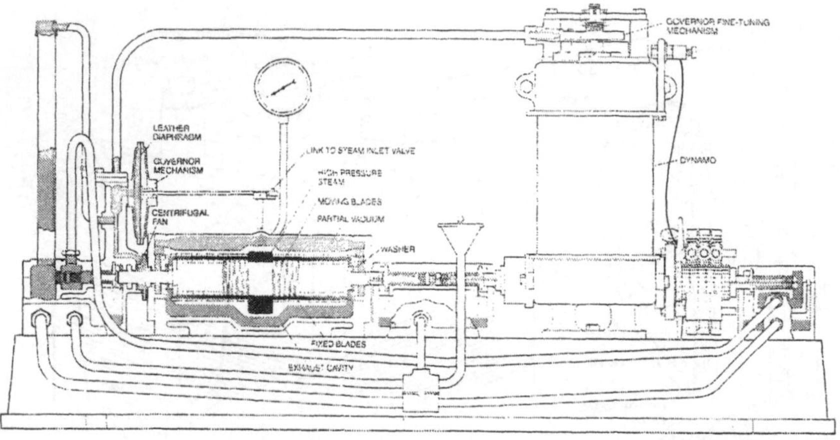

Figure 41. Parsons' prototype turbo-generator 1884, from Scaife (1985).

difficult, but the need for such was removed quite simply by having two complete sets of blading. The steam was supplied to the centre of the shaft and travelled outwards in opposite directions, to an exhaust at either end, in a so called 'balanced flow', which neatly cancelled this force (see figure 41). This arrangement remains a feature of many steam turbines to this day. Should the shaft for any reason tend to move axially, it was arranged that a suitably situated ridge in the casing would reduce the area available for the exhaust steam. As a consequence, if the shaft moved toward one end, the steam pressure at that end of the rotor would rise towards the inlet pressure, and push the rotor in the opposite direction. Though not listed as one of the claims, this was the germ of the idea for a balance piston that would later prove its value. Because of the high shaft speeds it was not practical to prevent steam escaping into the engine room where the shaft emerged into the open air by using conventional seals with greased packing. Instead, spaces were provided at these points and these were connected by ducting within the casing to the exhaust pipe. By arranging for some steam to be ejected by a nozzle into the exhaust duct, the pressure there was reduced below the ambient. This caused a current of air to be drawn in, so preventing any steam from escaping into the engine room. It was a small detail but it demonstrates how complete was the design even after less than one year's work. Once again this approach features in modern steam turbines.

The specification points out that if such a machine is driven by some external source of power it will act as a compressor. It then goes on to suggest the now familiar gas turbine cycle.

> *For this purpose I employ the pressure producer to force air or combustible gases into a furnace (into which there may or may not be introduced other fuel, liquid or solid). From the furnace the products of combustion can be led in a heated state to the multiple motor* (i.e. turbine) *which they will activate. Conveniently the pressure producer and multiple motor can be mounted on the same shaft, the former to be driven by the latter but I do not confine myself to this arrangement of parts—In some cases I employ water or other fluid to cool the blades either by conduction of heat through their roots or by other suitable arrangements to effect their protection.*

Clearly this reflects his experience with the torpedo experiments in which the power was provided by the products of combustion of gunpowder. Cryptic and undated entries in his notebook suggest that he may have carried out experiments with fans, furnaces and gas turbines. The final specification contains an additional comment that must also be related to these earlier experiments. It also shows that from the beginning he had marine applications in mind.

> *A compound motor such as described may be applied to marine propulsion, but as the velocity is necessarily high, it will be advisable to place several fine pitched screws on the shaft, in order to obtain a sufficient area of propeller blade: and one screw may be prevented from interfering with another, or others by suitable guide blades or other means.*

But all this ingenuity was of no avail unless a way was found of absorbing the power output which was developed at an extraordinarily high shaft speed. The design faced the same dilemma as de Laval's version of Hero's aeliopile. Parsons' inspired solution was to build an extraordinary dynamo. In later years he said that the achievement in designing this had given him even more pleasure than the turbine. Electric generators based on Faraday's researches required that conductors should cut magnetic fields. At first the magnetic field was provided by permanent magnets, but these were relatively weak. A major advance was made in 1866 when Charles Wheatstone and Werner von Siemens simultaneously described machines in which the magnetic field was created by currents produced by the generator itself. This meant that machines could now be built in ever larger sizes. They worked best at 'high' speeds, around a

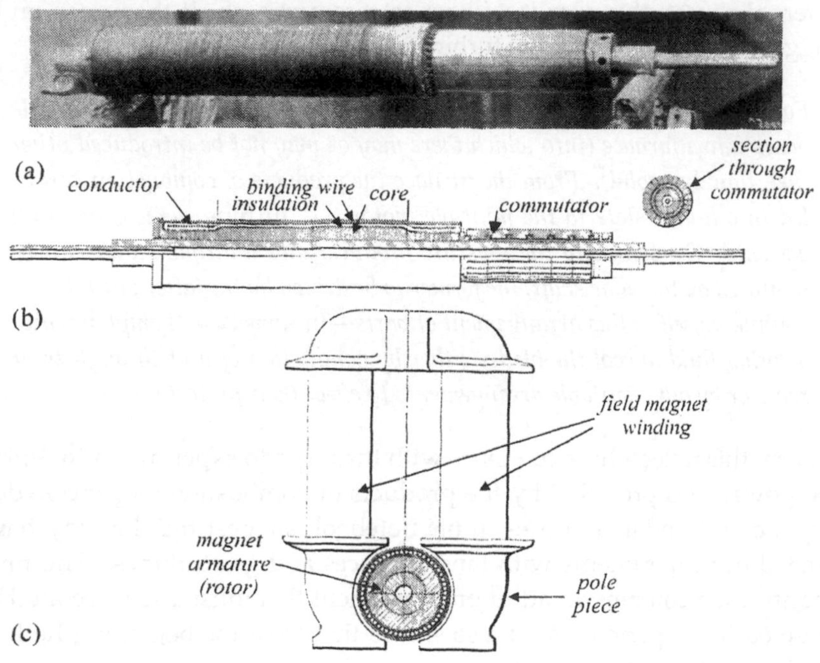

Figure 42. The dynamo of the prototype turbo-generator (1884): (a) photograph of rotor, (b) longitudinal section of rotor and (c) field magnet with a section through the armature (Richardson 1911).

thousand rpm or so, but to adapt such a machine to run directly coupled to a steam turbine at 18 000 rpm was as formidable a challenge then as it would be today. Quite apart from the fact that the shaft presented the same problems of vibration as in the case of the turbine, the conductors that are carried on the rotating armature, or 'keeper', of the dynamo's magnet are subject to the same huge centrifugal forces as the turbine blades (see figure 42).

In many ways patent 6734 is strange. As it points out, most of the details presented in it are not novel, and relate to existing Siemens generators. The claims relate to the application of forced lubrication to bearings, the use of a turbine drive and the use of anti-vibration bearings, restating much that has already been described in patent 6735. What was critical to the success of the machine was not claimed as novel at all. For example the armature is constructed by threading iron laminations, washers 0.015 inches thick, along the steel shaft. These were insulated from one another by paper and clamped together by a brass nut. It was necessary to do this to avoid currents being caused to flow in the iron as well as in the conductors. Such currents cause wasteful and unwanted heating. The moving conductors were laid on the surface of the armature in the manner of Friedrich von Hefner-Alteneck's invention of 1872 (Siemens 1892). They were laid axially, and held in place by tightly wound phosphor bronze wire in a way which is reminiscent of the use by Armstrong of tightly wound wire as a way of strengthening the barrels of his guns. It was found that bronze was liable to fracture at soft spots and so it was later replaced with steel piano wire. This, despite its magnetic properties, proved fully satisfactory and the method remained in use for many years to come. With the anti-vibration bearings, wire binding of armature conductors was a key element in the success of the first turbine-generator.

The magnitude and sign, positive or negative, of the voltage induced in each wire as it traces a circular path through the magnetic field varies with time in a sinusoidal manner. It is an 'alternating' voltage. However the earliest applications of electricity had evolved from the use of batteries, that generate a voltage of constant polarity, and create a 'direct', or steady, as distinct from an alternating current. And so provision has to be made for changing the output to DC. This was achieved by means of the commutator. The 50 conductors carried on the armature were all connected together in such a way that the voltage induced in each was added together. Each join between two wires at one end of the armature was connected to a segment of the commutator that was fixed around the shaft. The commutator was a very complex item with half as many segments as there were conductors, and each segment consisted of up to eight pieces. Each segment was insulated from its neighbours and from the shaft by using asbestos millboard. The segment assemblies were held together by dovetailing steel rings. To withstand the huge centrifugal forces at work, the whole assembly was placed between two threaded collars on the shaft. The greatest voltage across a conductor exists when it cuts the magnetic field at right angles, and the least when it is moving parallel to the field. Connection to conductors was achieved by a pair of stationary brushes made of bunches of brass wire. By locating the brushes at a point where the voltage in the conductors is least, it is possible to sustain a steady voltage of unchanging polarity, as well as minimizing the danger of sparking when a brush overlaps two adjacent commutator segments. Because the surface of the segments became worn in service, the whole assembly was designed so that it could be removed as a unit for servicing.

In order to achieve the desired precision with the existing machine tools, the rotor shaft was made as a composite structure. It was extended in length by two stubs. These had a hole bored through them, and were a push fit into the main body that was also bored out. In this way it was practical to make a passage for the lubricating oil to pass along the inside of the shaft from the outboard, generator, bearing to the point where it was coupled to the turbine shaft. Here it fed the journal bearings of the generator and the turbine at the centre of the machine. The temperature of the rotor rose significantly by virtue of the very high current density in the conductors, as well as due to the effects of very rapid movement of the magnetic field relative to the iron components. It was envisaged that the oil flow would provide some additional cooling. The drive connection from the turbine to the rotor was accomplished by a coupling. The ends of the two shafts were given a square cross section, and each protruded into a short cylinder with a bore that was also of square cross section. This coupling transmitted torque but left the two components independent of one another as regards any tendency to bend, which was beneficial in handling shaft vibrations. It was an important feature of later machines, and the details of the coupling were modified several times.

The magnetic circuit of the dynamo was constructed with two cast iron legs that each carried field windings. The path of the magnetic field was completed by two shaped cast iron pole pieces that guided the flux lines through the rotor, or armature that ran between them. It should be remembered that the scientific design of magnetic circuits was not established until 1886 (Hopkinson 1886); indeed Edison's early dynamos of this period were defective in this regard. Yet, as Gerald Stoney commented after Parsons' death (Stoney 1938)

> *Few better examples of Parsons' intuition can be given than the fact that the proportions of his first turbo-dynamo, designed 53 years ago when electrical knowledge was so rudimentary, could, as far as I know, hardly be improved upon today.*

Some questioned the use of cast iron for the magnets since wrought iron was more usual. But Parsons explained that there was a good reason for the choice. Wrought iron had been tried but it was found that self-excitation was not achieved until very close to the rated speed, and then there was a danger of a serious overswing in voltage which would burn out incandescent lamps. Cast iron on the other hand retained a significant permanent magnetism, and so some voltage was generated from the moment rotation began. Besides, the energy consumed in exciting the cast iron magnets was less than 2.5% of the output. Settling a rating for such a machine is a rather difficult matter, but the actual output was variously stated to be 6 HP, 10 HP, 4.5 kW and 6 electrical HP, while the steam supply was given as four times atmospheric, 60 psig or 75 psig. (Scaife 1984). Until widely agreed standards for acceptable temperature rise had been evolved any figure for continuous output was bound to be largely arbitrary.

Perhaps the neatest feature contained in the specification for patent 6734 is the way the magnet of the dynamo was combined with the pneumatically operated governor to create a servomotor with a superb performance (figure 40). A servomotor is a device for amplifying the power available to move a mechanism, and nowadays it is very commonly found in control devices. Steam, and then hydraulic power, had been used to provide the considerable effort required for operating gun turrets and controlling rudders of the new iron ships. The name had been coined by a French inventor J J L Farcot in his book '*Le Servo-Moteur ou Moteur Asservi*', published in 1873 (Bennett 1979). The use of the servo principle in Parsons' governor appears to have been one of the first applications of the principle to industrial use. It has been mentioned that one of the ways in which the speed of the turbine could be adjusted was by leaking air into the bellows that controlled the steam valve. When the bellows pressure was close to atmospheric the steam valve opened, and as it fell below it, it closed. This fact was used to refine the speed control of the bellows mechanism as follows. Some of the magnetic field of the field coils leaked from the cast iron components, and a pivoted iron lever was mounted on the top of the field magnet casting, where it tended to align itself with the field (see figure 40). As the speed of the dynamo increased, so did the strength of this magnetic field. The consequent movement of the lever was restrained by an adjustable coiled spring. The iron lever carried with it two brass fingers with a space between them. The tips of either of the fingers could close off the mouth of the tube that connected with the suction side of the bellows. As speed was building up, the lever moved, and this allowed air to leak into the bellows space, so causing the steam valve to open. Should the speed build further the second finger began to close off the leak again and so reduced steam supply, until an equilibrium position was achieved. The power for adjusting the steam valve was provided by the suction fan, but control of this power was achieved by the electromagnetic speed sensor. Although the arrangement is shown in detail, and its working is explained in patent 6734, it does not form one of the claims in that patent. A separate patent 14723 of 7 November 1884 includes claims for the servo motor

Figure 43. Parsons' prototype steam turbine rated at 6 kW at 18 000 rpm, by courtesy of N E I Parsons.

design. It also provides for the possibility that the soft iron sensor could be made sensitive not to the voltage of the dynamo but, if desired, to its output current. Another innovation was the fail-safe feature that was built into the arrangement. If by chance the dynamo failed and in consequence the iron lever swung back completely, the first brass finger moved to close off the mouth of the tube so causing the steam inlet valve to shut.

The performance of the whole arrangement was superb. It was possible to throw off the full electrical load of the machine without the speed, and therefore the voltage, varying by more than 1 or 2%. This carried a big bonus. The new incandescent lamps were very expensive: each cost something like the equivalent of a ton of coal, and their life was rapidly shortened by any excursions of voltage above the design value (Parsons N 1987). Therefore such close control of voltage greatly extended their life.

At Clarke Chapmans as at Kitsons, Parsons would have had access to some of the most skilled workmen, but his role in the manufacture of the prototype was not confined to the drawing board. It was remarked by E Kilburn Scott who had worked at the Victoria Works during that period, *'the Hon. C.A. Parsons is not only a clever engineer in theory and design but also an expert handicraftsman. Many parts of the first steam turbines were made with his own hands, and I have heard expert armature winders acknowledge that he could beat them at their own trade'.* (Parsons C A 1904a). Parsons took care to use the best equipment available to him. The shafts of the turbines were found to be noncircular when they came off the turning lathe and so they were ground on the latest Smith and Coventry automatic grinding machine. *'When the spindle had been turned and ground as nearly true as possible, and had been covered with rings of blades, the complete moving part was then ground in the case with fine emery'* (Parsons C A 1888a). In this way he was able to reduce the blade tip clearance to 0.015 in.

The complete specification of Patent 6734 is dated 24 November 1884, which suggests that the final details of the dynamo were the last to be settled. Just when the complete machine first ran is not known, but it must have been run during 1884 when the inventor was still only 30 years of age. He had become the first person to create a steam turbine with a really useful power output. It represented the fruit of an inspired and concentrated burst of creativity, that is reminiscent of the way that great scientists often display a peak in their powers at just such an early age. The surviving photographs of Charles Parsons show him either as a toddler at Birr, or as a middle aged man. Regrettably there are none dating to this period.

8.3 A note on governors

Control systems were developed along two parallel paths that seem not to have intersected until some years after this. On the one hand engineers had sought to perfect the governor for use on steam engines. It was a need that became more pressing as engines were used not just to pump water, but to turn machinery of increasing sophistication. Gradually the tendency for the action of governors to cause the speed of the engine to overshoot and so give rise to continual hunting was mastered, but the theoretical understanding of their performance was very limited. Astronomers meanwhile had an interest in devices that would turn an equatorially mounted telescope at exactly the correct speed to counteract the rotation of the earth. Fraunhofer's large telescope at Dorpat was controlled by clockwork in 1824, but its performance was marred by the presence of small oscillations in speed. In 1846 William Siemens described a much improved governor that he hoped to apply to steam engines. It was very quick in reacting to speed changes, and although not suited to use with engines the idea was very appropriate for telescope drives. Once photography became a realistic option it became important not just to keep the speed constant, but to keep the angle turned through exactly in step with the earth's rotation as measured by an accurate clock. At the summer meeting of the Institution of Mechanical Engineers in Dublin in 1888, at which Parsons read a paper describing his turbine, there was also a paper by Howard Grubb, the Dublin telescope manufacturer. It was entitled 'On the latest improvements in the clock-driving apparatus of astronomical telescopes'. Laurence Parsons during the discussion mentioned his experience with a clock drive that he had made for his own equatorially mounted 18 in reflector telescope at Birr. He had also struggled to fit a clock drive to the 36 in telescope in its new equatorial mounting. During the British Association meeting in 1861 James Clerk Maxwell with William Thomson and William Siemens formed a committee to determine the unit of resistance, the ohm. It was decided to make measurements with an electromagnetic machine that was run at a very accurately controlled speed. Arising from this project Maxwell published his paper on 'Governors' in 1867 (Maxwell 1867). Mathematicians struggled to understand the circumstances in which such devices would be stable. The Canadian born E J Routh at Cambridge University pursued the matter further, and in 1877 he won the Adams prize for his essay on the critieria which must be met for a system to behave in a stable manner. But such knowledge was not shared widely outside a small group of mathematicians and scientists. In England, especially, the gap between theory and practice remained wide (Bennett 1979). Just as in the matter of lubrication of high-speed shafts and the construction of dynamos, Parsons' technology was ahead of the theoreticians.

Chapter 9

Clarke Chapman and Parsons, Partners

9.1 Marketing the turbo generator

One characteristic that marked out Parsons from many other inventors was his grasp of the commercial realities that he must face. He knew that the steam turbine required a great deal of development before it could realize its true potential. Even with his own considerable private wealth it would be quite impossible to fund the necessary experimental work from his own resources. Therefore he had to sell machines to secure sufficient income to allow him fund incremental changes in design. The early machines were real 'steam eaters'. Compared to conventional steam engines they had a very heavy steam consumption. The prototype consumed up to 200 lb of steam per kW h of electrical output. This compared with as little as 40 lb of steam per kW h quoted around this time by Willans for reciprocating engines operating under similar conditions. Parsons well knew that the turbine would only show its best when built in the largest sizes, but as yet there was little demand for really large electric generators. He was therefore making a huge act of faith that the demand for electricity would increase enormously in the not too distant future. Just how confident he was about the future opportunities for his turbines can be judged by the fact that he ensured that his patents covered Belgium, France, Germany, Italy, the United States, Sweden and Canada.

Development of public electricity supply in Britain was in its infancy and growth was slow. In 1878 the *Electrician* magazine complained that while most other capital cities were lit by electric arc lamps, London still had none. Already where the newer science based industries were concerned British industry was finding that European and even American competitors had taken over the lead. Also in Britain the gas industry was long established, and this made the task of selling electricity for lighting much more difficult. In the United States this was not the case, and the expansion of the electric lighting industry was much more rapid. The early lighting installations all used direct current, with storage batteries to even out the demand on the generators. Swan had given Britain a lead in lamp manufacture, but it was Edison in America who set out to provide all the components required to complete the supply from producer to consumer. It took time for such items to be devised and for some degree of standardization to emerge. During 1881 and 1882 there was huge public interest in an industry that seemed to offer competition for what had, by now, become a somewhat complacent gas industry. But the reality was that it would be some years before electricity could compete with gas in price. Meanwhile the government passed the Electric Lighting Act in 1882. This was designed to avoid problems that had arisen among private companies involved in the gas industry during the early years. It gave local authorities the power to break up streets for the laying of mains, but it also gave them power to buy out private companies after 21 years. This latter feature discouraged investors. At the same time gas companies fought back by increasing their efficiency and keeping costs down. And so the early surge in demand for electrical equipment was quickly damped out.

The early supply systems that employed direct current were limited to an area that could be supplied from a given generator without an excessive voltage drop being caused by the resistance of the supply cables. This voltage drop of course increased with the amount of current supplied. As a consequence generating stations were built in locations which were far from ideal. When alternating current was used for the first time to supply the arc lamps at the Grosvenor Gallery in London in 1883, a step was taken that would have profound consequences for the future of the electricity supply industry. An alternating voltage was used in place of direct current and this made it possible to employ transformers designed by Gaulard and Gibbs. If power is transformed to a higher voltage the corresponding current is reduced, and with it the voltage drop experienced along the supply cables so that much greater areas can be supplied from one location. The adoption of alternating currents and transformers opened the way to rapid growth in the demand for electric power. In 1884 at an exhibition in Turin, Lucien Gaulard won a 15 000 lire prize by successfully mounting a demonstration of long distance transmission of power (Vincze 1985). A 30 kW steam driven generator supplied power over a looped circuit of 80 km. Around this time other pioneers were laying foundations for the application of alternating currents. In Hungary transformer design was perfected by engineers of Ganz and Company, and Nicolai Tesla, the Serbian engineer, built his first induction motor in 1883. When word of the work of Gaulard and Gibbs reached George Westinghouse in America, he set William Stanley to work, and by the end of 1886 Stanley had succeeded in putting a complete high voltage distribution system into operation. Westinghouse purchased the rights to the Gaulard and Gibbs patent in 1886, and to Tesla's motor in 1888, and he became the champion of alternating current in opposition to Edison who had espoused direct current.

If Parsons wanted to find customers it was essential that he should capitalize on the turbine's unique advantages. Mention has already been made of the benefit gained in terms of the extended lifetime of costly incandescent lamps when they were free from voltage surges. Also the quality of the light given out by lamps supplied by turbo-generators was remarkable for its steadiness. This contrasted with annoying fluctuations of brightness in the lamps that were supplied by slow speed reciprocating engines which, despite their large flywheels, were subject to appreciable cyclic fluctuations of speed. Gas light was still derived from a hot, smoky and flickering flame because the incandescent gas mantle was not invented by Dr Carl Auer von Welsbach until 1893. A further real benefit of Parsons' turbines was the possibility of running them without the need for an attendant to lubricate them or adjust speed as the electrical load varied. Lubrication of reciprocating steam engines involved getting oil to the pistons where it inevitably became mixed with the steam supply. Turbines, on the other hand, did not cause the oil and steam to come into contact, and so oil losses were greatly reduced with significant financial savings. Reciprocating engines created vibrations that were transmitted through their foundations to the ground, and constituted an annoyance at considerable distances. Finally the complete turbo-generator was relatively light and small, and therefore the capital cost was less than for equivalent reciprocating engines.

During the next four years machines were built for a variety of customers, for electric plating plants and for private lighting installations, but most were for use aboard ship. There was one place where steam was readily available, and where some degree of waste was tolerable, and that was aboard a steam ship. By 1884 Swan or Edison lamps were in use in over 150 steamers. Tyneside was already one of the largest centres for shipbuilding in the world. In February 1885 the *SS Earl Percy* became the first of several ships of the Tyne Shipping Company to be supplied with a steam turbo-generator for lighting. It was rated at 2 kW. The Chilean

Blanco Encalada, a vessel of advanced design, was the first battleship to be so equipped when it arrived at the Elswick shipyard to have new guns and boilers installed. It was obviously an advantage during this period that Parsons had decided to manufacture on Tyneside and not inland at Leeds for example. It must also have helped that the firm in which he had served his apprenticeship, Armstrongs, had recently merged with the Walker shipyard of Charles Mitchell and Company.

In 1884 Parsons wrote to Joseph Swan (C A Parsons 1884)

Launder Grange
Corbridge on Tyne
March 27

Dear Mr Swan,

It will give me very great pleasure and I should be very much interested in coming to see your house and the electric installation. There have been great strides made since I saw you last, some of the most successful seem to have been on board ship. I came over from America last summer in the Arizona lit by your lamps most successfully. I do not think there was a failure or a flicker during the whole passage. I quite agree with you that we want a thoroughly reliable and economical high speed engine. Brotherhoods is too wasteful in steam yet it has many points to recommend it and it keeps its position at present with the Admiralty. I know little of the practical workings of Willans' engine, certainly the bath of oil ? round the working ? parts is almost indispensable for continued working. We have found this ? necessary when working at the high temperature of 212° and have applied a small oil pump to keep up a constant circulation and keep down the wear which it does most successfully.

We have as a firm largely connected with shipping been thinking much on the subject of ship lighting. Such plant would harmonise well with our other business and might well be carried on with some special advantages... Kitsons have the management of my (epicycloidal engine) patents and I think they will succeed with what they are now supplying for dynamo driving.

I should like very much to hear your views as to the function of electric lighting when we meet.

Yours very truly,

Charles A. Parsons.

The pace of activity for Charles Parsons was intense. His wife Katharine gave birth to her first born child, Rachel Mary, on 25 January

1885. He now had a family to consider, but pressure of work must have left little time for him to enjoy his infant daughter. Like his father he had an instinct for publicizing his inventions, and the Inventions Exhibition which was due to open in London in May of 1885 offered a perfect chance to achieve this. His eldest brother, Laurence Lord Rosse, continued a family involvement in the running of the exhibition, which was opened by his erstwhile fellow footballer, Edward Prince of Wales. A machine was prepared for display, not the prototype but a machine with a speed of 12 000 rpm, most likely number 5. Essentially its design was the same as the prototype except that it ran at two thirds of the speed of the first machine and the number of stages was increased correspondingly from thirty to fifty. Castings were substituted for some components that in the original had been machined from the solid. Whereas the jury of judges awarded a gold medal for the epicycloidal engine, the steam turbine was deemed worth just a silver medal (Anon 1885a). Only the most perspicacious observers were able to sense its true potential at this juncture.

The lengthy account of the exhibition in the May issue of *Engineering* gave much attention to electric lighting and the related equipment. Detailed descriptions of both the epicycloidal engine and the steam turbo-generator were given, generously illustrated. Both were highly praised, the turbine especially so. It was possible for visitors to the exhibition to see not only the bank of incandescent lamps that lit the refreshment rooms and grill bar, but the machine itself in operation. The high pitched noise that it emitted was described by one observer to make it '*a veritable siren*'. But this was to some extent masked by the din of other machines nearby, so that at first sight the machine seemed to be at rest. There was a complete absence of those pulsations caused by the reciprocating engines, and that could be felt underfoot for appreciable distances.

The invention had received a good launching, but the summer was saddened by the death of Mary Dowager Lady Rosse on 22 July at the age of 72 years. Her health had been declining in recent years, but her death was unexpected. Katharine Parsons had only a rather brief acquaintance with her mother-in-law, but she had been brought to meet her by Charles when he was in London. Mary Rosse was buried at Parsonstown in a vault in the graveyard of Saint Brendan's, the abandoned Church of Ireland building. The coffin was brought by train to Parsonstown and after a short time resting in the Castle hall, the cortege formed up to take her coffin to the graveyard. Among those walking in the procession on a day of extraordinary heat were her sons Laurence, Randal and Charles, her son in law Sir John Conroy and daughter in law Mrs Randal Parsons, but not

Cassandra Lady Rosse. The relationship of the latter with her mother-in-law had been difficult. The sons paid their own tribute to the memory of their mother by commissioning C E Kemp to design superb memorial windows for the parish church at Heaton (Davison 1989). Among a number of souvenirs that she had treasured and that have been preserved in an ornamental cabinet at Birr Castle is a packet labelled 'Charlie's hair'

9.2 Turbo-generator improvements

The program of experimentation continued apace. Charles had learned from his father that careful measurements were the key if technical progress was to be maintained. Measurements were carried out on the dynamo to assess its efficiency. The resistance of its windings was established. The extent of energy losses in the magnetic circuit was determined.

> *There are losses due to eddy currents in the core and wire of the armature* (rotor), *and to magnetic retardation resulting from change of polarity of the core. These losses have been ascertained by separately exciting the magnets from another dynamo, and measuring the change of steam pressure required to maintain the speed constant; the corresponding power was then calculated. The commercial efficiency of this dynamo has been found to be about 95%* (Parsons C A 1888b).

Every time that the magnetic field in an iron circuit is reversed in direction the process absorbs energy. This phenomenon of magnetic hysteresis was first observed by Kohlrausch in 1866, but had only recently been clarified by Professor Ewing. He had carried out research into magnetic phenomena while he was a member of staff at the Imperial College of Engineering at Tokyo. Obviously this source of energy loss was of much greater significance in the new machines in which reversals of direction occurred at a rate which was up to ten times greater than previously experienced. But it would appear that Parsons had direct access to excellent scientific and technical advice, for when he read his paper describing the turbo-generator to the Summer Meeting of the Institution of Mechanical Engineers in Dublin in 1888 he thanked '*Mr Cross of Newcastle, Mr William Anderson of Erith and Professor FitzGerald of Dublin.* The last named was the Registrar of the Engineering School at Trinity College Dublin, and it was he who would, in 1889, propose the so called Fitzgerald–Lorentz effect. This was a hypothesis to explain the negative result of the experiment of Michelson and Morely to measure the velocity of the ether. He had published a note on a 'non-sparking dynamo' in June (FitzGerald 1888).

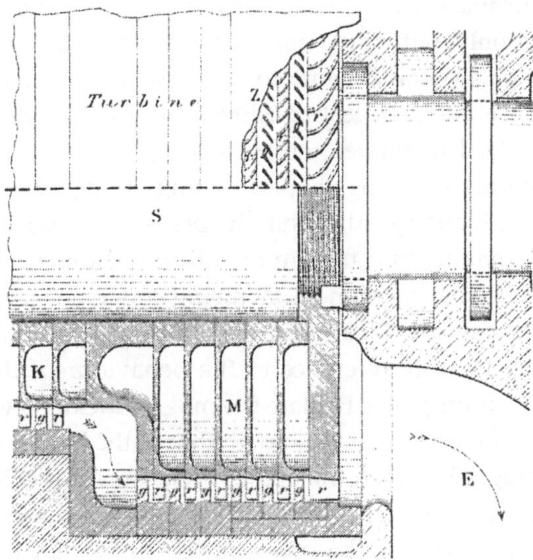

Figure 44. A sectioned view of the exhaust stages of a 32 kW turbine (1888). It shows the shaft S with cast discs K and M carrying the moving blades r. The cut-away at the top right shows how the front blade edges have been bent (Parsons C A 1888b).

During the discussion of Parsons' 1888 paper, Professor Forbes drew attention to the fact that the turbo-generator appeared especially suitable for operation as a turbo-alternator. Forbes had just tried an experiment with such a Parsons dynamo. He selected two conductors on opposite sides of the rotor and attached them together at one end and connected the other ends to two insulated rings on the shaft. These rings gave a voltage that changed its magnitude and direction at 8000 times a minute, the speed of the turbine. He found that despite his fears there was no tendency for the iron pole pieces to heat up excessively, and he thought that '*for alternating currents the turbo-electric generator seemed likely to become the best of all*'. Despite this it was not until 1890 that a turbo-alternator of Parsons' design was installed at the Forth Banks station of the Newcastle and District Electric Lighting Co., which he had helped to establish in the first place.

The blading of the first turbines was crude in the extreme, and despite the assertions in the patent that blade height would need to increase progressively toward the exit stages to accommodate the greater volume of steam as the pressure was reduced, the actual increase provided was minimal. He had of course employed blades made of sheet metal in his gas powered turbines that he used for the torpedo experiments, but his notebook deals only with their mechanical strength and gives no hint about their shape. When it came to the first steam turbine, the design of the blading was the simplest possible. Both fixed and moving blades were made identical in shape, and were simply formed by cutting slots in the circumference of a ring shaped casting. Using the same shape for fixed and moving blades ensures that the pressure drop is the same across both groups. It remained a feature of future Parsons turbines. It is not impossible that it arose from manufacturing convenience, but a glance at de Laval's design, figure 47, shows that it was not the only option open to Parsons. But the only reference to the behaviour of the steam in his notebook was an attempt to calculate the magnitude of its velocity, so as to ensure that it remained comparable with that of the blades themselves. So long as the passages between the blades were the same width and height throughout the machine, the velocity of the steam increased progressively because its volume rose as the pressure fell from inlet to outlet. Some mitigation of this effect was obtained by adjusting the angle at which the blades were cut, which achieved an increase in the effective area for flow as the steam expanded (see appendix 2). Parsons himself shed very little light on the processes by which he arrived at his designs but in his first paper describing the new machine, that was read at Newcastle in 1887, he stated that (Parsons C A 1888a)

> *the theory of the* (water) *turbine has been thoroughly understood... The important conditions for economical working being that the shape of the blades and the velocity of the wheel shall be so arranged that there shall be no shock or sudden change of velocity of the water in passing from the guide blades to the moving blades, and also that as the velocity or energy of the water on leaving the wheel is lost it shall be reduced to the utmost extent by the formation of the blades.*

The discussion on the paper was opened by Mr Weighton, chief draughtsman at the Hawthorn-Leslie's St Peter's works[1]. He said he was at a disadvantage, having to stand in at the last moment for Professor Garnett. He accepted that water turbines achieved high efficiencies, but he could not

> *see from Mr Parsons' paper that this machine could in its present form, attain to the efficiency which was there alleged. It might be so, theoretically.*

[1] R L Weighton was nominated to the newly created chair of Engineering and Naval Architecture at the Durham College of Science in 1891

> *But it was not shown, for instance, that the blades and guides had the proper curvatures to get the maximum effect out of the steam, which acted on the blades by impact. That was one point that struck him as being omitted. The difference of angle of the blades on the various wheels was not mentioned in the paper at all, he thought, and he first became aware of it by noticing it on the actual machine exhibited on the table.*

Parsons, in his reply to the discussion, noted that

> *Mr Weighton had drawn attention to some of the most important points, which enabled him to explain more fully some portions which he regretted had not been made clearer.*

But in fact he still did not throw any further light on the question of blade shapes and steam flow.

At an earlier date, in an effort to reduce the leakage of steam around the blade tips, the blades for the *Earl Percy* turbine had a shrouding silver soldered to their tips. It was not very successful and various strategies were devised to make the leakage path offer a greater resistance to steam flow, for example by arranging to create vortices in the flow. A model was made to test the idea, only using water instead of steam. This showed that there was little benefit to be gained. The next step was taken in 1888 (Parsons C A 1887). Blades formed of cast bronze were prone to fracture, so instead the discs were forged from delta metal which is a brass alloy containing some iron (Stoney 1933a). The front face of the disc was machined to undercut the leading edges of the blades and these were then bent with the aid of pincers, so that they were turned into a direction that was parallel to the shaft axis (see figure 44). As he had stated, in the case of water turbines optimum design requires that when the fluid emerges from the moving blades there should be no potential for doing further work, that is, there should be no component of velocity in the direction in which the blades were moving. In other words the steam should be flowing parallel to the shaft axis. The modification addressed this need, but it would only work properly if the pressure drop across the stage was appropriate. Nevertheless a 25% gain in efficiency was claimed for these curved blades compared with the original straight ones. At the same time it was also decided that the blade rings should be made in three groups with successive diameters increasing in the ratio of 1 : $\sqrt{2}$, as can be seen in figure 45. This went some way to achieving the need recognized in the patent 6735 of 1884, to ensure that as the pressure falls '*the space for the actuating fluid increases, either continuously or step by step*'. A machine of

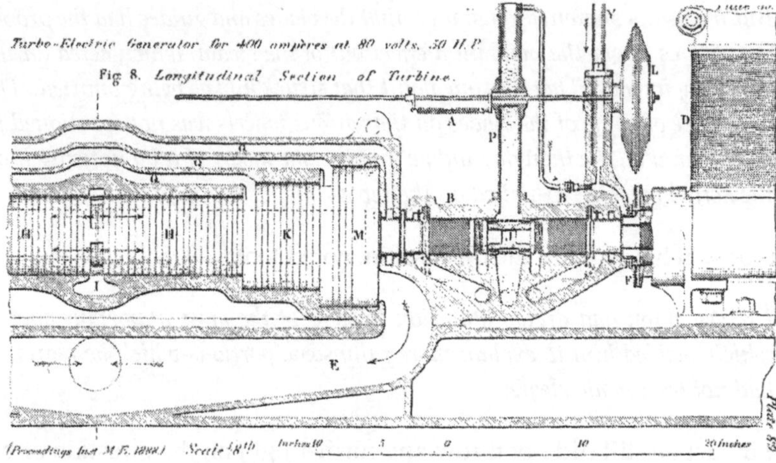

Figure 45. Sectioned view of 32 kW turbo-generator. There is balanced flow with three sections of blading to the right of the midpoint, each with increasing mean diameter. The inlets and outlets of each section are connected by pressure equalizing passages to the corresponding points on the left hand side (Parsons C.A 1888b).

32 kW output was built and the steam consumption was found to have been reduced to 58 lb of steam per kW h of output, less than one third of the prototype. The casings for these machines had become quite complex castings, with their multiple passages running from one end of the turbine to the other. The bearing pedestals were made as an integral part with the turbine casing, and this tended to cause problems as different parts of the casting acquired different temperatures during operation, and so became distorted.

It was not unusual at this time to mount a locomotive type boiler and a small steam engine on a (one) horse drawn cart of the sort intended for agricultural use. A couple of the first 12 000 rpm turbines were fitted out in this manner. In January 1886 the Swan Pond at Gateshead became frozen over after a severe frost. The Chief Constable of Gateshead planned to raise funds for the Royal Infirmary by using it for skating, and charging for admission. Parsons agreed to lend his turbo-generator, and lamps were supplied by Joseph Swan so that skating could continue at night. Over a three day period some £100 was raised, and of course the huge crowds could see for themselves this very visible, and revolutionary development!

9.3 Other new products

Parsons' inventiveness led him to explore quite unique designs, like the unipolar dynamo, which was based on Faraday's original magneto-electric experiment (Parsons C A 1922). This had used a copper disc rotating between the poles of a magnet, with the current that was generated being collected from one brush at the rim and another brush rubbing on the shaft. In 1887 Parsons modified his prototype. The dynamo was replaced by one that had a 3 in diameter soft iron shaft which passed through holes in new pole pieces fitted to the magnet, while copper sleeves shrunk onto the shaft replaced the armature windings of the normal dynamo. About 1 volt was generated, but the current could be 1000 amps, enough to weld together two $\frac{1}{2}$ inch diameter steel bars. In 1902 he built an even larger machine. Running at 4500 rpm it gave 10 000 amps at 5 volts.

Parsons had been appointed to head the electrical department of Clarke Chapman, and turbines were by no means his only concern. He had already experimented with arc lamps in connection with his epicycloidal engine. Methods for making the carbon rods for arc lamps had evolved gradually and now relied on the use of carbon obtained from gas works and its compaction with a variety of additives. He had tried inserting a core made of chalk to improve the light output. He now began to manufacture lamp mechanisms. As the carbon of the rods is burned away by the arc, some adjustment is needed to prevent the arc becoming so extended that it becomes extinguished. Great ingenuity had been displayed by different inventors during the previous 40 years of development of arc lamps (Carlson 1991). Parsons' design was straightforward. The upper carbon was supported by a cord connected to an iron plunger. The plunger was surrounded by a solenoid that carried the current feeding the arc. By suitable design, the plunger was arranged to draw the rods apart until the current in the arc reached the desired value, which was then automatically maintained. Shaping of the plunger ensured that the magnetic pull of the current diminished gradually as the rod burned away. The lower rod which burned more slowly was adjusted by hand. When supplied with direct current the top, positive electrode develops a crater in which temperatures of 3000 to 4000 °C are reached. It provides most of the light output. Arc lamps are several times more efficient than filament lamps, and when fed with direct current from dynamos are more efficient than when supplied with alternating current.

Parsons widened the scope of his lighting products by setting up the Sunbeam Lamp Company, with the help of Mr John W Edmundson and his partners Clarke and Chapman. Around 1885 he began making vacuum incandescent lamps in the ground floor of Park House. At first these were of small output, but since 1879 a legal battle had been going on which centred on the meaning of 'filament' used in Edison's patent (Josephson 1961). In 1889 the legal decision was made that patent protection applied only if the incandescent carbon of the lamp had a diameter of less than 1/32 in. In consequence the firm was restricted to making only lamps with larger diameter carbons. These had outputs of anything from 100 to 3000 candle power, the latter consuming 8 kW. At the Newcastle upon Tyne Royal Mining, Engineering and Industrial Exhibition that was held in 1887, several of these lamps were used to provide lighting, and were supplied by ten of Parsons turbines (Anon 1887). While efficient, the lamps had a relatively short life of 200 to 500 hours (Parsons C A 1922).

The method for manufacturing the carbon elements was to squirt a mixture of tar putty, anthracite and 4% zirconia into the shape of a rod. The zirconia reduced the resistance as the lamp grew older and the glass became darkened. Details of the process were kept very secret at the time. A son of John Edmundson described what he saw as a boy (Stoney 1938).

> *The carbons were made by an old man, and his apparatus was quite simple. The coal was broken up into a coarse powder in a mortar, and then ground in a coffee mill, after which it was treated with caustic soda and aqua regia, with prolonged washings. The resultant powder, after being dried, was ground in an agate mortar and sieved, after which it was subjected to a mysterious operation in a little cubby hole, kept carefully locked, and called by Sir Charles the 'Holy of holies'. What actually took place was that the old man weighed out some of the carbon powder and some icing sugar and ground them together in a mortar with a little absolute alcohol. The function of the sugar was to act as a further binding agent when the coal powder was mixed with the tar. It was kept in a jar labelled 'arsenic'. This was done by Parsons with the idea that the old man, or anyone else, would be chary of tasting a white powder so labelled. The old man always referred to the 'white powder' and the 'black powder'. When the production of lamps went beyond the experimental stage, and the methods of manufacture had to be altered to produce larger quantities, the old man was nearly heart broken.*

Next the filament rod was baked. After affixing electrical connecting leads and placing it in a glass envelope, the carbon element was heated by an electric current in an atmosphere of benzene. This caused a surface layer of carbon to be deposited. A vacuum was then drawn and the heating continued to drive off any gases. A Sprengel mercury vacuum pump was used, backed by a mechanical pump made to Parsons' design. This had a couple of cylinders of 18 in diameter with a 36 in stroke, and each was compounded with a smaller diameter piston and cylinder. John Edmundson recalled that

> *He* (Parsons) *and my father had been wrestling with the pumps ... I expect some mercury had got into the valves, and both had their hands good and black. Parsons whipped out a beautifully clean white handkerchief and proceeded to wipe his hands on it. One of the men said 'we often saw Mr Parsons and Tom Jefferson* (a fitter) *at work on the turbine which worked the generator for the lamp works, Mr Parsons with his coat off and his nice white shirt sleeves rolled up; I often wondered what his missus would say when he got home.'*

The approach to the manufacture of carbon rods for arc and incandescent lamps was by no means casual. In June of 1888 a paper by the Hon. Charles A Parsons was communicated to the Royal Society by his brother the Earl of Rosse (Parsons C A (1888c). It dealt with '*Experiments on Carbon at high temperatures and great Pressures, and in contact with other substances.*'

> *The primary object of these experiments was to achieve a dense form of carbon which should be more durable than the ordinary carbon when used in arc lamps, and to obtain a material better suited to the formation of the burners of incandescent lamps.*

He used a massive steel mould with an interior cavity measuring 6 in long by 3 in diameter. It was placed in a hydraulic press. A carbon rod $\frac{1}{4}$ in in diameter was placed in the cavity with an electrical connection at one end to the mould, and at the other to the spigot used to close the cavity. The spigot was electrically insulated with asbestos from the mould. The rod was surrounded by a variety of fluids all containing carbon such as benzene, paraffin, treacle and bisulphide of carbon. The top and bottom of the mould were securely sealed using an arrangement like the gas-check in guns, with which he would be familiar from his days at Armstrongs. The hydraulic press was then used to drive home the closure at one end and so generate pressures up to 2200 atmospheres. At the same time a dynamo was used to pass currents of up to 300 amps through the carbon rod. He noted that

> *In some of these experiments a considerable quantity of gas was generated,*

and the press had to be slightly slacked back during the experiment to accommodate it and maintain the pressure constant.

It was a remarkable technical achievement, to seal in gases and liquids and to provide effective electrical insulation for such heavy currents at such high pressures. But although a considerable deposit of carbon was formed on the rods, it proved to be soft and friable. Tests showed no improvement when the material was used in arc lamps.

He next tried surrounding the carbon rod with silver sand or silica instead of a liquid. Pressures were doubled to 4400 atmospheres, and currents up to 300 amps were used, creating conditions of very high temperature and pressure. This time he found that the density of the carbon had increased very significantly: from a normal 1.6, it reached between 2.2 and 2.4. Variations on the theme were tried using alumina instead of silica. The alumina was found to have melted to form a crust arourld the carbon, testifying to temperatures in excess of 2000 °C. Other experiments involved the use of layers of sand, lime and coke dust. From the melted remains, he recovered a very hard material which had the appearance of 'bort' (natural diamond fragments). The conditions generated in his apparatus were, he noted, not inconsistent with what appears *'to have existed in the craters or spouts of the Cape Diamond Mines at some epoch.'* In 1867 diamonds were first discovered in what became known as Kimberely in southern Africa, and in 1886 Cecil Rhodes had made the headlines by a proposal to secure a monopoly of the entire diamond industry (Pakenham 1991). As an aside it may be mentioned that one of Katharine's brothers, Christopher Bethell, who was acting chief of frontier police on the border of South Africa, was killed by the Boers in 1884 (Appleyard 1933).

In 1880 H B Hannay, a self-educated chemist, claimed to have created diamonds by sealing liquid mixtures, including hydrocarbons, in massively strong iron tubes that he raised to red heat[2]. S Marsden too, funded by a grant from the Royal Society, made a similar claim at this time. Although Parsons' experiments had a very practical objective, the manufacture of carbon rods for arc lamps, he was quick to grasp the possibility of achieving something much more ambitious, the manufacture of synthetic diamonds from graphite. But he recognized that the material that he had recovered was unlikely to be diamond. He set the problem to one side, returning to it 15 years later. It is interesting that the young engineer who was embarked on a career of invention and manufacture should have taken time to publish his scientific results in the

[2] For an excellent account of the history of attempts to synthesize diamond see Davies (1984)

Proceedings of the Royal Society. The influence of his father is certainly very apparent.

It was during this period that the first mirrors were made at Gateshead for use in searchlights and arc light projectors. Although Charles Parsons was not born until after the 6 ft telescope was built and the discovery of spiral nebulae had been announced by his father, nevertheless his childhood was filled with the activity of the astronomer assistants who taught him. After his father died he retained a close friendship with his elder brother who continued the operation of the Birr observatory. A full account of Charles' involvement with aspects of the optical industry is recounted later, in chapter 21.

On 19 September 1888, a couple of months after the meeting in Dublin of the Institution of Mechanical Engineers, an important addition was made to the staff that Parsons had gathered around him. George Gerald Stoney (1863–1942) was employed as an apprentice draughtsman at 10 shillings per week. He was the eldest son of Dr George Johnstone Stoney, one of the two brothers who had worked for William Parsons as astronomical assistants and tutors at Birr Castle. Although Gerald was nine years younger than Charles Parsons, there were strong links between the two families. Born in 1863, he was educated at home by his parents before entering Trinity College Dublin, from which he graduated as second senior moderator and gold medallist in experimental science in 1886. In the following year he also obtained his degree of BAI from the Engineering School, taking first place in all three groups of subjects. He started work under his uncle Bindon Stoney, who was chief engineer of the Port of Dublin. Like Parsons he was a keen cyclist; indeed while still a student he had published two papers with his father on bicycles and tricycles. As a youngster he had also learned how to silver mirrors from Charles Burton (1846–1882) who made mirrors for telescopes and worked both for Lord Rosse and at Greenwich Observatory (Burnett and Morrison-Low 1989). Stoney's later career was to be entwined with that of Charles Parsons in a very special way.

Chapter 10

A New Start—C A Parsons

10.1 Break-up of partnership

By the year 1889 something like 250 turbines had been built and sold. This was indeed impressive progress. The rating of the largest machine manufactured at Gateshead was 75 kW, but in January 1889 Parsons had Stoney working on drawings for a machine rated at 500 kW (Dowson 1942). It was envisaged as the combination of a standard non-condensing turbine with a second turbine of novel design, operating in tandem and which would exhaust to a vacuum. A number of supply companies were now employing alternating current, or at least planning to, so he decided to follow up the suggestion that the connections to the conductors on the surface of his standard generator armature could be modified simply to allow delivery of alternating current. The conductors were connected to form a coil in two parts joined in series, with the space between them occupied by wooden spacers. The commutator was removed and the two ends of the coil were brought out to slip rings. The arrangement can be seen in figure 46. The voltage generated by the alternator was some ten times higher than in an equivalent dynamo, consequently for a similar power output the current was smaller. This eased the problem of dissipating the heat generated by the current in the armature conductors. Despite this, the current density in machines of 150 kW output reached the extraordinarily high value of over 7000 amps in^{-2} (Stoney 1938). The 75 kW machine marked a very important step forward: it was the first ever high speed alternator, and the first 'turbo-alternator'. The turbine and the alternator are ideally matched. High speeds suit both devices, and today this combination provides the overwhelming bulk of electric power world wide. The next problem was to find a customer.

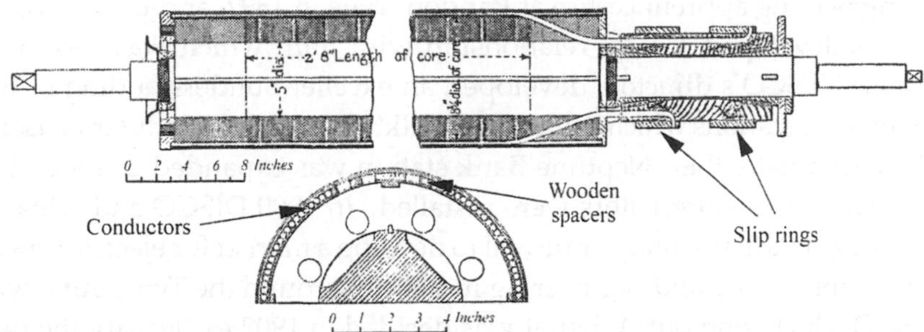

Figure 46. Rotor of a single phase alternator, showing the rotating armature rated at 150 kW and 4800 rpm (Stoney 1908).

In 1889 a group of businessmen set up the Newcastle upon Tyne Electric Supply Company (NESCO). They included J T Merz, a Mancunian of German extraction, R S Watson, a Quaker from Northumberland, and T G Gibson, one time mayor of Newcastle (Hore 1994). They built a power station at Pandon Dene in Newcastle and equipped it with alternators, but these ran at a slow speed, driven by conventional reciprocating steam engines built by Robey. Charles Parsons realized that it was not enough to rely entirely on special features to sell his turbines and so he decided on more direct action to secure a place for them. While still a partner at Clarke Chapman, he assumed the position of Managing Director of the Newcastle and District Electric Lighting Company, DISCO, which was registered in 14 January 1889. This partnership also included Lord Crawford who was chairman of the London Electricity Supply Corporation, a wealthy coal owner named Milburn and an industrialist, W M Angus, and was incorporated in 1889 (Hore 1994). Their station at Forth Banks was equipped with two of his new 75 kW turbo alternators built by Clarke Chapman. This marked a most important step, being the first ever instance of a central power station to be equipped with turbo-alternators. The Electric Lighting Acts of 1882 were amended in 1888 to extend to 40 years the period before which the local authority could take enterprises into public ownership. Nevertheless parliamentary bills still had to be promoted if permission was sought to extend the supply of power beyond the bounds of the local authority within which it was generated. As Lloyd George is reported to have said, electricity supply is *'not a matter of engineering, but of politics,'* (Hore 1994). Of the two supply companies, it was Parsons' rival NESCO which fared best in the long run. Charles Merz (1872–1940) the eldest son of J T Merz, began his engineering apprenticeship at Pandon Dene in 1892, and in later years Parsons developed a close relationship with him, which was fortunate because NESCO's directors developed an excellent understanding of the potential for stations which could offer 'bulk' electricity for industrial users. When eventually their Neptune Bank station was expanded in 1901, two of Parsons' turbo-alternators were installed. In 1900 DISCO and NESCO each sought Parliamentary approval to develop a market for electric power in the shipbuilding and engineering industries around the Tyne, but it was NESCO which won out. When it was decided in 1902 to electrify the two branch lines of the North Eastern Railway from Newcastle to Tynemouth, the first outside London, NESCO built the Carville power station. It was sited closer to the river than Neptune Bank. On the outbreak of war in 1914 the coverage of NESCO's system had grown from 11 square miles to 1400 square miles, and it supplied power to factories and shipyards as well as to one of the first electric railways outside London (Hadfield 1994).

Parsons knew that in order for the turbine to compete with the now highly developed steam engine, it still required considerable improvement and that it must be built in much larger sizes. Having publicly demonstrated the technical feasibility of the concept he was anxious not to lose his lead to a competitor. The 500 kW machine which he was planning would have been a huge leap forward, and it may have caused his partners to balk at the prospect of seemingly endless experiment and change, but a more serious matter, his involvement with DISCO, had upset his partners (Clarke 1998),

The Honble C.A. Parsons
Ryton on Tyne *8th February 1889*

Dear Sir,

We notice that you have accepted the position of 'Managing Director and Electrical Engineer' of the Newcastle and District Electric Lighting Company Ltd. As this is contrary to the provisions of our Deed of Partnership we must have at once, some explanation of a step which we entirely disapprove of.

In Our present relationship we think you will agree that it is better that any explanation should come through your solicitors.

Yours truly,

Clarke Chapman Parsons and Coy.

The tone of the letter suggests that relations were already somewhat strained. A letter to their legal advisers spells out the position more clearly,

Messrs Watson and Denby
Newcastle upon Tyne *14th February 1889*

Dear Sirs,

We duly received your letter of yesterday relative to Mr Parsons' position as Managing Director and Electrical Engineer to the Newcastle and District Electric Lighting Coy and in reply beg to say that the duties Mr Parsons has undertaken in our Firm occupy the whole of his time & therefore any duties he undertakes outside of the business, must be detrimental to it. It is clear that one or the other must be given up, but we leave it to Mr Parsons himself to choose which. Should he decide to give up those in our Works, no doubt an arrangement could be made as to his future position. Messrs Dees and Thompson say that Mr Parsons has been anxous to promote the interests of the firm. It seems to us that not only does the position he has taken place our firm in apparent hostility to the rival Company which should be our customer but precludes us from tendering to his own Company and thus instead of gaining by both we may lose both.
Yours very truly,

Clarke Chapman

This letter suggests that it was Parsons' failure to persuade NESCO to purchase turbines that triggered the crisis, rather than just poor relations with his partners. Whatever the exact reason, the partnership between Charles Parsons and Clarke and Chapman came to an abrupt end on the 31 December 1889 having lasted only six years. A document dated 22 June 1889 sets out an agreement drawn up between William Clarke, A H Chapman and J B Furneaux (a minor partner) and Parsons of Elvaston Hall, Ryton (Clarke 1889). It was agreed among other things that

(1) the partnership should be dissolved from 6th July, or on the retirement of Parsons, and it should be published in the London Gazette.

(2) There was to be no testing or tendering in the electrical lighting department without the partners' decision.

(3) Parsons would take fixed and moveable machinery and all patterns connected with the torpedo material in the electrical lighting department except the 8 H.P. gas engine, but also stores, stock, drawings of the electrical lighting department, work in progress etc and contracts for Parsons' turbo generators or electric lighting installations.

(4) He would also take the patents and patent rights and privileges (home and abroad) in respect of the invention of the Parsons turbo generators and arc lamps. Clarke Chapman and Furneaux would take exclusively all freehold works (including those of said electric lighting department)... together with 24 shares of £10 each in the Sunbeam Lamp Company Ltd., and the patent for Improvements in Incandescent lamps 15,118 of 1887 in the name of Clarke and Parsons.

(5) Parsons would also be entitled to correspond etc. relating to electric lighting... being temporarily placed in Park House. The value of the patents of (4) would be assessed by Gainsford Bruce esq. Queens Counsel as sole referee.

(6) Parsons is credited with £6,000 goodwill and £14,000 capital less £383-7-0 for reduction of capital, plus £800 for goodwill of torpedo business and £550 sums anticipating profit up to 3rd June 1886 (last audit).

(7) In order to afford convenient time for him to remove plant and allow him meanwhile establish his electrical lighting business, he is allowed to use the shop shown on the map for 5 months.

Evidently despite the disappointment that his partners must have felt at his withdrawal, they were still ready to facilitate him in many ways. During his lifetime the impression got about that Clarke Chapman had been unreasonable in their behaviour. After Parsons' death their legal representatives wrote to the *Engineer* and to the *Times* to refute this, and certainly the documentary evidence seems to support their contention. As the years passed, Parsons' reputation for 'fiery' conduct and impulsiveness became ever more firmly established. It would seem that these qualities also played their part in what proved to be a serious setback for him.

The arbitration on the value of the patents duly took place before Mr Gainsford Bruce Queen's Counsel and Member of Parliament. Clarke Chapman and Co. were represented by Mr Robson QC, and Parsons by Fletcher Moulton QC (Appleyard 1933). Clarke Chapman engaged the Irish born Sir William Thomson, later to be Lord Kelvin, and Gisbert Kapp, an electrical engineer, as their expert witnesses. For his part Parsons chose Dugald Clerk, an engineer who specialized in the design of oil engines, and J A Ewing (1855–1835). Ewing was an engineer, close in age to Parsons and a graduate of Edinburgh University. He had spent five years as professor of mechanical engineering in the Imperial College of Engineering at Tokyo. In 1890 he was appointed to the chair of mechanism and applied mechanics at the University of Cambridge.

Parsons' agreement committed him to paying seven eighths of the value of the patents if he wished to recover them. Substantial sums would still have to be laid out to pay the fees necessary to keep the patent in force and these could be set against the notional value. Dugald Clerk testified that considerable work remained to be done if the turbine were to reach the necessary standard of reliability and efficiency. He suggested a sum of a few thousand pounds would be adequate recompense. He listed earlier patents that in some measure anticipated the patents in contention,

12026 of 1848, Robert Wilson, Rotary steam and other engines.
1681 of 1863, Christian Schiele, Turbines.
5022 of 1863, W H Cutler, Rotary steam engine.
177 of 1881, John Imray, Apparatus for obtaining motive power from fluid currents.
2434 of 1887, W R Lake, Rotary steam engine.

Wilson's patent proposed *'combining together a succession of two or more rotary engines to be worked by the same steam in succession'*. The successive jets would be larger or else more numerous in order to allow for expansion. High speeds were envisaged.

Clerk's report stated that he had measured a steam consumption rate of 75 to 80 lb from an early version of Parsons turbine. This contrasted badly with the 22 to 40 lb per horse power per hour that current reciprocating engines consumed. He continued

> *The market for the turbo-generator on land is most limited—practically non-existent. There are conditions where it may obtain the preference on board ship, but unless present difficulties are overcome, even the existing sales will not be maintained. It is a most disastrous thing to experiment in public. This engine with so many points which although not new in the Patent Law sense, are new in the sense that little practical experience has been had with them, should be kept in the works and tested in every possible way for at least two or three years longer. I do not doubt but that Mr Parsons will overcome in time, and by dint of the expenditure of very great sums in experiment, the remaining difficulties, but he has not done so yet, and Messrs Clarke, Chapman and Co. have not done so. The* (turbine) *engine is much more difficult to manufacture than any ordinary steam engine, and when constructed at best it uses almost double the weight of steam of an ordinary engine for equal output.*

He spoke of *'an American competitor in the field with a radial flow turbine.'* This appears to have been a reference to C G Curtis (1860–1953) who in

1889 set up the Curtis Electric Manufacturing Company to manufacture motors and fans (Wood 1984). However it would it was not until 1896 that patents were filed by him. He concluded

The highest possible offer which might be got for the patents would not exceed £2,000 and this including foreign patents. Considering that over £2,000 still remains to be spent on fees to keep all the patents in force until the end of their term, this payment means a total expenditure, on the part of the person purchasing, of about £4,000, and no person in Britain would make a genuine offer of more than £2,000 in these circumstances.

Sir William Thomson countered by describing the steam turbine as a very well worked out realization of a bold idea that he considered to have been a great success. He mentioned that the lubrication system had worked so well at speeds of 8000 to 10 000 rpm it had run non-stop for months and years with little sign of wear and small expense. He admired the use of the semi-elastic bearing that permitted such high speeds, being an entirely new mode of applying previously known dynamic principles to the balanced rotation of a rigid body round an axis. He pointed out its suitability for use with a steam condenser. He thought the patents could be of 'infinite' value. While testimony like this from an expert of such ability was no doubt flattering to the ear, it was not what Parsons needed! Thomson turned to Wilson's patent which he thought would not work at all: in fact he claimed that the use of semi-circular shaped blades meant that it was impossible to say in which direction they would turn. It happened that one of Parsons' turbines had been installed for lighting of Lincoln's Inn Hall. Permission was obtained by Parsons to get access to its boiler for an experiment. He mounted a set of blades constructed according to Wilson's patent in a lathe and directed a jet of steam against them. No matter in which way the blades were set turning initially, they always responded to the flow of steam by turning in the same direction. Moulton asked Thomson, (Anon 1891),

Well Sir William, do you deny that the wheel went round in the direction in which Mr Parsons told you it would go?

I do not deny it.

Then how do you account for the fact that it does not behave in the capricious manner in which you said it would behave?

If Mr Dugald Clerk and Mr Parsons want any particular machine to go any particular way, it will go that way.

You have examined the machine and are you satisfied that it is as shown in the Wilson drawing in the patent specification?

Yes.

Moulton had won his point but it was incidental to the matter at issue. Robson offered on behalf of Clarke Chapman and Co. to retain the patents and to proceed on their own. Parsons decided to accept the offer, and he had now to set out to continue his work with all speed but without the benefit of the 1884 patents.

10.2 A competitor

Parsons had good reason to press ahead. C G P de Laval was making progress and had by now assembled the main components for his rival concept, an impulse turbine. It is interesting to consider how de Laval approached his design because this throws light on the significance of Parsons' own work up to this time. Whereas Parsons had chosen to reduce the pressure drop across each set of turbine blades, and so moderate the maximum velocity of the steam that had to be harnessed, de Laval had decided to extract the energy from the steam in one step. It had been long known that if a nozzle were shaped so that the cross sectional area of flow decreased continuously from inlet to outlet in a so called convergent nozzle, the maximum velocity of steam or other gas that was discharged, was reached with a very moderate ratio of inlet to outlet pressure, less than 2:1. The velocity of the jet was then equal to the velocity of sound in the particular fluid. In his British patent 7143 of 1889 de Laval described a nozzle in which the flow was made at first to converge but then was allowed to diverge, figure 47. In this way he was able to make use of much greater pressure ratios and to achieve much greater velocities in the steam, and therefore larger kinetic energy in the jet. To harness such velocities required the use of shaft speeds which were even higher than Parsons had employed.

De Laval carried out experiments at speeds of 40 000 to 60 000 rpm but they produced violent vibrations that destroyed the conventional bearings that he was using (Jung 1973). At last he arrived at a strategy which he tried out for the first time in January 1889. He used a long and very flexible shaft, in fact a length of cane, on which he mounted a heavy disc. This was set spinning in a lathe. At low speeds the out of balance forces caused the shaft to vibrate, but once it had passed a certain speed its motion became very calm and stable. A Swedish patent for the use of a flexible shaft was

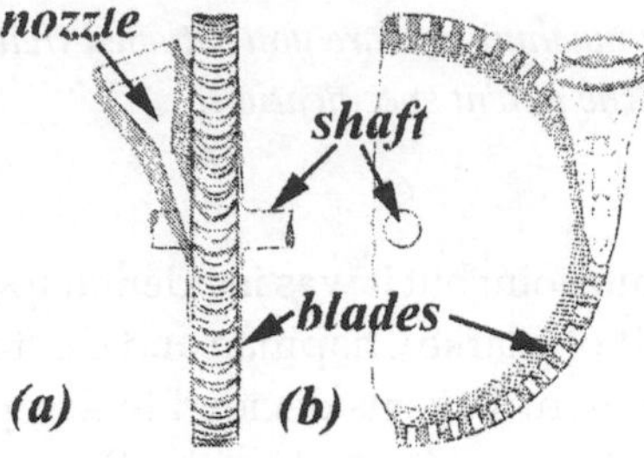

Figure 47. Details of de Laval's impulse steam turbine: (a) plan view of rotor disc with a section of the converging–diverging nozzle; (b) end view of the nozzle and blades, from de Laval's British patent 7143 of 1889.

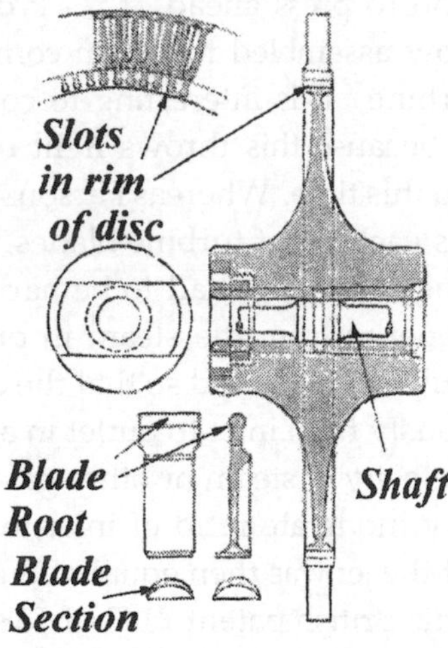

Figure 48. Details of a de Laval impulse turbine showing the massive hub of the rotor disc sitting on a slender shaft, and the method of slotting the crescent shaped blades around the disc periphery, from a Greenwood and Batley brochure.

awarded in May 1889. The combination of a disc with a massive hub and a long slender shaft resulted in a critical speed that was quite low, typically between one fifth and one tenth of the normal operating speed. Although this was a much greater separation than what was common in Parsons' machines, the actual concept of a critical speed does not appear to have been any more evident to de Laval than to Parsons at the time. In fact it seems to have emerged over a period of time during discussions involving officials in Patent Offices in different countries.

Like Parsons, de Laval incorporated some damping into the bearing housings. Blades were made of steel in a symmetrical, crescent shape, figure 48. This gave mechanical strength, and also ensured that the flow area at the inlet to the blade passages was equal to that at the exit. This latter point was important because it ensured that there should be virtually no pressure drop across the moving blades. The centrifugal forces acting on the blades were almost three times greater than Parsons had to cope with and so the turbine disc was made of forged steel. Around the circumference, holes were drilled parallel to the shaft, with radial saw cuts to allow a blade with its root to be slid into place. It is a technique for blade fixing that is still employed today. Because all the pressure drop occurred in the nozzle, the impulse design is free from difficulties to do with leakage around the blade tips. One last task remained, namely to reduce the shaft speed to a value where it could be coupled to machinery. From his earliest experiments de Laval had intended to use reduction gearing, but it was noisy and wore rapidly. In fact the development of quiet running gears with a long life proved to be a very difficult problem to solve, and it took him the best part of a further six years of experiments before he had solved the problem of making high speed gears (Jung 1968). Parsons' choice of a high speed dynamo got around this difficulty and had given him a lead of several years, though it probably would not have been suitable for direct coupling to an impulse turbine because of the even higher shaft speeds of the latter. The decision to rely on gearing did have the benefit that de Laval could use a governor mechanism and oil pump of conventional design because they could be driven from the output shaft that had been geared down to a relatively slow speed. The firm of A B de Lavals Ångturbin was established on 1 May 1891 to develop these ideas commercially. Although it would be some years before a truly practical turbo-generator was able to compete with Parsons' turbine, there was no time to be lost.

10.3 Establishing Heaton Works and radial flow turbines

In November 1889 after his break with Clarke and Chapman, Parsons crossed to the north bank of the river Tyne and with the help of friends he set up his own works at Heaton near Newcastle trading as C A Parsons and Co (see map of the Tyne) (Stoney 1931a). He brought with him a dozen or so of the team that he had assembled at Clarke Chapman. These included J H Armstrong, H Bishop, Francis Hodgkinson, R Williams, J B Willis, and Gerald Stoney. The name of Stoney first appears on the salary book on 23 April 1890, paid £2.10s.0d. per week. The workforce was built up to 58 all

Figure 49. Photograph of the radial flow steam turbine constructed by Professor Osborne Reynolds around the year 1875, opened out to show a and b, the two casing components, c, the rotor disc with two rows of blades, and d, the brake disc. The turbine is preserved in the museum of the School of Engineering at the University of Manchester. By courtesy of Professor J D Jackson.

told. Heaton at the time was a quiet suburb on the edge of Newcastle. A 2 acre site was leased close to Heaton Railway Junction and a workshop 50 ft wide by 140 ft long was built. Public transport at the time was by horse drawn trams and private transport was by horse and carriage or by bicycle which was coming into fashion (Anon 1949). With a considerable wage bill to be met, an immediate start was made with such products as arc lighting equipment and searchlight mirrors in order to produce some quick income, but the awful reality had then to be faced that access to all the features that were covered by the three seminal patents of 1884 was henceforth denied to Parsons. He seems never to have entertained for a moment the possibility of abandoning the development of the steam turbine. It was just a matter of finding other, alternative solutions to the problems with which he was now thoroughly familiar. One immediately thinks of his father's attitude when faced with the many reverses encountered in making the giant metal specula for his telescopes.

For his first venture he chose a reaction design but applied it to a radial flow configuration (Richardson 1911). It consisted of 13 stages or cells, each with a rotor disc fixed to the horizontal shaft. Between these rotating discs were diaphragms attached to the casing, split into two parts and with seals at the point where they were pierced by the shaft. Steam was led in at one end of the shaft and exhausted at the other, having traversed all the stages in turn. Each stage was designed like a Francis water turbine, with the fluid entering tangentially at the circumference of the rotor (see figure 50). The steam flowed radially inwards towards the shaft, and it was then turned and redirected through passages in the succeeding diaphragm to the circumferential region of the following stage. This was in fact precisely the arrangement that Professor Osborne Reynolds had proposed in his patent 724 of 1875, *'Improvements in apparatus for obtaining motive power from fluids, and also for raising or forcing fluids'*. The patent describes pressure compounding and deals with both axial flow and radial flow arrangements. Reynolds had made model water pumps as well as a model steam turbine (figure 49) (Gibson 1948)[1]. It is difficult to know whether Parsons was aware of Reynolds' 1875 patent; he never made reference to it in his papers. When his brother Clere read his paper on the efficiency of screw propellers to the Institution of Mechanical Engineers in 1879, the meeting was held in Manchester and Professor Reynolds contributed to the discussion. Clere could well have learned of Reynolds' patent and his turbine experiments, and could then have mentioned them to Charles during the period when both were working at Kitsons at Leeds. It was said that (Gibson 1948)

> *This* (turbine) *ran at 12,000 rpm and was probably the first practical machine of its kind. While it worked quite successfully, owing to its small size (the wheel diameter was only 6 inch) the leakage losses due to the clearance between the blades and the casing were relatively large and the steam consumption so high that he came to the conclusion that it could not be developed as a competitor of the steam engine and did not pursue the matter further.*

The only means of extracting power from the model was by a friction brake, so that it can hardly be described as a practical machine. What is interesting about it is that although it is a radial flow machine, the flow is in a direction that is *opposite* to that shown in Reynolds' patent drawing, namely outwards. Obviously inward flow does not work for steam. Parsons discovered this when his blading was destroyed after

[1] The machine is in the museum of the Department of Engineering at the University of Manchester

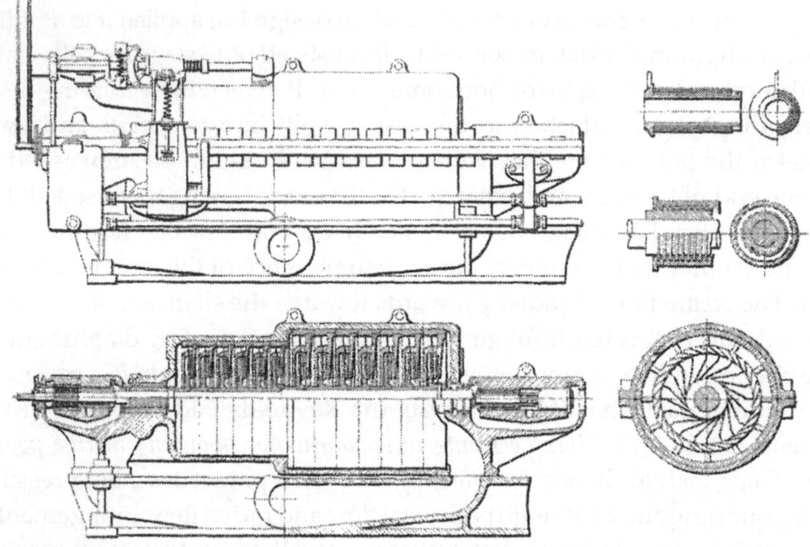

Figure 50. The first 32 kW inward flow radial steam turbine dubbed 'Jumbo', built in 1889. A side view and section are shown, and at the right hand side can be seen, at the top, the concentric tube bearings, next the thrust bearing, and at the bottom, the inward flow radial configuration of the rotor passageways (Richardson 1911).

running for no more than one hour. Water and bits of debris that were pushed towards the centre of the moving disc by the flow of steam were being thrown out to the perimeter by centrifugal force.

The first radial flow machine built by Parsons was quite large and complete. On this occasion he did not first try out the design principle on a small scale. The machine was dubbed the 'Jumbo', and was coupled to a 32 kW electric generator running at 6000 rpm (figure 50). The multiplicity of relatively large brass discs on a long shaft presented a quite new dynamic configuration. In order to achieve even a trial run he had to produce an alternative design to the compressed washers of the 1884 patent to support the bearings. This had to be able to absorb energy from any incipient tendency for the shaft to vibrate before this could build to a dangerous amplitude. His solution was to enclose the bearing journal in a nest of tubes. The inner, phosphor bronze sleeve that carried the shaft sat inside three, close fitting, steel tubes. The outermost of these was supported by the casing. Thin films of oil were forced between the surfaces of the tubes and developed strong viscous forces which reacted against any tendency for lateral movement. This was very effective in damping out any vibrations and in any case the assembly of washers previously used to support the bearings of the prototype had been found to be unsatisfactory because they cut into both the journal and the casing surfaces. Placing reliance on the oil film eliminated this mechanical wear and tear. For the next 20 years this was the preferred solution. This ingenious invention was absolutely essential in the period before the technology for balancing high speed machinery had been developed, but we have no insight into what led Parsons to devise it.

The fact that he could no longer balance forces by causing the steam to move in two, balanced, flows travelling in opposite directions created two new problems. First he had to find a way of making a seal between the shaft and the casing at the point where the steam entered at the high pressure end of the machine. For this he invented the highly successful labyrinth seal. Secondly he needed a way of resisting the axial force on the shaft. He achieved this with a thrust bearing contrived from a succession of collars arranged on the shaft which meshed against similar ridges in the casing.

Now that the speed of the main shaft had come down, Parsons could use a flyball governor. This was mounted on a secondary shaft running at a reduced speed, initially powered by a friction drive, but later gearing was employed. Once more he showed great originality in his approach to the control mechanism. The steam inlet valve was moved by an electrical solenoid and was caused to continually open and close so as to create 'gusts' of steam. By adjusting the ratio of the time open to the time closed the average steam flow could be regulated. As the flyballs of the governor moved outwards with an increase of speed, they moved a sleeve on the secondary shaft. This had a cam profile that was arranged to reduce the open/shut ratio of the inlet valve. Patent 1120 of 1890 covered this invention. It was found in later machines that the solenoid was too weak to control the steam valve position and so a servo motor was used to amplify the force available. At first compressed air was used in the servo, then after a further trial of an all-electric control, steam was substituted for compressed air (patent 15677 of 1892).

Throttle governing reduces power by lowering the mass flow through the machine. But by reducing the inlet pressure it also destroys some of the potential that is stored in the steam for doing useful work. It is the equivalent in a water turbine of wasting some of the 'head'. Beside this it has the effect of changing the magnitudes and directions of the velocity of the steam flowing through the blade passages, from the design values. In principle gust governing should have given better thermodynamic

Figure 51. Photograph of a disc with blades cut from the solid, for a radial, outward flow turbine, like the Scotland Yard machine 1890 (Richardson 1911).

efficiency than throttle governing because while the steam is flowing there is no impediment at all to flow. The governor valve is either fully open or fully closed. But the gust concept was not particularly efficient since the pressure, and hence temperature, of the steam were continuously varying. Moreover if the fluctuations in speed were to be kept small enough not to be noticed in the flickering of lamps supplied by such machines, it was necessary to use a relatively rapid succession of gusts, 200 to 400 per minute. An incidental benefit was that the mechanism was continually in motion, so there was no tendency for it to stick and its response to changes was very rapid and certain.

After the first run of the Jumbo, the discs were replaced with new ones made of cast iron, but still efficiency was poor and drainage presented problems, so inward flow was abandoned. A new machine dubbed the 'Mongrel' was built, in which Parsons reverted to the alternative that had been described in the patent specification of 6735. Speaking of the blading he had written

> *When concentric they* (the blades) *are arranged upon a disc or plate or on both sides of a plate or disc to balance the pressure, the first being near the centre, the second outside of it and so on, the vanes* (blades) *projecting from the surface of the disc. This disc works in conjunction with a fixed disc (or discs, on each side of it) which also serves as a casing. Upon this disc are arranged circles of fixed vanes which stand between the circles of*

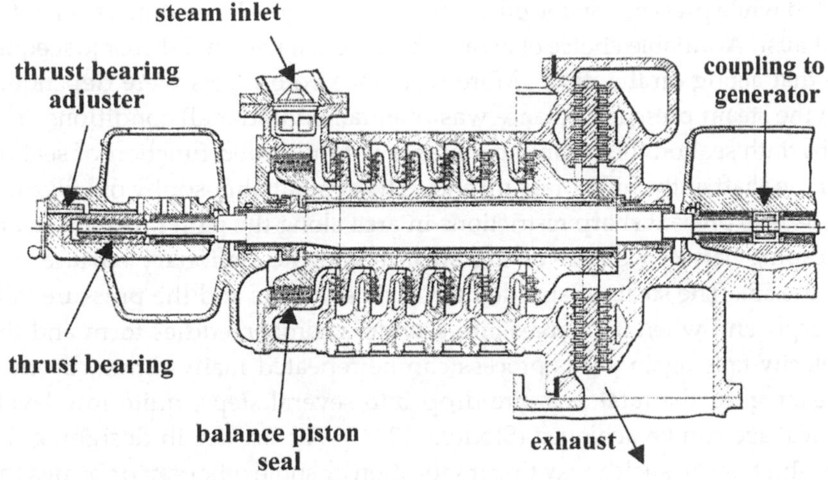

Figure 52. Longitudinal section through a 100 kW radial flow condensing turbine supplied to Cambridge Electric Lighting Co. built in 1891. A labyrinth seal is used to seal the balance piston (Parsons R H).

> *vanes on the first mentioned disc and serve to direct the fluid on to them in the required direction. As each circle of moving vanes is larger than the one within it, it follows that the actuating fluid will be expanded in passing from one to the other.*

As has been observed already, this configuration fortunately did not form part of the final patent specification.

Now more cautious, he used the casing and shaft of the Jumbo to first try out a single rotating element. This had a brass disc bolted to it, with blades that had been machined from the solid. A corresponding disc was fixed to the casing. Steam was admitted at the centre of the disc and flowed outwards through the rings of blades (see figure 51). Results were satisfactory and machines were made with a series of discs, each carrying either three or four circles of blades. Steam emerging at the circumference of the first disc was led down behind the back of the succeeding fixed disc to enter near the centre of the next moving disc. A 32 kW machine supplied to the Metropolitan Police at New Scotland Yard in London had five such discs. This construction involved large pressure differences between the front and back surface of the discs, and in order to counteract the resulting large axial force a further innovation was adopted. This was the balance, or dummy, piston (figure 52). On one face of the piston high pressure steam acted while pressure on the other face was kept low by a connection to the exhaust. A suitable choice of areas could yield a net axial thrust just equal to that acting on the discs. Moreover since both forces were dependent on the steam pressure, balance was maintained under all conditions. The labyrinth seal on the balance piston served the added function of sealing off the shaft at the high pressure end. The general philosophy of labyrinth seals is to present sharp restrictions in area along the path taken by steam escaping between the revolving surface and the stationary surface (see figure 55). The steam is speeded up at these points and the pressure falls sharply and when it emerges into a wider chamber, eddies form and the velocity falls again. This process can be repeated many times. By thus breaking up the total pressure drop into several steps, quite low levels of leakage can be achieved (Stodola 1933). The art lies in designing the constrictions in such a way that if vibration or some other factor brings the two surfaces together, no serious damage will be done. Stoney commented (Stoney 1933a)

> *It is curious how long it took to recognise that one or other of two elements passing one another at high speed, must have a really thin edge. It would be about 1897 that this was fully recognized and the present type of labyrinth packing adopted.*

Adoption of the balance (dummy) piston allows a reduction in the length of the machine, so reducing first costs, and is a common feature of modern turbines. This was a most significant invention. It was to be of special importance when eventually Parsons regained his axial flow patents. It allowed him to do without the split, balanced flow arrangement of the prototype. This meant that the blades could be made longer, and consequently leakage around the tips was relatively less important. It has been generally adopted by turbine manufacturers. Provision was also made for the clearance between the blade tips and the fixed or moving discs to be given a fine adjustment. This was achieved by making the small thrust bearing in such a way that the outer component, inside which the shaft rotated, could be moved axially relative to the casing.

A number of machines were made to this design, but other improvements were quickly incorporated. The steam tended to condense inside the machine, and the resulting accumulation of water reduced performance. To combat this the fixed diaphragms were cast with passages through them, which could be heated by a current of steam. More significant were the developments in blade design. As foreseen in the Complete Specification of 6735, but not identified as a claim,

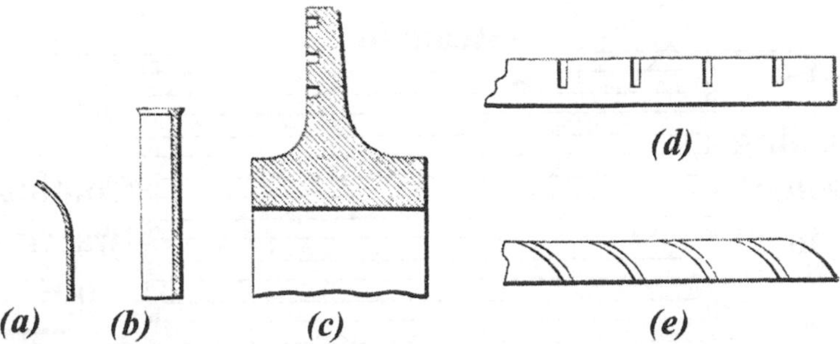

Figure 53. Blading for a radial flow turbine made from sheet metal: (a) cross section of a stamped blade, (b) showing 'upset' foot at top, (c) radial section through a disc, showing slots, (d) side view of a blade carrier, and (e) plan view of same carrier (Richardson 1911).

> *In some cases it may be found convenient to make the blades of sheet metal and to secure them in suitable grooves or recesses in the rings—in this way the blades may be made thinner and they may be accurately formed to any desired shape...*

Blades were stamped out of brass sheet, figure 53(a), and formed to the desired shape with a foot created by 'upsetting'. Upsetting is a forging process that causes the metal section to swell out. The blades were slid into slots in a carrier made of metal strip, (d) and (e), that was in turn secured in a groove cut in the face of the discs and diaphragms, (c). The latter were now made of cast iron.

10.4 Condensing turbines

The first machines of this general construction, exhausting to atmosphere, were supplied to DISCO's Forth Banks station when output was being increased. But the time had come to consider designing a machine that could exhaust into a condenser. Of course the earliest steam engines, Newcomen's atmospheric engines (figure 2), had relied for their motive power entirely on the vacuum created by condensing steam. James Watt made a great leap forward by arranging that condensation of the steam took place away from the piston and cylinder in a separate 'condenser'. Henceforward this became a normal feature of large engines that were

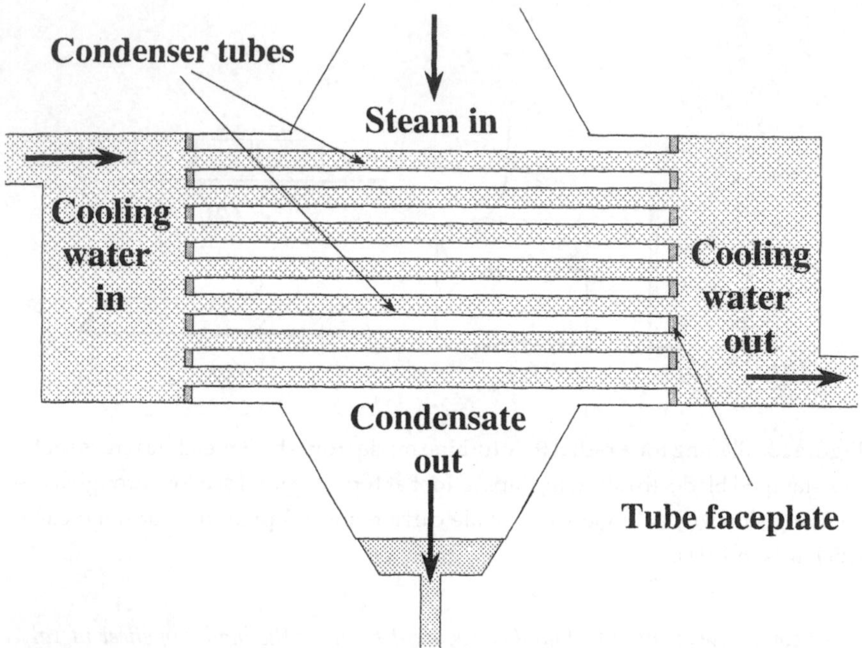

Figure 54. Schematic drawing of a steam condenser, showing the inlet for steam and the outlet for condensed water, as well as the inlet and outlet for the cooling water supply.

required to be highly efficient. As boiler designs improved, the pressure of steam increased, and some engines, like those in locomotives, dispensed altogether with condensers. For marine applications however, condensers were valuable not just for efficiency but because they allowed the boiler to run on distilled water and so avoid problems due to salt deposits on the walls. Early condensers were quite primitive. Surface condensers were simply air tight tanks cooled on the outside by a supply of water. In jet condensers a spray of water was used to condense the exhausted steam. Meanwhile improvements in manufacturing had made available thin tubes of seamless copper. These were arranged inside a metal tank between suitably drilled face plates that formed opposing walls. Chambers outside the tank fed cooling water to and from the tubes, while steam from the engine was led into the spaces between the tubes, as can be seen in figure 54. In this way very large surface areas could be made available for heat exchange. With condensers built to such designs, it was possible to achieve pressures as low as 3 to 10% of atmospheric. Since any trace of air

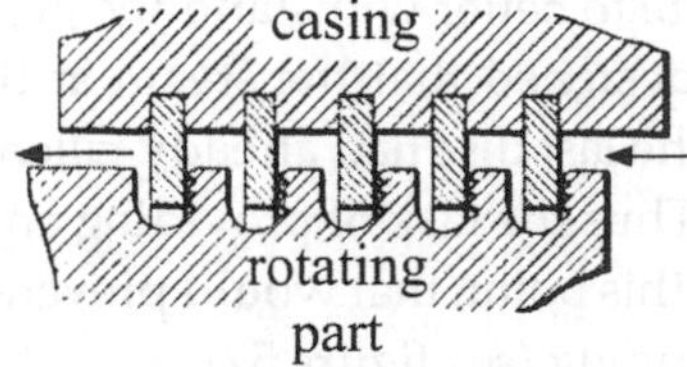

Figure 55. The principle of the labyrinth seal can be seen in this example of a later design for a dummy piston seal, which was employed for the 1 MW turbine built in 1900 for Elberfeld. The sectioned view shows brass rings which are fitted into the casing. These run in grooves cut into the periphery of the rotating piston, which have serrations cut in their right hand face. The arrows indicate the direction of steam flow. (Parsons R H).

in the exhaust steam spoils the effectiveness of the condenser, there was a need for pumps that would extract not only the water condensed but also any air. Earlier, while manufacturing incandescent lamps, Parsons had designed and built a quadruple stage pump for extracting air from the glass envelopes, and he had a good working knowledge of building such air pumps.

An engine with a single piston and cylinder can deal efficiently with only a modest ratio between the supply and exhaust pressure, so that it cannot take full advantage of a very good vacuum. This difficulty was avoided by compounding. Two, three or even four cylinders were placed in series, each passing ever larger volumes of steam to the next in line. Compounding of course requires the successive pistons and cylinders to be ever larger and more massive. This is where steam turbines enjoyed a major advantage because they are well suited to handling the expansion of steam down to the very lowest pressures without a great increase in size. There was another advantage. Because the lubricant in a steam turbine is kept separate from the steam, it does not foul the surfaces of the boilers or condensers. Steam engines on the other hand require some lubrication for the piston and cylinder surfaces, and this inevitably gets mixed with the steam supply.

Expansion of steam down to low, sub-atmospheric, pressures increases its volume enormously, typically by over one hundred times. To take care of this, an extra disc was fitted to the standard radial flow machine. As well as being of much larger diameter than the others, this was fitted with blades on both faces. The peripheral velocity of this wheel was 565 ft s^{-1}, an extraordinarily high value for the time, and it was made of mild steel boiler plate to better withstand the greater centrifugal forces. The machine was so designed that the steam at the point where it was entering the centre of the last disc had already fallen to a pressure that was close to atmospheric. Thus the task of providing an effective seal between the shaft and casing at this point, that would prevent the ingress of air, was made much less demanding (see figure 52).

Where the shaft emerges from the casing a tendency exists for air to be drawn in from the outside to mix with the steam, so spoiling the vacuum. To prevent this another invention was devised. A supply of steam, at a pressure slightly above atmospheric, was provided to a chamber surrounding the shaft. Of course, some steam flowed in each direction, and some escaped to the atmosphere and was lost. But the loss was small and the exclusion of air made this well worth while. The strategy for sealing off the shaft at the exhaust end, with the so-called steam packed gland, is still in use. It was covered by patent 5312 of 1887. Among later solutions to this problem was the adoption by General Electric of carbon rings made in segments, to form a snug fit to the shaft (Stodola 1927). Also, at Westinghouse, experiments were made using water in place of steam to exclude air (Hodgkinson 1939). In Parsons' machine the pressure of the steam did not fall to its lowest value until it emerged at the rim of the last disc where there was no sealing problem, and it moved thence via the exhaust duct to the condenser. The governor operated on the 'gusting' principle, with a solenoid that took its current from the exciter dynamo, controlling a steam servo.

As if such work was not enough to occupy him, Parsons decided for some reason to take another, serious look at Hero's aeliopile (figure 35). He constructed a single arm with a jet at a radius of 17.25 in, which was balanced with a counterweight. Turning at 5000 rpm in a good vacuum he found that it consumed 70 lb per kW h, a quite inadequate performance. A modification was made so that the jet played against cups set in a ring fixed to the casing. The jet itself also had three cups attached to it. The intention was that the steam would be turned back from the stationary cups attached to the casing and would impinge successively on each of the moving cups. It was an idea later perfected in the Reidler–Stumpf turbine (Stodola 1927), but this attempt was not successful. Next a full blooded reaction machine was built. The arms carried jets at each end, and were aerodynamically shaped and highly polished. Each set of arms was contained in a pressure tight cell. Steam was fed through three such units in succession before emerging into a double sided radial flow wheel. The design was covered by patent 8854 of 1893. Despite all the care taken with this machine, tests gave disappointing results. There were great frictional losses caused by the rotation of the arms at high speed in steam that was still quite dense.

In the course of these experiments, another test machine was built to make measurements on a single disc which was mounted on a relatively long and slender shaft (Richardson 1911). It was tried with blades on one face only and with blades on both faces, but it was found that the disc tended to oscillate laterally. The clearance between blade tips and the casing that was necessary to prevent these oscillations from causing blade damage, created unacceptable leakage losses. But it was found that when several such discs were mounted on a shaft of the same length, and bolted together, there was no problem. Clearly in this latter machine, the shaft had been stiffened sufficiently by the multiple disc assembly to raise the critical speed above the working speed. Coping with the dynamics of high speed shafts was still very much a matter of experience rather than of calculation.

Parsons' experience of working with turbines in a condensing mode inspired one further idea that was to prove valuable in future years. Steam that expands from a pressure of 200 psi down to a good vacuum of 1 psi increases its volume by up to 100 times, depending on the circumstances. Steam engines that were fed at pressures of several atmospheres and which exhausted to a condenser could cope with an increase of volume of perhaps 25-fold. As has been mentioned, the usual solution was compounding, with steam from one cylinder being fed to another cylinder of even larger volume. This led to very large diameter, and hence massive, low pressure cylinders. Parsons realised that the steam turbine which could quite easily handle the very lowest pressures without great increase in size, could be used to complement the reciprocating engine. A turbine could replace the cylinders that worked at the lowest pressures. His patent 367 of 1894 covers the combination of a reciprocating steam engine feeding into a low pressure steam turbine that exhausts to a vacuum. Although it was envisaged as suitable for use in propelling a ship, at the time at which the patent was filed there was still no turbine in existence which drove a propeller. Indeed the first commercial application was actually not built until 1908 when it was employed in the steamer *Otaki* for Messrs Denny of Dumbarton, after which it was adopted by several ship owners. The same concept is embodied in many modern turbo-charged motor car engines. In these the energy in the exhaust gases from the reciprocating engine that would otherwise go to waste, is used to power a high speed turbine (coupled to an air compressor). An engine of a given size develops much greater power when supplied with air that has been compressed in this way.

10.5 Dynamos and alternators

During the period after the break-up with Clarke Chapman, Parsons could not employ his own version of the Siemens dynamo. He turned instead to the Gramme, ring, design of armature. Whereas all the conductors in the Siemens machine lie close to the surface of the rotating armature, in the Gramme machine current from the outer conductors flows back in the opposite direction through additional conductors that lie inside the armature and close to the surface of its shaft. This is wasteful of material, it generates greater losses and makes cooling more difficult. Indeed the limitations on size were so severe that some 410 kW turbines had to be fitted with two Gramme dynamos mounted in tandem in order to supply sufficient current safely (Richardson 1911). Sometime around 1890 the construction of the commutator in his dynamos was also improved. During manufacture the many individual segments were first held in place around the shaft by twine, with mica sheet employed to keep each segment insulated from its neighbour. At either end of the assembly outer cylindrical layers of mica were tied in place with string, and steel rings, heated to red heat, were then slipped over them. As these cooled, the rings gripped the assembly firmly in place. It is an arrangement that was widely adopted. Stoney remarked how surprised he was that the idea was never patented by Parsons (Stoney 1938). Mention has been made of the fact that during the period before Parsons recovered his right to use patent 14723 of 1884, which covered the design of the dynamo, he had to adopt the Gramme construction. However the patent expressly stated that

> *I do not lay any general claim to the construction of an armature with a hollow axis on which metallic discs are threaded nor to laying of conductors in grooves or channels or the laying of conductors along and tightly binding them upon the armature core…*

So he was fortunate that there was nothing to stop him continuing to use a similar arrangement for alternators.

Although the total quantity of turbines sold between 1890 and 1894 was not great, notable progress was made in obtaining acceptance among public supply companies. During the first few months Charles Parsons was somewhat constrained. When writing on 28 May 1890 to his brother Laurence about the decision to install electric lighting in the Museum in Dublin, he said in a postscript (C.L.(1)),

> *Edmundson's had almost promised to put in turbo-generators, but now such an order would go to Clarke, Chapman and Co. We are sending out*

circulars to say we have disposed of the turbo patents and will shortly bring out a combination plant of motor and dynamo having many improvements. We have agreed not to advertise any new patent for six months, and one has to be careful.

No municipal electricity supply organization had yet taken the risk of installing steam turbines, but Cambridge Corporation did consider a 100 kW turbo-alternator running at 4800 rpm for its planned generating station. In 1891, one of the members of the Corporation, who was sceptical of the claims made for the turbine, asked J A Ewing who had already testified in support of Parsons at the patent hearing and was now Professor of Engineering at Cambridge University to carry out a test (Richardson 1911). Ewing had expected a poor performance, but his measured value for the consumption of steam was 37 lb per kW h with a modest condenser vacuum of 26.5 in He contrasted this with the performance of a *'good ordinary compound condensing engine'* of 36 lb per kW h. He was won over by what he saw as the machine of the future. Thus another milestone had been passed, parity of performance with the conventional steam engine. It was accomplished after only two years work on the new configuration, and eight years from the day he first set to work on the steam turbine.

On 18 June 1891 Parsons was able to write that the Cambridge town Council had approved the use of turbines for lighting the town (C.L.(2)).

Elvaston Hall
Ryton-on-Tyne
18 June 1891

My dear Laurence,
We have just heard that the town Council have approved of the Turbine System for the lighting of Cambridge. Their Committee carefully tested the merits of the question before deciding and they were of the opinion that it was the most economical altogether, and much cheaper in first cost. It seems that when condensing, and with intermittent admission of steam, it is much more economical at average loads than the best compound engines and dynamos—partly because there is less friction, and partly because there is no condensation when steam-jacketed, and the steam gets fairly expanded at all loads. I thought you might like to know, as we were talking about it last week. When Lord Rayleigh was here he said he would expect very good results from condensing; this appears to be practically possible and gives results up to theory. It has turned very warm here today.
Your affectionate brother,
Charles

In 1892 the Cambridge Corporation sold the right to supply electricity for a sum of £2040 to the Cambridge Supply Company, which was headed by Charles Parsons as Managing Director. Other members of the board included John Bell Simpson and the Rev. G B Finch, a fellow Cambridge graduate (Smith 1989). The following year the firm of C A Parsons was successful in its tender to supply the newly created Scarborough Electric Supply Company with boilers, generating sets and condensing equipment (Parsons R H 1940). Once more he had had to take shares in the enterprise to ensure a sale.

In 1893 Portsmouth became the first municipally owned power plant to use turbines. One of the largest of these machines, rated at 150 kW, was installed under the direction of Gerald Stoney. It was intended to run in parallel with several large Ferranti alternators that turned at a leisurely 100 rpm. The event was celebrated with a municipal banquet, after which aldermen and guests repaired to inspect the power station. This was lit by arc lamps fed by the turbine. Because it was not yet synchronized with the other machines, they were running at a speed that gave a slightly lower mains frequency. The effect was to make the Ferranti machines appear to be turning very slowly, in reverse, to the astonishment of the well dined guests (Parsons R H 1940). In fact the two quite different kinds of machine ran quite satisfactorily in harness with one another despite fears to the contrary.

Stoney's account of an accident that occurred to a turbine gives a vivid impression of of the situation in those days (Dowson 1942).

> *Radial turbines were installed at Cambridge, Scarborough etc., and were for an output of 150 kW, 2,000 volts, single phase alternators, 4800 rpm. The h.p. discs were of cast iron, and the final, l.p. disc, which was bladed on both sides, was of mild steel boiler plate 27 inches in diameter. The peripheral speed was 560 ft per second. This would give a stress of 12 tons/in^2 at the hole, not counting the load of the blades, but fortunately none knew at that time how to calculate stresses in a disc, or they would have hesitated to put 12 ton/in^2 on 28 ton material. One of the Scarborough machines ran away on account of the armature bursting due to the binding wire going and the square on the coupling* (to the turbine) *shearing off. The governor valve was leaking and thus failed to stop the turbine. The driver ran away also, but finally steam was shut off the boiler.*
>
> *The turbine was practically wrecked, some of the h.p. cast iron wheels burst and the l.p. disc expanded about equally at the hole and at the rim, the former from 4 inches to $4\frac{1}{4}$ inches and the latter from 27 inches to $27\frac{1}{4}$ inches. Of*

course, all the blades had gone, and the balancing holes for passing steam through the disc were oval. These l.p. discs used sometimes to become slack on the spindle, they were then bored out and bushed and rarely became slack again. Truly we were brave people in those days!

One of the Cambridge machines, figure 52, also overspeeded due to a damaged armature winding, causing the final disc to expand appreciably. Fortunately it seized up without disintegrating (Barker 1939). On considering the fact that the bore of the disc, once stretched, did not change if subjected to an overspeed a second time, it was decided that all production discs should be overspeeded before being finally assembled. The process of stretching the bore of a cylinder, such as a gun barrel, beyond its elastic limit is known as 'autofrettaging'. It was patented in Switzerland in 1937 (Gibb 1947).

Parsons' efforts to market his machines were considerably assisted by his appointment of a young Scot, Archibald Campbell Swinton (1863–1930), as his agent in London (Swinton 1930). His father was Professor of Civil Law at Edinburgh University. During his school years he showed unusual technical ability, experimenting with photography and the Bell telephone that had been invented just two years before, but he received no formal third level education. Instead, at the age of 19 he joined Sir W G Armstrong and Co as an apprentice, paying a premium of £600. He developed a strong interest in electrical engineering and while still a draughtsman, not yet 21 years of age, he wrote a text *The Principles and Practice of Electric Lighting*. He sent a copy to the directors, and was asked by one of them, Sir Andrew Noble, to look after fitting out ships with electric light. He pioneered the use of lead sheathing as a protective covering for the electric cables that were insulated with gutta percha. When he left Armstrongs in 1887 he set up as a consultant in London. He had seen Parsons' turbines in 1885 at the Gateshead works, and they were to become a close friends. Although he was described as possessing a natural diffidence, his strength of character carried him through reverses of fortune. He had a gift for working with other men and his common sense, sound technical ability and business ability made him a valued board member. It was through his good offices that Parsons eventually recovered the rights to his 1884 patents on 16 December 1893 for a reasonable sum; £1500 instead of the £98 000 originally demanded (Stoney 1931a). C Turnbull, who was at that time an apprentice, but in later years chief engineer of the Tynemouth Corporation, remembered (Turnbull 1939)

Figure 56. Algernon (Tommy) Parsons (1893) standing beside a helicopter powered by a light weight steam engine. The sail at the right hand resists the tendency for the propeller to spin the vehicle. © Birr Scientific and Heritage Foundation by courtesy of Lord Rosse.

> *One morning he* (Parsons) *came into the works in great form and met me walking along the shop, when he said that he had recovered possession of his patents. I was surprised at him telling me this, as I did not expect that he would bother with an apprentice, but I had no idea what it meant to him. Later I saw that the whole question of the turbine was his life to him—he lived for it.*

Considering what a significant event this was in the affairs of Charles Parsons, it is strange how little detail about the negotiations seems to have survived. True he had made a success of the radial flow machines, and it may have seemed that he could have been successful with them if he had to, but once he recovered the rights to the original patents he reverted without a second thought to what the passage of time has shown to be distinctly the better option.

10.6 Experiments with flight

In the summer of 1893 Parsons also became interested in flying machines (Parsons C A 1896). Almost as a hobby he constructed a small steam engine of novel design. Steam engines are affected by the condensation that takes place in the cylinder when the steam cools as it expands. To counter this tendency Parsons placed the boiler tank *around* the cylinder of the engine. Perhaps it was an idea suggested by his experience with radial flow turbines in which he had used steam to heat the special passages incorporated into the fixed iron diaphragms. At any rate he was so pleased with the results that he used the design in an engine for a helicopter. The boiler was a tube 14 in long by 2.5 in diameter, and was turned from the solid, with walls just 10 to 15 thousandths of an inch thick. Its manufacture severely tried the patience of the foreman who was entrusted with the job (Stoney 1938). The engine was arranged to drive a twin bladed, horizontal helicopter propeller. A large sail at the end of a boom served to oppose a tendency for the vehicle to rotate. The boiler was placed over a flame until the pressure reached 50 lb in^{-2} at which point the engine was started and the craft was allowed to rise into the air. It rose to a height of 12 ft vertically. The engine, running at 1200 rpm, developed $\frac{1}{4}$ horse power. Including 3 oz of water, but excluding the weight of the burner, the complete engine weighed just $1\frac{1}{4}$ lb. This represents a weight of 5 lb per h.p. and may be compared with the Wright brothers' first engine that weighed 16 lb per h.p. There is a photograph taken around 1893 of his seven year old son Algernon standing beside the helicopter (figure 56). Later in life Algernon served for a time with the Royal Flying Corps. Charles Parsons retained his interest in aeronautics. In 1913, as President of the Northumberland and Durham Aero Club, and of the North East Coast Institution of Engineers and Shipbuilders, he arranged for the two bodies to merge (Clarke 1984).

He turned next to construct a monoplane using a frame made from cane with a covering of silk. The wing span was 11 ft, with an area of 22 square feet. There was a tail but no undercarriage. The total weight was 3.5 lb. It achieved a flight of 80 yards reaching a height of 20 ft. We have no information about its aerodynamic stability. A rather blurred photograph of the machine in flight is the only illustration that we have of the aircraft. A different strategy was also tried that involved filling the boiler with methyl alcohol instead of water. As the alcohol boiled and the vapour was exhausted from the engine, it was burned and used to heat the boiler. A very high evaporation rate was achieved, equal to 120 lb of water per square foot per hour. However there were problems in maintaining a flame in the face of the strong air currents caused by the propeller. To set this episode in context, it may be noted that the Smithsonian Institute in America was funding Samuel Pierpoint Langley at the time, to allow him to develop an aircraft powered by a steam engine (Crouch 1989)[2]. Over a period of

[2] Langley was the same person as had worked on lunar temperature measurement

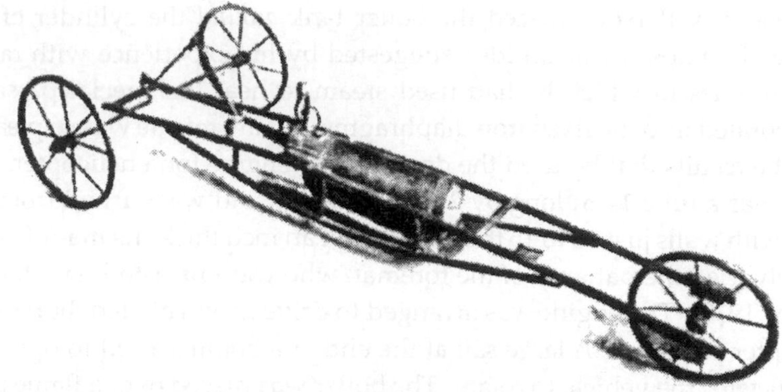

Figure 57. The Spider, a self-propelled toy made from the steam engine, but with methylated spirits for both the working fluid and fuel (Stoney 1938).

eight or nine years more than one hundred models were constructed, but it was not until 6 May 1896 that his efforts were crowned with success when his 'Aerodrome 5' flew for 90 seconds and covered some 3000 ft. Orville Wright in his account of *'How we Invented the Airplane'* includes Parsons among those who had worked on the problem before, though without success (Wright 1988).

Before leaving this brush with aeronautics, mention may be made of a prescient contribution that had been made by a past president, Jeremiah Head, during the discussion in 1888 of Parsons' paper to the Institution of Mechanical Engineers in Dublin (Parsons C A 1888b).

> *Without having worked out any calculation, he ventured to think the compound steam turbine gave a greater power for its weight than any kind of engine known... It had always been said for a long time past that it would be impossible to propel machines in the air because of the great weight of any engine which would be required to do it.*

He went on to consider the problem posed by the need to carry both fuel and water. He suggested the use of petroleum which had half the weight of coal for a given evaporative effect, and went on to say

> *But if the petroleum could be burnt in the engine itself, as was already being done in the cylinder of the petroleum engine, it would save the water altogether... Whether the products of combustion of petroleum could be applied to such a machine he did not know.*

The photographs that illustrated Parsons' description in the article in *Nature* were taken by Gerald Stoney, who was by now very closely associated with his experimental work. Writing after Parsons' death, Stoney recalled that the little engine was modified yet again. It was fitted with wire wheels as a tricycle and was run on the lawn for the amusement of the two young children (figure 57). A smaller version was used inside the house in the library, and he remembered Parsons and the children running after it, stamping out the flames as the leaking methylated spirits ignited on the carpet.

It appears that Parsons was considering another change of residence from the summer of 1892. He settled on Holeyn Hall which was located at Wylam on Tyne, almost twice as distant from Heaton as Ryton. After February 1894 his notepaper is headed Wylam on Tyne. He equipped his new home with a workshop and spent much time in it working on ideas that interested him.

Chapter 11

A Marine Prototype

11.1 Developments in shipbuilding

In 1914 when Charles Parsons was given the freedom of Newcastle upon Tyne, he said (Parsons C A 1914)

> *When it* (the turbine) *had beaten a compound engine driving a dynamo, my old friend Dr John Bell Simpson said to me one day when we were out shooting: 'Why not try it at driving a ship?' To which I replied that I thought the time was ripe for the attempt.*

This might seem to suggest that the possibility only dawned on him rather late in the day, in 1891. But the wording of patent 6735 of 1884 made it quite clear that thoughts of applying the steam turbine to marine propulsion had been in Parsons' mind from the beginning, and Richardson states that (Richardson 1911)

> *Mr Parsons began the study of the marine problem at the Gateshead Works...* (where) *his first design of a machine for driving a propeller was... of the tandem type, of 200 H.P. with two separate casings... This machine however was never made.*

It seems most likely that Parsons had early formed the judgement that the first priority was to develop the turbine principle on a scale where it could give a useful amount of power, and that this could be achieved with least cost by employing the turbine to drive an electrical generator. As regards marine applications he would be well aware, from his brother's experiments on screw propellers, that there were special problems to be faced in securing an efficient design for a propeller even when it was running at speeds of one or two hundred revolutions per minute. Since the speed of small turbines must necessarily be high, of the order of thousands of revolutions per minute, a huge hurdle had to be surmounted before their output could be successfully harnessed for propulsion. It is clear that he never doubted that this could be done when the time was ripe. His sights were set on using turbines to propel the largest vessels, indeed vessels that were significantly larger and faster than any in existence at the time, but he was neither aware of the full extent of the obstacles before him, nor of the means that he would discover for overcoming them. One is reminded of his father's certainty when he pressed on to the construction of the world's largest telescope, even when he was facing what appeared at the time to be insurmountable difficulties in the fabrication of mirrors.

In the first half of the century American shipbuilders with access to plentiful supplies of timber had, for a time, challenged Britain's dominance in shipbuilding. Their elegant and fast clippers began to circumnavigate the world, joining the east coast to the west coast of America. But the second half of the century saw the most incredible technical advances. The first iron-clads were wooden ships with iron armour but gradually the use of wood for the hull of a ship was dispensed with altogether. Slowly cheap iron, and then mild steel from the Bessemer and Siemens processes, began to become available, and the picture changed rapidly (Pollard and Robertson 1979). Iron ships of a given size could carry one third more cargo by weight, and one half more by volume, than a wooden ship. When steel began to replace iron in the 1870s, the weight of a hull of a given size fell by a further 15%. The early, low pressure, steam engines freed vessels from a dependence on favourable winds, but they were inefficient. Their engines consumed 10 lb of coal per IHPh, and their range was necessarily modest given the relatively small capacity of their bunkers. But the adoption of higher pressures and the use of compounding in engines cut this consumption to one quarter. Ultimately in the early 20th century a further reduction was achieved, down to $1\frac{1}{4}$ lb of coal per IHPh, or just one eighth. So even on long voyages the clippers could be challenged.

By the year 1890 Britain had become responsible for 80% of all the world's shipbuilding (Pollard and Robertson 1979). Of this almost one half came from the north east coast, that is from the yards on the banks of the rivers Tyne, Tees and Wear. The output of the shipyards of the north east coast reached a peak in 1883, comfortably surpassing the yards on the river Clyde. Although this was followed by a sharp cyclical decline, the region retained its dominant position. A number of factors helped to create this situation. The area was rich in stocks of coal and iron ore and the export trade in coal had encouraged shipbuilding. Also the construction of railway steam engines and other similar heavy machinery had already created a large pool of skilled mechanics. Management of the region's rivers by straightening, embanking and dredging had transformed them. Eventually the Tyne became suitable for the launch of battleships and even giant liners like the *Mauretania*. When Sir W G Armstrong merged with Charles Mitchell, a firm was created that could not only build ships from its own supply of steel, but also arm warships with the latest designs of guns and hydraulic machinery. This attracted customers from all over the world, from Argentina, Australia, Austria, Brazil, Chile, China, India, Italy, Japan, Norway, Portugal, Romania, Spain, Turkey, and the United States. As long ago as October 1872 a high level delegation of Japanese officials had toured the Armstrong works and other industries in the North East of England. In 1883 the first warship was launched for delivery to Japan (Conte-Helm 1989).

Doubtless the reliability, light weight, and freedom from vibration of the turbine suggested that ultimately a market could be found in the propulsion of larger, relatively high speed passenger ships that were beginning to emerge; still it required gifted foresight to be able to anticipate the progress that the next 20 years would bring. Before the necessary giant turbines could be designed and tried in service, an immediate customer was needed who would value high speeds, without being unduly concerned with efficiency. Such a customer was the Royal Navy. When Whitehead had developed the locomotive torpedo as a naval weapon he created a demand for very fast, unarmoured vessels that could close quickly with the enemy, launch their missiles at short range, and then speed away. Driven by a strong spirit of competitive rivalry, two small shipbuilders in particular were designing such vessels in England that pushed performance to the limits of technological possibility. They were J I Thornycroft[1] and A Yarrow[2], while in France J A Normand was an able competitor. In his paper to the Institution of Civil Engineers in 1881 Thornycroft described the design of two high speed vessels, the yacht *Gitana* built for Baroness Adolphe Rothschild and what was termed a 'first class torpedo boat'. The latter vessel was some 100 ft in length and 40 tons displacement. The maximum speed attainable by a vessel with a steam engine of a given power output depends on the shape of the hull, but also

[1] Sir John Isaac Thornycroft (1843–1928) served as a draughtsman before studying at Glasgow University under Rankine and W Thomson. He later attended the School of Naval Architecture at South Kensington (D.N.B.e).

[2] Sir Alfred Fernandez Yarrow (1842–1932) was of Scottish and Jewish descent. He left school at 15 and served a five year appreticeship with a firm of marine engine builders (D.N.B d)

on its wetted surface area, which is related to its displacement and hence its total weight. Consequently the naval architect shared the same concerns as the modern day aircraft designer, and was always watching to avoid unnecessary additions to the all-up weight. To achieve lightness, the design of the boilers was changed, and was modelled on railway locomotive practice, where combustion gases passed through a multiplicity of fire tubes immersed in water. In addition, the rate of combustion per square foot of grate area was greatly increased by the addition of forced draught to the fires. As a further measure to minimize weight, the engines were designed to run unusually fast, at between 400 and 600 rpm. Speeds of up to 22 knots were reached, but it was noted that vigorous vibrations were excited at certain engine speeds. There was a strong demand for this new class of vessel world-wide and large numbers had already been purchased. The need to protect battleships from this new menace led to the development of a class of larger, fast, torpedo boat destroyers which could sail in all weathers.

The author Rudyard Kipling was introduced to Thornycroft who invited him to witness speed trials of a 30 knot destroyer. He described the experience in a vivid letter. Although it relates to a somewhat later period, it is worth reproducing because it conveys so graphically what was involved in the design and testing of the latest warships (Kipling).

> *Well Sir, this is about all there was to the boat. She was 19 ft. beam 7 draft aft and 5 forward and 210 overall. She was filthy black—no bright work anywhere; and covered with oil and coal dust—a turtleback forward to turn the worst of the seas: a conning tower plated with $\frac{1}{2}$ inch steel to turn rifle fire: but her skin was 3/16 of an inch everywhere else!... We pulled out of the Medway into the mouth of the Thames at an easy twelve knots to get down to our course... A lumpy sea and a thirty knot breeze. Then I was introduced to George Brown—Thornycroft's head man, who had attended more than 2,000 trials! He had a goatee beard and a head like a Yankee: a born engineer. We talked about steam trials.*
>
> *'Yes' said George Brown, 'we've had every damned thing happen to torpedo boats that could happen. We've shed our propeller blades: we've twisted our rudders off: and we've next to waltzed the engines off the bed-plates.'*
>
> *By that time we'd freshened up to 17 knots—jogging along easily. They wrapped my neck in a comforter, gave me heavy oilskins, and tied a sou'-wester over my ears. The wind was pretty keen and now and again the top of the sea came aboard... They began to rig up the indicators, to know how many revolutions we were doing and I went into the engine room. Two*

engines of 3,000 h.p. apiece were making about 230 to the minute. Our stoke hold was open.

Then I heard someone say 'We'll shut down as soon as you say sir' and they screwed down the stoke-hold hatches and a fan (700 rpm) began to pump forced draft into the fires. Then the Captain said 'Let go' or words to that effect, and do you know the feeling of standing up in a car when the thing starts up quick. I nearly fell down on the deck. The little witch jumped from 22 to 30 like a whipped horse—and the three hour trial had begun.

It was like a nightmare. The vibration shook not only your body but your intestines and finally seemed to settle on your heart. The breeze along the deck made it difficult to walk. I staggered aft above the twin screws and there saw a blue-jacket, vomiting like a girl.

All we could do was get under the lee of the conning-tower and hang on while the devil's darning needle tore up and down the coast ... The wake ran out behind us like white hot iron; the engine-room was one lather of oil and water; the engines were running 400 to the minute; the gauges, the main steam pipes and everything that wasn't actually built in to her were quivering and jumping; there was half an inch of oil and water on the floors and you couldn't see the cranks in the crankpit.

Just for the fun the skipper said 'we'll take her over the mile', that is marked by the two red Admiralty buoys—and is the official mile for all the ships of the Navy. The first time we had the wind at our back so I wasn't blinded... we covered the mile in 1 50.5 or something over 32 knots to the hour.

Then we turned round. We faced into that 30 knot gale and for the honour of the thing I had to stay up on the bridge. This was pure hell. The wind got under my sou'-wester; and I was nearly choked by the string round my throat. But we did the mile in the face of wind and tide in 2.5 - 6 or 8, the timings did not agree. Then we went on till we all turned white with fatigue. At last those awful 3 hours came to an end, but not before the speaking tubes to the Captain's bridge had been smashed off by vibration.

Well then we jogged back to Sheerness at 20 knots an hour. We were all as black as sweeps; and utterly played out. It took me two days to get the 'jumps' out of my legs.

11.2 Planning a turbine powered ship

Parsons faced two tasks in trying to develop marine applications for his invention. First he had to design a steam turbine that was suitable for driving screw propellers efficiently. Then it was necessary to make an appropriate vessel in which to install it. One obvious possibility would have been to have Yarrow or Thornycroft design and build the hull and boilers for him. But their shipyards were far to the south, on the Thames. In the end Parsons took on himself the whole task of designing a suitable shape of hull, and overseeing its construction on the Tyne. This greatly facilitated carrying out the various modifications that would inevitably prove to be necessary.

Rather than deal with shipbuilders he chose a firm of sheet metal manufacturers, Messrs Brown and Wood, to build the hull. Their site at Wallsend was conveniently situated a couple of miles to the east of the Heaton Works and it had a slipway on the river bank. Manufacture of components that could not be made at Heaton was subcontracted out, like the boiler drum that was ordered from Hawthorn Leslie and Co. Parsons' interest in the sea was reflected in his choice of friends. Among these was a wealthy young landowner from Northumberland, Christopher John Leyland (1849–1926). He was born C J Naylor and was five years older than Parsons. When he inherited the wealth of the Leyland family of bankers, he adopted Leyland as his surname. As a youngster of 13 he entered the Navy in 1862. When his naval 'education' was completed, he retired as a sub-lieutenant in 1872 having already seen service in China and the Pacific. Their friendship was close and he was chosen to captain the test vessel when it was built.

No doubt Parsons had been giving thought to suitable shapes for the hull of a high speed vessel for a number of years. While he was a child, the racing performance of the *Titania* had been the subject of much discussion during family holidays on the Isle of Wight with its designer, the renowned naval architect, Scott Russell. Judging by Parsons' remarks to John Bell Simpson, design work must have begun in earnest around 1892. Ideally, from cost considerations, his boat had to be no larger than was absolutely necessary to demonstrate the concept effectively. Yet it had to accommodate a bulky coal fired boiler. The smallest turbine that was likely to demonstrate a reasonable efficiency was quite large; in any case the power required to propel a boat rises roughly in proportion to the cube of its speed. The vessel that he planned was to be broadly the size of Thornycroft's torpedo boat, but would be very much faster. As a first step in design, he built a number of models for a hull (Stoney 1931). He was evidently very familiar with the work of the Froudes, father and son. Besides he had had his own experience of building the little boat powered by a form of air turbine (figure 32),

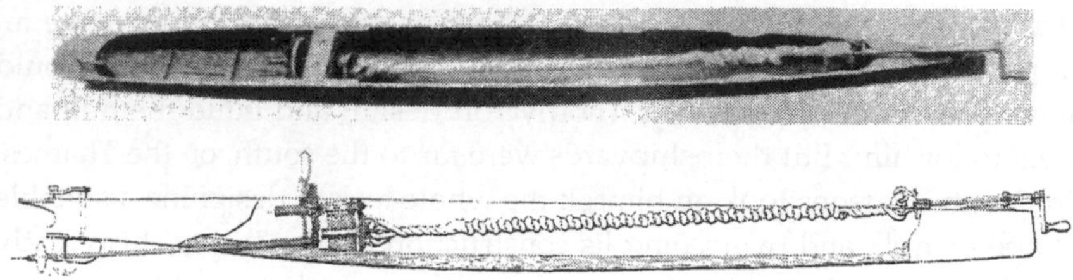

Figure 58. Photograph of 6 ft long model for *Turbinia*. Below it there is a sectioned view showing the rubber motor, the gearing and the propeller shaft. Note the distinctive shape of the stern (Richardson 1911).

and of carrying out tests on his own torpedoes some years before. The Scottish shipbuilders William Denny and Sons of Dunbarton had built the first commercially owned test tank for models in 1881, based on Froude's design, but Parsons had to make do with something much more modest. His first small model was 2 ft long, which represents a scale of roughly 50 to one, and is held in the Museum at Newcastle upon Tyne.

The first mention of tests is of those that were conducted in 1893 in a pond in a quarry near his home in Ryton (Richardson 1911). Here he towed the model with a fishing line. With the objective of minimizing the wave resistance encountered at high speeds he chose a very slender hull form, with a length to beam ratio of around 11:1[3]. The stem had a knife-like slenderness, and the cross section filled out at mid-ships to a U shape, before tapering again to the stern.

One of the first results to emerge from his tests was the discovery that a rounded stern was unsatisfactory, and the model had a tendency to 'sit up'. As a result he changed to the flat stern adopted by Thornycroft, perhaps at the prompting of Leyland. He was not satisfied merely to tow his models. He also equipped them with clockwork gearing, that drove a propeller having a diameter of $\frac{1}{2}$ inch, with a pitch of the same amount. The power was supplied by strands of rubber, following the example of contemporary model aircraft experimenters (see figure 58). Self-propelled models are commonplace now, but at that time they were a novelty.[4]

[3] Jung points out that *Turbinia* was the first ship ever to attain a Froude number greater than unity. This meant that at full speed she moved on the crest of a wave system that, from one crest to the next, was six times longer than the length of the hull (Jung 1982).

[4] Before his death in 1883, Tideman, the Chief Constructor of the Royal Dutch Navy, had used an old torpedo motor powered by compressed air to drive his models (Moore 1984).

Froude had demonstrated that a model would generate waves of the same shape relative to the hull as the full scale vessel, if it were driven at a speed proportional to the square root of the model scale. This model could reach 6 knots at a propeller speed of 18 000 rpm, which therefore represented a real speed of approximately 42 knots. It was reassuring that a screw propeller could function effectively at such an incredible speed, even if it was of miniature size. On the other hand it may have been responsible for misleading Parsons in regard to the way the full scale propeller would behave. The Admiralty tank had been transferred to Haslar Creek near Portsmouth in 1886, but Parsons at this time had no direct access to this facility which was intended to serve only the Navy's demands. Nevertheless his extraordinary experimental skill allowed him to extract quantitative data from his own models. He was sufficiently pleased with the 2 ft model to construct one 6 ft in length. This gave sufficient accuracy to guide his interpretation of the results of trials of the full sized boat. The rubber motor of the larger model drove the propeller at 8000 rpm. He was able to establish its power output, net of friction losses, by replacing the screw propeller with a fan. The fan was then surrounded by an enclosure which was dragged around by the resulting current of air. The torque on the enclosure could be measured by a lever arm, and from this and a knowledge of the speed, an estimate of power was made. To obtain a measure of the speed of the motor he made use of an observation which he had made. He had noted that as the rubber motor was wound up, a series of knots appeared along the length of the rubber. Further twisting caused 'double' knots to form, and when these had extended the full length, 'triple' knotting developed. He counted the number of revolutions which the propeller made as the rubber bands unwound from a state of 'triple' to a state of 'double knotting' and noted that these states could be reliably observed and that the number of revolutions made during this interval was quite repeatable. He could then compare the distance travelled by the vessel through the water during this interval with the pitch length of the screw propeller multiplied by the number of revolutions made. The difference between these two lengths gave an estimate for the 'slip' encountered by the propeller[5]. This is an important indication of the behaviour of the screw.

At some later stage, between 1895 and 1896, he also used a pond at the Heaton Works to carry out towing tests. These were a good deal more accurate than the ones carried out at Ryton. The model was towed by a

[5] See also appendix 1.

length of salmon fishing line which was wound onto a pulley. The pulley was driven by a falling weight. As the weight descended, its supporting line unwound from the shaft, so rotating the pulley attached to it. Two weights were used: one which had a short drop brought the model up to speed quickly, the other weight powered the main run. Two metal tags were placed 30 feet apart on the tow line. By timing their passage past electrical contacts, the speed could be measured. Essentially the same arrangement had been used by Tideman in 1875 to test a model of the *Atjeh* at the Amsterdam Dockyard (Moor 1984).

While his test program was still not complete, Parsons was sufficiently satisfied with the lines of the boat that he began work on the detailed design of the hull, its motor and boiler. In each of these he followed existing developments, though pushing them to new limits. The hull was built of thin steel plate which ranged from 3/16 in (4.75 mm) for the bottom, to 1/16 in (1.6 mm) near the stern. It was divided into five spaces by watertight bulkheads (Parsons C A 1897a). Although the radial flow design of turbine would be abandoned as soon as he recovered the 1884 patent rights, still it was a proven design, and, more to the point, it had already been adapted for use with a condenser. For marine applications a condenser is essential, because it preserves the supply of fresh water needed for use in the boiler. Besides the multi-disc design of the radial flow machine had other advantages. The mean diameter of the rows of blades was larger than the early axial flow designs, so that a lower shaft speed could be used without letting the blade velocity fall too low relative to steam velocity. In addition the rotor assembly was much stiffer than the earlier, and more slender shafts of the axial flow machines. This kept the critical speed well above the working speed and simplified the task of keeping the clearance at the tips of blades to a minimum. Model tests suggested that a speed of 30 knots would require an effective power of 820 h.p. As a rule of thumb, the propulsive efficiency, that is the ratio of useful propelling power to engine output, was taken to be 50%. The boiler and turbine were therefore designed to deliver 1650 h.p. at a shaft velocity of 1600 to 1700 revolutions per minute (Cook 1939). This was easily the most powerful steam turbine yet built. In metric units, its output was 1.24 MW. To assist him with this venture he was able to use employees who had already proved their ability while working with him, men like Robert Barnaby who carried out the design of the hull, and J B Willis, head draughtsman, and A Wass who designed the turbine. Barnaby also took charge of the construction when it began (Richardson 1911). Charles justified his decision to take the whole responsibility for the design on himself in a letter to his brother (C.L.(10)),

Elvaston Hall
Ryton on Tyne
2 Feb 1894

My dear Laurence,

In regard to the matter of associating engineers and shipbuilders in the experimental enterprise I think in regard to the latter that with the staff we have we shall not make any serious blunders on the hull, at least I hope not. Our foreman and leading draughtsman have both been employed in similar work before, and I am able to check their calculations of weights and strengths. In regard to the engineering part, outside the motor, air pumps etc there only remains the boiler, screw propeller and shaft. The first we are following Yarrow's most recent practice which is almost identical with Gurney's steam carriage boiler of 1850, or about then, and subsequently patented by several different persons.

Then as to the screw, conditions are somewhat uncertain, but we are following the lines only making it a wee bit finer pitch. The screw shaft is also run under unusual conditions of speed but I hope the ordinary bearings lengthened will fulfil the necessary conditions. If not we can put elastic cushions around them. When, if successful we go to a big Co^y^ there will require to be a large skilled and highly paid staff.

Your affectionate brother,

Charles

Parsons worked through the design calculations for the hull in a notebook which has survived (Parsons C A 1893). To quote Osler and Grieve (1980a),

> *In conclusion it must be said that the overall impression is one of a man who is learning a new art by the simple expedient of going over the work of others, either to follow their reasoning or to satisfy himself of the validity of the design. This impression comes from the fact that so many of the calculations are continued long after changes have taken place in the design. One can almost hear Sir Charles Parsons saying 'Where did he get that from?' and 'How did you go about it?' before returning to the office to work out the results for himself. Bearing in mind that the hull was just a vehicle for his engine the record of dimensions, weights and form kept in the notebook show that he was totally absorbed in the overall project to the last detail...*

They also note that *'All calculations were reduced by the simplest means, there are no logarithms in the book'*.

There are few dates but what there are range from 3 October 1893 to 15 December 1893. Two variants were worked up, each for a length of 100 ft, but one having a beam of 8 ft and the other 9 ft. In the end it was the latter which was used. Meticulous calculations of weight were made. As well as estimating the weight of the steel plates and frames, and of machinery and fittings, an estimate was even made of the weight of wood, of different densities, and of the paint required. From these the displacement, the wetted surface area, the centre of gravity and the centre of buoyancy were derived. In her final form, Parsons itemized the total displacement of $44\frac{1}{2}$ tons as made up of 3.65 tons due tọ the engines, 18.35 tons for the remaining machinery, boiler, screws and shafting etc, 15 tons for the hull and 7.5 tons for coal and water (Parsons C A 1897a). In addition to the notebook, a large number of drawings have survived (TWAS 927). As originally built, a single motor drove a single propeller of 20 in diameter. The turbine generally resembled the condensing machine tested by Professor Ewing for Cambridge, but it had ten times the power output and only one third of its speed. The reduced speed allowed the fly ball governor, the air pump and the boiler fan all to be driven from the turbine shaft, either by gearing or, in the case of the fan, directly.

Parsons had had a great deal of experience building and designing steam turbines and air pumps. Besides, Gerald Stoney was already an able lieutenant who was by now quite capable of putting Parsons' ideas into practice. But the construction of a boiler was less familiar territory. Both Yarrow and Thornycroft had moved away from 'fire tube' boilers in which combustion gases passed through tubes immersed in water, to the 'water tube' type in which the gases passed around the outer surface of water filled tubes. The latter contained far less water and this made for quick starting, but care was required to ensure that there was a balance between the water available in each tube and the heat in the combustion gases flowing round it. In locomotive practice a blast of steam in the chimney was used to draw air through the boiler furnace. The resulting loss of pure water in the steam jet was not acceptable at sea, and so a fan had to be used to produce the same effect. The fan raised the pressure in the entire stokehold by several inches of water gauge. Since the fan was directly coupled to the turbine, the faster the vessel went the more intense the draught became and as a result during high speed runs the stokers struggled to maintain a sufficient supply of coal in the grate (Parsons C A 1894). Originally a superheater was located in the funnel to raise the steam temperature above boiling point,

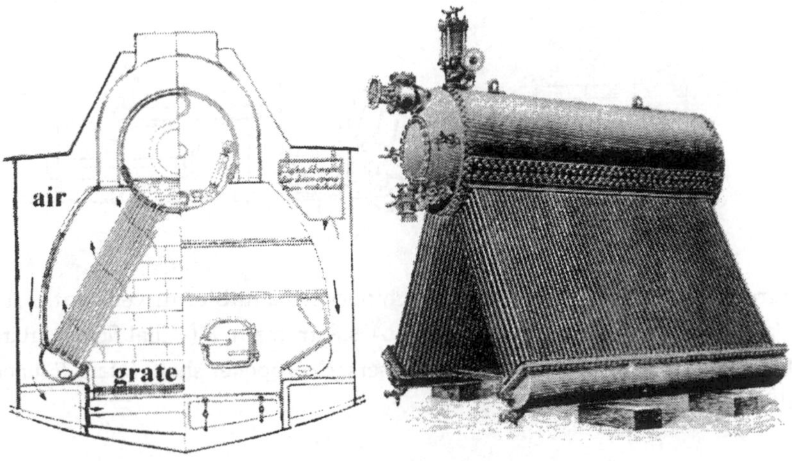

Figure 59. A Yarrow water tube boiler with on the left a cross section and to the right a view of the drums and tubes awaiting installation. Air is led down to the grate through a space in the casing. The combustion gases traverse the two inclined banks of tubes in which water is vaporized. Steam is drawn off from the header drum at the top where it accumulates (Perry 1900).

but it had to be removed because the temperature of the combustion gases was found to vary too widely under different conditions. During high speed runs flames streamed from the funnel, scorching the surrounding paintwork.

He had just begun to build high performance condensers for his land based turbines, and there was still much to learn about the correct design principles. One advantage was that the sea provided a copious supply of cooling water for the condenser tubes. Whereas in land based installations pumps are needed to circulate cooling water, this complication was avoided at sea by using an adjustable scoop mounted beneath the hull and facing in the direction of travel, to perform the same duty. This could be adjusted to reverse the direction of flow should it be necessary to remove any fouling or debris. There is a fragment of an undated letter to Laurence which must have been written before February 1894 (C.L.(17)),

Elvaston Hall
Ryton on Tyne
late 1893 or early 1894

... *experiment with a centrifugal pump with a 10" fan* (i.e. impeller) *at*

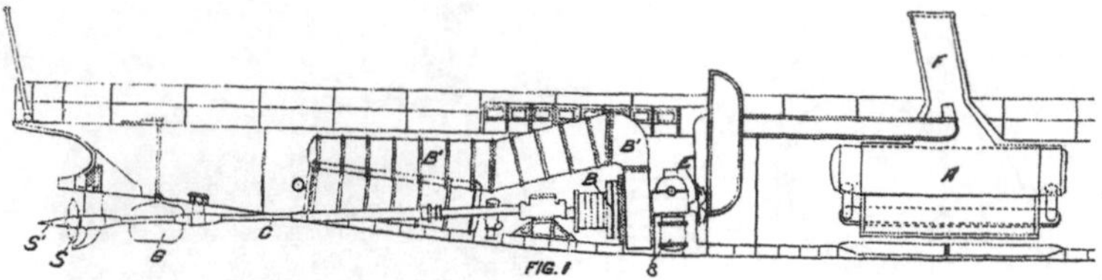

Figure 60. General arrangement of machinery for a turbine driven boat as shown in Parsons' patent 394 of 1894: *A* boiler, *A'* steam main, *B* (radial flow) turbine, *B'* exhaust ducting leading to twin condensers, *C* propeller shaft, *E* fan, *F* funnel, *G* rudder and *S* screw.

> *4,500 revolutions which works perfectly and gives a very high efficiency (viz. 63 percent) for so small a pump. It seems probably that the vacuum difficulty will not turn up and that most of the pressure of this blade on the water is done by pressure and not by suction.*
>
> *I think I mentioned in my former letters that we expect to get 30 to 40 percent more work out of the steam than with ordinary engines.*
>
> *There will be 8 miles of 5/16" condenser tubes and about 2 miles of $\frac{1}{2}$ brass and steel tubes in the boiler.*
>
> *There is nothing novel about the boiler or condenser though they are both somewhat modified to suit circumstances from the ordinary torpedo boat practice.*
>
> *Your affectionate brother,*
>
> *Charles*

The trial with the centrifugal pump brings to mind Clere's paper on the subject. Stoney recalled that while he was observing such a test, so high was the pressure which was developed that the pump casing burst and he was drenched in water (Stoney 1938). In the letter, Charles was comparing the pump impeller with a screw propeller where the suction at the inlet is comparable with the pressure rise at the outlet.

In January 1894 he applied for a patent, number 394, for '*Improvements in mechanism for propelling and controlling steam vessels*'. It contains, among other material, details of the boat design up to that moment. Among the figures is one that shows a general arrangement of an installation which is essentially the one which was used for the planned vessel (figure 60). The sectioned view shows the boiler and funnel amidships with the turbine aft of this. Not shown are the coal bunkers located at either end of the boiler. Ducting, left and right, leads the exhaust steam to the condensers which are located on either side of the propeller shaft. There is a single propeller and an offset rudder. The 41 ft long propeller shaft was 3 in diameter, which is relatively slender. He planned to drive it directly by the turbine, so that its speed would reach perhaps 2500 rpm, which was very fast for such a long shaft. As has been observed Thornycroft's vessels had engine speeds that were only a fifth of this. From his letter of 2 February, he obviously anticipated that the shaft might have a tendency to 'whirl', but he had remedies in mind to deal with this possibility. Many other features are dealt with in the patent, such as suggested arrangements for reversing thrust to allow the vessel to go astern. In fact the first turbine, as installed, had a set of blades mounted on the rim of the last disc, with nozzles positioned in the casing in such a way that when supplied with steam, the propeller was turned in reverse. In normal use there was no supply of steam to the nozzles and these blades would rotate in what was effectively a vacuum, so there was little loss of power. The patent also contained a proposal for a shaft with four propellers mounted one behind the other, to deal with any tendency to '*produce a vacuum behind the blades*'. He evidently had misgivings about propeller performance even before tests were begun. He also describes a variety of damped, spring, supports to deal with shaft vibrations.

A major problem in screw propelled ships of all kinds is the need to handle the thrust created by the propeller and very massive bearings were required to absorb the large forces. This patent proposed that some of the thrust could be balanced by the net steam pressures on the discs and the balance (dummy) piston of the steam turbine. The complete specification was delivered in November 1894, which was after he had recovered his rights to use the 1884 patents, and patent 394 also contains details of an alternative, axial flow turbine.

Even as he was planning a suitable shape for the hull in 1893, Parsons was taking steps to assemble the capital needed to allow a demonstration vessel to be built. Although the construction of land based turbines was making progress, that side of the business was not capable of subsidizing the marine venture. From the start he sought the help of his elder brother in recruiting share holders in the company which he was planning and Charles asked him to act as Chairman of the Board (C.L.(4)).

Elvaston Hall
Ryton on Tyne
27 March/93

My dear Laurence,

All the people asked have now agreed to join the board, George Clayton has agreed and takes 20 shares all the others take 10 each. George Clayton you perhaps remember was a lawyer and came in to 2 million or more a couple of years ago. His joining is I think very satisfactory, his opinion carrying considerable weight.

I have spoken to Simpson since writing to you and he will with much pleasure act if you wish as your representative as chairman in your absence. He is a man you can implicitly trust. You know him as a genial person ready with his jokes, but he is a sound common sense businessman of great firmness when he thinks it necessary and would be a most suitable person for the position.

I have in all cases pointed out that there is considerable risk, but that I consider the matter as sound as can be seen without trial on full-sized scale, so that no one can blame if anything should happen.

They are all apparently very keen and think the venture at the present time (when fast speed is the rage) very opportune.

We hope to have a meeting in about ten days time to settle the prospectus to send to certain people and not the public—to ask for the 250 less the 70–180 shares yet required. It would seem there will be no difficulty now in getting the rest taken up. I must say that the way people have taken up the venture is a surprise compared with (electric) *lighting companies I have had to do with, where the directors have grudged taking shares to the extent of £250. At Cambridge we fixed £500 as the Directors qualification and we were obliged to reduce it to £250 in order to get Cambridge directors.*

In Britain at least the demand for electricity was still weak, but anything relating to naval activity was a quite different matter! The second half of the century saw a great surge in colonial adventures throughout the world. In response to this the British Navy, which was already the world's greatest in terms of size, was being given an extensive upgrading. Newspapers of the day regularly carried letters from correspondents who feared that the 'continental powers' were overtaking the British lead, and called for increased investment in the navy. In 1891 Lord Brassey read a paper to the Institution of Naval Architects on the 'Future policy of warship building' in which he stated that official government policy was to keep the

British Navy at a strength at least equal to that of any other two (continental) powers combined. While the tables which were appended seemed to show a satisfactory position, *'the British Navy at the end of the 19th century though numerically a very imposing force, was in certain respects a drowsy, inefficient, moth eaten organism'* (Marder 1952). The navy scares in the 1880s led the government to embark on a huge shipbuilding program which had its beginning in 1889 when the Naval Defence Act was passed by parliament. One individual who was to figure prominently in this was Jackie Fisher who was moving rapidly upwards in rank. In 1886 he had been appointed Director of Naval Ordnance and in 1890 Rear Admiral. He was absolutely dedicated to the Service, and when head hunted to join Sir J Whitworth and Co, he turned down the offer of a directorship writing to a friend, *'If you want to know what God thinks of riches—look at the people he gives them to'*. In 1892 he joined the Admiralty as third Sea Lord responsible for all matériel. He encouraged Yarrow to *'develop* (torpedo boat) *destroyers which would be able to handle the swarm of French torpedo boats'* which he expected to be facing in the event of armed conflict. He sought to replace Lancashire boilers with water tube (Belleville) designs to improve fuel economy and be capable of a speedier start up from cold. A campaign was mounted to get politicians to support the building of seven new major ships in order to maintain equality with France and Russia.

11.3 The Marine Company

Such was the picture when Parsons established the Marine Steam Turbine Company in January 1894 (Richardson 1911). The board of directors which was headed by Lord Rosse comprised

- N G Clayton of Chesters, Northumberland
- Christopher Leyland of Haggerston Castle, Beal, Northumberland
- J B Simpson (deputy chairman) of Bradley Hall, Wylam on Tyne
- A A Campbell Swinton of 66 Victoria St, Westminster
- C A Parsons (managing director).

H C Harvey was solicitor for the Company, and assisted in its formation. At a later date Norman C Cookson, an industrialist of Oakwood, Wylam on Tyne, also joined the board. The authorized capital was £25 000. A private issue of 480 shares of £50 each raised £24 000. A prospectus was issued which set out the aims of the company.

> *The object of the Company is to provide the necessary capital for efficiently and thoroughly testing the application of Mr Parsons' well known steam*

turbine to the propulsion of vessels. If it is successful, it is believed that the new system will revolutionise the present method of utilising steam as a motive power, and also that it will allow much higher rates of speed to be attained than has hitherto been possible with the fastest vessels.

Up to the last five years it has been found impracticable to obtain economical results from a motor of the steam turbine class, though such motors, on account of their light weight, small size and reduced initial cost present great advantages over ordinary engines for certain classes of work.

Recently—and more especially within the last two years—the steam turbine has been developed and improved. It has further been adapted for condensing, and results have been attained which place its performance as regards economy among the best recorded. Reports have been made by the following authorities:- Professor J.A.Ewing F.R.S., Professor A.B.W.Kennedy F.R.S. and Professor George Forbes F.R.S.

It is confidently anticipated that with turbines of, say, 1,000 horse-power and upwards, having a speed of revolution of about 2,000 per minute, the consumption of steam per effective horse-power will be less than with the best triple compound condensing engines.

The initial cost of a steam turbine will be very considerably less than that of an ordinary marine engine of the same power. To these advantages must be added the consideration that the space occupied by the turbines will be very much less than that occupied by ordinary engines, thus largely increasing the carrying capacity of the vessel. The reduction of the amount of vibration admits of a diminution of weight of the hull, which under the present system must be built stronger and heavier than will be necessary under the new system, in order to resist the effects of the vibration of the present class of marine engines.

Another important feature is the reduced size and weight of the shaft and propeller. This will not only facilitate duplication and repair, and enable spare parts to be carried to an extent not hitherto practicable, but will also admit of screw propelled vessels being used for navigating shallow waters, where at present only paddle steamers can be employed.

The merits of the proposed system may be summarised thus:- Increased speed, increased carrying power of vessel, increased economy of steam consumption, reduced initial cost, reduced weight of machinery, largely reduced vibration, and reduced size and weight of screw propeller and shafting.

The efficiency of the screw propeller, the arrangements incidental to the adoption of higher speeds, the best form and the proportions and mounting of the propeller, the material of which it should be made, and other points, can only be decided by investigation and practical experiment; and it is to provide funds for the complete and exhaustive testing of the new system in these and other respects that the Company is being formed.

The new company was quite distinct from Messrs C A Parsons of Heaton. It was therefore essential that the former should receive an exclusive right to use, for marine purposes only, the patent 6735 of 1884 and other related patents which now belonged once more to Charles A Parsons. In consideration of this, the new company allotted him £9000 worth of fully paid up shares. His intellectual property therefore made him the largest shareholder, and constituted 3/8 of the total nominal capital of £25 000 (Anon 1898). At the same time as the Company was being launched, Parsons had of course taken out patent 394 of 1894 which dealt with many features associated with ship propulsion.

As a member of the House of Lords, Laurence was well placed to approach other members of the nobility with capital to invest (C.L.(7)).

Elvaston Hall
Ryton on Tyne
26 Jan /94

My dear Laurence,
It would be admirable if you could get Lord Brassey and his brother to join the C^o^. Also it occurs to me that the Duke of Devonshire who is largely interested in the Barrow Docks and shipping would be a great acquisition.

At the present times there are about 64 shares remaining but it is always open to the directors to refuse applications should that be necessary in order to take in some necessary or desirable persons who would benefit the C^o^ as a whole by their influence.

The C^o^ was registered yesterday. I enclose several copies of the circular in case you may want some. The plates and angles are all in now and the building operation will soon commence.

Your affectionate brother,

Charles

Lord Brassey was an Oxford graduate, DCL, and held various government appointments, many of them associated with the navy. From 1893 to 1896 he was President of the Institution of Naval Architects, it being a quaint custom of that Institution that it chose for the office of president men who were not themselves full members, i.e. naval architects, but were nonetheless all of them peers of the realm (Barnaby 1960). Another letter followed quickly (C.L.(8)).

Elvaston Hall
Ryton on Tyne
28 Jan^{y} / 94

My dear Laurence,

There are now 434 shares applied for and in addition Johnstone Stoney has applied for 10. I think it is suitable for him to do so and I said so to Gerald Stoney. He would probably be offended if we refused his application but we might allot him a less number he having applied for the stated minimum.

It is rather curious that John and Ted Purser and Griffith have applied. It appears that old Purser left a considerable sum of money and Griffith being G.Stoney's master at North Wall told them of the scheme. I think it would be a good plan for the future C^{o} which we hope to form later to have Ld Ardilaun now and especially Ld Iveagh now. The fact of the Pursers and Griffith going in to the thing should carry weight with them. I do not think Bramwell would go in and I think we better keep clear of Ld Kelvin, he might be a bother at the present time.

In some of our original patents lately bought back, subdivision of the motor into several or separate shafts is described[6]*, and I think claimed, and one of the two patents taken out this month is primarily for the combination of a high pressure ordinary engine working one screw, the steam exhausting to a low pressure turbine working another screw on its way to the condenser. These arrangements seem more suitable to a large ship than a small boat and when we come to large ships as we hope to do, they should apply largely.*

I have several times tried to arrange turbines working opposite ways. In about 1866 (?1886) a man contacted us to make one work with compressed air for a Whitehead torpedo[7]*. It looked rather complicated when drawn out. The fact of the case having to ? revolve with tubular shafts and concentric stuffing boxes makes things appear more complicated.*

Perhaps if you thought fit you could ask Ld Ardilaun and Ld Iveagh.

Your affectionate brother,

[6] Patent 6735 of 1884 *'The motors or successive portions of the compound motor, may be arranged either upon one common shaft or upon different shafts'*.

[7] The Tyne and Wear museum possesses a small geared turbine said to have been intended for torpedo propulsion (figure 88).

Charles

I sent you a couple of letters to 15 Jermyn St. I hope you got them.

Lord Iveagh and Lord Ardilaun were both members of Guinness, the Irish brewing family. Johnstone Stoney was the father of Gerald Stoney, and a man with a considerable scientific reputation. Griffith was the assistant to the Chief Engineer of Dublin Port at North Wall, who was Bindon Stoney, a brother of Johnstone Stoney.

The reference to arranging '*turbines to work opposite ways*' relates no doubt to the wish to have contra-rotating propellers, which are more efficient and have less tendency to set the torpedo spinning in the opposite sense to the propeller. There is a notebook which covers a number of topics, among them calculations for a 14 in Whitehead torpedo to be driven by a compressed air powered turbine (Parsons C A 1881b). The entry is undated.

Three patents were applied for in January 1894. Number 367 was for the combination of a standard reciprocating marine engine and a low pressure turbine. At that point in time of course there was no evidence at all that a steam turbine could successfully drive a screw propeller efficiently. Number 394, as already stated, covered the application of the steam turbine to ship propulsion as realized in *Turbinia*. It contained 27 claims in all. Another patent, 394A, described the use of a centrifugal fan attached directly to the turbine shaft, and the application of the resulting forced draught in various ways to a steam raising boiler. A couple of days later Parsons wrote on company paper (C.L.(9))

C A Parsons & Co.
Electrical Engineers
Heaton Works (London Office: 66 Victoria St)
30 Jan 1894

My dear Laurence,

It seems to me that it may be as well as you say to have Ld Ardilaun and Ld Iveagh for the present. There are several people, in fact the majority about here who have not been approached in any way and there will be a few possibly who may wish for shares and we may take in later.

In regard to the boiler this has been very carefully considered in conjunction with the available data I can find, and I have carefully calculated the proportions of steam and water in the tubes nearest to the fire and find that this cannot be less than 1/3 of the volume (of the) water under utmost forcing. The same calculations give for the Admiralty boilers of ordinary construction 1/7 water ?only? near the tube plates moderately forced. I do not anticipate any expansion from difference of temperature in the several rows of tubes.

Yarrow's boilers are all straight tubes and he is building a large number of them and they have given good results.

The difference in temp're in the ?? condensers amounted to from 32^0 to 212^0 in winter weather according to the load, and starting and stopping. This gives stresses beyond the elastic limit of copper and hence the rupture.

In the case of a boiler with good circulation the temperature will remain approximately equal throughout.

In Lancashire boilers there is no circulation and the bottom often remains cold after steam has begun to rise but when evaporation is set up by drawing off steam, all is mixed up. Heavy strains are set up but no trouble practically is experienced.

I consider curved tubes should be avoided in a water tube boiler as being difficult to clean and replace. However Thornycroft has patented a curved tube boiler for which he claims good results, but Yarrow's boiler appears to have given better results for the same weight, which I think is the point we have to consider at the present time. And I think we should also avoid as far as we can all experiment excepting the main issue of increased speed. It appears Yarrow's form of boiler dates back before 1865, and Gurney in his steam carriages had a similar boiler but without a surface condenser the small tubes soon burn out.

Your affectionate brother,
Charles

It is blowing hard today. I tried the model in a quarry close by, with comparatively big waves it cut through them like a knife. The screw did not appear to draw in any air, it seemed to go first rate and nearly as fast as in smooth water.

Around this time he wrote to R E Froude to tell him of his worries that cavities would form in front of the propeller blades, that is on the front surfaces (Goodall 1942). Froude in his reply agreed that cavitation and the possibility of air being drawn in to the propeller were dangers to be guarded against.

Business was now picking up, and he was looking into ways of increasing his manufacturing capacity quickly. Evidently a small turbo-generator had been built for Birr Castle but had been returned to Newcastle for attention (C.L.(11)),

Holeyn Hall
Wylam on Tyne
4 Feby /94

My dear Laurence,

Your little Turbo is now gutted and a new inside is being put into it. Work is slow as we have more than we can get through in the time, indeed we have some (£)19,000 worth which ought to be out by July but our present capacity means some 8 to 10 months work. There are nearly as much again of orders which would be given if we could meet them. We are therefore looking about to get some arrangement to increase production. I have at present two alternatives before us,

(1) To get machines made for us at Greenwood and Batley of Leeds.

(2) To take over Palmer's Ordnance works at Jarrow on Tyne and start a limited Coy to run the whole business.

No 1 is the least trouble but No 2 is the bigger scheme. The Ordnance works were built 4 years ago and partly fitted up with magnificent tools which would mostly work in for Turbos. They have made only one small gun (the whole having been mismanaged). If Palmers will agree reasonable terms we could put some £20,000 worth of Turbos in hands there and about 1/3 occupy the Works. We are trying to get the place on a rental for say 3 years with the option to purchase at a fixed price. Of course the place is worth only scrap value and the lathes which are all new would in part, work in. However the matter has not yet proceeded very far and requires further investigations - the Leeds labour market being more favourable than in the Tyne.

Your affectionate brother,
Charles

Charles Parsons could hardly have stated his objectives and intentions more publicly than by his action in persuading so many very prominent men to support his plans financially. Those plans were very clearly set out in the company prospectus for all to see. Failure would be all too visible should it ensue. Reflecting on his position in 1894, one is struck by the strong parallel with that of his father when his plans for the building of the 72 in Leviathan were carried in the September issue of the *Illustrated London News* of 1843. At that stage, no more than his son, could William Parsons have foreseen the success that would eventually reward such a huge gamble.

11.4 *Turbinia* launched

During 1894 construction work proceeded swiftly. On 2 August 1894 the *Turbinia*, as the vessel was named, was launched with some difficulty, from the slips at the Works of Brown and Hood. C J Leyland recalled (Leyland 1935)

> *Turbinia showed no anxiety to get a move on when we essayed to launch her and she had to be pushed off the slips. However, later on, she made amends for her unwillingness to take to the water.*

At 3 ft $5\frac{1}{2}$ in draught, her displacement was estimated at 44 tons.

There are two main sources of information about the trials performed with the *Turbinia*. One is Parsons' own notebook (Parsons C A 1897b). We are also fortunate that a considerable number of the letters from Charles to Laurence Parsons during this period have been preserved. In them he describes progress made from day to day in dealing with the difficulties which he faced in transforming his ideas into reality. The letters reveal much about the man and his approach to problems, and they will be quoted extensively. He wrote of the initial run (C.L.(12))

> *Holeyn Hall*
> *Wylam on Tyne*
> *3 Oct 94*
>
> *My dear Laurence,*
>
> *We had a preliminary run with the boat yesterday to Tynemouth and back. The boat travelled very smoothly indeed and with 60 lb^s steam and 9 lb^s vacuum the speed appeared to be between 18 and 20 knots. We could only run for a few minutes at a time owing to want of draught. And in so short a space of time the main air pump could not sweep out the air thoroughly.*
>
> *So far the screw appears to behave perfectly—that is up to 1,200 revolutions. The (boiler) draught will soon be remedied and pending some alteration to the fan we are putting a temporary steam jet in the chimney. We are also doubling the speed of the main air pump by a quicker pitch worm.*
>
> *The condenser and pipes are now nearly air tight and I hope soon to get a proper vacuum of $14\frac{1}{4}$ lb^s. Theoretically the power we developed was some 300 horse and this agrees with the corresponding speeds obtained with ordinary boats with similar powers. I think on the whole the first run is quite satisfactory. The thrust of the propeller appeared to be about balanced by the steam thrust and there was almost complete absence of vibration.*

We shall now work things gradually up and so far there appears no reason to expect much further expenditure in the experimental stage, though it may be too soon to give a reliable opinion.

Your affectionate brother,
Charles

Events were to show that his optimism about avoiding the need for extra expenditure would prove to be badly misplaced. There is a fairly straight stretch of the Tyne between Wallsend and the river mouth. Although the legal speed limit for steamships was 7 knots, permission was kindly given by the Tyne Improvement Commissioners for its use in runs at speeds at up to 31 knots when the weather precluded runs on the nautical mile in the North Sea (Parsons C A 1914). One month later he reported (C.L.(13))

Holeyn Hall
Wylam on Tyne
6 Novr 94

My dear Laurence,

I found that the boiler tubes were perfectly choked with soot, it not having been swept since the start and steam had been got up 8 times—this was enough to stop the draught. We shall have another preliminary run this week after a run at anchor tomorrow.

I am arranging to double the speed of the main air pump and add a compound cylinder of larger size to the auxiliary air pump. These latter alterations will take a fortnight to three weeks, but meantime with wider (fire) bars and a clean boiler I suspect we shall get a good run this week.

As far as can be estimated we got 18 knots with 300 to 400 horse, which agrees closely with results of ordinary torpedo boats. The boat having lain for 3 months since the launch, has long weeds on the bottom reducing the speed probably 2 knots.

Your affectionate brother,
Charles

Speed trials were begun on 14 November. He wrote (C.L.(14))

Holeyn Hall
Wylam on Tyne
26 Novr 94

My Dear Laurence,

Many thanks for wishing me to come and I should very much like to do so if I could get away but we seem to be getting busier every month with Turbine electrical work, and now have just twice the number of hands we had last year, and I fear there is not much prospect of getting any distance at present. We rather expect to get several 120 h.p. turbines ordered for Guinness Brewery. I might go over later to see Geoghan.

We propose to have a director's meeting to consider the Foreign Patents for the Boat and some other matters of minor importance on Decr 4th and I hope to have all present alterations completed by then, and to have run at over 20 knots. And if all goes well we may get 30 knots or more on 1,000 to 1,200 h.p. When we get all the minor parts right we shall then be able to get the power developed-

We have only burned 12 lb^{s} of coal (per hour) per square foot of grate on a draught equal to a 30 foot chimney only. We shall work up to 40–60 and perhaps 80 lb^{s} per sq^{r} foot on locomotive draught. I believe priming is the only limit to the steaming powers of the water tube boiler.

You seem to have done very well in Woodville and also the grounds. I was out shooting at Thomas Spears on the N.Tyne last Saturday and we got 95 pheasants, 40 rabbits and some grouse and black game and 2 wood cock. I think it is the best I have had this year.

Your affectionate brother
Charles

P.S. I fear I shall not be able to get to uncle Laurence's funeral, the family will be adequately represented without me.

'Priming' refers to conditions in which the throughput of steam is so great that the separation of steam and water in the boiler is incomplete, and some water is carried into the turbine with obvious dangers for the blading. It is clear from the catalogue of slaughtered wild life that the pleasure that he experienced in hunting as a child still remained his principal form of relaxation. Laurence was his father's brother. A couple of days later he wrote (C.L.(15))

Holeyn Hall
Wylam on Tyne
28 Novr 94

My Dear Laurence,

In regard to the boat, we propose to have a meeting on Dec 4th. I presume that only Swinton and Simpson will be present as Leyland will be in Scotland. Simpson wishes for the meeting and the business will be to confirm the arrangements we have made for taking out Foreign Patents which includes all the most important Foreign Countries.

The alterations which I hope will be completed by Decr 4th are,

1. Working the main pump independently by adding a 6 inch steam cylinder[8]*. This will give the pump a speed of 350 revolutions per minute against 36 average on the last trials.*

2. A new fan of larger capacity to deliver air into the stokeholds and air pipes for delivery into each stoke hold.

The air leaks are small imperceptible leaks all over the exhaust pipes and system, and not so much from the glands round the spindle. It is difficult in a light structure to keep all the joints quite tight and I think we originally speeded the pump to(o) *slow.*

All these alterations are very small matters but it is difficult to get many men working on them at once.

Your affectionate brother

Charles A. Parsons

P.S. I think we may shortly anticipate a good speed. Enclosed is a cutting which appeared last week written by some busy body.

This letter contains several interesting pieces of information. It makes clear the importance which Parsons attached to securing the protection of patents world-wide. Also we can see that the launching only marked the start of a program in which the whole system, turbine, condenser, boiler, ancillaries and propellers were to be fine tuned. The boiler had a fire door at each end and there were two stokeholds. At full speed the task of keeping the fires adequately fuelled was massive. This was the first time ever that a condensing turbine was mounted aboard a ship, which brought its own problems of access etc. Finally it makes clear how essential it was that Parsons' research and development activity should be nestled within

[8] That is instead of driving the pump directly by a geared drive from the turbine.

a nurturing manufacturing environment, which could provide without delay the necessary resources, like a steam cylinder to drive the air pump independently of the turbine. This could only have been achieved, in England at least, by Parsons' personal ownership and direction of his own firm of C A Parsons & Co, Electrical Engineers.

Progress was reported again to the Chairman of the Boat Co. (C.L.(16))

Holeyn Hall
Wylam on Tyne
6 Decr 94

My dear Laurence,

The alteration to the inside of your little turbo is now set out and will be sent to you. We are arranging it with two diameters of turbines or compound, and expect it will take much less steam than before. A thrust bearing with collars at the end of the spindle will prevent liability of endwise wear.

We had a Boat meeting on Tuesday, Leyland, Simpson and Swinton attending.

The patents were discussed and the arrangements confirmed, they cover the principal foreign countries. Nothing else was done beyond signing the original contracts between C.A.P.C^{o} and the Boat C^{o} which were drawn some 8 or 10 months ago and for no reason left unsigned. I hope to have a trial run tomorrow or Saturday, and possibly Leyland and Swinton will be there on Saturday. We should now get a good vacuum and a good draught.

Your affectionate brother

Charles

The drawings for changes to the little turbine had been completed. While the winter weather slowed work on the boat, the demand for land based turbines was building up (C.L.(18)).

Holeyn Hall
Wylam on Tyne
9 Feb^{y} /95

My dear Laurence,

There is a Bachelor's ball in Newcastle on the 14th for which Lady Bazington and daughter are coming, so I shall probably be at home then and not able to come South.

If you have time to come we shall be delighted. The Boat is quite at a standstill. No work is possible in the frost and all the painters on the Tyne are laid off. I have been engaged in the last few days in negotiations for increasing our output of Turbos either by getting Greenwood and Batley of Leeds or some good firms to manufacture for us, or take a lease with option to purchase Palmers Ordnance Works at Jarrow which have been idle for 4 or 5 years. The latter case would be preferable if Palmers agree to reasonable market values.

Your affectionate brother,

Charles

If you come to N.C. we will call a boat meeting and report progress and go into costs and discuss further progress-

Another, undated, letter seems to relate to this period (C.L.(42)).

Holeyn Hall
Wylam on Tyne
Sunday (?spring 1895)

My dear Laurence,

I hope you are better and all right again. I was in London last week and heard from Chris (?Leyland) *that you were better. I hope that we shall get another trial of the boat this or next week as the weather is now more favourable. There is still a good deal of ? about which is not very nice with our 20 knots with a light hull and thin skin.*

We got a large order for Turbos from the Metropolitan C^o last week making a total of 5400 H.P. for them.

Your little machine is getting on and will I suspect soon be finished. We are getting a Newcastle firm Messrs Hawthorn Leslie and Co to help us manufacture Turbos, they are old friends and glad to get manufacturing work.

When do you think you may be over again

Your affectionate brother,

Charles

Charles Parsons evidently spent a good deal of his time travelling the 300 miles between Newcastle and London. After ten years of unremitting effort, the market for steam turbines had become firmly established. He had recovered the rights to use his patents of 1884, he had built his own works at Heaton and as yet he had no serious commercial rival. But the reference to the fate of Palmer's Ordnance Works serves as a reminder that even a firm like Palmers which had been building ships since 1852 was not invulnerable to commercial failures. It is interesting that Greenwood and Batley did eventually take up the manufacture of steam turbines, but it was to the design of the Swedish rival, de Laval.

Several years before this, in 1888, the records at Haslar show that Clere Parsons had received copies of many letters and papers by William Froude on the subject of ship forms (Goodall 1942). By then Clere had left Easton and Anderson, and was working at Kitsons in Leeds. Charles Parsons had evidently also done business with William's son, R E Froude, receiving copies of his recent papers and, in return, supplying him with a small motor,

Holeyn Hall
Wylam on Tyne
2 Jan 1895

Dear Mr Froude,

I hope the little electric motor we sent you some time ago is giving the results you require. In regard to the kind assistance you gave me some time ago in sending me copies of yours and your father's papers on the screw propeller, I may say we have been hard at work since then in adapting the steam turbine to marine propulsion with satisfactory results as far as we have gone. Numerous details have had to be dealt with occupying much time and we have so far realized a speed of 20 knots with a fraction of full power, the coefficients of the screw appear to hold good as far as we have gone and I am testing a slightly modified pattern at the same time as getting details of forced draught, etc, into good working order preparatory to full power trails.

Have you, may I ask, made any progressive trials on resistance of a 100-ft boat and 9-ft beam at speeds from 20 to 30 knots? I have been reading Yarrow's Expts in Naval Architecture 1883, and his experiments seem to indicate a considerable drop from the 'cube speed' curve above 21 knots.

I wondered if any of your experiments went to show that this drop is continued.

From all I know at present I am unable to judge this, but perhaps your experiments may fully clear up the question and I should be very obliged if you can enlighten me.

Yours very truly,

Charles A. Parsons

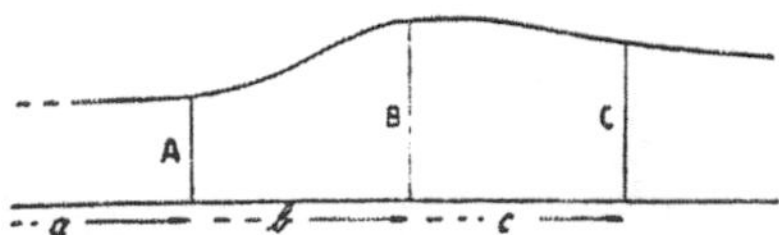

Figure 61. Image from letter 7/1/95 from R E Froude (Goodall 1942).

Parsons received a prompt reply (Goodall 1942)

7th January 1895

Dear Mr Parsons,

Many thanks for yours of 2nd. I am much interested to hear of the progress of your steam-turbine torpedo boat.

We have made experiments throughout the whole range of speed for a great number of forms of torpedo boats and 'destroyers' though none so narrow in proportion as you quote, viz 110 ft × 9 ft, 10 beams is the maximum length ratio. The features of the curve of resistance are much the same for all forms, but more emphatic the greater the ratio of displt/length. Taking the sample form in which the ratio is smallest, if we plot Res/speed^2 (or equally H.P./Speed^3) as ordinates to a base of speed, we get such a curve as is shown on the next page (figure 61). *For 100 ft length, if we take A as equal unity, B and C would equal about 1.7 and 1.5 respectively, the speed represented by a, b, c, being respectively, say 11.5, 18 and 24 knots.*

The summit at B is well rounded, the descent towards C afterwards fairly straight, with a tendency towards concavity upwards. In some pinnace forms in which we have reached an even higher speed relative to $\sqrt{(\text{length})}$, we have found this concavity very well marked, but we have never reached the end of the declining gradient and I expect it would continue, though declining in intensity, to a good deal over 30 knots for a 100 ft boat.

With best wishes of the Season, Yours very truly,

R.E. Froude

P.S. The motor still remains inactive, the job for which it was required still remaining in abeyance. R.E.F.

Parsons' attention was turned on the design of the screw propeller in an effort to improve the speed. So far he had not exceeded 20 knots which was far short of his target of 30 knots. A fragment of a letter to Laurence seems to relate to this period; it reads (C.L.(18))[9]

[9] There are two letters marked as 18.

Holeyn Hall
Wylam on Tyne
1895

I note what you say about the other points which shall be attended to—

I think the boat would be better without the false wooden extension and that the vibration at the stern is due to the water from the blade hitting the stern. There is however an irregularity of from 1/16" to 1/8" in the inclination of the blades of the present screw. Whether due to the accident or to error of machining I have not been able to discover. I expect the 4 bladed screw will have much less slip and work more smoothly. I think we may reasonably expect to soon get a higher screw efficiency and also develop larger powers. The resistance also goes up in a slower ratio after 18 knots.

Your affectionate brother,

Charles

The false wooden extension seems to refer to a wooden fairing that was bolted on to the flat face of the stern (Osler and Grieve 1978). Also it is clear that it was written during a period when in the course of a total of 31 trials, two bladed, three bladed and four bladed propellers were all tried, as well a combination of up to four propellers on the same shaft (Cook 1939).

The design of a screw propeller is complex[10]. The function of a propeller is to create a sharp increase in pressure of the fluid which passes through it. This pressure when applied across the effective area of the propeller yields a thrust. As the thrust required to move the vessel increases, so too must the magnitude of the pressure jump. Half of the pressure jump comes at the entrance to the propeller where a fall in pressure, a suction, which has been created by the acceleration of the water stream as it approaches the screw disc, is cancelled by a relative rise in pressure. The other half comes as a further increase of pressure in the fluid leaving the propeller. Of course if too much power is demanded of the propeller, the suction will become so great that the pressure will reach the evaporating pressure of the water, which will, in effect, boil. In practice, problems occur long before this condition is reached. The maximum 'effective area' is that of a circle of equal diameter to the propeller. But if the blades are slender, and few in number, the effective area may be only a fraction of this. Also Rankine's simple momentum model indicated that the most efficient scheme was to cause the propeller to act on a large volume of fluid, imparting only a relatively small increase in its velocity.

[10] See also appendix 1.

This calls for a large effective area of propeller. Finally the significance of the pitch of a propeller, whether coarse or fine, is similar to the pitch of, say a metal screw, turning within a nut. So a screw of fine pitch must rotate many more times than a coarse pitch screw in order to advance a given distance. The speed of the turbine, and hence propeller, for a given speed through the water, will be linked with the pitch of the propeller. Finally it was known from experience that a real propeller advanced by a smaller distance through the water than the number of revolutions multiplied by the pitch would suggest. This phenomenon, called slip, is inescapable but should amount to not much more than 20% perhaps.

In the next letter Charles gives his reasoning (C.L.(19)).

Holeyn Hall
Wylam on Tyne
26 March 95

My dear Laurence,

I have just got your letter on returning from 3 days in London where we are just getting an order for 4 Turbos of 1,000 horse power for the City of London Co to replace Brush plants or Thomson Houston plants.

The Turbinia

We will have the log move out in the future and test it against the mile as a check. I understand the posts on the mile outside (the mouth of the Tyne) *are shortly to be repaired by the other ship builders for their trials.*

We find on examination the (furnace) *bars are seriously burnt. They are the usual cross section and so far have not had hard work, so it must be the pyrites in the coal. We shall try the best North Country Coal at 11/- per ton on Thursday next.*

I find on checking the speeds and revolutions that we have almost exactly 50% slip against 33% calculated, there therefore appears to be vacuum behind the blades, which a larger area and the new screw will correct in all probability.

I think the apparent high (shaft) speed comparatively at the lower (vessel) speeds is due to two causes.

1st increased slip due to formation of vacuum when forced, so that though the (shaft) *revolutions go up rapidly the* (vessel) *speed goes up in slower ratio.*

2nd the increased ratio of resistance at from 12 to 18 knots

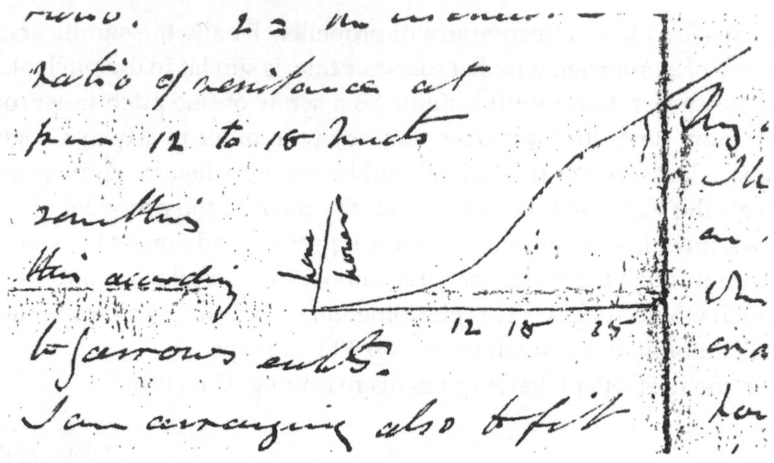

ratio of resistance at
from 12 to 18 knots
something
thus according
to Yarrow's results.
I am arranging also to fit

Figure 62. Image of part of letter 19 (C.L(19)).

> *something thus* (figure 62) *according to Yarrow's results I am arranging also to fit a second screw in front of the rudder—to get more blade area—without too large a diameter or too small pitch ratio.*
>
> *We are increasing the diameter of the fan and hope to realise the condition of blowing cinders up the chimney as in locomotives working full power, which have a suction up to 10" of water in the smoke box. Air pressure now is 3".*
>
> *Excuse this long letter.*

A further page is really not decipherable. We know that these letters were often written after a long and tiring day's work, and they are in consequence not always as legible as would be wished. The reference to a vacuum 'behind' the blades must mean 'behind the leading edge' of the blades. When comparing the graph in Froude's letter with that in Parsons' letter, remember that it plots resistance as an ordinate, while Yarrow's graph used in Parsons' letter plots power, which equals resistance multiplied by speed. The reference to 50% slip could relate to trial No 18, which is the second, undated, entry in Parsons' notebook. This trial has a note which states that the tests were done with a 24 in diameter, 27 in pitch propeller. The first entry trial No 17, also undated, records *'Rosse, self, Barnard'* as present. Robert Barnard who was the manager of the project generally acted as steersman. Parsons has identified his problem as the excessive value of slip, experienced due *'to formation of Vacuum when forced.'*

The correspondence, as the next letter shows, was two way although we do not possess any of Laurence's letters (C.L.(20)). It was written on the same day as the trial.

Holeyn Hall
Wylam on Tyne
28 March 1895

My dear Laurence,

As to the screw and your suggestion of one blade, I fear that the front edge would give most of the thrust and so be apt to bend the shaft and stern frame, but there appears to be no objection to a two bladed screw with great lengths of blade, say $\frac{1}{2}$ turn to each blade, thus covering the whole disc area of the screw with blade surface. This arrangement will eliminate the cutting or parting resistance due to a thick blade as the length ÷ by thickness will be large- one would like to make it with increasing pitch backwards. I make out that at 18 knots the slip of the present screw is 54% whereas 33% is that calculated, there appears to be vacuum and increased blade area will cure this. I will try a screw like the one on the model and also put one in hands for the boat.

We tried today some Durham coal which Smithson recommended, and it appears to give about 10% less steam than the Welsh we have been using, and curiously seems to burn the bars in the same way. Now we are only burning from 33 to 40 lb^s of coal per sq^{re} foot of grate, and locomotives go up to 60, 90 and 120 lb^s per sq^{re} foot. I am looking into the bar question to see what is the matter. There appears to be something wrong here and I will consult Hawthorns and if necessary W.H.White on the subject.

I took the feed today exactly, by timing the donkeys (pumps) *and it showed 8040 lb^s per hour with 75 lb^s steam* (pressure) *on the motor. This is a very small consumption indeed and much less than calculated owing I believe to good superheating.*

As I want to get the evaporation up to 2,000 lb^s we will on the next opportunity when the boat is slipped gauge out...

This letter has no signature and a portion of it may be missing. It was written on the same day as the trial run. The suggestion of a single bladed screw seems very strange, and it is turned down by Charles, but his manner is not dismissive. Clearly Parsons had easy access to other experts at places like Hawthorns, the nearby firm of shipbuilders who were already helping out in turbine manufacture, and to naval architects like

W H White who now worked for the Admiralty as head of construction, after a period at Elswick[11]. There were a number of ways of estimating the steam consumption of the turbine. If conditions were stable and the amount of water in the boiler was steady, it was easier to measure the water input feeding the boiler than the corresponding steam output. Since water was supplied by a donkey engine which delivered a fixed volume per stroke, it sufficed to count the number of strokes per minute. The figure of 2000 lb^s for the evaporation rate should perhaps be 20 000, because later tests, admittedly on a different turbine, found a figure of 30 000 lb per hour at maximum output. Some light is shed on the way data was occasionally acquired, by a story told by Frank Hodgkinson (Hodgkinson 1939).

Sometimes, however, improvisations were carried to extremes. During the early trials of Turbinia, the steam consumption was estimated by the speed of the reciprocating feed pump. After the boat was tied up, the fireman was instructed to set the feed pump running at the same speed as during the full power run, which he did by guess, not by counting. In response to remarks on my part, Sir Charles said that he had occasionally checked the fireman and found him quite accurate.

An incomplete letter which seems to date to sometime in March touches on other concerns affecting the two brothers (C.L.(22)).

Holeyn Hall
Wylam on Tyne
1895

You will find it a relief to have an energetic man instead of Coghlan though at first it may be some bother. The man you want is difficult to find as he has to have a general knowledge of mill wright work and not be a simple fitter or turner.

We are just arranging to supply 4 1,000 H.P. turbines for the City of London Co, this is in addition to the 6 more 700 H.P. plants for the Metropolitan Co

Private

We are hoping to sell our various patents to Turbos for electricity to the General Electric Co of America for £10,000 down and £15,000 over a term

[11] William Henry White (1845–1913) was not only a very competent naval architect and administrator, but personally resourceful. On 20 December 1886, he with Lord Charles Beresford inspected the French submersible *Nautilus* at Tilbury docks. During a dive the boat became stuck on the mud. The air was being rapidly depleted when White finally succeeded in dislodging the vessel by systematically moving the passengers from one side to the other (Manning 1923).

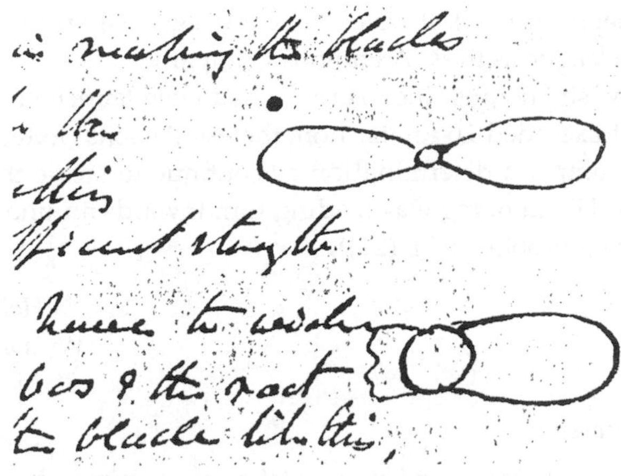
in making the blades
like
this
fficient strength
have to widen
boss & the root
the blade like this,

Figure 63. Image of part of letter 22 (C.L.(22)).

> *of years depending on the output, several of their people are over here now negotiating.*
>
> *P.S. The difficulty with us in making the blades like this,* (figure 63) *is getting sufficient strength We have to widen the boss and the root of the blade like this and I need to put in breadth where it is not so effective.*

The reference to Coghlan is of course to William Parsons' leading mechanic who had joined his team at Birr in 1841. After a long life of service he had retired. He died the following year on 18 March 1896, at the age of eighty years (Parsons L 1896).

The interest expressed by the American General Electric Co. was an indication that the steam turbine was now an established commercial reality. In the end the General Electric Co. was persuaded by their brilliant but conservative young engineer, C P Steinmetz, not to purchase Parsons' steam turbine patents (Kline 1992). Although he considered this field *'very promising'*, he did not *'think much of Parsons' steam turbine'*. He thought *'a simpler design, even if not as efficient in steam consumption, would be desirable.'* He recommended that they should *'delay the closing of any agreement with Parsons until we have satisfied ourselves with regard to our own design'*. Steinmetz was born in Breslau and educated at the Technische Hochschule in Berlin, but his political opinions caused him to emigrate to the United States where he joined the Thomson–Houston Company, which was at the time a close rival to the Edison Electric Light Company. As the bankers got to work, Edison's company became Edison General Electric, and in 1892 this was merged with Thomson–Houston to become the General Electric

Company (Josephson 1961). It was a huge and aggressive business, not at all in the same league as the C A Parsons Company.

It was now six months since *Turbinia* had been launched and still its speed had not exceeded 20 knots; nonetheless Parsons' letters show no sign of panic, simply a determination to continue to tackle the problem systematically. His thinking was leading him towards an understanding of the underlying problem (C.L.(21)),

Holeyn Hall
Wylam on Tyne
30 March 1895

My dear Laurence,

If you do not wish to be bothered by speculation on screw propellers put this in the waste paper basket.

For a speed of 30 knots a thrust of about 6 tons appears to be necessary and assuming the screw 28" dia^r this gives a mean pressure of 21 lb^s over the disc area of the screw (the present screw is 32" dia^r but the blades cover only about 1/3rd of the disc area).

Now it seems to me that water flowing into such a column of 28" dia^r will necessarily part company and vacuum spaces be formed. For a screw 28" dia^r the mean pressure works out 21.1 lb^s per sq^r inch and it would appear that very little vacuum would be produced. A larger screw of the Archimedean type would give excessive friction. I therefore propose to make an Archimedean screw 28" dia^r and on working it out it will probably have a moderate slip of 20%.

To assist this screw if necessary I propose to place in front of it at a distance of about 4 feet a screw of 24" dia^r archimedean type and of the less pitch of 19" (the speed of 30 knots and 2,400 rev^{ns} corresponds to a pitch of 15" exactly). This front screw will do half the acceleration and the aft screw (which would be reduced to 24" dia^r if so assisted by the forward screw) would do the remainder. Besides this the velocity added by the front screw will be much dissipated before reaching the back screw.

The skin friction of the two screws (the after one being reduced to 24") will be about the same as a single 28" screw.

Your affectionate brother,

Charles

In this letter Parsons spells out with figures the basis for his conclusion in the letter written on the 26 March. He has realized that his problem relates to the effective pressure jump associated with the propeller. His instinct tells him *'that water flowing into such a column of 28" dia^r will necessarily part company and vacuum spaces be formed.'* But his calculation gives a pressure of *21.1 lb^s per sq^r inch* (which is twice 10.6 lb per square inch) *'and it would appear that very little vacuum would be produced.'* Bearing in mind that a full vacuum represents a suction of 14.7 lb per square inch, his logic seems unassailable. But of course the calculation assumes that the 6 tons thrust was spread uniformly over the entire surface of the propeller disc. In practice the area of the actual blades when projected onto this disc was much less. As he stated, *'the blades cover only about 1/3rd of the disc area* (of the 32 in diameter screw)'. It would be some weeks before he appreciated the full importance of this. But in an effort to bring down the size of the pressure jump necessary to achieve the required magnitude of thrust, he considers increasing the effective area of the screw. An Archimedean screw is one in which the blades of a two bladed screw are broadened and spiral for more than half of one pitch length (see figure 67). The problem with such devices is that, with the increase of surface area, there is a rise in the associated surface friction. In fact Charles, like his brother Clere, seems to have been too concerned with the importance of losses due to friction, being still unaware of the nature of the other problems which could be caused by the pattern of water flow over the blades.

Chapter 12

Obstacles Overcome

12.1 Pursuit of missing thrust

In the spring of 1895 the solution to the problem of propeller design was close at hand. It seems that Parsons had become convinced that the inability to reach high speeds was due to a failure of the propeller and not to any inadequacy on the part of the turbine. To put the matter beyond doubt he devised an instrument that could be fitted between the turbine shaft and the propeller shaft (Richardson 1911). This, 'dynamometer' would measure the force tending to twist the shaft, and so, when taken with a measurement of the shaft speed, an estimate could be made of the actual power which was being transmitted. In figure 64 it can be seen that the turbine shaft and the propeller shaft each have a claw shaped device attached. The claws engage so that rotation of the one effects the rotation of the other. The claws are kept from touching by four spiral springs which can be seen in figure 64(b). As the springs become compressed by increasing torque, a bell crank lever, visible in figure 64(a) and (c), moves. Sharp steel points make a permanent record of the movement on a piece of soft wood inserted for this purpose. It would appear that like so many of Parsons' ideas, this was the first use of such a device to monitor the power actually being delivered to the propeller shaft of a vessel while it was under way. He lost no time in putting it to use. The following letter appears to have been written at the end of March (C.L.(24)).

Holeyn Hall
Wylam on Tyne
Wednesday (1895)

My dear Laurence,

You will be expecting to hear the results of the dynamometer tests. The

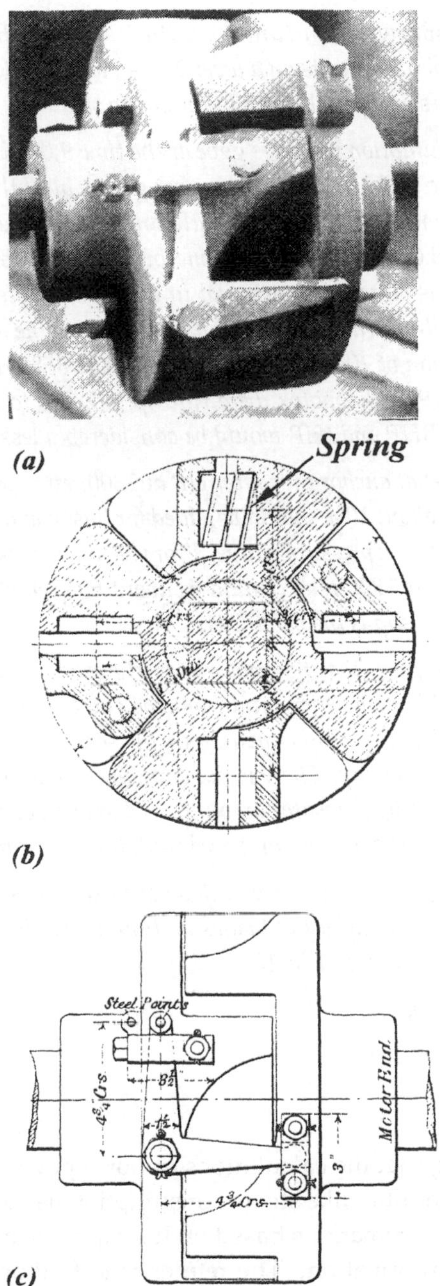

Figure 64. Shaft dynamometer constructed for the *Turbinia* tests: (a) photograph; (b) cross section; (c) side elevation. The turbine (motor) shaft is to the right hand side (Richardson 1911).

deflection of the springs when running with 100 lb^{s} *on the Deck* (steam) *gauge are identical with 6 cwts on a lever* $2'11\frac{3}{4}''$ *which at 1,900 revolutions works out 869 horsepower delivered to the screw shaft.*

The (steam) *consumption readings gave at the time 9,000 lb*^{s} *per hour with 50 lb*^{s} (per square inch) *on deck* (pressure) *gauge, but I think the water in the* (boiler water level) *gauge fell slightly and that we shall be very near a true measure of 10,000 lb*^{s} *per hour. Numerous tests of turbines have shown the consumption is closely proportional to the absolute steam pressure, so this gives 17,700 lb*^{s} *per hour with 100 lb*^{s} *pressure on the deck gauge. This gives a consumption of 20.4 lb*^{s} *per brake H.P. or at 85% ratio to indicated 17.34 lb*^{s} *per I.H.P. If the motor were run up to 2,500 to 2,800 (rpm) the consumption per BHP and IHP would be considerably less.*

On the Towing test at anchor the water HP at 1300 rev^{n} *was 522, one screw was damaged and about 10% should be added for this, making 574 water HP with 100 lb*^{s} *on the deck pressure gauge. From Monday's tests the calculated dynamometer HP at 1800 rev*^{ns} *would be about 695 which would seem to show that excessive slip is the chief loss.*

In conclusion it appears that the motor is fairly good and at higher speeds will make a very high efficiency, and from the known resistance of similar boats it appears that the loss is entirely in the screw and is chiefly, if not wholly due to excessive slip. There is not enough evidence yet to show what the resistance of cutting the water amounts to, but at all events, the loss due to it, whatever it be, is small as compared with the loss from excessive slip.

The trial of the larger (3′-6″) *screw with sword like blades will throw more light on this. We will make the blades as thin as possible and cast them forward at an inclination of 1 in 10.*

Best wishes to you all

your affectionate brother

Charles

It is interesting that the calculations of horse power are fairly rough. His estimates of quantities like steam consumption per brake horse power involve several approximations based on his experience, but in his hands these gave useful information. The reference to IHP or 'indicated horse power' relates to steam engine practice in which engine output was inferred from 'indicator diagrams'[1]. The actual shaft output of a steam

[1] An indicator diagram plots pressure in the cylinder of a steam engine against the piston position, and hence the volume of steam, at a given point in the piston cycle.

engine was less than this by a factor close to 0.85. Presumably the towing test at anchor involved measuring the restraining force required to keep the vessel at rest. Of course the flow of water past the propeller would not be the same as in a speed run. Another, undated, letter which must have followed close on the heels of the one above, reads (C.L.(23))

Holeyn Hall
Wylam on Tyne
Sunday 24th March/6th April? 1895

Important

In further reference to the propeller, the matter is now I think quite cleared up by a paper, proof of which I got this morning by Thornycroft. He makes out from the trials of the Daring and two other very high speed boats that if the mean pressure of propulsion over the blade area exceeds $11\frac{3}{4}$ *lb*^{s} *then a vacuum, or as Froude has termed it "Cavitation" is set up, the slip goes up enormously as well as the power required for a given revolutions.*

In our present screw we had some 60 lb^s *mean pressure on the blades, and therefore enormous "Cavitation" set up. I think the best course will be to follow out our patent and put two screws one abaft the other, the front one say 19″ pitch and the after one 22″ pitch so as to gradually accelerate the water column in two stages, also make the screws of the Archimedean form with maximum blade area and keeping the diameter down to 28″.*

This is all very reassuring as even in the Daring trials where the screw was working close to the commencement of 'Cavitation', an increase of 45% in blade area put matters right and raised the speed from 24 to $29\frac{1}{4}$ *knots with the same H.P. We may expect a much greater increase in our case. If two screws are not enough we will put on more.*

Yours

The paper to which Charles referred was read to the Institution of Civil Engineers in London on 2 April 1895 by J I Thornycroft and S W Barnaby, and was entitled 'Torpedo-Boat Destroyers.' It is largely devoted to details of design, but at the end it was noted that *'some difficulty was at first experienced with the propellers'*. Tests on the *Daring*, in June the previous year, gave values of slip in excess of 20% and even reaching 30% at 24 knots. A variety of diameters and pitches for the screws were tried, but it was only when the effective blade area was increased that a real improvement in maximum thrust was obtained. It was postulated that the onset of 'cavitation' occurred when the mean pressure jump exceeds $11\frac{1}{4}$ lb per square inch. The suction component was taken to be 0.6 of the total pressure jump, that is $6\frac{3}{4}$ lb per square inch, which is far short of 14.7 lb per square inch which would represent a complete vacuum. The authors were able to verify from their records that no previous vessels built by them had exceeded this threshold. However they could offer no reason for the existence of this particular limit. We can now see how close Parsons had come to the truth in his letter of 30 March. Had he allowed for the factor of 1/3 in his calculation in March, he would have arrived at a figure close to the '*60 lbs mean pressure*' which, as a result of the clear exposition given in the paper, he now recognized as being close to the real situation. Thornycroft, it is true, had the advantage of having built a great many such vessels, whereas the *Turbinia* was Parsons' first and only such experience.

12.2 The study of cavitation

So far the phenomenon of cavitation in screws had not been visually observed. Parsons now set out to explore the problem with models using scientific principles. Cavitation occurs because voids open up within the liquid when the pressure locally falls to a value at which boiling takes place. The first approach was to vary this pressure by heating the water. At 100° C water boils at atmospheric pressure, and so close to this temperature a very small degree of suction is sufficient to cause bubbles to form. A second approach was to artificially reduce the pressure at the propeller by creating a degree of vacuum above the water surface with the help of a pump. The test rig which was devised went through a number of versions. The Hon. Geoffry Parsons, son of Lord Rosse, served an apprenticeship at the Elswick works from 1896 to 1899, and visited Holeyn Hall during this period (Parsons G L 1939). He was told that the first step had been to borrow a large saucepan from the kitchen. A suitable packing was devised to allow a shaft carrying a model screw to enter at the side. The shaft was spun and when the water temperature neared boiling point, cavitation was observed to set in. A more elaborate rig was devised which would allow close visual observation. The Captain of the Turbinia, C J Leyland, described his experiences acting as an assistant (Leyland 1935).

> *In those days, after trying her* (Turbinia) *at sea, Parsons and I were often up until the housemaids were stirring in the morning, trying propellers in the little experimental tank. This was hard labour. I had to look after the boiler that drove the propeller, the lamp for boiling the water in the tank, and almost incessantly heaving round the air pump to maintain the vacuum.*

Figure 65. Small scale cavitation tunnel for propeller experiments, showing the synchronous rotating disc used to 'freeze' the propeller image (Richardson 1911).

I suggested to Parsons making use of electricity, and then we worked both the propeller and the air pump by that means.

Our earliest observations were taken by flashes, just like lightning, to fix the vision in our eyes, but later Parsons improved on that, and by means of a band on a cone, we reduced the apparent revolutions so that the result was more easily noted.

The 'band on a cone' seems to refer to a form of stroboscope arrangement. Improvements were made, and photography was used to make a permanent record.

These first photographs were taken when the propeller was revolving, I think, at some 950 revs. We did it on a Bank Holiday as we had borrowed one of the best portrait cameras in Newcastle for the job.

Parsons contrived a shutter heavily loaded with lead, and with two short rubber bands attached to the floor to pull it down. It was held by a mousetrap arrangement and one of the children was told off to release this.

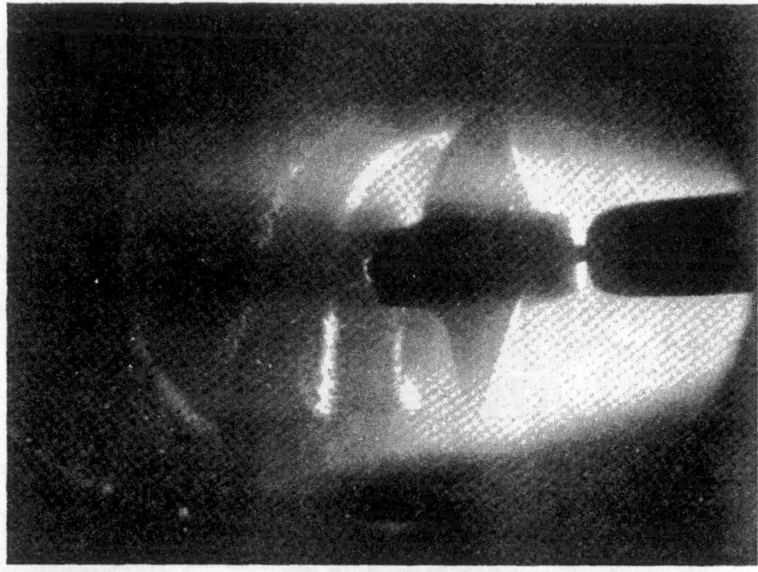

Figure 66. Advanced stage of cavitation; propeller speed 1800 rpm (Richardson 1911).

> *The light was obtained by reflecting the sun from a searchlight mirror, and as our glass in the tank was small and only held in position by rubber bands, the beam of light had to be carefully controlled.*

This example shows once more how Parsons turned to experimentation when he needed to understand some phenomenon. There was a real flavour of 'string and sealing wax' about the arrangements, and yet in his hands they yielded valuable information. When he moved house to Holeyn Hall he had provided himself with a workshop where he could experiment, away from the Heaton Works. These accounts throw some light on the domestic scene, with the kitchen being raided and the children partaking in the experiments. At this time they would be ten and eleven years old, and Charles was repeating his own childhood experience in his father's workshops in Birr.

Just observing the propeller in bright illumination would certainly show the development of bubbles as cavitation set in, but stroboscopic illumination was needed if details were to be made out. With the help of a rotating shutter synchronized with the propeller, its motion could be 'frozen' (figure 66). It was evident that cavitation first appeared at one location on the propeller blades, then gradually extended until their entire

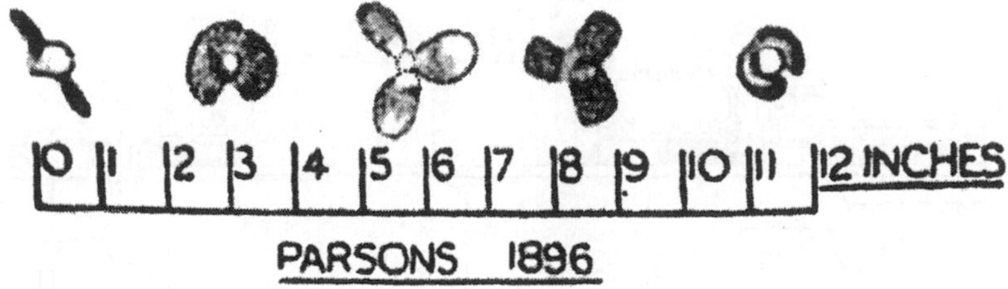

Figure 67. Miniature experimental screw propellers described as, from left to right, '*sword like, Archimedean, elliptical three bladed, wide three blade* and *Archimedean*' (Goodall 1942).

surface was enveloped. In figure 66 the bubbles can be seen as a light coloured helix streaming from the tips of the propeller blade. Further refinements were added to allow the thrust to be measured. Eventually adoption of an electric arc removed the need to depend on sunlight for illumination. Quite early on, confirmation was obtained of Thornycroft's experience. In the case of a propeller in which the blades covered an area equal to only 25% of the disc, the thrust fell off sharply when it was brought to a state of cavitation. But when a propeller with blades covering 70% of the area was run under identical conditions, the loss of thrust was quite small. A photograph of a collection of miniature, 2 in diameter propellers, dating to 1896, was printed in Sir Stanley Goodall's memorial lecture in 1942 (figure 67) (Goodall 1942).

12.3 Axial flow with three shafts

By this time Parsons had tested seven different arrangements of propellers and had tried nine different sets of propellers (Parsons C A 1903). The next step was a major one, and there must have been intense discussion among the Board members about it. Unfortunately we have no certain date for it. It was decided to adopt a radical solution to the cavitation problem and in future the power was to be spread over three independent propeller shafts, instead of one, so reducing the duty on individual propellers. Now that he had recovered the use of his 1884 patents he was able to take up some design concepts which he had been developing just before the break with Clarke Chapman. The single, radial turbine was replaced by three turbines in tandem, that is in series. The high pressure, intermediate pressure and low pressure turbines would each take an equal share of the total power

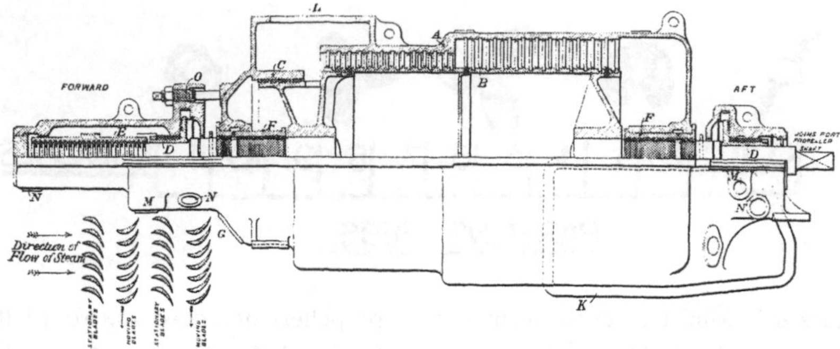

Figure 68. *Turbinia*'s intermediate pressure turbine, with steam inlet at *L*, blades at *A* and *B*, a labyrinth seal for the dummy piston at *C*, a thrust bearing at *E*, steam glands at *F*, journal bearings for the stub shafts at *D* and the exhaust at *K*. This overall architecture was used for most subsequent marine turbines. (Richardson 1911). Compare this with figure 45.

and would each drive one of the shafts. This brought an extra benefit in that only one casing was subjected to relatively high temperatures and pressures, while another dealt with the quite different problems of low temperature high vacuum conditions. Besides, because of the big jump in the power of this machine compared to previous axial flow designs, and the consequent increase in steam flow, he could afford to mount his turbine blades on the surface of a drum of relatively large diameter, instead of on the rather slender shafts used with the smaller machines. In order to allow even taller blades he chose not to split the steam flow but cause it to travel in one direction only. The resulting end thrust was balanced with his recently developed 'dummy' piston. Several of these features were used in the most recent machines which he had just supplied for the Manchester Square power station in London (Parsons R H 1936). A fourth, reversing, turbine was planned to share the central shaft with the low pressure turbine, although it was not fitted initially.

The design of these turbines established the main features of direct drive marine engines for many years to come. Figure 68 shows how the rotor drum was carried at each end by spider castings. These were bored out to take stub shafts. The spider casting was heated so that it could slide over the shaft, and then allowed to cool to create a tight fit. With this construction reasonable blade velocities could be achieved even with shaft speeds as low as 2200 rpm. Also the drum construction was so stiff that there was no risk of whirling, and the blade clearances could be kept small. Considerable work was needed to remove the radial turbine and install the two extra shafts. Fortunately the condensers could be left as they were. The decision taken to construct the hull locally under Parsons' own supervision now paid a dividend. A partially dated letter contains some domestic news and also refers to the new developments (C.L.(41)).

Holeyn Hall
Wylam on Tyne
5 Feb^{y} (1896)[2]

My dear Laurence,

I am sorry you have trouble with the supporting bricks of the boiler. I hope otherwise it will prove satisfactory but when you get it done I think you will find the boiler a comfort as compared with the wretched vertical type which only evaporate about 5.5 lb^{s} *against 8 of the Cornish.*

I presume I shall have to go to Guinness Brewery when they get the 1st *plant connected up and under steam. They are waiting at present for the pipes which they are doing themselves. They hurried us tremendously for the plant and when it arrived our men after erecting it had to come home to Newcastle till they had the pipes ready.*

I shall be delighted to run down (to Birr Castle) *for a day or two and will let you know.*

The report on the alterations to the boat has now been signed, also J Leyland and Swinton, and Simpson has given his approval but forgot to sign. Cookson (a new member of the Board) *has seen the plan of the new arrangement and is pleased. The three motors lie close to the bottom and almost entirely below the water line. Calculation seems to show that we may expect 40% more HP per lb of steam at the same revolutions. In addition we have been testing some blades of more perfect shape and finish (in the test house trial machine) which appear to give a much higher efficiency than we have yet used and these* (we) *will put in the boat motors (the new construction of blade is also cheaper).*

We shall have a run next week with the boat having 4, 3 bladed screws and two guide or deflecting plates, as found to answer best on the model. We shall not try and force the boiler higher than before, though the larger fan and larger air ways are finished, as it would be a pity to risk doing damage until the screws and new motors are fitted. We always know we can probably draw upon the boiler for some 30% more steam than we have yet done—but

2 From the contents, the year must be 1896.

this should I think be left to the last. I propose therefore to run the boat with 100 lbs on the deck gauge as before and take the time (for the mile).

This will give us an idea as to the best arrangement of screws on the other two shafts which we will soon have to get made. The patterns for the (castings for) *three motors are now finished, and one is gone to the foundry. Directly after our next run we will remove the present motor when laying at Wallsend and get the seatings ready and everything possible done before slipping the boat again to fix in the 2 additional stern tubes and brackets for carrying the screw shafts.*

Your affec^te^ brother

Charles

P.S. You will have seen that Cambridge Co is paying $3\frac{1}{2}\%$. *There seems reasonable prospect that this next year it will pay* $4\frac{1}{2}\%$ *or 5%*

P.P.S. As far as I can make out by calculation in two ways, the new (turbine) *blades appear to give a result closely approaching perfection, after allowing for the inevitable loss due to windage past the ends—previously we had always found a missing quantity of some 20% which has puzzled us for years, and it seems now that it has been due to bad shape, irregularities in form, and surface polish. Stoney is checking my figures and we shall repeat the tests, and also putting the new blades into 2 of the 4 100 HP plants we are making for London, one lot against the other. If the saving is really as great as it now appears, it will be an immense help to the turbine manufacture.*

The boiler mentioned was a replacement for the one which supplied steam for the engine in the workshops at Birr. It is clear that while Parsons carried the responsibility for developing *Turbinia*, he had to get the formal approval of his fellow directors for a major decision. Their financial interest was substantial and it would have been difficult if not impossible to proceed without their backing. Mr N C Cookson of Oakwood, Wylam on Tyne was an industrialist and neighbour of Parsons, and had joined the Board after the formation of the company. The mention of work on the design of turbine blades using a specially devised machine for tests, reminds us that *Turbinia* was only one of several projects which demanded attention.

In February 1896 while the new machinery was being prepared, trials were resumed (Parsons C A 1903). It is probable that the *'run next week'* is the third entry in Parsons' notebook, trial No 19 on 12 February 1896. The entry states that *'Bernard, Stoney and Self'* were present. The shaft carried four propellers, three which were three bladed had a diameter and pitch of 18 in. A fourth with four blades had a diameter and pitch of 16 in.

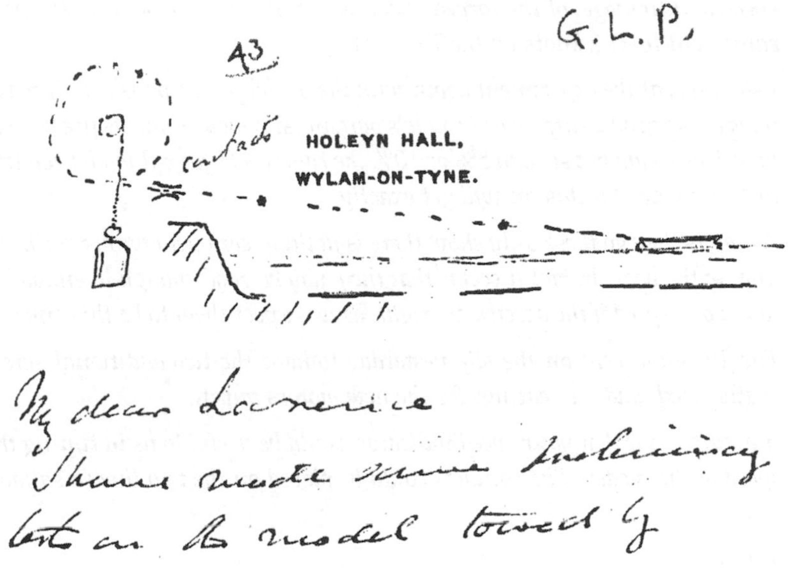

Figure 69. Image from letter 43 (C.L.(43)).

It was fitted forward of the others. Between the first and second screws and between the third and fourth screws, deflecting plates were fitted on brackets. These deflectors are reminiscent of Clere Parsons' proposals in his paper of 1879. A time of 3′45″ for the nautical mile was recorded, equivalent to 16 knots. Ignoring the tide, the slip was very high at 41.5%. *'The air pump stuck for a time and worked the auxiliary. Prevented run back on mile'.*

It would appear that he also concentrated on model tests during this period. An undated letter reports on progress (C.L.(43)).

Holeyn Hall
Wylam on Tyne
Summer 1896?

My dear Laurence,

I have made some preliminary tests on the model towed by a weight and wheel and axle, and then deduced the resistance on Froude's method. I have not Froude's exact coefficients, only those given by W H White in his book which is superficial. I am however getting the old numbers of the Naval Architects which have his papers in full. So far the resistances work out as on opposite page (see below) *and they agree closely with towing tests made by Yarrow in 1882 and also with HP curve reduced to the same equivalent speed and tonnage, of the torpedo boat destroyer Sohol, 30 knot max speed equivalent to 21.2 knots on the Turbinia.*

I shall repeat the experiment again with the model french polished. It was I imagine as good a surface as Froude's parafin (sic) *models and as previously mentioned I am not sure to 5% or 10% the right coeff for 6 ft model relative to 10 ft model but this we will get exactly.*

As far as they go these only show there is nothing seriously wrong with the lines of the boat. In fact it seems that they may be good enough to enable us to get our speed if the screws are right, as we expect them to be this time.

The Turbinia went on the slip yesterday to have the two additional screw shafts fitted, and the seatings for the new motors put in.

It occurs to me that your speed indicator would be useful to us in timing the speed of the model. The contacts could be placed on the tow line (a salmon line).

Yours

Charles

Speed in knots	Thrust tons		IHP
10	0.2		60
15	1.5		260
20	2.6		590
21.25*	2.8	Turbinia	700
		Yarrow	600
		Sohol	500
25	3.67		1030
30	4.8		1650
35	5.79		2550

Note: Thrust HP assumed 60% of IHP which is usually assumed by the chief builders

P.S. I can either put on the indicator we have at Wallsend or your new one as you think best.

It was important that the surface of the model was quite smooth. Froude made models from wax which was carefully cast and burnished after being shaped. The set up for towing the model was fairly standard. The tow line was reeled onto an axle which was driven by two falling weights. One had only a short distance to fall, and gave an initial acceleration up to the desired steady speed. The other exerted a constant force, see figure 69. It would appear that Laurence had some sort of timer, perhaps designed for use with the telescopes at Birr, which could measure the time interval between the passage of two electrical markers on the line. Laurence experimented with a number of clockwork mechanisms in his astronomical work, and no doubt his timer would have been based on one of these. The *Sohol* was a 30 knot destroyer built by Yarrow for the Russian Navy in 1894. When we compare these estimated figures for the resistance of the full sized boat with the estimates which Froude obtained later with his 10 ft model, we find that Parsons' figure is 15% higher at 30 knots (Parsons C A 1903). Given the uncertainties mentioned in the letter, and the modest resources available to Parsons, this is not bad agreement. The letter reminds us of the difficulties in acquiring items from the technical literature in days before photocopying. This phase of the project was very like the series of tests which Charles had personally carried out on the rocket propelled torpedoes. C Turnbull, who was an apprentice at that time, recalled that (Turnbull 1939)

On one occasion, when he was running the Turbinia model on the cooling pond in the dinner hour, he saw me looking at it and he conveyed to me that my presence was not acceptable, but it was done with courtesy.

On 21 May 1896 a provisional patent application, 11 086, was submitted covering 'Improvements in propelling vessels by steam turbines' which described the new tandem arrangement. Between Spring and September of 1896, there was a gap while the new turbines, shafts and screws were being made and fitted. The next letter to Laurence that we have was written after Charles had returned from a short vacation (C.L.(25)).

Holeyn Hall
Wylam on Tyne
2 Septr 1896

My dear Laurence,

I got back to Newcastle early this morning after spending 4 days with Cooksons. On the day after we arrived I got 5 salmon from 9, down to 5 lb^{s} and in the afternoon had a stalk after a stag which I was lucky in bagging, they said it was the best sport in one day for a long time. The other days we were not so lucky, on the Monday two salmon and on the Saturday a *high flood. Brawless Trent comprises 50 square miles and seemed to swarm with deer—no grouse or other shooting.*

The boat is now about completed and we hope to put steam on for a preliminary turn round on Monday. The new arrangement looks much better than the original one, with more room to get about the engine room. The draught of the boat seems to be the same as before—but the C.G. (centre of gravity) *is lower. We are sending you the old speed recorder, I presume you do not require the batteries but the contact switch is being cut (not the one for time, but the keys). This with the two boxes is I presume all you require, if not let me know.*

Your affectionate brother

Charles

P.S. Have you settled when you come over—

The next entry in the *Turbinia* notebook is for trial No 20 which took place on 17 September 1896 with *'Barnard Cross and Self'*. The centre shaft had four screws as before, but the guide plates had been removed. The two side shafts had three screws each. All were 18 in diameter and pitch, except the first on the middle shaft which was 16 in. There is a note *'Bottom dirty, weeds 3″ long 2 runs on Hartley mile, rather choppy, slip mean 29%.'*

The next letter we have was written on the day of the trial (C.L.(26)),

Holeyn Hall
Wylam on Tyne
17 Septr 1896

My dear Laurence,

We got 21 knots today steaming easily. I expected 24 or 25 and was surprised to notice a sluggishness which seemed to diminish after starting. I at first attributed it to the new shafting being stiff but had misgivings as to this. On returning we examined the bottom of the boat as far as we could reach with the hand and found to our astonishment it was coated with small muscles (sic) *of about this size, spaced at from* $\frac{1}{2}''$ *to 1″ pitch. But at the bow and to some 8 ft aft they were almost entirely swept off by the water—I also came across a barnacle about this height* (figure 70). *We shall have the bottom scrubbed and the fire grate and bars rearranged by the middle of next week-*

Today was only a trial of the new engines which passed entirely satisfactorily. We had about 95 lbs mean (steam pressure) *on the engines and hope to work up 150 next week. Have you any data you know of as to the resistance of a surface covered like this* (figure 71).

Figure 70. Image from letter 26 (C.L.(26)).

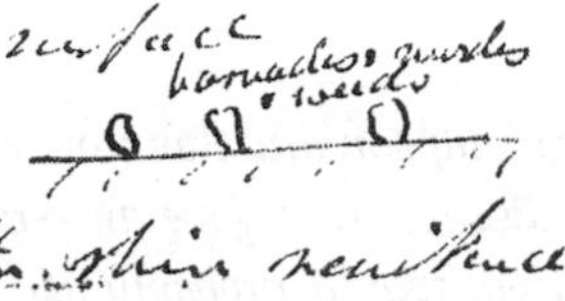

Figure 71. Image from letter 26 (C.L.(26)).

If we assume skin resistance slightly more than double that of a varnished surface it brings today's results in accord with calculation

	Rev^{ns}	*Slip*
Press at	*HP 2000*	*mean*
HP motor	*IP 2016*	*33%*
110 $^{sq.in}$ abs.	*IP(?LP) 1880*	

The calculated slip was about 20% , doubling skin resistance increases total resistance in the ratio 5:8 according to Froude's method. We must I think wait till say next week to estimate the speed we shall reach. We shall now in any case drive the machinery up to its utmost limit which we have hitherto not I think nearly reached. The new machinery works with much less vibration than the old.

Your affectionate brother,

Charles

P.S. If the screws give an efficiency of 65% we ought to get 1650 IHP with about the same quantity of steam as we had when you were aboard and the new motors viz. from 18,000 to 20,000 lb^s per hour.

A major change had been made to the turbines and propellers and yet 33% slip was being experienced. The period of inactivity while changes were being made had caused deterioration of the surface of the hull, and characteristically he satisfies himself that the effect was what could be predicted. There is no sign of anxiety, just a matter of getting problems sorted out. Shortly afterwards he reported on a Board meeting which had been held (C.L.(27)),

Holeyn Hall
Wylam on Tyne
28 Sept /96

My dear Laurence,

At the meeting today were Simpson and Swinton, matters were talked over and the present position discussed. I gave an estimate that the cleaning and smoothing of the bottom would probably add some two knots at the same power and that the new centre shaft would make about $\frac{1}{2}$ knot, and the feeling was to get on as fast as possible to a somewhat higher speed, and at the most reasonable cost.

The painting to be properly done means scraping off the old filling (done before she was launched) which has scaled in places leaving edges and roughness. Re-filling, painting and varnishing all this will take at least a fortnight to be properly set before placing in the water. The replacing (of) *the shaft and brackets is a simple job. It will lighten the stern by about $\frac{1}{2}$ a ton. The friction of the present shaft is so great that it takes 2 men on the fan to move it round. The other motors and new shafts can be easily turned by a man and a spanner, so that a great deal of friction will be saved.*

The old screws go on the new shaft when bushed. We were out again this afternoon to try a run on the knot at sea, but it was too hazy to see the post—this has been the first smooth day at sea.

We had out Stoney, Cross and Barnard, and got on a straight run measurements of temperatures and water pressures at the scoup (sic)*—also a measurement of the feed to the boiler. We are still bothered by the burning of the fire bars—now confined to the centre of the grate only, and we are also agreed that it is due to the scouring action of the air along under the bars. It is primarily due to the boiler being placed rather near the bottom of the boat to lower the centre of gravity. But we are also agreed that by a modification of the air pipes, and by admitting the air on all 4 sides of the grate viz on the sides as well as the ends, and also equally at the ends, that the bother will be overcome. It has not hitherto hindered the progress but now we find it serious, as when burnt the bars become clogged and the fires checked. We were all down in the stokeholds for some time today watching, and there seems no doubt as to the cause and the remedy.*

We shall want the boat to steam for at least an hour at full speed, to show her off to advantage to Admiralty people. I think you will be pleased with the improved absence of vibration, both the new motors and the air pump work very much better than when you were aboard.

The time of the run on Saturday was taken by Stoney as 2′28 2/5″, and by myself 2′28 1/5″, and we ascertained on stopping that there was no tide flowing, there was no motion relative to the piles on the quay. We also made a run with 63 lbs and got $20\frac{1}{2}$ knots, against 115 lbs and 24.3 knots. This is a long letter I fear. We are looking out for lodgings for Geoff. Hope he will be here before he commences work.

Your affectionate brother

Charles

The run mentioned at the end of the letter was evidently made on the straight stretch of the Tyne known as the Northumberland Dock mile, where he had permission from the Commissioners to experiment. The notebook shows this was trial No 21. It is clear that being a member of the board was no sinecure. All at one time or another participated in test runs, trial No 22 had *'Rosse, Stoney, Barnard, self'* present, while on No 25 *'Rosse Swinton Stoney and Self'* were present when a speed of 29.5 knots was achieved. No date is given for the latter. Operating the vessel was now sufficiently controlled for the more adventurous women members of the family to come aboard. Trial No 26 had among other objectives to allow Parry to take photographs of *Turbinia* while she travelled at 30 knots. Accompanying Parsons and Stoney were Mrs Parsons and Miss Bethell, presumably a sister. On other occasions Rachel and Leyland's daughter enjoyed the thrill of high speed travel. Though they were just as likely to end up soaked as the men, they made no complaint on that score.

The reference to the need for two men to turn the central shaft by the fan is explained by the fact that the forced draught fan for the boiler was driven directly by the low pressure turbine and the fan impeller could be used to obtain leverage. When the new turbines were fitted, the old central shaft was initially left in place. Of course its replacement had to deal with only 1/3 of the power for which the original was designed, and so it was much lighter. Geoff was Laurence's son.

In 1882 W H White had been enticed by Lord Armstrong to join Sir William Armstrong, Mitchell and Co. as Warship designer and Manager of the warship-building branch at a salary of £2000, three times what the Admiralty paid him (Manning 1923). In 1885, facing the threat of war with Russia, Lord George Hamilton prevailed upon him to return to the

Admiralty with a brief to introduce many reforms. For some time Parsons had been in close contact with Admiralty personnel, including White, about his plans. One obsession of the navy was the need for vessels to be able to go astern quite fast if occasion demanded. This was becoming less necessary as warship designs improved, but old traditions were hard to disregard. With the current reciprocating engines, reversing was a simple matter and almost full power astern was readily provided. In the old radial turbine, a row of blades for reversing was made an integral part of the low pressure disc. But at first there was no provision for reversing in the multi-shafted arrangement. He takes the matter up in a letter which must have been written in 1896 (C.L.(28)),

Holeyn Hall
Wylam on Tyne
28 Dec (1896)

My dear Laurence,

Many thanks for your letters. I am now again in correspondence with W.H.White with the object of endeavouring to get their rule of $\frac{2}{3}$ full speed of revolutions astern modified. It means no increase of cost or weight or complication whatever with common engines, but it does mean all three with the turbine engines. We might very well come to a compromise of say 10 knots astern in exchange for $\frac{1}{2}$ knot more ahead. This would I think suit us and them.

I think, with you that the larger C^{o} ought to pay well and your action and offer should encourage others. We are getting particulars of several sites with plenty of room for engine works and a certain amount of initial shipbuilding. I was not clear about the land. It appears the rental varies from £50 to £70 per acre and the cost of freehold from £1,000 to £1,500 per acre or so. This is for a site, on the water and 'on' the railway. Good sites for works are nearly as much.

Your affectionate brother,
Charles

I have been comparing results with the model resistances, and to my surprise find that in the case for the turbines speed and steam pressure at 29.6 knots, the resistance corresponds to about 28.1 knots. On thinking it over it appears that from the capilary (sic) *action in the model the front waves cling to the side and so greatly increase the wetted surface. At moderate speeds this is not so much the case. This effect was not mentioned by Froude. We shall be better able to say how the comparison is when we get a complete set of runs at increasing speeds on a fine day.*

I think our next (?) attention will be increased churning and air (?) ways. Also putting on new propellers 24 pitch against the present 18″ pitch. I think the turbines must be up to full speed at 29.6 knots viz 2400 revolutions, and their efficiency will not be reduced by slightly slowing them, and the screws will be much improved thereby. We may perhaps get 33 or 34 knots. The new screws were put in hand a couple of months ago and are nearly finished. They are more nearly of the usual proportions.

Many thanks for the paper which I have returned. The use of accumulators for day, and after midnight, load has been a good deal used in this country. It has advantages, and against there are the cost and loss in charge and discharge. At Cambridge a small plant is put on (at times of light demand) and one man minds it.

The air pump on the Turbinia is compound, 18″ piston below making it a Smeaton pump and above is a single acting one 6″ dia^{r}, and the same stroke. On the opposite end of the shaft is the steam cylinder 6″ dia^{r} and 6″ stroke. Last day the reducing valve was sticking and the pressure was varying from 50 to 120 lb^{s}

It is clear that thoughts for the future of the marine business were in the air. At that moment the possibility of engaging in shipbuilding had not been ruled out. The intense industrial activity along the banks of the Tyne made land very expensive. This letter reminds us that all during these speed runs, Parsons was relying on the measurements he had made with his model to give him the estimates of power which would be necessary to propel *Turbinia* at any particular speed. As later events were to prove, these estimates were remarkably accurate. His reference to increased wetted surface shows the close attention which he paid to the behaviour of his model.

The early electric lighting schemes, whether public or private, faced the problem of meeting demand late at night and by day, when the level was too low to justify running a steam plant. By opting for a direct current system, rechargeable batteries could be used. The only other way out of this dilemma was to build up demand to a level where it was always economical to run at least a small machine. The level of demand was dependent on the legal regulations affecting the running of public main cables. This determined the maximum area which could be served. It was also important to encourage the development of industrial customers.

In both respects the electricity supply industry in Britain was greatly hampered compared with that in Europe and America.

12.4 Fine tuning performance

As the year 1897 opened, the performance of *Turbinia* was approaching the targets that had been set, after more than two full years of trials. Parsons was confident of ultimate success, and steps were being taken for the establishment of a successor to the Marine Steam Turbine Company, which would exploit the achievements of the program of invention and experimentation. Meanwhile a new setback was reported which required the boat to be taken from the water (C.L.(29)),

Holeyn Hall
Wylam on Tyne
5 Jany 1897

My dear Laurence,

The boat is going to the slip the end of this week. We had a short run last Thursday and measured the thrusts. There is a very heavy overbalance of steam (thrust) *over screw* (thrust) *on the low pressure motor. Some mistake appears to have been made in our drawing office, I cannot otherwise account for it as I gave the Chief Draughtsman the steam pressures and thrust I wanted taken—which if carried out would have been nearly right. The intermediate* (pressure motor) *is also to a small extent overbalanced and the HP motor is about right. We are taking the motors out and adjusting them—We also took the revolutions which were*

LP	*IP*	*HP*
2450	*1450*	*1650*

The first is above the maximum to which it was tested before being put on board and I do not like to run them faster in regular work. The mean slip is now down to about 20%.

I have been puzzled by the smallness of the thrust observed, considering the enormous HP delivered, and I put it down at first sight to the screws not being the best obtainable, now that we have 9 instead of 3 only, as when they were found best for the single motor. I have been making calculations of skin friction of blades, and comparisons with the data of other screws and also

Froude's papers etc, and have made out a balance sheet of the horsepower developed and expended. There is necessarily some guess work about it but I think that it is substantially correct. It shows the motors to consume only 10 lb^s of steam per effective or Brake H.P., that is some 45% of the whole developed power viz 1860 is going in blade friction against the water and 20% in slip leaving 850 for useful propulsion or thrust horsepower, which agrees with resistance curves. This great waste in the skin friction arises from the very wide blades covering 0.6 of the disc area (they must be wide to prevent vacuum) and the fine pitch, 18" pitch, 18" dia^r. Now that we have got the revolutions up to 2450 or say 2500 mean we can cut them down to say 1900 without incurring a loss of more than, 5% to 7% in the turbine. This being so, increasing the pitch to say 24" will at the same speed reduce the skin friction in the ratio $(\frac{3}{4})^3$ or by more than $\frac{1}{2}$.

I do not like to go beyond this at present, so I have put in hands another set of 9 screws exactly the same as the present, only 24" instead of 18" as at present. The extraordinary economy of the engines of 10 lb agrees with the theoretical efficiency of 70% (which I have in my notes as 10.4). In smaller engines we have reached 60% of the total energy of the steam, so it is not at all improbable—The theoretical efficiency of the boat turbines at 2500 is about (?) 90% leaving 22% for losses. It is possible that the actual steam being used is somewhat above 18600 calculated but it cannot be very much so. In torpedo boat engines the consumption is about 17 to 18 lb^s (of steam) *per IHP or about 20 per Brake HP. I am inclined to think that Yarrow tried 24 sets of different screws and raised his speed from $20\frac{1}{2}$ to 23 knots thereby. I think we have a greater rise in store for us with the same steam consumption we had in the last trials.*

It would be very satisfactory if we can get 4 knots more, the saving of some 500 HP would about do this.

We have one very favourable enquiry for a boat 104 ft long 14'0" beam 3 feet draught for passenger traffic, by the Youghal and Blackwater Steamboat C^o Ltd of Youghal C^o Cork, speed about 18 miles per hour. We have 5 or 6 enquiries not suitable. There appear to be two suitable sites at Wallsend and Hebburn and we are getting particulars etc.

We are testing some reversing methods—one suggested by Swinton looks very promising.

Your affectionate brother,

Charles

Harvey is busy with the prospectus and I have sent a long and buttery letter to W.H.White (which should satisfy Swinton as to its tone). Your little turbine is waiting a new commutator which will soon be finished, when it will be tested and sent off.

Because the three axial flow turbines shared the same steam flow and were now in series, some delicate adjustments were required to the designs to ensure that each turbine developed roughly the same fraction of the total power. As well as this, the balance between the thrust on the bearing due to the propeller and that due to the net steam pressure in the turbine had also to be attended to.

The boat now had nine screws mounted on three shafts. They were of 18 in pitch and 18 in diameter. The turbines had reached higher speeds than intended. At this time Parsons reckoned that of 1860 HP produced by the turbines, 45% i.e. 837 HP, was wasted in skin friction at the propellers. A slip of 20% meant that 80% of the remaining 1023 HP, or 850 HP (actually 818 HP), was left to propel the boat. Put another way the 'propulsive efficiency' was around 45% . Adopting a longer pitch of 24 in would allow the same boat speed to be held with a much lower turbine speed. This lower speed of rotation would also help to reduce the friction loss caused by the wide propeller blades.

Already there was enough public interest in turbine propulsion for Parsons to receive specific enquiries for manufacture, and he has been looking for sites along the river for a new factory. The reference to a *'buttery'* letter to W H White *'which should satisfy Swinton'* is a reminder of the episode mentioned earlier in which it was Swinton's counsels which succeeded in ending the acrimonious breach with Clarke Chapman, and achieved the amicable recovery of Parsons' valuable patent rights. Harvey was the solicitor who had been charged with the preparations for the launching of the original, experimental, company. He was now preparing the launch of what was to be known as Parsons Marine Steam Turbine Company

Laurence ventured some suggestions to his brother. One was inspired by the milk separator which had been developed by C G P de Laval, in which the bowl containing milk and cream is mounted at the top of a long vertical shaft. This and other proposals are treated seriously, but the arguments against are stated in a reply (C.L.(30)),

Holeyn Hall
Wylam on Tyne
5 Feby 1897

My dear Laurence,

I have been reading over again your letter re turbine engines after consideration.

First as to taking the weight off bearings by spring blades pressed upwards with a force equal to the weight carried.

I think something might be done on this principle with advantage. It has been considered but never tried.

I do not think (?) *shafts placed on end would answer practically. Doubtless milk separators work well so placed, running on two hard steel balls. Stoney tried balls like this in a turbo but they would not stand constant work and eventually seized and melted, the collar thrust bearing works better but I think it would give trouble placed on end, that is in ordinary use, and besides the weight cannot all be taken by the steam pressure which is variable.*

Stoney also tried last year, ball bearings but they would not stand the speed.

As to end balance in boat motors we have in our patent a device the same as you suggest.

I do not think anything beyond what we have in the Turbinia is necessary, for when we ran on the mile there was 5,200 lb^s^ on the thrust of the low pressure motor viz the screw thrust was about 3000 lb^s^ and the (?) *sternward steam pressure 8200 lb^s^ leaving 5200 for the thrust* (bearing) *to carry, this as you know occurred through some mistake—the steam balance can be got to within 500 lb^s^ of the* (propeller) *thrust in ordinary practice easily, and the thrust bearing can carry 1000 to 1500 lb^s^ easily I think.*

Then as to the self contained oilers for the bearings, I think we must try something of this sort. If it could be got to work satisfactorily it would be a great advantage undoubtedly , and expense in construction also-

In our large turbos we are now putting the oil under 10 to 20 lbs pressure to ensure better lubrication—same as we have in Turbinia—but for small plants a method of the kind you suggest, if successful would be a great advantage.

Boat The new screws 2nd set will be on Monday. She will be launched on Tuesday and the machinery put on board again, and I hope to have a run again in 10 to 14 days.

W.H.White is coming on Monday to look over her. I think there is business coming in this quarter. We have got out a list of machinery wanted, some £4,000 to £5,000 worth, so as to be ready to order it without delay. Tool makers are busy now and this will be the largest part of the programme for the start. Buildings will go up quickly. We are also setting out shops suitable for the Works. Negotiations are still in hand for the land. On the whole (?) Manducs at Wallsend seems most suitable so far.

Returning to the W.H.White visit (his letter only reached me this evening) I presume he will examine carefully and see there is nothing of special refinement in the screws, shafting, stern tubes, glands—boiler, engines, nothing in fact to require nearly as much skill as the management of a set of modern torpedo boat engines—all this I will endeavour to point out. Also that the boiler tubes show no signs of having been forced. Personally I think we can get some more steam out of it, and this added to the clear advantage of the change of screws ought to give us the speed we want of 32 or more knots.

The larger chimney will, by the way, have a double advantage, it will of course improve the draught and also, should a boiler tube burst, as sometimes happens, all the steam will go upwards freely.

It rather seemed to me that it might be well to get some outsider like Prof. Weighton Prof of Engineering at the Durham College to make steam and water tests. I think I shall ask W.H.White what he thinks. You see if we got someone of sufficient standing from London or at a distance, he might be days or weeks waiting for a smooth enough sea, and it would be a question if he could wait. Besides this, it is always better I think, to get a local man if possible. And we are not going to ask for the money in London.

Forgive this long letter.

Your affectionate brother,

Charles

The plans to arrange for independently witnessed tests to be carried out is proof that he was sure that success was close. W H White's encouragement has given him enough confidence to press on with the planning of the new factory. His letters show that he had great confidence in the resistance measurements which he had made with his own models, but he suspected that more reliable data might be available from the Admiralty research establishment. As already mentioned, early in 1895 he had written to E R Froude to know if any tests had been carried out at Haslar on models of the form of *Turbinia*. He was told no, and that such tests would require Admiralty permission. Yet on 15 and 16 February 1897 preliminary tests were in fact carried out, with a second set between 18 and 21 May. It seems that Parsons knew of these, and they may well have been the result of the close interest exhibited by senior Admiralty staff such as Sir W H White, who incidentally had recently been conferred with a knighthood in recognition of his work. The mistake which had been made with the thrust balance of the low pressure shaft did at least have one compensation. It showed that the thrust bearings had plenty of overload capacity. In a later letter to Laurence he writes (C.L.(31))

Holeyn Hall
Wylam on Tyne
16 Feb[y] 1897

My dear Laurence,

You seem to have made a satisfactory purchase of the 'Hargreaves Estates'. No doubt near towns much is done in the way of buying when opportunity offers and the price is low as compared with many building sites, £500 per acre is often given on the Tyne for ordinary building land.

There is no particular news here. Katie has been away for a fortnight in Torquay and is now in London with Maurice, and returns here on Monday in time for Nansen's lecture. I saw Geoff (son of Laurence) *at the Club on Saturday. He went to our neighbours the Fenwicks for Sunday, and came with us to hear Nansen. Leyland is still on the sofa. I rather think he must be worse than they give out. He writes every week, but beyond what he says himself, I hear very little.*

We are thinking of a Board meeting about 26th to talk over details and try and get things fixed up, as we hope to get some longer runs next week weather permitting.

Sir W.H. White turned up unexpectedly last week and spent all one morning till 1.30 looking at the motors and the boat on the slip. He examined everything closely and expressed himself pleased, and I gathered he was convinced that we could go astern fast if we wanted by a reverse turbine on the Low Press. Shaft. The only thing he suggested could be improved was the stern of the boat, which he thought should be built up square as shown (figure 72). *He thought that if water came over the top of the present tail it might suck violently and retard the boat. I myself do not think it comes over enough to do this, but quite agree what he suggests, though ugly, would be a practical improvement-*

W.H.White quite agreed with me that the new screws (both sets) ought to be superior to those on at the last trial. I have got the communication for the Naval Architects completed as far as possible pending further trials.

Your affectionate brother,

Charles

Figure 72. Image from letter 31 (C.L.(31)).

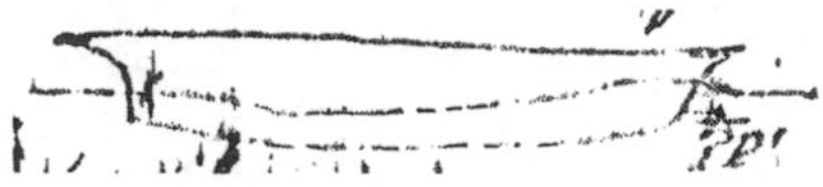

Figure 73. Image from letter 31. PP' is a pressure sensing tube. (C.L.(31)).

P.S. I mentioned to White that you had made some speed recording instruments. He at once exclaimed that they had all, including Froude, been on the problem for years and had tried every imaginable device, but the real difficulty was that the float was dropped into disturbed water and failed to tell the true vel^{y} *of the ship—no apparatus had so far given reasonable results.*

He thought the pressure type of tube, projecting was the most hopeful (figure 73)*, but this was affected by the pressure wave of the hull. He said they had thought they had found a place not so affected, but had been mistaken and it was a most difficult problem. I imagine that they have been aiming at extreme accuracy, but he did not mention the extent of the errors they had observed.*

The most accurate measure of speed was to time runs in both directions past two accurately located mile posts out at sea. Other means had to be resorted to on occasion. Laurence appears to have been advocating a pitot type of instrument. Frank Hodgkinson recalled his part in these tests (Hodgkinson 1939).

It was not always convenient to go out to the measured mile; so arrangements were made to measure the speed by means of a fishing line with a piece of wood at the end. At two appropriate points a piece of copper wire was wrapped around the line, and the line ran out between electrical contacts, Sir Charles delegated his brother, Lord Rosse, to obtain a good reel and line, but he only purchased a line. It was my job to attend to the speed measurement. I coiled up the line in the cabin in such a way that, with hope, it would pay out freely. The line got snarled up and was broken. Sir Charles was disgusted, not because of me, he said, but because of the parsimoniousness of his brother.

The communication for the Naval Architects refers to a paper which he was invited to present by Sir William White, 'On the application of the turbine to marine propulsion.' (It was read to the Institution of Naval Architects on 8 April 1897.) Apparently the astern turbine had not yet been fitted. His next letter followed after a week (C.L.(33))

Holeyn Hall
21st Feby 1897
Wylam on Tyne

My dear Laurence,

We are preparing to have a boat meeting on the 26th to talk over matters a little further, but seeing that we shall only commence the more complete trials of the boat on Wednesday next, we cannot do very much beyond talking matters over. We hope this and the following week to get some further measurements of speed, water and coal, weather permitting. The river course is too short to get accurate measurements.

Leyland is better but not yet able to get about.

Your affectionate brother,

Charles,

Leyland's illness continued to keep him from participating in the trials. A week later Parsons wrote again (C.L.(32))

Holeyn Hall
28th Feby 1897
Wylam on Tyne

My dear Laurence,

It has been blowing a whole gale for the last 10 days. We went out on Friday morning in the midst of it, one of Palmer's 30 knot destroyers went out at the same time for a preliminary run. We steamed round her, she got going at 15 to 20 knots. We got drenched and had to turn home on account of a blowing joint and losing fresh water, and also to get back in time for the meeting at 2.

We talked matters in general and the impression is now not to build hulls at all, but to sub-contract them and fit them up. We also thought we should approach Ewing (Prof. of engineering at Cambridge) who made the last report we have had on the turbine—to ask him to test the Turbinia in conjunction with Prof. Weighton, Prof. of engineering at the College in Newcastle. The latter to do the details of the trials.

Figure 74. Guests grip the rails at the stern of *Turbinia* during a run at 30 knots, courtesy of the Discovery Museum, Newcastle upon Tyne.

> *We shall be out again on Wednesday, weather permitting, and I hope to get some further results of a favourable kind.*
>
> *Your affectionate brother,*
>
> *Charles*

This run was No 27, carried out on 26 February in a '*heavy W gale*' with '*Stoney, Barnard, Self*'. The measured mile ran roughly N.N.W by S.S.E. The small size of *Turbinia* and her low freeboard made her very vulnerable in bad weather, especially to winds from such a direction. Steaming at full speed, the seas broke over the deck and neither passengers nor crew escaped a soaking. On returning to land Parsons had right away to face in to a business meeting. It is clear from this letter that Parsons' intentions were to pursue manufacturing as a means to developing the steam turbine and not as an end in itself. He was not interested in creating a massive shipbuilding organization just as a commercial proposition. His thoughts on testing changed. He had had an excellent report from Ewing on the turbine for Cambridge, and besides as Professor of Engineering at Cambridge Ewing's views were likely to carry more weight with the Admiralty.

The pressure to make progress was now intense and the notebook records five trials between 26 February and 1 April; in each case those present were Stoney, Barnard and Parsons. In the next letter Parsons reports (C.L.(34))

Holeyn Hall
Wylam on Tyne
7 March 1897

My dear Laurence,

We went out in Turbinia on Thursday, but it was blowing hard with sleet. The wind had shifted from W to SSW so it was too rough outside. We got a run on the river $2'1\frac{1}{2}''$ *and a bad start. We estimated that we reached 31 knots on part of the run. Today we luckily got a smooth sea with a long swell, and got a long run and measured the* (boiler) *feed water. Altogether we covered 30 miles at* $25\frac{1}{2}$ *knots. Two runs on the mile N and S gave a mean 28.12 knots, and feed was 20,095 lbs per hour at 26 knots.*

With the present screws, No 2 set 24″ dia^r 24″ pitch 6 propellers, the revolutions maximum were 1900, and our fan draught was only $5\frac{1}{2}''$ *and 4″ (water gauge) on longer runs which was enough to hold 100 lb^s on the* (steam pressure) *meter steadily. The stoking was very easy. The performance of the screws, which as you know were put in hand 5 months ago, and before the last trials (at which you were present) seems to confirm our conclusion that smaller diameter and longer pitch ratios will give better results, and that our No 3 set of 9 screws 18″ dia^r and 24″ pitch will be much superior.*

The bottom of the boat is getting slimy and this will, with the roughness of the sea account for about $1\frac{1}{2}$ *knots. At times on riding over a swell some 20 feet of the bow was clear of the water, but there was no racing of the screws. The thrusts were quite cool, a good vacuum and everything worked without a hitch.*

We are going to slip the boat this week and put on No 3 set of screws which are ready, and I expect we shall reach 32 knots and steam steadily at 30 to 31.

Prof. Ewing of Cambridge and Prof. Weighton of the Newcastle College are willing to come and make a joint report.

In regard to the water consumption the measurement agrees very closely with the calculated consumption (within 5%) and the water meter agrees with the theoretical discharge of the pump, assuming about 15% slip (a moderate allowance for a duplex pump). All therefore goes to confirm that in the long run the engines were consuming about 15 lbs per brake HP, and giving 1340 BHP. The slip was about 17% and the IHP from the † model expt[s] at the above speed should be 1075. If we divide 1340 by the ratio (0.85) of Brake to IHP, we get 1586. There is thus about 500 HP unaccounted for, which I think is due to the skin friction of the (necessarily) abnormally wide blades of 1 pitch ratio, which at 2000 revolutions works out at over 800 HP. I hope the No 3 set will reduce this by more than one half, or save 500 HP at 30 knots.

Going back Comparing the No 1 and No 2 sets. †

The pitch ratios of these sets are both unity, but No 2 blades are narrow and the screws of more ordinary proportions. The results of No 1 and No 2 are very nearly the same, but slightly in favour of No 2.

I discussed the screws fully with W.H.White and so far as he went, he agreed with me, but he said 'you cannot tell till you try', but from what we now know and which is really dependent on ordinary coefficients, skin friction etc (nothing abnormal) it seems to me that it is quite clear that No 3 ought to be much better than No 1 or No 2 set. The steam consumption of 20,075 with 1556 IHP (P6) (note that page 6 of the letter is marked thus †, see above) *gives 12.6 lbs per IHP, which is very low indeed for any marine engine.*

Your affectionate brother,

Charles

I hope you will excuse the scrawl, fingers are cold and raw with runs.

That last postscript says so much. The writer had just spent two consecutive and physically tiring days at sea in icy weather, and yet the letter is one of the longest of the series. Obviously he is delighted that things are making sense at last, and he has virtually reached his target. It is not enough just to report that they exceeded 30 knots, but he is keen to tell how all the data fit together. The lengthened propeller pitch had produced the desired effect and reduced the maximum speed of rotation of the turbine, without a serious loss of thrust. The adjustments to the three turbines had been effective and none of the thrust bearings was overloaded. We know from his notebook that the reversing turbine had been made operational. On 4 March it notes *'astern revs 288.'*

It seems likely that the 'model experiments' referred to were those made by himself and not the ones made by Froude on 15 and 16 February.

The facilities for measuring the feed water supplied to the boiler had been improved by the adoption of water meters. Characteristically he compares the new results with the older method to satisfy himself that he had not been misled by relying on counting the strokes of the feed-pump. From now on he commands a firm base for future designs. He lost no time in preparing for the definitive trials, delaying only long enough to fit the No 3 set of screws. On 30 March the notebook records '*no wind long swell*'. Stoney recalled such a day (Dowson 1942),

> *We went out to make for Professor Ewing a consumption test at 12 knots. There was a nasty swell on, which seriously affected most of the crew, though Mr Parsons himself seemed proof. We went up and down the coast over the Hartley mile, and we rolled, and rolled, and rolled. I was taking the times and as I sat on the deck I wondered whether the mile posts would come along first or whether the sea would take its toll. The mile posts did come along, however the sea took its toll, and I did not read the stop watch for the next five minutes.*

Froude replied to a recent letter (Goodall 1942),

> *30th March 1897*
>
> *Dear Mr Parsons,*
>
> *Many thanks for your letter which suggests several interesting points for investigation. There is no doubt that we are sadly in want of more experimental information on some points, now that the question of cavitation at high speeds has put a new aspect on the problems by putting a maximum limit on the pressure per unit area.*
>
> *However we can do nothing in the matter here immediately, so I will wait to give fuller consideration till I have had the advantage of reading your I.N.A. paper. I am most interested to hear of the hot water experiments. I wish we could boil our tank and cool it again afterwards. ...*
>
> *In haste, yours very truly,*
>
> *R.E.Froude*

12.5 Objective achieved!

On 1 April Parsons was out again with Barnard and Stoney. On the last run he reached 31.01 knots and collected the data which he needed for his paper to the Institution of Naval Architects in London on 8 April. Parsons lost no time in announcing his success. The paper that he read to members of the Institution of Naval Architects carried the results of the run made only one week earlier on 1 April (Parsons C A 1897a). He was able to announce that a mean speed of 31.01 knots on the measured nautical mile had been achieved, with a maximum of 32.6 knots. This occasion must have felt something like the meeting of the British Association for the Advancement of Science held at Cambridge in 1845, at which his father was able to pass around the pencil drawing which he had made of M51, the first time that a nebula had ever been shown to have a 'spiral' structure, the first fruits of the Leviathan. Charles for his part had not only met the challenge of adapting the steam turbine to marine propulsion, but in doing so he had reached a record speed. For this paper Parsons quoted data taken by Stoney and himself, and calculations of turbine power were based on measurements made on his own model hull. The paper contained an account of the stages in the development of *Turbinia*, including the first ever account of a study of cavitation made on a model propeller with heated water and stroboscopic illumination. In the discussion afterwards it was noted that *Turbinia* had exactly equalled the record set by Normand's *Forban* on her trials.

The very next day, 9 April, Parsons was back in Newcastle, and joined by Barnard, Stoney, Ewing and his assistant Stanley Dunkerley for trial No 32. Great care was taken in measurements of steam consumption, using calibrated water flow meters for the boiler feed, as well as steam pressures gauges at the boiler and the turbine inlet. Similarly the number of revolutions made by each propeller shaft was recorded. There was no provision for measuring power directly, and so in writing his report Ewing had recourse to the results of model tests. He does not reveal in his report the source of this data, but we know now that it was sent to Parsons on 13 April by E R Froude (Goodall 1942). Data were collected over a range of speeds, from 6.75 to 31 knots. Ewing later recalled that (Ewing 1931)

> *After we had been cruising for several days at various speeds over a measured mile on the north east coast, observing the relation of steam consumption to speed in weather which was too rough to allow the engines to be worked at their full power, we were returning up the Tyne at the modest pace allowed by local regulations. The river, as it happened was nearly empty, the tide slack and the water smooth. Passing the posts of a measured mile on the river bank below Wallsend, Parsons was tempted, and said 'What about a full power run here?' to which I replied 'She's your ship'. In a few minutes we were speeding through the water at a speed which in those days was a record for any vessel.*

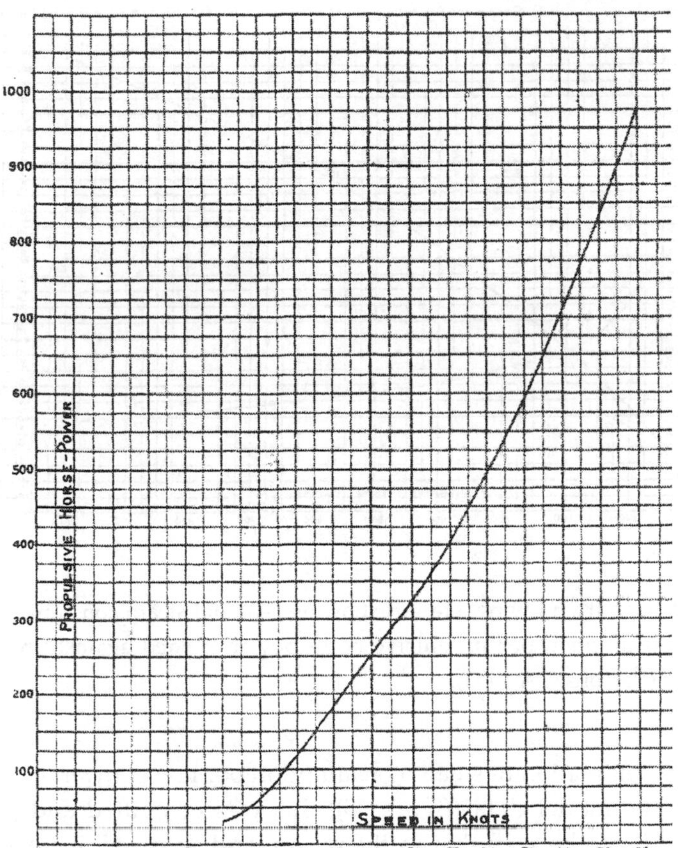

Figure 75. Estimates of propulsive horse power from Froude's experiments on a model of *Turbinia* (Parsons C A 1903).

A formal full speed run, in both directions, on the measured mile was duly carried out. As Ewing states in his final report (Parsons C A 1903),

> *The speed reached in this trial, 32.76 knots in the mean, is, I believe, the highest recorded for any vessel. It is greatly in excess of the speed hitherto reached in boats so small as the Turbinia.*

Apart from the maximum speed achieved, other parameters were crucially important. In particular some estimate of fuel economy was needed. Assuming that competitors would use boilers of similar efficiency, this

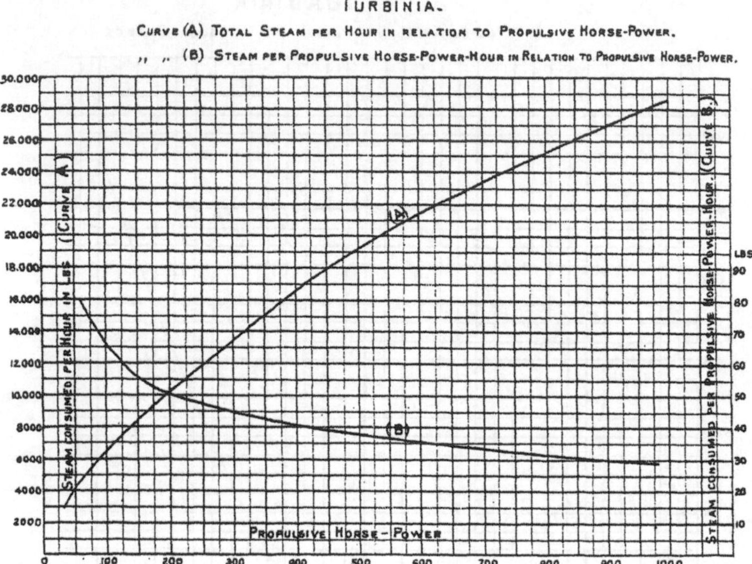

Figure 76. Graph from Ewing's report showing (A) total steam consumption and (B) steam consumption per propulsive horse power, plotted against propulsive horse power (Parsons C A 1903).

required the estimation of steam consumption per horsepower.

Using the model data (figure 75), this was computed and plotted (figure 76). It can be seen that as the power, and so speed, increases the steam consumption per propulsive horsepower falls steadily. Propulsive horsepower is less than turbine output power because the overall propeller efficiency is less than 100% and because of losses due to slip. By making the assumption that, as in similar craft, 50% of the engine output was converted to useful propulsive horsepower, Ewing concluded that the turbines delivered 2100 HP for a steam consumption of $14\frac{1}{2}$ lbs per hour, which was a very competitive achievement. Ewing noted other features,

> *The turbines are remarkably handy, and allow sudden starts to be made with impunity... Starting from rest a speed of rotation corresponding to 28 knots was acquired in 20 seconds after the signal was given to open the stop valve. A conspicuous feature is the absence of vibration, due to the absence of reciprocating parts, and the perfect balance of the turbines. The contrast in this respect with ordinary high-speed boats is most striking. An*

incidental merit of the turbine is that it needs no internal lubrication, and the oil which circulates through the bearings has no opportunity of mixing with the steam. The advantage of having the condensed steam free of oil will be particularly felt in cases where water-tube boilers are used.

He went on

The general impression I have formed from the trials is entirely favourable to the prospects of this novel method of marine propulsion.They (turbines) *have a distinct advantage over ordinary engines in the first cost, in probable cost of maintenance, and in the cost of attendance, as well as in bulk, weight and in freedom from vibration.*

Armed with such a testimonial, Parsons was now in a position to press the Admiralty for business. At the highest levels there was encouragement from men like Jackie Fisher, third Sea Lord at the Admiralty, and Rear Admiral Sir John Durston, Engineer in Chief. Nevertheless there was still much resistance in the Navy among serving officers, and something dramatic was needed to capture the public's attention. An opportunity was found in the Naval Review at Spithead which was to be held on Saturday 26 June. It was Sir John Durston who suggested, and obtained permission for, *Turbinia* to be present at the Diamond Jubilee Review (Osler and Grieve 1980c). On the journey south from Newcastle Parsons himself travelled on board.

12.6 Diamond Jubilee Spithead review

By the year 1897 Queen Victoria had ruled the United Kingdom for 60 years and for the last twenty years she had also ruled as Empress of India. During her reign the synergy of her country's bustling manufacturing industry, and the growth of an empire which extended to every continent, had propelled Great Britain to unparalleled heights of prosperity and influence. Essential for this was Britain's maritime prowess. Her shipyards supplied a large fraction of the world's mercantile fleet. Her own fleet of merchant ships brought cheap goods from the underdeveloped world, and carried textiles and other manufactured goods to customers all across the globe. The part played by the Royal Navy was central. Even though many technical innovations were pioneered abroad, the strength of British industry had helped it to maintain its superiority. This was evident when the French stole a march with the first ever ironclad, the *Gloire*, in 1858. They were quickly overtaken when *HMS Warrior* was commissioned in

1860, the first warship to be constructed all iron, with an internally divided hull. But as the turn of the century approached, the United States and the European 'powers' were enjoying the fruits of their own industrial development, and were ready to challenge Britain. This trend could be seen in the increasing attention being paid to the construction of ever larger battleships, heavily armoured and heavily armed.

In the months before the Jubilee celebrations, the *London Times* carried accounts of fund raising efforts being made to alleviate famine in India, and reports of fighting at Epirus in the Graeco Turkish war. On 30 April the Queen returned from Cherbourg on the Royal Yacht escorted by two warships *HMY Osborne* and Trinity Yacht *Irene*. On 1 June an unprecedented Levée was held at St James's Palace at which 600 people were presented to the Prince of Wales. There were reports from places throughout the kingdom and abroad of the plans being made to celebrate the Jubilee. The *Times* of 12 June carried a map showing the berthing arrangements for the Naval Review that was to be held at Spithead in the Solent, between Portsmouth and the Isle of Wight. It brought together a great number of vessels, over 150 in total, of which a number had been built on the Tyne. When marshalled in four main lanes they stretched for a total of 25 miles. Of these, 22 were battleships armed with 13.5 in calibre guns. Ten of them displaced 14 000 tons and belonged to the *Majestic* and *Royal Sovereign* class, with the latter having the capability of reaching 17.5 knots. The ships made a brilliant display, with buff coloured funnels and their upper structure painted white; the hulls were black above a red waterline. Flags and bunting fluttered from stem to stern. Two further lines were reserved for special merchant vessels and for warships representing 13 countries. Among the latter was to be the *König Wilhelm* with Prince Henry of Prussia aboard.

On 22 June the Queen processed in Imperial state around London. At last on Saturday 26 June the Naval Review began in fine weather. The festive atmosphere is well captured in the painting shown in figure 77. Some smaller vessels had been given the job of keeping order, and ensuring that the lines were kept clear for the Royal procession which was headed by the Trinity House yacht *Irene*. Next, on the Royal Yacht *Victoria and Albert* was Edward Prince of Wales, representing the Queen. Two more vessels carried royal guests. After them came the Lords of the Admiralty with their guests in the *Enchantress*. On board was Sir William H White, the Admiralty's director of naval construction. Finally came the *Danube* with members of the House of Lords, the *Wildfire* with colonial premiers, the enormous liner *Campania* with members of the House of Commons and,

Figure 77. A painting by Eduardo de Martino of *Turbinia* at Spithead on 26 June 1897. The Royal Collection © 1999 Her Majesty Queen Elizabeth II.

completing the procession, the *Eldorado* with foreign ambassadors. There, waiting for the procession of dignitaries to end, lay *Turbinia* with steam up and under the command of her Captain, Christopher Leyland. Charles Parsons as usual was at the engine room controls and Gerald Stoney by his side. As passengers, they had Lord Lonsdale and Sir George Baden-Powell aboard. *Turbinia* moved out into the lane between the battleships and cruisers, accelerating rapidly until she reached the astonishing speed of 34.5 knots. The impact which this display made on those present is perhaps best conveyed by the report of the special correspondent of the Times newspaper written on 27 June aboard *HMS Powerful*.

> *During the passage of the Royal procession the lines were kept creditably clear by the vigilant and ubiquitous patrol-boats told off for the purpose, but in spite of all their efforts some small craft and steam boats managed to defy their authority. Among these was the now famous* Turbinia, *the fastest vessel in the world. At the cost of the deliberate disregard of authority, she contrived to give herself an effective advertisement by steaming at astonishing speed between the lines A and B shortly after the Royal procession had passed. The patrol boats which attempted to check her adventurous and lawless proceedings were distanced in a twinkling, but at last one of them managed by placing herself athwart her course to drive her out of the lines astern of the French cruiser* Pothau. *Her speed was, as I have said, simply astonishing, but its manifestation was accompanied by a mighty rushing sound and by a stream of flame from her funnel at least as long as the funnel itself. Unless these commonplace, but very serious defects can be corrected, it is manifest*

> *that the system of propulsion devised by Mr Parsons cannot be applied to torpedo boats for whose operations, silence, secrecy, and invisibility are indispensable. The Turbinia again made her appearance as the Royal Yacht was weighing her anchor and assuming a position, rather slowly and with much backing of her engines accompanied by the setting of a jib, which would enable her to steam through the lines. The Turbinia waited astern of the Powerful until the Victoria and Albert was well under way, and then followed her, at first at a moderate speed, but gradually quickening up until the sea in her immediate wake was churned into a mass of white and seething foam. Probably she overtook the yacht within a short distance and passed her at full speed, and perhaps her lawlessness may be excused by the novelty and importance of the invention she embodies. But visitors to the Jubilee Review of 1887 will perhaps remember that a prominent feature of that occasion was the appearance of the Nordenfeldt submarine, or rather submerged boat, just as the appearance of the Turbinia was a prominent feature of the present occasion. Little or nothing has since been heard of the Nordenfelt boat as a practical invention. Absit Omen. Everyone would regret if the Turbinia, after her brilliant but unauthorised exhibition of yesterday should turn out only a similar nine days wonder.*

The suggestion of cheeky illegality which this report conveyed enhanced Parsons' popular esteem, but it was quite erroneous. It seems Prince Henry of Prussia, younger brother of Kaiser Wilhelm II, who had been watching the earlier display from the deck of the *König Wilhelm*, had made a request that *Turbinia* should come alongside his Flagship and show a turn of speed. Permission was given for this by an admiral and conveyed to Leyland, who recalled (Leyland 1935)

> *I saw a vedette boat following, and as I surmised it would have a message for us, I stopped. The Lieutenant told me that Prince Henry of Prussia wished to see me alongside his Flagship. The Prince asked me to show a turn of speed... we steamed to westward, shoving on the coal. We then passed between two ships and met a vedette dashing through to head us off. It was a close shave. I just had space to avoid her. Her crew dashed into her bows and her Lieutenant unbuckled his sword, expecting to swim. He evidently spoke to me and I said something to him, but as we passed at nearly 45 knots, it may have been just as well that our impromptu remarks did not carry.*

This account was confirmed in a letter to the editor of the *Times* from Baden-Powell, who added that a speed of 34 knots had been reached *'with an entire absence of vibration'*. It is also said that, on the evening of the

Review, Parsons received an invitation from the Prince to go aboard the *König Wilhelm* to discuss the future potential of the marine steam turbine. This was a significant meeting because the turbine was later adopted by the German Navy on the urging of the Kaiser himself against much opposition.

As Leyland observed, other boats were not '*quick enough to give us right of way, whilst some lost their heads completely.*' Stoney remembered (Dowson 1942)

> *Shortly after, we were going quietly along towing a boat from Captain C.J.Leyland's yacht, which was our house boat. Just as we went past the bow of a battleship, a French yacht appeared. There was no time to stop, but Leyland who was our captain signalled full steam ahead. I promptly sat down on deck, as the acceleration of 2500 s.h.p. on 40 tons is pretty big. We shaved the bow of the big yacht but the tow rope broke and the boat behind (in which was Leyland's Scotch skipper) went bang into the side of the French Yacht. I just heard a volley of French and Scotch from the two skippers.*

Despite the unusual handling characteristics of *Turbinia*, Leyland was quite prepared to take chances to display his craft to the best effect. Fortunately nothing worse than 'close shaves' resulted, but had a collision occurred the consequences for Parsons would have been very serious indeed. The fact that Parsons was prepared to take such a chance shows, once again, that courage was one quality which he had in abundance.

It was one thing to make runs off the mouth of the Tyne, but the journey south to the Isle of Wight had been something else. Like all steam powered vessels of the time *Turbinia* had, among its ten or 11 crew members, several stokers. While she may have been superbly designed for high speeds, her long narrow shape guaranteed that in heavy seas she gave a most uncomfortable ride. John H Barker was appointed by Parsons as Engineer to the Cambridge Electric Lighting Company in 1893, and he recalled (Barker 1939)

> *My only association* (with Turbinia) *was to send a fireman* (stoker) *from Cambridge to take the ship home from the Diamond Jubilee Naval Review at Spithead. The stokehold crew were drunk with success and good cheer; they mutinied or were frankly incapable.*

Turbinia became a great attraction. Leyland recalled a couple of the visitors to whom he had been asked to demonstrate what she was capable of (Leyland 1935).

> *Among those we took to sea off the Tyne was the then Chief Engineer of the Navy. It was rather rough at sea, and after passing the piers I asked for*

Figure 78. *Turbinia* travelling at speed; note how the bow has risen out of the water; courtesy of N E I Parsons.

> *instructions and was told to run her at full speed, which we did to northward. The Chief Engineer of the Navy was sitting on an after skylight. I went to him on several occasions, and suggested he should come into the wheelhouse, but was always received with a negative shake of the head, although he was being drenched. At last he said, 'If you will stop her I will move, but not till then.'*
>
> *We also took out a Russian Naval Attaché who came on board in patent leather boots, and although I offered him a spare set of waterproofs and sea boots, which I kept on board, he would not have them. When we reached the mouth of the Tyne he asked me to turn back as the seas were breaking over us, and then confided to me that his wife had made him promise not to go outside. I was somewhat annoyed.*

In the intervening years, the design of *Turbinia* has been the subject of much attention. The data contained in the surviving notebook and drawings, as well as the records of the tests carried out by E R Froude at Haslar, have been used with computer programs to obtain an up to date, objective, assessment. In a way this can be likened to the recent physical reconstruction of the giant 72 in telescope at Birr which has demonstrated that the mechanism of the telescope is a successful engineering construction, though it has not yet been possible to extend this judgement to include the optical performance. On the occasion of the centenary of *Turbinia's* record breaking run, her design was assessed by two naval architects with regard to the hull shape, and to the propulsion package, and it was then considered in the light of a systematic review of international developments in the period 1905 to 1982 (Mackie and Hutchinson 1997). Using the most modern procedures, they concluded that at 34 knots an effective propulsive power of 698 kW for the hull alone was required. This was compared with a modern equivalent and they

> *concluded that the equivalent form tested at the David Taylor Model Basin in the 1950's and 1960's was unable to better the resistance characteristics achieved by Parsons more than half a century earlier.*

A similar analysis of the performance of the turbine driven propellers was carried out, using data for modern propeller designs. It showed how the change from the single shaft three screw configuration, to three shafts with three screws on each, brought the propeller efficiency up from 40% to between 60 and 70%. They summarized by saying,

> *So as with the hull form, the propulsor package developed by Parsons was near optimum for the ungeared direct drive propulsion system that he was seeking to demonstrate.*

Chapter 13

Turbine Blades and Propellers

In the course of his struggles to develop the steam turbine to the point where it could be a commercial challenge to the existing, conventional steam engines, Charles Parsons fought on two fronts. For land use he perfected the turbo-generator as a source of electric power for a supply industry which was still in its infancy. At sea the struggle was to match the divergent demands of the turbine on the one hand and the screw propeller on the other. He was successful in both battles after a period of only 13 years, years that were packed with innovative creativity of a high order. Central to much of his work during this period was the science of fluid mechanics, which describes how fluids, liquids and gases, interact with solid structures. The foundations of that subject had been laid long before by mathematicians like Daniel Bernouilli (1699–1782), but experimentation continually revealed deficiencies in the 'model', friction-free, fluids which were postulated. For example, theory suggested that a steamlined body like a fish should not experience any net force when moved through a liquid. The reality with actual fluids is of course quite otherwise. Experimental facts were being established by the observations of engineers like the Froudes, father and son, and Osborne Reynolds, and these were still being explored during the late 19th and early 20th century. Concepts like streamline flow and boundary layer were not yet familiar to engineers. This was just the period when Charles Parsons was himself struggling to perfect machinery such as turbines, compressors and propellers that involve fluid flow. Because he was in some respects so uncommunicative, the only way to establish how his thinking evolved is by examining closely each step in development. This is well illustrated by following the way that the shape of his turbine blades evolved.

Earlier, in chapter 9, an account was given of the progress which was

Figure 79. Part of a ring of blades cut out and formed from a strip of delta metal (Richardson 1911).

made between the construction of the first turbine and the point at which Parsons broke with his partners. After his break with Clarke Chapman in 1889, when Parsons was forced to develop radial flow machines, the blades continued to be machined from a solid brass casting, which was now in the shape of a disc (see figure 51). Since this was clearly expensive to manufacture he turned to an approach first adumbrated in patent 6734 of 1884,

> *although I have shown straight blades cut from the solid... yet any other forms of blades may be employed. In some cases it may be found convenient to make the blades of sheet metal and to secure them in grooves or recesses in the rings. In this way the blades may be made thinner and they may be accurately formed either before or after insertion in the rings.*

A study of water turbines had shown that an optimum design requires that when the fluid emerges from the moving blades there should be no potential for doing further work, that is, there should be no component of velocity in the direction in which the blades were moving. In other words the steam should be flowing parallel to the shaft axis. Furthermore Parsons' design required that the steam should be accelerated by causing it to pass through passages of reducing cross section as it traversed the successive rows of blades. Simply curving thin blades is enough to ensure that the width available for flow between the walls decreases. This can be seen in figure 80, in which two equal diameter circles have been placed to the left of the row of fixed blades to assist in gauging the distance between the walls when measured at right angles to their surface. This diagram is drawn using a ratio between steam velocity v_1 and the blade velocity v_b, which gives optimum performance. The dashed arrows show the direction of steam flow relative to the blades when the pressure drop is appropriate. The velocity v_1 of the fluid leaving the fixed blades has a large component

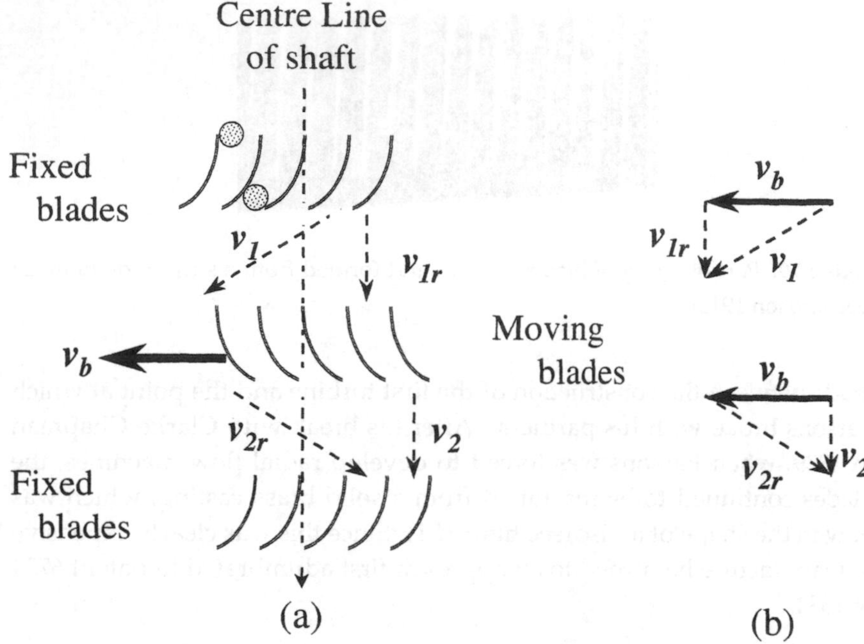

Figure 80. A diagram showing the velocity of steam as it traverses fixed and moving blades. (a) A view of rows of fixed and moving blades seen from above. The dashed arrows indicate the magnitude and direction of fluid velocities relative to the fixed blades, v_1 and v_2, and relative to the moving blades, v_{1r} and v_{2r}. (b) How the blade velocity and steam velocities combine.

which is at right angles to the axis, and v_1 combines with the large blade velocity v_b in such a way that, relatively, the steam appears to enter the moving blades in an axial direction v_{1r}. Finally, because the volume of the steam expands significantly as the pressure falls, the geometry of the blading, in particular its height and its mean diameter, needed to be varied continuously if the steam velocity were to be kept in the same proportion to the blade velocity throughout the turbine. When Parsons reverted to axial flow arrangements he went some way to meeting this requirement by grouping the rows of blades in units of the same height and diameter in what he described as 'expansions'. This went some way to achieving the need recognized in the patent 6735 of 1884, namely to ensure that, as the pressure falls, '*the space for the actuating fluid increases, either continuously or step by step.*

When the parallel flow patent rights were recovered in 1894 the blade

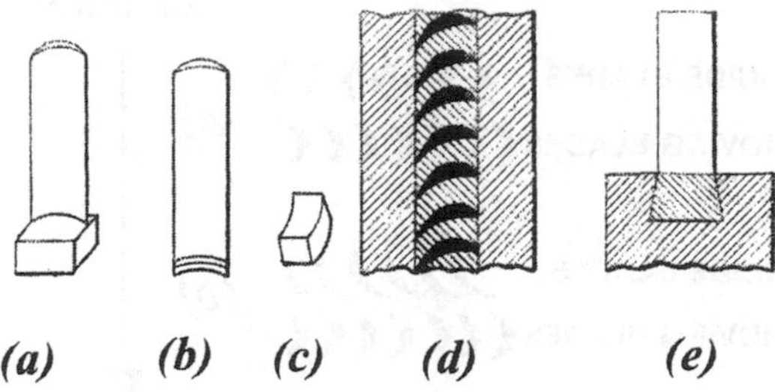

Figure 81. (a) A blade and spacer stamped out in one piece, (b) a stamped blade and (c) a separate spacer; (d) a plan view of blades and spacers in place and (e) an elevation of blades and spacers assembled in a slot (1894) (Richardson 1911).

section was changed and a degree of thickening was introduced near the mid-point. Initially an attempt was made to cut out a group of blades from a strip of delta metal, like the teeth of a comb (see figure 79). The strip could be bent round into a ring, ready for fitting to the rotor or the casing. However there were difficulties with this and eventually individual blades were made from brass strip which was drawn through a die (figure 81) with separate distance pieces as (c). A plan and an end view can be seen in (d) and (e). As he noted in his letter to Laurence, this approach achieved a much smoother surface finish (C.L.(41)). The blades and spacers were located in a groove and caulked in place, (d) and (e). Stoney recalled (Stoney 1933)

> *... the blading problem then was to get the blades and sections to stay on the grooves until they were caulked. At first they were smeared with a sticky and messy mixture of bicycle cement and oil but someone in the shop... discovered that by driving them circumferentially they would stay until caulked between and the problem was solved. At first the grooves were dovetailed, but the blading was apt to work out due to differential expansion and serrated grooves were adopted. The slope of the sides of dovetailed grooves must be more than the angle of friction, a point not realized at first.*

The thickening of the mid-section of these blades which can be seen in (d) was in part, as in the de Laval design, intended to stiffen the blade, but this was not the only objective. Parsons' design was arrived at after a period of intense research. It is important to remember that Curtis and Rateau were following de Laval's blade design, which followed quite closely water turbine practice as exemplified in the 'Pelton wheel', but Parsons

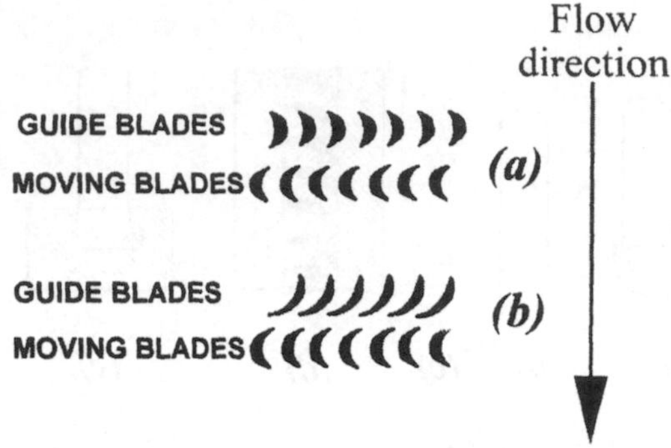

Figure 82. Two of the blading schemes which were tested in a special water tunnel (Richardson 1911).

was breaking quite new ground, with no precursor to guide him. Jumbo, the machine that had originally been made for experiments with radial flow, was again pressed into service and modified to act as a test bed on which to try out different ideas for blade shapes and dimensions, and to assess them for efficiency and reliability. More important perhaps was the construction at this time of a water tunnel. The idea was to mimic the flow of fluid along the passages between turbine blades in a way that could be directly observed. Pieces of wood were shaped like a turbine blade in section, and were sandwiched between two sheets of plate-glass. A current of water was made to flow past the 'blades'. By injecting colorant into the flow of water, the paths taken by the flow were traced out and were visible to the onlooker (Richardson 1911).

Tests were made with this water tunnel to compare blades with a shape like that used in de Laval's impulse turbine with ones of Parsons' own design (see figure 82). When water was caused to flow across an arrangement using only impulse type blades as in (*a*), the flow broke down and did not follow the direction of the blade surfaces. Of course a real impulse turbine employs fixed blades or nozzles, which have a quite different profile to that of the moving blades, so this set-up did not correspond to a real machine, and this behaviour was therefore not surprising. But more important, it was also verified that when the blades of Parsons' 'reaction' design were used as the fixed guides or nozzles, to feed a row of impulse type moving blades as in (*b*), then the flow remained orderly. This diagram is the only surviving evidence of these experiments, but many other arrangements must have been tried, leading eventually to a characteristic cross section for which the point of maximum thickening is further forward than the mid-point. In his study of various strategies for checking steam leakage in labyrinth packings, Gerald Stoney had earlier used a model in which water replaced steam, though there is no suggestion that in that case the flow was made visible (Richardson 1911). In 1896 a final check on the new blade shape was made when an order for four 75 kW machines was received for the Hotel Cecil in London. The first two machines were given blades of delta metal (figure 79), the third had blades like those in figure 81(a), but the last had the drawn brass blades of figure 81(b)–(d). This latter machine proved on test to be clearly the most efficient.

Reynolds made his classic study of the flow of water in a pipe in 1880. By injecting a dye into the flow he was able to observe directly the onset of 'turbulence', as the velocity of the water was gradually increased. The same idea was also adapted by Hele-Shaw to illustrate 'stream line' flow, that up to this time had existed as a mathematical concept propounded by writers like Rankine, but had never before been made visible (Hele-Shaw 1898). In his paper, which was illustrated with photographs, Hele-Shaw makes it plain that the technique required the greatest possible skill and attention to detail if reliable results were to be obtained. His work was carried out during 1897, but neither he nor Parsons make any reference to Parsons' own work that preceded it. The use of a water tunnel was a most significant step because it allowed Parsons to see for himself how shape and flow interact, and it greatly facilitated his intuitive reasoning. This use of models to study problems of fluid mechanics coincided with similar experiments which he was making on propellers. Parsons was one of the first to apply such sophisticated methods of flow visualization as an aid in the design of machinery. What is most puzzling is the fact that the only record of these critical experiments is the reference contained in Richardson's book.[1]

At much the same time as this, those studying the flight of heavier than air machines were making progress in a quite different way. They were performing tests to measure the lift and drag forces created by the flow of air over aerofoils of different shapes that were being tried out for wings. In fact Horatio F Phillips had taken out a patent in 1884, to cover his designs of aerofoils that were curved on both the top and bottom surfaces (see for example figure 83) (Gibbs-Smith 1985). Years later, in 1913, Phillips wrote to Parsons claiming that he had stolen his ideas for aerofoils from

[1] A fragment of a manuscript version of page 47 of Richardson's book is undated but written on notepaper headed '24 Grosvenor Gardens SW' (PAR 30/1).

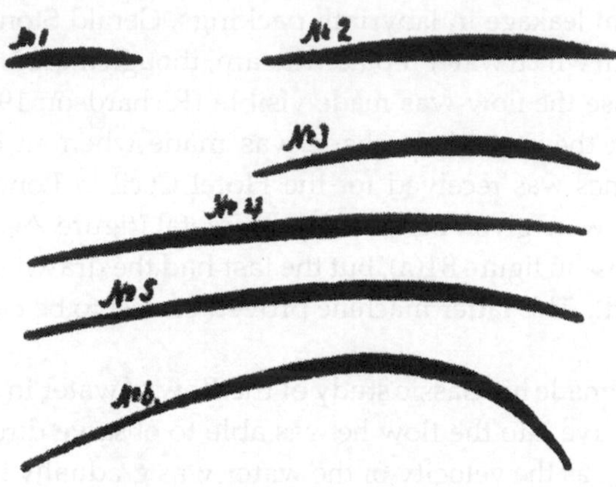

Figure 83. Aerofoil sections from Horatio F Phillips' patent 13768 of 1884.

him after a conversation, and had used them for his turbines. Certainly Parsons' interest in aviation would have made him generally aware of developments in wing design, but he flatly denied the accusation[2],

> *I therefore regret to have to state emphatically that proposals for the thickening of backs generally of turbine blades were not new to me at the time of the interview, nor has it had anything to do with the progress of steam turbines as developed by our Companies, that is of the so-called Compound Steam Turbine.*

Certainly there is no evidence that Parsons viewed the behaviour of blades as being 'winglike'.

By 1894 the essential features of Parsons' 'reaction' blading had been established. First the shape was aerofoil in section, with a thickening that was greatest towards the leading edge. Then the blades were mounted with the inlet pointed close to the direction of the shaft axis, and the outlet surface making an angle of roughly 70 degrees to it. Finally, the fixed and moving blades were of identical shape. That did not mean that the design remained static. Over the years Parsons accumulated a bewildering variety of shapes for land and marine turbines as he sought continually

[2] Parsons C A, undated letter from Holeyn Hall to H F Phillips PAR44/2.

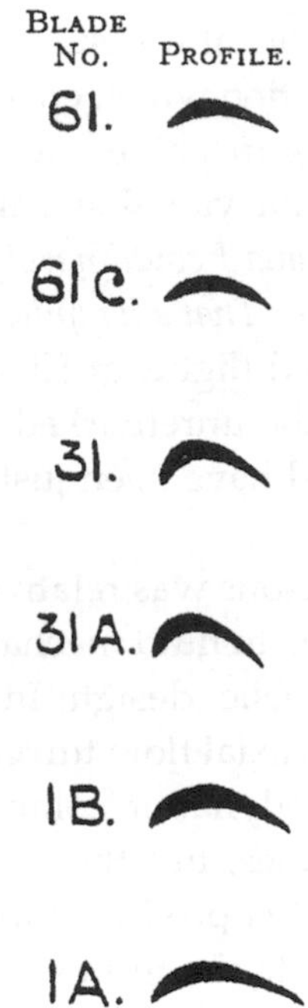

Figure 84. Some blade profiles used between 1894 and 1904 (Dowson 1936).

to improve the performance of his machines, as can be seen for example in the selection of figure 84 (Dowson 1936). Many writers have drawn attention to the remarkable similarity between the shape arrived at by Parsons and that of more modern designs. There is a suggestion that it was the result of a special inspiration that allowed him to make the necessary leap in thinking, but it seems much more likely that it was Parsons' special gifts as an experimenter and his scientific assessment of results of tests on models that explain his success. Even during his student days he had used models to assist his reasoning as a designer. His method of working is well documented by Richardson in his description of Parsons' tests on the 6 ft model of *Turbinia* (Richardson 1911), but it required just as much skill to make the flow past blade arrays that are in relative motion visible to the observer. However because he spoke little about these experiments and because nothing like his notebook on Gas Engines has survived, the importance of this work has passed unremarked. Nevertheless his later success was very dependent on having arrived at a correct solution to the problem. Just a few years later, in 1901, the

Wright brothers conducted critical measurements on a variety of airfoils, accumulating precise information which enabled them to chose a shape of wing suitable for their flying machine. In their hands, the simple but elegant equipment poured out valuable data in a short space of time. Orville remembered '*Wilbur and I could hardly wait for morning to come, to get something that interested us. That's happiness!*' (Crouch 1989). While it was the drama of the powered flights at Kitty Hawk which commanded public admiration, without the unremarked series of earlier model tests their attempts at flight would have been just as unsuccessful as those of their predecessors.

It may be asked why Parsons was relatively successful with his early turbines although his attention to fluid mechanics was so scant, and yet he had so much trouble with propeller design. In his 1884 patent, Parsons had foreseen that if he worked his axial flow turbine in reverse, it would act as a compressor, a so-called aerodynamic compressor. Such machines have much in common with turbines, but there is one significant difference. The build-up of pressure that is produced in a compressor comes about as the high velocity imparted to the fluid stream by the moving blades is eventually reduced. This is achieved by widening the passage through which the fluid flows. If this widening takes place too abruptly the flow ceases to be guided by the walls and becomes disordered. In the case of a propeller there is only one guiding surface and the tendency for fluid to part company with it under certain conditions is if anything greater. In the turbine, on the other hand, fluid is speeded up as pressure falls, and in this case the fluid is pressed against the walls as they converge and the flow has less tendency to become disrupted. Moreover if turbulence does develop in the fluid of a turbine, and energy is degraded to heat, some of this heat can still be rescued as work in later stages of the machine. The penalty for poor blade design in turbines is therefore very much less severe than in the case of compressors.

Propellers can be considered as a special kind of compressor with just one set of moving blades. Professor Rankine at the University of Glasgow had put forward a mathematical model of a propeller. It treated the propeller as equivalent to a disc which created a jump in the pressure in the fluid passing through it. While this was a valuable insight, it gave very limited assistance to the designer who was concerned to chose a suitable pitch, blade width and shape of blade. Eventually the aeronautical engineer F W Lanchester, in 1907, set out a mathematical theory which accounted for Phillips' experimental observations. Even then it required the clearer exposition of Ludwig Prandtl of Göttingen before the theory

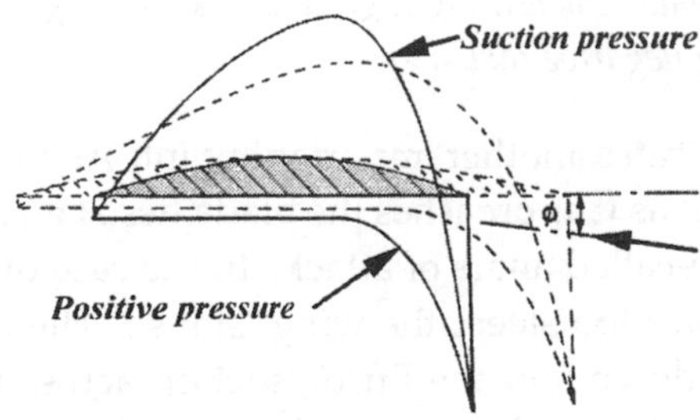

Figure 85. Distribution of pressure across the chord of a propeller blade which is shown in section. The direction of the fluid flow is shown by the arrow inclined at an angle of ϕ. The dotted figure relates to a blade of wider chord but carrying the same load (Goodall 1942).

began to impact on the practical design of aircraft. What emerged from the theory was the not altogether surprising fact that the level of suction pressure on one surface, and of positive pressure on the opposite surface of a blade, varied from the leading edge back to the trailing edge, generally as in figure 85 (Goodall 1942). This shows that it is not correct to treat the propeller blade as if the pressures acting on its surfaces were uniform everywhere. It also demonstrates how a wider blade reduces the peak suction pressure. While Parsons had no way of measuring the pressure at different points on the blade surface, his cavitation tunnel did allow him to vary the speed of the propeller shaft and this allowed him to observe behaviour as the screw encountered ever increasing cavitation. He described his experiences as follows (Parsons C A 1897a).

> *The screw was illuminated by light from an arc lamp reflected from a revolving mirror attached to the screw shaft, which fell on it at one point only of the revolution, and by this means the shape, form and growth of the cavities could be clearly seen and traced as if stationary. It appeared that a cavity or blister first formed a little way behind the leading edge, and near the tip of the blade; then as the speed of revolution increased, it enlarged in all directions until, at a speed corresponding to that in the Turbinia's propeller, it had grown so as to cover a sector of the screw disc of 90°. When the speed was still further increased, the screw as a whole revolved in a cylindrical cavity, from one end of which the blades scraped off layers of solid water, delivering them on to the other. In this extreme case nearly the whole energy of the screw was expended in maintaining this vacuous space. It also appeared that, when the cavity had grown to be a little larger than*

the width of the blade, the leading edge acted like a wedge, the forward side of the edge giving negative thrust.

Figure 85 illustrates another important point. As shown by the arrow, the path of the fluid as it approaches the blade lies at an angle to the plane of the blade, the so-called angle of attack. In the case of an aircraft wing, if this angle becomes too steep, the wing 'stalls'. The orderly flow over its surface breaks down and the lift or suction across its upper surface collapses dramatically. In the case of the propeller, this angle becomes larger if the amount of slip is increased. Figure 85 represents a section of blade at a particular radius. At other radial positions, closer to the hub say, the velocity of the blade surface relative to that of the water will be different. Clearly, even with the aid of his cavitation apparatus, it required inspired intuition to arrive at a suitable design of propeller. The task of course was made so much more difficult because the speeds of rotation that Parsons was forced to use with turbines were five to ten times what was normally used in steam powered vessels at the time. Some years later Parsons spoke again on the subject of propeller design (Parsons C A 1911c).

I would remind mathematicians that water is not a perfect fluid, that the actual lines of flow round a moving sphere cannot, I think, be represented by a source and sink, for they are turbulent and complex. In the rear of a sphere moving through water there is a long trail of swirls and vortices, and a similar condition of flow prevails around aeroplane wings when set at a large angle to the direction of motion. ... I remember... some years ago the flow of water round the rudder of a ship's rudder was illustrated by a lantern slide in which there was a thin film of viscous fluid flowing past an obstcale, and Professor Dunkerley remarked that if those were the lines of flow round the ship's rudder, the rudder would not turn the ship. Mathematicians, I fear postulate conditions that do not exist, and forget sometimes that water is in reality not the perfect fluid of the mathematicians... and that its motion... can only be understood by means of experimental observation and determination.

Looking back over this early period, it is possible to distinguish two quite different classes of advance in the course of innovation. In one case a number of ideas are conceived together, but are not necessarily implemented at once. Thus among the suggestions in patent 6735 of 1884 is the proposal '*to make the blades of sheet metal and to secure them in suitable grooves or recesses in the rings*', And again it is suggested that '*a compound motor such as described may be applied to marine propulsion, but as the velocity is necessarily high, it will be advisable to place several fine pitched screws on the shaft'*. Foreseeing an axial flow compressor he says that *'Motors according to my invention... if such an apparatus be driven... becomes a pump... and can be used for producing a pressure in a fluid.'* He continues, proposing the now familiar gas turbine cycle,

> *Such a fluid pressure can be combined with a multiple motor according to my invention so that the necessary motive power to drive the motor for any required purpose may be obtained from fuel or combustible gases of any kind. For this purpose I employ the pressure producer to force air or combustible gases into a close furnace of any suitable kind... into which furnace there may be introduced other fuel (liquid or solid). From the furnace the products of combustion may be led, in a heated state to the multiple motor, which they will actuate. Conveniently the pressure producer and multiple motor can be mounted on the same shaft, the former driven by the latter.*

The first of these proposals was put into effect in 1891 (figure 53). The second was achieved over a period from 1893 to 1897. The third proposal for a compressor was never really successfully implemented. As for the gas turbine cycle, he recognized that it would not be practicable, at least during his lifetime, due to difficulties in finding materials capable of sustaining the high temperatures generated in the furnace.

All of these ideas were set down around the same time in the great burst of imaginative invention around 1884. But there was another class of ideas that were defined only in the course of experimentation. A much earlier example of this second class can be found in his patent 4266 of 1878. This sets out arrangements for forced lubrication of high speed bearings, in particular the boring out of passages in the crank shaft to provide an oil supply to the big end bearing of the connecting rod, as shown in figure 30. This feature had been quite absent in the first patent 2344 of 1877, and was introduced directly as a result of practical experience with the first machine. Another such example was the characteristic, aerofoil shape of his turbine blades, which took ten years to emerge. It is interesting that the work on propeller design followed on the heels of the seminal work on turbine blades, and that both relied on experimental studies involving flow visualization techniques. There is no doubt that Parsons' understanding of the principles of mechanisms was greatly furthered by the teaching at Cambridge, but it may be argued that if he had been given the opportunity to study under Professor Rankine at Glasgow, he would have set out with a better grasp of fluid mechanics.

By 1895 de Laval had successfully marketed his turbine using mechanical gearing to reduce the rotor speed of up to 24 000 rpm down

Figure 86. Launch number 2 at full speed (Richardson 1911).

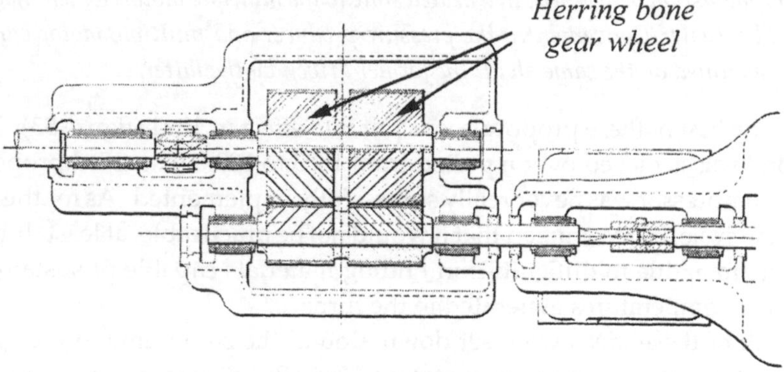

Figure 87. Herring bone, helical gear drive for a 150 kW alternator (1898) (Richardson 1911).

to a tenth or less. His gears were still very noisy, and they could handle only relatively small amounts of power. From his diaries it appears that he probably considered applying a steam turbine to ship propulsion as early as September 1886. However the coupled pair of turbines for travel ahead and in reverse were based on the Pelton water wheel, and were intended to drive simple spur gears (Jung 1968). In 1892 he actually built an engine to propel a ship, comprising a pair of 15 h.p. impulse turbines. These were of the same basic design as figure 47, joined back to back and coupled to a two stage helical gear box (Jung 1982). One turbine was used for forward and the other for reverse motion. However, like Osborne Reynolds' turbine, this project failed to reach commercial production. The reason appears to have been the fact that the largest turbines which he had been able to make developed only a few hundred horsepower, which was insufficient to compete successfully with reciprocating machinery. Parsons

Figure 88. (a) Compressed air powered turbine, running at the extraordinary speed of 40 000 rpm, and rated at 40 HP. It was built to propel a 14 in torpedo. The top cover is removed to show the turbine shaft with a herring bone helical pinion at the right hand, which drives the reduction gearing to give an output at 99.5 rpm. Discovery Museum at Newcastle upon Tyne.

was unaware of this venture, but he was very conscious of the potential that gears offered him if he could only overcome their current limitations. He took his first steps to explore the practical difficulties in 1897, when he obtained an order from Mr F Buddle-Atkinson to provide a turbine drive for a launch that was required to service the latter's yacht *Charmian*. The launch was only 22 feet long and was to be capable of a speed of around 9 knots. Parsons designed a 10 HP turbine that ran at 19 000 rpm. It drove twin propellers via a helical gear with the rather large reduction ratio of 13.5:1. By cutting the teeth in the form of a helix, somewhat smoother and less noisy operation was possible. The small pinion gear on the turbine shaft meshed with two much larger diameter gear wheels, one on either side, each driving a separate propeller shaft. This helped to balance the sideways thrusts from the gear wheels on the pinion, but because the gear

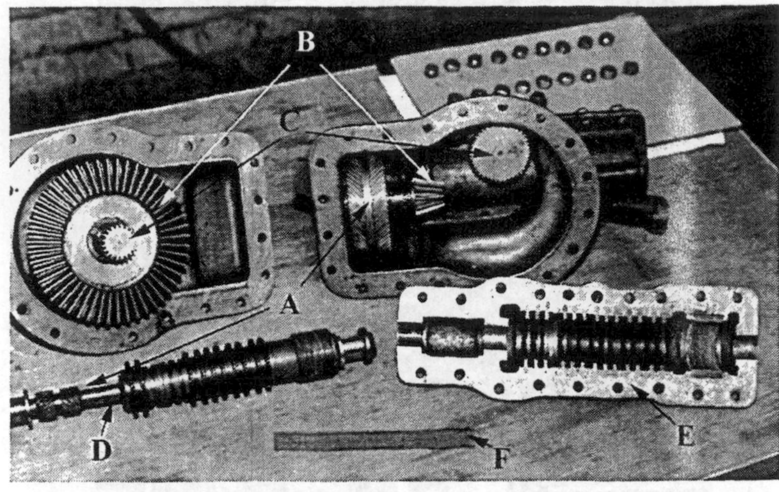

Figure 88. (b) Details of the gear train, A herring-bone helical pinion, B bevel gears, C spur gears, D turbine rotor, E turbine casing and at F is a 6 in rule.

teeth were cut at an angle, there was still an axial thrust on the turbine shaft that was unbalanced. When the engine was examined in 1980, it was found that this had resulted in considerable wear (Jarrett 1980). In a response to a question during the discussion of his paper to the Institution of Naval Architects which raised the problems caused by the very high propeller speeds, Parsons said (Parsons C A 1897a)

> *The speed can certainly be very much reduced, especially in ships of larger beam… By increasing the beam, provided the horse-power is considerable, the speed of revolution will be diminished in direct proportion as the diameter of the turbine is increased… If gearing is used it would be applicable to all classes of ships. But we have always looked with considerable amount of distrust on gearing. For small horse-powers we have tried gearing and found it fairly satisfactory, but for large powers it would be a very questionable experiment to try.*

In 1898 gearing was also used on a much larger turbine, driving an alternator of 150 kW rating that was supplied to the Newcastle and District Electric Lighting Company. The output speed was reduced from 9600 to 4800 revolutions per minute. In this case the gears were cut in 'herring bone' style (see figure 87), an approach adopted by de Laval sometime after 1891. This avoided the problem of unbalanced axial thrust on the shafts which had been a feature of launch 2 because the teeth on each gear were cut in two sections and were so slanted that this axial thrust cancelled itself out. This unit was given heavy usage, but the results were not encouraging; the noise *'was insupportable, and the cost of upkeep abnormally heavy.'*

One other encounter with reduction gearing dates to 1897 when a motor was constructed to drive a 14 in Whitehead torpedo (calculations for a balanced, axial flow, 30 000 rpm turbine, driven by air at 400 p.s.i., are contained in a notebook (Parsons C A 1881b). The turbine was supplied with compressed air and coupled to a propeller through an elaborate three stage reduction gear box (see figures 88(a) and (b)). The double herring bone pinion on the turbine shaft A meshed with a wheel on a shaft that also carried a bevel pinion gear. The gear wheel was fixed to a shaft which carried a simple spur gear C. This meshed with a wheel driving the output shaft. It appears as if this project was not pursued any further. There were more pressing matters to be attended to at that time, and the development of gearing was put to one side for another ten years.

Chapter 14

International Recognition

14.1 Establishing a home market

When Katharine Bethhell met her future husband in 1882 she was only 23 years old herself. She quickly realized that he was an unusual person, *'a strange and weird young man'* as she herself put it. But her own character was remarkable too. She was prepared, after the briefest honeymoon, to accompany her husband and his mechanic on winter mornings and watch while they experimented with gunpowder fuelled torpedoes on Roundhay Lake. The tour which the young couple made to America in 1883 may have made amends to her for this rather unromantic start to married life, but, as the years passed by, the strain of being married to an inventive genius and businessman became worse if anything. While she was carrying her first child, Rachel Mary, he was giving life to his greatest achievement, the steam turbine. When their second child, Algernon George, was born in the following year, it seemed as if Katharine would emulate her mother in law Mary Countess Rosse, who bore 11 children. But it was not so, and these were the only children of the marriage. It is true that family sizes were starting to decline in the population as a whole as childhood mortality, though still high, began to fall, but two of Charles' brothers both had more children: Laurence had four and Clere six. We cannot be sure, but the combination of a feisty and independent mother, and a workaholic father, possessed of what could at times be a fiery temper, may have led to a domestic climate which transferred the energy of sexual drive into other channels. His friends recalled that he enjoyed children, and he obviously encouraged Rachel and Tommy, as his son was known, to come into his workshops at Elvaston and later at Holeyn Hall. In this he was emulating his own father's behaviour. Look again at figure 23(a), which is one of the photographs used to prepare the illustrations for the paper on the giant telescope for the Royal Society in 1861. It shows not only Lord Rosse and his assistants, but two women and his two youngest children, Clere and Charles; figure 23(b) is a detail from one of the engravings made from such a photo and is not at all the style one might expect for a publication of such an august learned society! (Parsons W 1861).

One of Charles Parsons' failings was a tendency to absentmindedness. On one occasion his wife waited at his office for him to accompany her home, while he attended to a problem in the works. But she found herself left high and dry after everyone had left and he had set off on his own, quite oblivious of his commitment. Some idea of how Katharine must have dealt with these lapses is revealed in what she told a friend late in life (Redmayne 1942). On an occasion when they were travelling abroad he '*blew up*' she said, and '*I walked away and didn't own him*'. She made no attempt to alter his behaviour, but just accepted it as unchangeable, while protecting herself as best she could from its consequences. One of the staff at Gateshead recalled (Hodgkinson 1939)

> *I remember one occasion, I was working at a drawing-board in one of the rooms at Gateshead Hall, and was alone. Sir Charles came in, and I hopped off my stool and handed him a pad to sketch on. He paid no attention, merely looked at me with a glassy expression in his eyes. He looked under a table, lifted up some drawings and looked beneath them, pulled some envelopes out of his pocket, and enquired: Hodgkinson, have you seen my overcoat anywhere?*

He was largely uncaring of his personal appearance, and was quite ready to roll up his sleeves to sort out a problem with a recalcitrant piece of machinery, sometimes soiling his clothes in the process. On one occasion he managed to step into a bucket of lubricating oil. When he came ashore at Wallsend after a run in *Turbinia*, he would often climb on his bicycle and set off for his office at Heaton to deal with correspondence, with no thought to change out of his wet clothes.

His job at Clarke Chapman had required him to run a successful business. This involved manufacture not only of steam turbines, but of electrical machinery, arc lamps and searchlight mirrors. Meanwhile he had to push the sales of his inventions, which meant participating in exhibitions, presenting papers and negotiating with patent agents. He was able to do this and continue his programme of development only because he brought together a very effective workforce. Whatever about his quick temper and autocratic style of management, he obviously inspired great

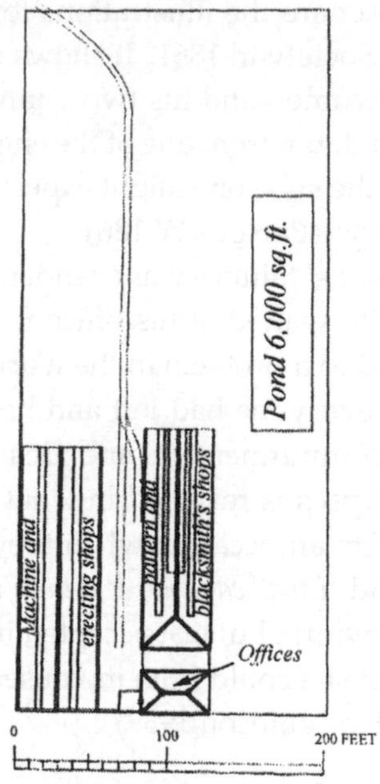

Figure 89. Map of the original Heaton Works in 1889 (Anon 1949).

loyalty among his staff, which became evident when he eventually left Gateshead to establish his own works at Heaton. The dozen or so who followed him formed the nucleus of the new organization and included several who played an important role in the later development of the turbine. It should be remembered that they were, all of them, facing a very uncertain future. The right to use the patents of 1884 was reserved to Clarke Chapman and Co. and there was no certainty that Parsons would find a way around this difficulty.

A completely new factory had first to be built (Anon 1949). The layout is shown in figure 89. The pond which can be seen was 155 ft long by 40 ft wide, and was the one in which some of the *Turbinia* model tests were carried out. During the period when he was developing the alternative, radial, flow design he must have faced severe difficulties meeting his financial commitments, while at the same time he had to put in place a suitable administrative organization. Yet Gerald Stoney reflecting on the period remarked that '*such a thing as discouragement was unknown*' (Stoney 1939). Something of the character of the man is to be seen in a letter written to Laurence (C.L.(1)),

Figure 90. Sketch of dioptric glass in letter of 28 May 1890 (C.L.(1)).

Elvaston Hall
Ryton on Tyne
May 28th 1890

My dear Laurence,

You must have had a very interesting party at Lord Rayleigh's. I think Swinton expected to be a bit out of the running with Sir W Thompson (sic)[1] *and the others.*

We have splendid weather, Kath and the children are at Sidwood for 10 days and I have had 11 days of country and some fishing—Gerald Stoney is quite a keen sportsman and devoted to bicycling which however does not interfere with his work.

I am feeling the relief of having my arbitration settled, it had been going on so long that I was getting used to it, but now one finds the difference—I think the other side made a bad bargain as we shall make a better and cheaper compound steam turbine and dynamo outside the old patents-

I hope before long to try the iron back covered with speculum metal for you, it will be interesting even if no use-

We are getting our glass mirrors much more perfect and with a better polish.

Sir J Duglas wants us to take up the manufacture of side lights for ships. At present dioptric glasses by Chance are used for the best quality. I think we shall try a v plain glass as in projectors

Your affectionate brother,

Charles A Parsons

[1] Sir William Thomson's name is incorrectly spelled, a not uncommon mistake then as now.

John William Strutt, third Baron Rayleigh, was a Cambridge graduate two years younger than Laurence, and with William Thomson, later Lord Kelvin, he was one of the leaders of British science. It was no wonder Campbell Swinton felt somewhat overawed in their company. It is clear that Stoney's relationship with Parsons was quite intimate, though Stoney never ever presumed to take advantage of it. How calmly Parsons greets the loss of his patents! With much experimentation still before him, he is nevertheless calmly confident of ultimate success. William Parsons' legacy of speculum making still provides a subject of mutual interest for the two brothers. A fragment, post script of another letter, reinforces this point, while reminding us of the occupational hazards of a mechanical engineer who kept in touch with the shop floor (Appleyard 1933),

> *P.S. I am at home today from a nasty spill in our test-house yesterday. I stepped backwards off a place three foot high, forgetting the drop, and fell on the corner of a bed-plate and nearly broke some ribs, luckily not quite, so I shall be about again tomorrow. Sir George Stokes was asking me last week how you supported the 3-feet mirror, whether on triangles or hollowed plates.*

It may be that his spill had longer term consequences because for a period around 1904 several correspondents refer to a bout of 'sciatica' which kept Parsons from his desk.

By 1890 the first radial flow turbo-generator had been delivered, and in 1891 the first turbo-generator exhausting to a vacuum condenser was completed. Sales began to pick up and orders were placed for deliveries to London, Middlesborough, Bradford, Blackpool, Birmingham and Dublin. In 1893 the main shops were extended, and in the following year a boiler and chimney were built to facilitate in-house testing (Anon 1949). In that year the largest radial flow machine ever was built for the *Turbinia*, at the same time as the right to use the original patents was recovered. Development of the original axial flow concept was again pressed forward. In 1895 the pressure of work was such that thought was given to sub-contracting the manufacture of turbines to Greenwood and Batley, or of purchasing Palmer's works at Jarrow. In the end it was decided to extend the Heaton works. In 1899 the site area was increased from two to eight acres and the machine shop and erection shop buildings were extended. A test building was constructed as well as new offices. The pond was moved and increased in size to 32 000 sqft. Export markets were opened. By 1896 two 75 kW machines were ordered for Calcutta, and a 120 kW turbo alternator was supplied to Victoria in Australia. The first, 60 kW,

Figure 91. View of one of the original workshops at Heaton in 1896, showing turbo-alternators being erected. Workbenches run along the left hand wall; numerous machine tools are powered by belt drives from a gas engine located at the back left. The armature shop is on the right, with the coke fired cylindrical drying oven clearly visible (Anon 1899).

turbine was installed in Russia in 1897. All of this was happening while the problems with the *Turbinia* were being successfully dealt with.

Around this time an opportunity arose to demonstrate in a very public way the special merits of the turbine. The Manchester Square station of the Metropolitan Electric Supply Company in London opened in 1890 with ten alternators driven by the industry standard Willans engines running at 350 rpm. Complaints from neighbouring householders came in almost at once complaining of vibrations. The station had been built on a bed of spongey clay. In all £2000 was spent trying to support the bed on which the engines sat, but all to no avail. In April 1894 an injunction was taken out against the company. The Chief Engineer Mr Frank Bailey knew of the Cambridge machines and approached Parsons for help. The latter agreed to supply a machine of 350 kW which was designed for axial flow, and was one of the largest turbine driven alternators built until then. It proved entirely successful in removing the cause of complaints, and further machines were ordered for the Manchester Square and also the Sardinia

Street stations (Parsons R H). C A Holbrow was the resident engineer at Manchester Square and had frequent contact with Parsons at that time (Holbrow 1939). He recalled that the

> *station was at that time equipped with ten Willans, central valve engines... It was no easy matter to dismantle these sets, one by one in a very limited space, and to make matters worse the station was severely overloaded... This entailed these two* (turbo) *sets doing practically all the station output, and undoubtedly many of the serious troubles that occurred could have been avoided if it had been possible to shut down the plant when indication of blade or armature troubles first appeared.*
>
> *On many occasions, Sir Charles would come into the station at a period of heavy load, and we often discussed what must have been to him rather galling, keeping the plant running. But it was either that or failure of supply, and no one could have been nicer than Sir Charles, or more helpful at these critical times. Indeed, on several occasions, he worked all night with his men and usually came to my room for breakfast, and gave me explanations of the trouble that had occurred and how to remedy it.*
>
> *In all my fifty years' experience of power station work, I never knew anyone who faced up to his difficulties—and there were many—like Sir Charles.*

14.2 Turbines in the United States of America

The steam turbo-generator was gradually establishing itself. In 1893 de Laval had exhibited a small electric generating set at the World's Fair in Chicago (Hodgkinson 1939). Mention has already been made of the visit which was made to the Heaton Works by a delegation from the American General Electric Company in 1895. This had proved abortive but was quickly followed by another expression of interest from the United States. It was sent by George Westinghouse (1846–1912), a prolific inventor who had made his fortune with his braking system for use on the burgeoning railway network which was developing in America. He had travelled frequently to England and other European countries in connection with this business. With his team of young engineers he had begun to develop a serious interest in electric power and was able to fulfil his contract to provide lighting for the great World's Fair at Chicago in 1893, despite strong opposition. In 1896 the Westinghouse alternating current generators, which were installed on the water turbines at the Niagara Falls, were used to supply power to Buffalo, twenty miles distant. By 1895 over 400 Parsons turbines had been built, and news of the progress being made with *Turbinia* caught the attention of Westinghouse. He realized that the steam turbine was a possible choice as a motor to drive his alternators and he moved decisively. As Frank Hodgkinson recalled (Hodgkinson 1939),

> *A single interchange of letters indicated that Sir Charles was willing to consider a license agreement with the Westinghouse Machine Co. Thereupon Mr Westinghouse delegated his Vice-President, Mr E.E.Keller, to proceed to England to investigate this type of turbine. It was characteristic of the ability of Mr Westinghouse to delegate authority, that his instructions to Keller were that if he saw value in this turbine development, he was to acquire the American rights, and to pay whatever considerations he thought proper. This culminated in an exclusive agreement being consummated in the later part of 1895 for the building and sale of Parsons steam turbines in the United States, exclusive of their application to marine propulsion.*

The agreement required the payment of £10 000 for a five year license, and it gave rights to several United States and Canadian patents (Prest 1931). During this time the licensees were required to build a minimum quantity of machines as judged by their total kW output, but they never exceeded this minimum, and the agreement was not renewed. In negotiating the deal, Parsons was assisted by Stanley Faber Prest (1858–1931), an engineer, educated at Glasgow University and apprenticed at W Hawthorn's Works in Newcastle. The deal also called for someone to travel to the United States to convey the necessary design 'know-how'. Francis Hodgkinson (1863–1949), a young Tynesider who had first worked on turbines in the days of the Gateshead Works, was chosen for the task. He stayed for 40 years with Westinghouse, until his retirement in 1936. A condensing turbine coupled to a 120 kW single phase alternator was dispatched from Heaton, accompanied by Prest. This machine was something of a triumph of the foundryman's art, with multiple passages in the casing arranged to balance steam pressures, but it never went into commercial service. The first machine for which Hodgkinson was responsible also failed to measure up commercially, and was returned by the customer. Meanwhile the Edison Company of New York had imported two 300 kW de Laval turbines for two of their plants. In fact it was 1899 before a successful installation was achieved at the Westinghouse Brake Company's plant at Wilmerding in Pennsylvania. The long distances separating Newcastle upon Tyne and Pennsylvania obviously made Hodgkinson's task more difficult. He corresponded with Parsons and in 1898 he wrote (Hodgkinson 1898)

> *My dear Sir,*
>
> *I have for a long time been hoping to receive a letter from you with reply to mine asking questions with regard to the possibility of using solid bearings in turbines.*
>
> *I am getting on very well. We have 3 300kW turbines nearing completion and have a few castings for the 1500kW. Those experiments with packing rings to replace the Dummies turned out a failure. Mr Westinghouse now has a new scheme for this purpose which we shall be trying out in a few days*

Writing on 28 April he says '*Glad you like the 1500kW designs. The first 3, 300kW ran OK.*' On 1 May 1899 he wrote a long letter acknowledging receipt of letters of 2 and 17 April from Parsons. He gave details of experiments on shaft seals and went on '*I enclose a photo of a new form of balancing machine I have devised. We have had several enquiries lately for turbo fans... I believe the necessary calculations are not very straightforward*'.

In the following year, on 3 April Hodgkinson wrote with news. The 1500 kW machine was found to suffer from vibration, demand for turbines was slow and, he added,

> *George Westinghouse certainly is a very wonderful man. In spite of the enormous amount of work he gets through he is continually making suggestions for the improvement of various products—unfortunately his ideas sometimes are only 'half hatched' and he is inclined to think an idea is good merely because he says so. In spite of all this I personally get on very well with him... salaries don't go for much here, living costs are at least twice as much as in England.*

It might be thought that Hodgkinson while in America had absorbed a degree of directness which would not perhaps have been manifested were he still working in Newcastle. Was it true that Parsons' success arose from his lucky chance of being first in the field, and that any able man could have done as well? Some answer to this question can be gained from the realization that, as his later career proved, Francis Hodgkinson was a very able engineer, and George Westinghouse had a good record as an inventor. Nevertheless initially they both found it difficult to make a success of turbine development. This fact highlights the importance of the skills that Parsons had acquired, and the firm grasp that he had established of the relevant technology.

While Stoney was holidaying in Scotland he wrote to Parsons about an idea which would eventually be taken up widely, but which Parsons resisted for a long time. It was to combine the de Laval and Parsons principles in one machine (Stoney 1897).

Melrose
Sept 27 1897

Dear Mr Parsons,

I have been thinking a good deal over what Hodgkinson has said about combining Laval and Parsons turbines. I doubt if the radial form he proposes will work except for very big powers as the blades are so short, but a combined Laval and Parsons parallel flow seems to me promising. My idea is a parallel cylinder like the boat turbines or the one at Clara pit, only the first row of blades would be Laval acted on by a ring of jets. In the Laval there might be about two expansions leaving the lower expansions for the Parsons turbine. This gets rid of the leakage of the high pressure part, and makes the whole very short and compact. I have worked out roughly and largely from memory, as I have nothing here but Molesworth, a 150 kW condensing type for 4,800 revs and also a 350 kW type for 3,000 revs. They both want only about 14–16 rows of turbines after the Laval ring, and being so very short the clearances could be very fine, probably the 350 kW would run easily at .02″ clearance. The advantages I see at first glance are

Getting rid of leakage of H P part

Whole turbine very short. Therefore clearances less, also weight and cost and space occupied.

Would the combination of Laval and Parsons be a good patent? I enclose the calculations for these two turbines in case you might like to look over them.

If the weather is fine we hope to ride to Moffatt on Wednesday, but all depends on weather, but will let the firm know my movements. We are having a delightful time here exploring the country.

Yours very truly

Gerald Stoney

14.3 Parsons Marine Steam Turbine Co. Ltd. Ships for the Navy

The business of supplying land based turbo-generators had flourished, and now the time had come to reap the rewards of the *Turbinia* experiment. In 1897 the company which had funded the construction of the *Turbinia* was superseded by the Parsons' Marine Steam Turbine Company Limited, PMSTCo. The directors of the new company consisted of those who had formed the original board, with the exception of N G Clayton who had recently died and who had been replaced by a neighbour of Parsons, N C Cookson. PMSTCo. had a much larger nominal capital of £500 000, which was represented by 5000 shares. Of these 800 were issued fully paid, and a further 1600 were 80% paid, totalling £208 000 (Anon 1900a). Parsons himself held a minority interest with 300 shares. In a deal made on 27 September 1897 the license, the *Turbinia* and all effects were purchased from the syndicate, which was paid £30 000 in cash and £80 000 in shares, together with '*certain royalties and certain rights to further shares in the event of further issues of capital. Of this consideration the Patentee* (Parsons) *received for his interest in the first Company* (The Marine Steam Turbine Co. Ltd) *£9,000 in cash and about* £30 000 *in shares*' (Anon 1898). The Marine Steam Turbine Co. Ltd did not cease to exist but continued to enjoy an income based on the royalty and license rights which it still retained (PAR 38/12).

The propulsive machinery for *Turbinia* had been manufactured at Heaton, but now a separate works was needed which would be dedicated to marine work. A suitable site covering 23 acres, with a frontage of 900 ft, was found on the banks of the Tyne at Wallsend, some four miles downstream from Newcastle. In 1897 a two bay shop equipped with a 40 ton crane was erected. A third bay with a crane capable of 60 ton lifts was added in 1906 and in 1908 a second crane was added to give a combined capacity of 120 tons. The factory was known as the Turbinia Works. Initially electric power was generated in house but later a supply was taken from the nearby Carville station. Some of the larger machine tools were built at Heaton. A boiler house and condenser were installed to permit completed turbines to be tested before despatch.

In 1892 the Admiralty had initiated a construction programme of torpedo boat destroyers, and shipbuilders on Tyneside had been responsible for building many of these. Discussions with Sir William White and other Admiralty staff had been going on for some time, and as a result a design was prepared for a turbine driven torpedo boat destroyer. It was modelled closely on the current 30 knot vessels with reciprocating engines, being built for the navy. A tender was submitted in July 1897, and, after further discussions, an order was placed for the *Viper* in the following January (Parsons C A 1903). The vessel was to be 210 ft in length, 21 ft beam and of 370 tons displacement. It was built under sub-contract by Hawthorn Leslie and Co., with turbines supplied by PMSTCo. Just before this Sir W G Armstrong Whitworth and Co. ordered turbines for a very similar vessel which had been designed by Philip Watts and was being built at Elswick. It was called the *Cobra*. Despite the success of *Turbinia*, the Admiralty was not going to risk taxpayers' money, and it insisted that Parsons should give a guarantee of £100 000 in case *Viper* was a failure. This was an enormous sum, but his good friend Christopher Leyland came to the rescue. Seeing the crestfallen look on Parsons' face when he heard the terms, he turned to him and said '*It is all right Charley, I'll back you*' (Stoney 1939).

Establishing the PMSTCo. as a limited company caused Parsons to think again about the Heaton Works. Although the latter was entirely owned by him, others had a legitimate interest in its future should anything befall its creator. He set out his thoughts in a letter to his brother (C.L.(35)),

Holeyn Hall
Wylam on Tyne
8 Decr 1897

My dear Laurence,

As I shall not see you before you leave for India, I am arranging to make Heaton Works a private Limited C^o for two reasons.

(1) That if anything happened to me, it would be extremely awkward for the Marine C^o and it could be awkward for my executors.

(2) The present staff including Stoney who have grown up with the place have justly vested interest which should be recognized and protected.

I propose to have partly different executors for Heaton and my interest in the Marine C^o, from those of my other property. Those for the first I suggest, are yourself, Gerald Stoney and Harvey, if you will do me the great honour to act.

It is very desirable that the management of Heaton Works should act well with the Marine C^o, as there will necessarily be some give and take, and both Foster and Harvey are very anxious to have you. Harvey and Stoney would do the work, both are interested in the Marine C^o and Stoney will be in both Companies, and is a most upright fellow.

The first signatories of Heaton would be myself and the permanent staff viz 7. The only charge on the place is the mortgage of £9,000 at 5% , and the capital stands at over £40,000 which includes nothing for patents which have all been paid out of current monies, as well as experimental work, so that it is financially in a first rate condition.

In regard to my other property I am proposing yourself and Katie. In regard to Heaton Works £25,000 in cash was put in viz. £20,000 in Clarke Chapman's time and £5,000 since.

This is all that has gone in, and what has been drawn out has been approximately 3% over the whole time + about £3,000 invested in Cambridge, Newcastle and Scarborough. This is in short the whole case as it stands. Perhaps you will like to think it over on your voyage and I should be greatly obliged if you can see your way to act, but if you had rather not, do not hesitate for an instant to decline.

Your affectionate brother,
Charles

14.4 Patent extension

In fact, for reasons that are not known, the decision to transform Messrs C A Parsons and Co. into a private, limited, company was not implemented until 1913 (Bedford). This letter makes it clear that although Parsons had not made a huge fortune, his personal wealth had enjoyed a respectable growth over the last 15 years. But as the negotiations with the American firms had clearly shown, much of his potential wealth lay in his patent rights, and some of these were very near the end of their normal life. The five year period during which Clarke Chapman owned, but failed to develop, the axial flow designs represented a serious loss to Parsons. Besides, the recent success of *Turbinia* had demonstrated for all to see that his invention was indeed of major significance. And so on 22 October 1897 Parsons decided to seek the rare privilege of an extension of patent 6735 of 1884. It is true that James Watt had obtained an extension to the life of his 1769 patent from 14 to 31 years by an Act of Parliament, but this was most unusual.

Parsons' petition came before the House of Lords Judicial Committee of the Privy Council for hearing on 19 April 1898 (Anon 1898). While the earlier arbitration proceedings had not given the result that was hoped for, the exercise had been a useful encounter with the legal process and this helped to ensure that the petition was made in the most efficacious manner. In the course of the proceedings the argument was made by Dugald Clerk that the application to marine propulsion had been foreseen in patent 6735, but the possibility of exploiting this commercially did not exist until the PMSTCo. was established in 1897. At this point the patent had only nine months further life, and the new Works at Wallsend had yet to be built. This was a clever approach, because the success of *Turbinia* had shown the enormous benefits that could follow if the turbine were to be adopted by the Royal Navy. Public agitation for a massive programme of warship construction was in full swing at that time, and obviously any decision which set back such a programme might prove harmful to the national interest. This had nothing to do with patent law, but the possibility might be expected to be present at the back of their Lordships' minds. The case was presented at a hearing on 19 April 1898, by Fletcher Moulton Q C and A J Walter. It was opposed by the Attorney General Sir Richard Webster and Mr Sutton, acting for the Crown. The petition first set out the contents of patent 6735 of 1884, and this was accompanied by suitable models. It went on to describe the up-hill struggle there had been to obtain orders. *'The machines were exhibited and lectured on, and articles were written upon them in the technical press'*. The temporary loss of access to the patents was described, as well as the *Turbinia* developments leading to the establishment of the PMSTCo. Financial details of this and of the Westinghouse deal were set out. Three expert witnesses gave evidence: Dugald Clerk gave details of the *Turbinia*, Parsons described the nature and solution to the problems associated with the screw propellers and William Thomson, now Lord Kelvin, spoke in support, *'the success of the turbine is a great step in steam navigation, one of the greatest steps since the introduction of steam navigation, nearly 100 years ago.'* William Swan presented accounts which showed that a sum of £7031 13s.9d. had been spent, and that at 7%, the Capital employed would have earned £8139 7s.7d. This represented a net loss of £1107 13s.10d. But the Attorney General objected that *'When it suited the Petitioner he contended that all the merit was due to the 1884 patent, but when he had to deal with large sums of money he contended that they had to be apportioned to other patents than the 1884 patent'*. In giving his judgement on 14 May 1898, Lord Macnaghten noted that the accounts for the foreign patents also showed an overall loss, of £202 4s.11d., and he concluded that' ... *weighing all the other circumstances of the case, their Lordships are of the opinion that Mr Parsons has not been adequately remunerated, and they will consequently make their report to that effect. They have already intimated that the patent should be prolonged for five years'*.

14.5 First warships

The *Viper* was launched and a start was made with her trials. She was twice the length of *Turbinia* and had a displacement of 370 tons. Her engines, No 693, like those for the *Cobra*, No 687, were made at Heaton and their design resembled that of *Turbinia*, though there were now two complete turbines, each of which had a high pressure and a low pressure cylinder. These four cylinders powered four shafts, each of which carried two screws. The total output was estimated to be 12 300 IHP. The low pressure turbines drove the inner shafts, and each also had an astern turbine coupled to it. When not in use the latter ran in a vacuum and so absorbed very little power. The pumps which were used to extract air from the condensers were powered by their own, geared, turbines. The boilers were of the Yarrow type. *Viper* set a world record with a maximum speed on the measured mile in excess of 37 knots. This was achieved with almost complete absence of vibration (Richardson 1911). Maximum propeller speed had been reduced to barely 1200 rpm, which was a large drop compared to the *Turbinia*. She could make $15\frac{1}{2}$ knots when going astern. When steaming at a steady 31 knots she consumed 2.38 lb of coal per IHPh. The *Cobra* achieved a speed of 34.6 knots on a three hour run.

The *Cobra* had been completed before *Viper*, but was rammed by a collier while it was tied up at the Turbinia Works. This delayed her trials significantly. Charles wrote to his brother about the *Viper* (C.L.(36)),

Holeyn Hall
Wylam-on-Tyne
21st November 1899

My dear Laurence,

The Viper had her first preliminary run today and reached 32 knots with about $\frac{3}{4}$ full steam and 3″ water gauge air pressure (in the stokehold).

The fan shields (Hawthorn's work) somewhat checked the outlet and prevented them getting up 5″ (boiler forced draught) *which we had in the Elswick boat (i.e. Cobra). This will be remedied. A silent blow off valve (own work) seized and stuck open, and could not be shut, wasting about 12% of the whole steam.*

It was a very rough day. 3 Govern' inspectors were out and also Leyland (started at 10 a.m.. Stopped fast running at 2 p.m.) The inspectors seemed greatly pleased and said they considered the performance as most satisfactory in every way, and that it was the roughest sea trial of a destroyer they had been on. Even Leyland in the conning tower got wet through.

The present screws should bring her up to $33\frac{1}{2}$ or 34 knots on the next trial—and after a modification she should do 35 easily—She seems to be about $1\frac{1}{2}$ knots faster than the Elswick boat (Cobra) and is 33 tons lighter.

I was much pleased with the views of the inspectors. They remarked on the complete absence of vibration, and the arrangement of machinery. There was absolutely no trouble with the main engines—beyond the heating of the thrust bearings while working up. As you remember they have not been sufficiently relieved by steam balance on either vessel, but this can be done later on. Meantime they have to be watched.

I have no doubt after a few trials the Gov[t] will order some boats.

Your aff[te] brother

Charles

The Elswick boat goes out with modified screws next week, and the Viper goes out again with the same screws as at present, in about a week to 10 days.

The progress was very reminiscent of the experience with *Turbinia*, though now much swifter. Parsons himself attended most of the trial runs of both the *Viper* and the *Cobra*.

In 1899 a key addition to the staff of PMSTCo. was made in the person of Robert John Walker (1870–1936), a naval architect (Anon 1936). As always Parsons sought out the best man available: Walker had served his apprenticeship with W Doxford and Sons on the river Wear. While working in the design office he studied for the examination in Naval Architecture of the Department of Science and Art at South Kensington, and in 1896 obtained first class honours. He was promoted head of the design office and was involved in the design of two destroyers for the Admiralty. Though described as *'of a modest and retiring disposition'*, he was appointed General Manager four years after joining the PMSTCo. In 1906 he became Managing Director, jointly with Parsons himself.

While the building and commissioning of the destroyers was progressing, it was decided to bring the *Turbinia* across to France to the exhibition being held in Paris during 1900. With 22 souls aboard, Leyland set out from Newcastle. He recalled that the weather was atrocious, everyone but himself was seasick and the card in the compass was lifted from its pivot. The stokers had difficulty fuelling the boiler (Leyland 1935). Much water had been shipped and the bows were low. He decided to turn back, and with good luck steered safely to her berth. He resumed the journey and steamed a couple of hundred miles to Yarmouth where they took on coal. When they arrived at Dover the weather was again very bad and he feared that the sea would come down the funnel, or even wreck the hull. In the light of subsequent tragic events such a fear could be seen as entirely reasonable. He continued

> *Crossing the Channel a fire occurred in our bunker, but we put it out the morning after our arrival at Havre. Barnard then went down the funnel to look at the boiler. I was standing on the wharf at the time and a Frenchman asked me if we always went down below that way. I assured him that at sea it was the one and only way. We had two days running at Havre. The first day for the French Naval Authorities and the second for representatives of steamship and railway companies.*
>
> *Parsons crossed to France for these runs and then left me to take the Turbinia to Paris. Arriving at Rouen, we passed a Newcastle collier and were hailed with 'Go it Turbinia—Good old Wallsend!'. This occurred unfortunately during the time of the Fashoda incident*[2] *which rather strained relations between the two countries. The berth indicated for us to occupy at the wharf was between two steamers, the stern of one of which was but little clear of the other. Our pilot could not manoeuvre the Turbinia in and wanted the boat to lay out warps. At last I gave him up as a bad job and took control, and by using full power ahead and astern quickly put her in. The pilot was afraid, dashed his cap on deck and danced about calling, 'Mon Capitaine, comme vous êtes fou.' ('My captain how mad you are').*

Leyland had problems at Customs which were cleared up, and he continues,

> *The Seine Navigation Department was nearly as bad and it was almost dusk when the committee arrived to inspect and measure the Turbinia, arguing that in case I sold her at the Exhibition they must be able to verify her. I had steam up hoping to get away that evening, so I took them first down to the engine-room and then personally blocking the exit, I turned on steam ahead in two turbines and astern in the other, and whilst they were in a state of terror as she jumped and vibrated, I dinned into them what an exhibition of power I was demonstrating, and I kept them there until they promised to sign all documents if I let them out in a couple of minutes. All arrangements were completed on the wharf.*
>
> *The new pilot for the up-river journey was nervous and left me to take her out into the stream, but after we had passed the first bridge he came to the conclusion 'Qu'elle dirige bien,' (How well she steers) and took the helm himself and promptly fell in love with Turbinia…*

[2] An imperial conflict between France and Britain in central Africa in 1898.

We tied up at the Exhibition and except for two men and myself the crew went home... Later in the summer there was an international meeting of Naval Architects. The Chief Engineer of the French Navy wrote asking me to come over and run Turbinia in the Seine as a side show. I replied that it was out of the question. He then telegraphed imploring me to come.

All our staff was engaged going around in the 'Dreadnought' (not the turbine powered vessel of course) *to Portsmouth, so I wired a friend at the exhibition for his aid, requesting him to get me four stokers from the Nord Channel Steamer if possible. This he did, and with three friends of mine who were at the exhibition, I made her ready just in time.*

R.Barnard, and some others who came over from Portsmouth, went down the Seine to the Basin Meulan where the run was to take place. The Chief of the Seine navigation, and one or two other officials, together with the river pilot, accompanied us. On arriving at the reach of the river selected for the run on the following Monday I was asked to go full out. This I did although we were not ready for a speed run.

Leyland had been advised by the Chief of the Navigation department at Meulan to dispense with the pilot, and to practice by himself on Sunday *'—unless my English Sabbatarian education prevented me...'*

On Monday I took Turbinia up to the locks where the excursion party met her. They were like so many ants coming down one ladder to get on board to listen to a hurried explanation prior to being hurried up another ladder. The excursion steamer was allowed to get well ahead and then we followed, getting our fires up with forced draught... R.Barnard came to me reporting a big head of steam so I told him to keep stoking. I had the chief of the Seine Navigation, a Russian Naval officer and the French Chief Engineer with me but Frank Barker (a director of Parsons Foreign Patents Co) *preferred the other steamer. The water was too shallow for a really good run, and instead of divergent waves we were driving rollers in front of us, sucking all the water away from the banks abreast and followed by great waves of replacement. Frank Barker told me afterwards that the French Chief Engineer decided that I was really mad and that the Turbinia would jump the river banks into the fields. I doubt if I made over 28 knots. My passengers were enthusiastic.*

In the hands of the flamboyant, but able, ex-Royal Navy officer, the *Turbinia* proved to be a magnificent advertisement for the steam turbine.

Around this time Charles was able to report on the *Cobra* (C.L.(37)),

Holeyn Hall
Wylam-on-Tyne
22 June 1900

Dear Laurence,

The Cobra had a semi-official trial today, and the mean speed of 6 consecutive runs was 34.89 knots, and the mean for 3 hours not allowing for turns was 34.32. She also went astern at 13.24 knots with under $\frac{1}{2}$ full astern power. At this speed the stern wave reached the deck.

The sea was absolutely smooth. The above speeds represent about 12,000 I.H.P. ahead—or some 2,500 H.P more than the contract. This power being required owing to the excess of weight, chiefly the hull.

The Admiralty and Armstrong's officials expressed themselves very pleased and reported the trial as entirely satisfactory. I suppose she is now as good as sold to the Admiralty. The next fastest vessel in the Service is Thornycroft's Albatross 31$\frac{1}{2}$ knots.

It would now appear that the Viper will reach a mean speed of close on 36 knots on her official trials, or 5 knots above her guarantee.

Leyland wires from Paris this evening that it is reported that the Turbinia has got the Grand Prix. She has reached Paris safely. I think the trial of the Cobra is the most important event for the Marine C° since the trials of the Turbinia.

The Admiralty officials were very much impressed with the fact that the engineering staff had nothing to do practically. The Cobra went out at 401 tons displacement, and consumption of coal was estimated at about 2.5 lb^{s} *per I.H.P.hr., or within the usual guarantee.*

Your aff^{te} *brother*

Charles

P.S. The fact of the machinery and boilers being passed beyond the designed power somewhat increases the consumption of coal per I.H.P..

A later letter that summer continues the story of the trials (C.L.(38)),

Holeyn Hall
Wylam-on-Tyne
27 Aug 1900

My dear Laurence,

Many thanks for your long letter telling me all the news. I fear it is impossible for me to get away at present. There are only 3 or 4 days before the trials of the two boats, and things have to be arranged between, so that it will probably be the end of Septr before I am reasonably free.

The Viper did well on her coal consumption trial at the guaranteed speed of 31 knots, being 2.38 lb^s^ per I.H.P. which is among the lowest consumptions of the latest destroyers. We should however do better on a repetition of the trial. Next Friday she goes for another coal consumption trial at $32\frac{1}{2}$ *knots at the request of the Admiralty. The Cobra has only one more trial, being a coal consumption at 15 knot, no guarantee being attached. The Viper thus has two more and the Cobra one more trial before being handed over.*

It looks as if Wallsend would become very busy this autumn and begin to earn good dividends. In regard to your Heaton property, it has just occurred to me whether it would be possible to convey the various small sites to a middleman in bulk, and so save you the cost of so many small deeds. I merely mention it for what it may be worth—lawyers are so greedy.

Our standard types of turbines with the help of special machinery which we are getting to work, is apparently cheapening the cost of production, and I can see at Wallsend that we shall be able to turn out future sets of machinery at much lower costs. We only want the start which we expect this autumn.

Leyland has been working very hard at the business, and constantly goes to London and Portsmouth.

Your aff^te^ brother

Charles

Parsons participated in eleven of the dozen trials of the Cobra (Appleyard 1933). The next letter was written in Portsmouth (C.L.(39)).

Kepples Head Hotel
Portsmouth
14 Sept 1900

My dear Laurence,

The paragraph in Engineering seems to have been composed by a rival or enemy, he has only told half the story. The Viper's coal consumption is below the average of 30 knot destroyers reckoned per I.H.P.. The Viper was only working 2/3rd full power at 31.118 knots. Had her engines been made for say 32 knots maximum, the coal would have been less. Then the Albatross won't go more than $31\frac{1}{2}$ *knots and she has all kinds of refinement to economise coal, such as feed-water heaters and compressed air jet over the fire to improve combustion. Also it is the opinion of the Admiralty that though Thornycroft boilers are more economical than Yarrow, yet the former have grave defects and cannot be forced to the same extent as the latter without burning the tubes. In reality it appears that at 31 knots the turbines of the Viper are more economical than the* (reciprocating) *engines of the Albatross. In further boats there is no doubt that we can improve on the Viper to the extent of 15% or 20%, and get the same high rate of speed.*

I think the Admiralty are very pleased with the results of the Viper—She goes out on Monday for the usual coal consumption trial at 15 knots. The Cobra is now definitely bought by the Admiralty, and our Japanese destroyer is nearly settled now—

I do not think the paragraph in Engineering will affect orders for passenger vessels.

Your aff[te] *brother*

Charles

A letter to the editor of *Engineering* had already been written on 10 September by the PMSTCo. arguing that the comparison with the coal consumption tests made on *HMS Albatross* was not fair.

A story which was quoted by Lord Rayleigh (S.P.(2)), throws some light on Parsons the man. It concerned the official Admiralty Trial and was told by James Denny.

A preliminary trial trip of the Viper was made early in the day, and the bearing of the engineers was ominous. They differed with Mr Parsons as to the trial trip rate of wages, and, as the latter knew his own mind, the engineers walked off the ship. Everyone thought the day's proceedings must end there and then, but Mr Parsons thought otherwise. He turned on his apprentices to do a journeyman's work, picked up some men off the quay, borrowed some more from Messrs Hawthorn Leslie and Co., who had the contract for the hull and the boilers, and made all into a scratch crew for the trial trip. Under these extraordinary circumstances the Viper ran her trial, and on that day did the unparalleled speed of 37 knots. When Mr Parsons emerged from the engine room, dirty and warm, all crowded around him to congratulate him, but he took the whole thing as a matter of course.

Around this time Parsons wrote to give his brother news of a plan which aimed to get a toehold in the market for fast passenger ships like the mail boats which crossed the Irish Sea (C.L.(44)).

Holeyn Hall
Wylam-on-Tyne
Sunday
(?winter 1900/spring 1901)

My dear Laurence,

I have seen Lord Stalbridge with Leyland[3]*. He said he would be prepared to consider the fitting of the Violet ($18\frac{1}{2}$ mail boat) with new boilers and turbine machinery on the principle of no cure no pay.*

The vessel is of no use to them at present and he says (it) *is of suitable lines for a faster speed—so I shall run over to Holyhead* (port on the Welsh side of the Irish Sea) *on the first opportunity and see Cap. Binny the superintendent, and look over the vessel and get the matter in shape in case we all think it desirable to go in for the work.*

The weather is still against trials. Not a single day for the last 2 months—

No time is being lost with the Viper which is being finished off to proceed South in about a month. Meanwhile we shall give her another full power run and try for $36\frac{1}{2}$ knots.

By the way would it fit in with all your arrangements to see us for a week at Easter.

Your aff^e^ brother

Charles

It seems that he did finally get an extended family holiday at Birr Castle, the first for many years (Barker 1939).

The progress of these two revolutionary ships was being eagerly watched by a wide audience. Amidst dispatches from South Africa on the progress of the war against the Boers, the *Times* newspaper carried a report that on 11 July 1901 a question was asked in the House of Commons (Anon 1901a). In reply Mr Arnold-Foster stated that

No conclusion has yet been reached as to the adoption of the turbine system of propulsion of vessels of His Majesty's Fleet. Trials are now in progress with the Viper under service conditions and in comparison with other destroyers. Much valuable information which will assist in reaching a decision as to further applications of the turbine system will thus be obtained. Experience

[3] Lord Stalbridge (1837–1912), Richard de Aquila Grosvenor, a son of the Marquess of Westminster, was a Cambridge graduate and Chairman of the London and North Western Railway 1891–1911.

with the Viper has shown that a higher speed can be obtained with her turbo-motors than can be obtained in a destroyer of similar dimensions and form fitted with ordinary engines. Vibration is practically avoided. It does not appear correct to say that turbo-engines occupy less space than ordinary reciprocating engines, but they require less vertical height, and can therefore, as the hon. member suggests, be more readily kept below the waterline—an undoubted advantage. The total engine room and boiler room complement of the Viper is larger than required in 30 knot destroyers fitted with ordinary engines, but the horse power developed at her maximum speed is greatly in excess in the former. Proposals, with outline designs, for fitting turbo motors in 3rd class cruisers and new types of destroyers are now under consideration. No decision has yet been reached regarding vessels in this year's programme. It is not contemplated at present to adopt the decision for battleships or large cruisers.

14.6 Disaster

Both the *Viper* and the *Cobra* had completed their contract trials, high speed as well as low speed. It seemed as if success had been crowned and future business looked to be assured at last. But it was not so. On 3 August 1901 the *Viper*, now in the hands of the Royal Navy, was sent to reconnoitre the Channel Islands, Aldernay and Guernsey. Steaming at 20 knots she arrived at 2 p.m., only to encounter mist which quickly became dense fog. At 5.25 p.m. she struck rocks and was abandoned by the crew one hour later. At least there was no loss of life. Clearly the turbine machinery was in no way implicated in the disaster, but it was a severe setback to the cause.

The following month, on Tuesday 17 September, the *Cobra* set out from Newcastle at 5 o'clock in the evening, *en route* to Porsmouth. On board there were 54 naval personnel, one representative of the Elswick shipyard, 22 employees of Parsons' and a couple of catering staff. After reaching Flamborough Head some 130 miles or so due south, she set course for a sea run of 60 miles. It was dull and there was rain. The '*sea was rather rough*' and severe rolling was experienced. The forward boilers were shut down and the speed was reduced to 10 knots during the night. At 7 a.m. the next morning the crew felt a shock and the vessel immediately began to break in two. Of some 79 men on board only 12 survived. Survivors told their story in the *Times* newspaper. There was difficulty in getting the boats launched. One boat, a whaler, was overwhelmed as 40 or 50 tried to board her. A dinghy avoided this fate because it had its canvas cover still in place. The survivors had to swim over to her. Some got aboard at once, others had to cling on for an hour or more until the weather made it possible to haul them aboard. *'One man who belonged to Messrs Parsons Works was heard to cry, just as the waves closed over him "Oh my God, my wife and my children"'*. The 12 survivors reached the ship's dinghy and safety, under the care of the chief engineer John Percey. When they were picked up almost 12 hours later by a steamer, they were in dire straits without water or proper clothing. Among those drowned was Robert Barnard, the manager from PMSTCo. who had so often acted as steersman on the *Turbinia*. *'That bright and hopeful young man who contributed no little towards her* (Turbinia's) *success'*, as the writer in *Engineering* described him. The following day, Thursday at 10.30 a.m., there was a phone call from Wallsend to Heaton with the news of the disaster. John H Barker, manager at the Heaton Works at the time, found Parsons in the drawing office. He called him out to give him the news (Barker 1939).

> *We went to his office, his first thought being for Mrs Barnard, whose husband was lost; next how news should be broken to the staff at Heaton Works before the news appeared in the press. On the invitation of the builders of the boat I was present on the last trial the Cobra made. Sir Charles remembered this and asked what I observed. He also knew that I had been with shipbuilders and to sea. I told him that I had taken particular note, when sitting on the engine-room skylight, how the deck plates, to use an expressive term, 'hogged', when reaching the crest of the wave and 'hollowed' when over the trough. The pause between the two motions was distinctly visible and sounded as if a tea-tray were pressed and released. I remarked on this to some official, I forget which. 'Oh she is only breathing was the reply.'*

Parsons was deeply affected by this accident and the loss of life. The men were sent home for the rest of the week. He retreated to his office that day and remained there in silence with his thoughts for several hours (Smith 1996). For the rest of his life he was concerned for the dependants of those drowned (Bedford). No doubt the thought must have struck him that just such a disaster could well have cost him his own life, or that of other senior staff during the numerous trials which they had undertaken in foul weather in the North Sea during the last seven years.

Following the sinking of the *Cobra* a Court Martial was convened to inquire into the loss. On 16 October 1901, at the end of a six day hearing, the finding was given (Anon 1901b).

> *. . . The Court has come to the conclusion that His Majesty's ship Cobra did not touch the ground, nor meet with any obstruction, nor was her loss due*

to any error of navigation, but is attributable to the structural weakness of the ship. The Court find that the Cobra was weaker than other destroyers, and in view of the fact, it is to be regretted that she was purchased into His Majesty's service.

The Court exonerated Chief Engineer Percey and the other survivors, and commended them for the way they had conducted themselves after the shipwreck.

The navy was clearly concerned at this finding, and on 5 November the 'Committee on Torpedo-Boat Destroyers' was set up to report on the construction and strength of the *Cobra.* Tests were carried out on another vessel of similar construction, the *Wolf*, both at sea and in dry dock, but no conclusive evidence was found for the likely cause of the disaster. More recently the opinion has been expressed that '*The Cobra was known to be in shallow water at the time of her loss, and it certainly seems probable that the fracture was initiated by the bow touching the ground during a heavy pitch*' (Barnaby 1960).

Chapter 15

Competitors and Licensees

15.1 A presence on the continent

Turbo-generators designed by both de Laval and Parsons were being produced in quite large numbers by the end of the century, and yet there was relatively little interest in mainland Europe. This was to change in 1900 when, on the advice of its English consulting engineer W H Lindley (1853–1918), the City Corporation of Elberfeld, in north-east Germany, installed two Parsons turbo-alternators each of 1 MW rating[1]. These were the largest such machines made anywhere up to that date. Partly because of difficulties in securing large enough castings, the turbine was constructed as a tandem machine, like the second *Turbinia* turbine (Parsons R H). That is, the separate high pressure and low pressure turbines, the one feeding into the other, and the alternator, were all mounted in line with their shafts joined by couplings, see figures 92 and 93. The speed was 1500 rpm. Steam conditions were to be a pressure of 11 atmospheres absolute (140 psig), and a temperature of 183°C plus 50°C of superheat (452° F). The machines were designed to run with a cooling water supply of 18°C. Strict requirements were laid down in regard to steam consumption and governor behaviour. The alternator output from the rotating armature had a frequency of 50 Hz, single phase at 4000 volts. An overload capability of 1.25 MW was specified. The City Corporation appointed a Commission to certify performance, consisting of Professor of Physics H F Weber of the Zurich Polytechnicum, M Shröter Professor of Engineering at the Munich Polytechnicum and W H Lindley.

Preliminary tests were carried out at the Heaton Works, observed by Lindley and Weber. In the course of these, an accident occurred in which

[1] Elberfeld is now a suburb of Wuppertal.

Figure 92. One of the Elberfeld 1 MW turbo-alternators (1900), with Mr J Eskdale who was responsible for erecting it (Richardson 1911).

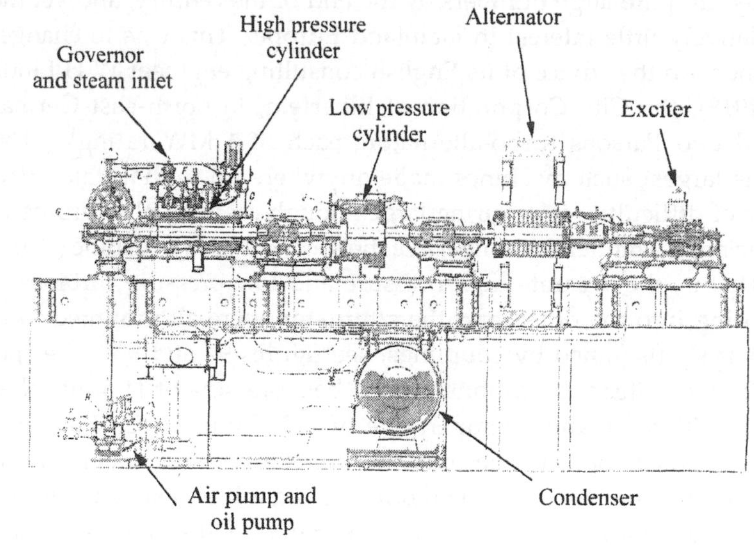

Figure 93. Sectioned view of the Elberfeld turbo-alternator mounted on its massive bed-plate, above its condenser (Parsons R H).

the blades of the high pressure cylinder of the first turbine were all stripped. Stoney described it as *'one of the biggest knock-out blows that Mr Parsons ever received'*. One can imagine the dismay that this caused, but repairs were effected with such speed that the tests were completed within a week of the incident, and the psychological damage was largely countered. The cause of the accident lay in the faulty operation of the makeshift superheater. This was used to raise the temperature of the steam leaving the boiler above its boiling point. The temperature of steam in a plain boiler is the same as that of the water being evaporated and cannot rise without the whole mass of water being heated. On the other hand in a superheater the temperature of the steam is determined by a balance between the temperature of the heating gases *and* the quantity of steam actually flowing in the tubes. It is easy to see how the steam temperature could rise out of control in a test-bed situation.

In following the progress of the turbine from the original prototype stage, one can see that at certain points an advance was made by a change to the configuration, as for instance by the adoption of a 'dummy' piston or of labyrinth glands. In other cases the inherent characteristics of the materials being used could be controlling factors. As the temperature of steam employed steadily rose, two phenomena were encountered which were not yet understood. The first concerned brass. The alloy material, α brass, had been adopted for blade manufacture because its ductility made it easy to form to the desired shape and it was resistant to corrosion. What was not known at this time was that, at temperatures above 200 °C (392 °F), it gradually became brittle when slowly strained (Greenwood *et al* 1954). De Laval did not encounter this difficulty because he used steel blades in order to cope with the extremely high centrifugal forces. Another materials problem concerned the use of cast iron, used largely for the casings. This was found to have a tendency to 'grow' at temperatures above around 230 °C. This resulted in distortion in service which could cause damage. Besides if the heating of a casting was uneven, as a result of the presence of steam passages for example, the casting could become warped, again causing damage.

The results obtained during the tests were significantly better than guaranteed, even though the actual superheat available at the time was only 14.3 instead of 50 °C (Parsons R H). There was one exception. At quarter of full load, steam consumption was 9% above guarantee. When corrected for the rated steam conditions, the consumption at full load was 8.63 kg per hour per kWh (14.3 lb per hour per HPh). This was unequalled by any engine at that time. The machine was controlled by a 'gust governor' which performed excellently. On taking up full load from zero, the speed fell by 3.6%. Also two thirds of the full load could be rejected suddenly without the speed rising by more than 1.9%.

The laws of thermodynamics indicate that the efficiency with which a turbine or other engine converts heat to useful work should increase as the temperature of the heat source is raised. This accounts for the increased efficiency of engines supplied with steam from boilers at higher pressure, and therefore higher temperature. The tandem construction of the Elberfeld machines allowed Parsons to experiment with a technique that is now quite commonplace, reheat. After expanding in the high-pressure cylinder, the steam becomes cooler. He therefore led it through an adapted locomotive boiler before continuing its expansion in the low pressure cylinder. The loco boiler added heat which raised the steam temperature to near its original value. Although there was quite a loss of pressure in the makeshift boiler set-up, a small reduction in steam consumption was achieved, something less than 5%. Because the gain was small, and because of the attendant complications, the idea was not pursued again until 20 years later.

15.2 Variants of de Laval's impulse turbine

Over the previous ten years de Laval had succeeded in demonstrating an alternative to the approach chosen by Parsons for steam turbine development. His machines were effective in a certain range of applications, but this range was restricted. Speed reduction gears were essential for his design, and they still could handle only a few hundred horsepower. Using a single stage of expansion they were unable to make the best use of the higher temperatures and pressures in the steam supply which were becoming available. At about the same time, two inventors, Charles Gordon Curtis in the United States and Auguste Camille Edmond Rateau in France, each found their own way to overcome these limitations by compounding de Laval's impulse turbine. Neither relied on gearing.

Curtis (1860–1953) who was born in Boston in Massachusetts, practised as a patent lawyer (Somerscales 1991). He had an interest in manufacturing and he owned a firm that made motors and fans. In 1896 he took out his turbine patents which covered both pressure and velocity compounding. Velocity compounding is achieved by adding an extra, stationary, set of blades to an impulse machine, so arranged that the steam emerging from the rotating blades, 'buckets' as they were termed, is redirected. When it emerges from these stationary blades, it travels in roughly the same direction as when the steam left the inlet nozzle in the first place. It can then be made to act on a further row of moving blades fixed to the rim of the same disc (see figure 94(b). Thus the kinetic energy of the

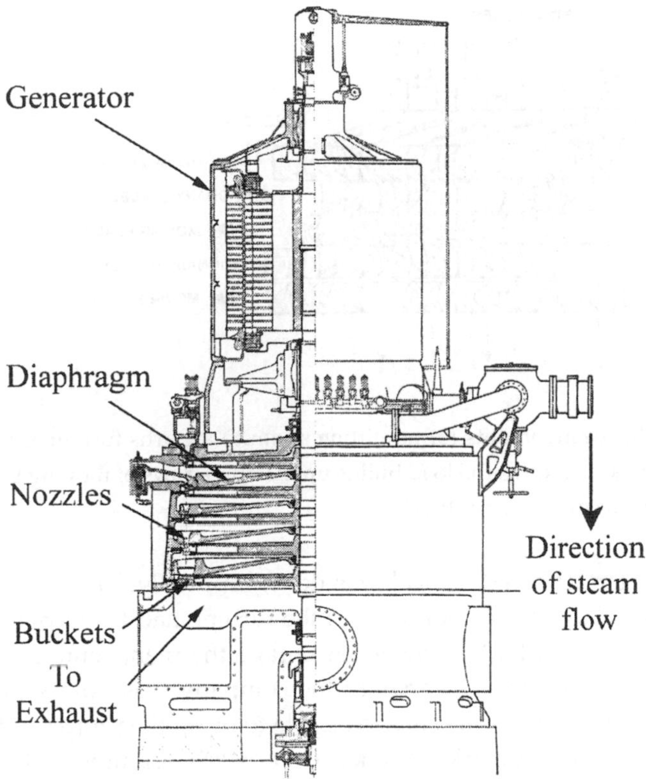

Figure 94. (a) Curtis turbine, 2.5 MW at 900 rpm, manufactured by the General Electric Company of Schenectady. Note that the shaft is vertical. There are four pressure steps, each with velocity compounding (Ewing 1911).

steam jet is extracted in two, or even three steps instead of one. Further, as in Parsons' design, several such stages were placed in tandem i.e. pressure compounding was used. Like de Laval's turbine, once the steam leaves the nozzles its velocity falls but its pressure does not change. One or two small machines were built along these lines but there was difficulty in finding a partner interested in commencing manufacture. Eventually an agreement was made with the General Electric Company, and Curtis began to experiment at Schenectady in 1897 or 1898. Since the abortive visit by GE engineers to Parsons in Newcastle in 1895, their rivals the Westinghouse Machine Company had already made a start, manufacturing turbines to Parsons' designs. GE were required to make the decision on whether or not to purchase Curtis' patent rights by 1900. Despite some doubts about

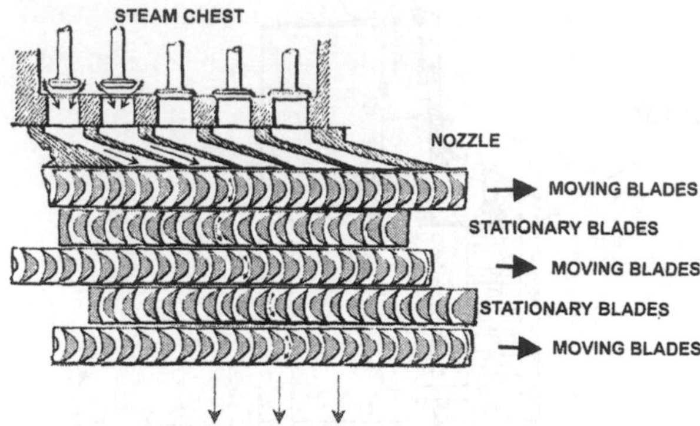

Figure 94. (b) Arrangements for admitting steam to a Curtis turbine by opening inlet valves in sequence as the load builds up. All three rows of moving blades are fixed to the same disc (Ewing 1911).

progress so far achieved, a deal was done, reputedly for the huge sum of $1500 000. Enough development of the turbine had taken place in the hands of inventors like Parsons to encourage the management of GE to face the inevitable risks. Moreover the company was now sufficiently sophisticated in a technical sense to be able to bring the project to fruition quickly. Development work was taken over by W L R Emmet (1859–1941). After a further four years the first successful commercial machine was installed at the Fisk Street power station in Chicago. It powered a 5 MW alternator (see figure 94(a)). In one leap the GE company had placed itself in the forefront of turbine builders. While Curtis was a brilliant inventor, it has been noted that *'he was not the one to make his invention into a commercial reality'*.

A C E Rateau (1863–1930) enjoyed the classical education of a French engineer and was a student at the École Polytechnique in Paris (Dickinson 1939). In 1882, having graduated at the top of his class, he progressed to a three year course at the École Supérieure des Mines. After this, while holding professorial appointments, he researched the subject of fluid mechanics, and experimented with various rotary flow machines. Between 1896 and 1898 he patented his multi-stage impulse turbine. Essentially this consisted of a succession of de Laval turbines on a common shaft, the exhaust of one becoming the supply for the succeeding stage. In this way the maximum steam velocity could be significantly reduced, and hence also the speed of shaft rotation. The nozzles of his machine were constructed as passages, formed by blades located at the outer edge of a diaphragm.

At the point where the shaft passed through this diaphragm, a labyrinth seal was placed to withstand the pressure drop across the nozzles. The moving blades were mounted at the rim of a disc that was mounted on the relatively slender shaft. Like all impulse designs, it was possible to admit steam to the inlet stage through one, two or several segments around the circumference. These could be brought into use progressively to increase steam flow, see for example figure 94(b). This removed the waste of energy which occurs when control of the quantity of steam is achieved by throttling the pressure, as had been done in Parsons' prototype. Experiments on turbines were carried out for Rateau by the firm of Sauter, Harlé et Cie from 1894 (Rateau 1904). According to S F Prest, who was negotiating at the time with Westinghouse on behalf of Parsons, Rateau offered his patents to Parsons, but the offer was refused (Prest 1931). Rateau established his own company, La Société Rateau, in 1903, with a factory at La Corneuve in Paris and at Maysen in Belgium. He quickly developed designs that were suitable for marine propulsion, and Yarrow was an early customer.

In Switzerland the long established firm of Escher Wyss specialized in building water turbines. A graduate of the Polytechnicum in Zurich, Heinrich Zoelly (1862–1937), who had joined the company in 1888, began to experiment with steam turbines working on the impulse principle. His first machine in 1903 was modelled somewhat on the Pelton wheel water turbine, but he quickly moved to the pressure compounded concept adopted by Rateau. He mounted his blades on more substantial steel discs than Rateau and was able to use higher blade velocities. With the technical assistance of Professor A Stodola (1875–1942) he rapidly raised the efficiency of his machines. He made license agreements with several firms including Schneider at Le Creusot in France.

In Sweden Birger Ljungström (1872–1948) and his brother Fredrik (1875–1964), two very prolific and resourceful inventors, created a successful variant on the reaction turbine (Hansson 1955). They had visited Parsons' works in 1896 while they were in Newcastle in connection with a high-pressure boiler which they had designed. This visit led to a novel design for a radial flow machine somewhat along the lines that Parsons had pursued, but in their case there were no stationary blades. Two facing discs rotated in opposite directions, with the rings of blades in the two discs so interleaved that steam passed from one ring to the next as it expanded radially outwards. This gave a very elegant and compact design. It was efficient and reliable and found favour for marine use. Several firms, including Mitsu Bishi, took out licences to manufacture. However there was an upper limit to the physical size of such units and

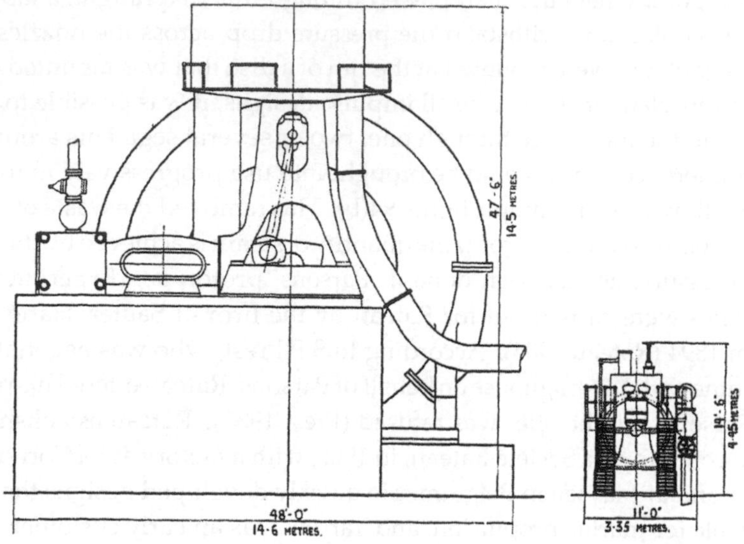

Figure 95. Figure showing the relative sizes of a reciprocating steam engine and a turbo-alternator, both rated at 3.5 MW, emphasizing the compactness of the latter (Parsons R H).

ultimately the design fell out of favour. In a letter to Parsons dated 14 March 1912 which accompanied a copy of the pamphlet announcing the new design, Ljungström wrote '*I could not deny myself the pleasure of giving you first an account of the invention which was only possible through your pioneer work and should be glad if this letter led to a commercial and technical co-operation*' (PAR/5). It appears that Parsons did not respond to this offer.

Quite suddenly, as the new century got into its stride, Parsons no longer had the field virtually to himself. In England, with the possible exception of the North-East, public supply of electricity developed slowly. Its progress was hedged in by political restraints. Municipal authorities generally lacked sufficient capital to invest, and manufacturing industry was slow to modernize. In 1900 the editor of *Engineering* writing on 'Electric energy in bulk', said that '*British engineers are often forced to admit with shame that they are inferior to the engineers of America, Germany and Switzerland in respect to practical knowledge of electrical matters*' (Anon 1900b). But in the United States, especially, the demand for electric power was growing very fast. In 1902 there were 12 turbines supplying a total output of 5.3 MW. By 1907, there were 629 turbines with a total output of 1009 MW (Sommerscales 1991). The General Electric Company wasted no time, and in 1902 it licensed the Allgemain Elektricitäts Gesellschaft, AEG, in Germany, and British Thomson Houston in England to build their Curtis turbines. Manufacture of the giant reciprocating steam engines for electricity was now in terminal decline. One reason, their great size, is well illustrated by the drawing in figure 95. Other advantages of turbines were their freedom from vibrations, reduced maintenance requirements, and their ability to handle higher inlet temperatures and to take advantage of higher degrees of vacuum. A great number of firms took up the manufacture of turbines from this point onwards, but for the most part their designs were based either on the Parsons reaction machines, or on impulse turbines pioneered by de Laval, Curtis or Rateau. Among the names of other ingenious designs, none of which could compete successfully in the long run, were Seger, Electra, Terry and Riedler-Stumpf (Stodola 1927).

15.3 Establishing the principles of design

There was a large increase in published material on steam turbines. In June of 1904 a meeting devoted to steam turbines was held in Chicago by the Institution of Mechanical Engineers jointly with the American Society of Mechanical Engineers. Papers were read by Frank Hodgkinson of Westinghouse, W L R Emmet of General Electric, E S Lea and E Meden of the de Laval Company at Trenton in New Jersey and A Rateau. Rateau's contribution is interesting because it was the only paper to show a diagram of steam velocities superimposed on sections of blading. A glance at the printed discussions shows that, as the number and variety of steam turbines multiplied, there was an increased demand for quantitative data about their construction and performance. The *Transactions of the Institution of Engineers and Shipbuilders in Scotland* for the 1905–1906 session carried papers by E M Speakman and R M Neilson which went a long way to meeting this demand.

When discussing turbine performance there are two 'efficiencies' which are employed. The thermodynamic efficiency expresses the actual mechanical work output of a turbo-motor as a fraction of the thermal energy that is supplied to the boiler in the form of fuel. During the early years this was sufficient when comparing turbines with reciprocating engines, and was often expressed as lb hr^{-1} of coal per IHP h. But since coal can have a variable energy content it became the practice to make comparisons between the mechanical output of a turbine for a given flow of steam and what an ideal steam engine should achieve with the same flow.

If we take a fairly typical value of boiler pressure, 160 lb in^2, and a condenser pressure of 0.75 lb in^2, the ideal is 7.6 lb HP^{-1} h^{-1}, or 10.25 lb $kW^{-1}h^{-1}$. But there is another efficiency, sometimes called the diagram efficiency. This assumes the existence of an ideal, friction free, turbine. It compares the useful work which should be available when steam of a given velocity is supplied to turbine blades having a given velocity, with the actual work achieved. The ratio depends on the geometry, that is on the angles of the blades, fixed and moving, which in turn depends on whether the machine is based on the impulse or reaction principle. Figures ranging from 65% to 75% were being achieved. Putting these two concepts together, a turbine with say 70% efficiency, working with the steam conditions already stated, should consume 10.8 lb of steam per hour for each horsepower output, or 14.5 lb $kW^{-1}h^{-1}$.

When Parsons and Stoney read a paper on steam turbines to the Institution of Civil Engineers in 1905, there was a section 'General Theory'. In fact it was a purely verbal discussion in the most general terms (Parsons and Stoney 1905). In the discussion that followed it was left to Professor Donat Banki of Budapest to introduce a mathematical analysis based on the geometry of both blade and steam velocities. By now there was a steady stream of research and publication on phenomena related to turbines in the German language. The classic book *Steam Turbines*, by Professor A Stodola of the Zurich Polytechnicum, which was quoted by Banki, was already in its third edition. It was left to Professor Andrew Jamieson to introduce the thermodynamic properties of steam as contained in tables, and in charts like the, by now well established, temperature–entropy diagram. After testing his mathematical analysis of the Parsons turbine against published data, a writer in the journal *Engineering*, commented that (Anon 1905a),

> *A careful examination of the Elberfeld data, however, leads to the conclusion that the bucket* (blade) *speed is too low. In view of the great experience of the makers, there is no doubt some temerity in advancing this opinion; but there is a great deal of information available pointing to the conclusion that these turbines have been designed to a large extent by rule of thumb.*

What provoked this observation was the fact that this turbine could readily drive a 2.5 MW alternator, even though it was coupled to one rated at 1 MW. This meant that normal running was at partial load, with a corresponding loss of efficiency. There had also been unnecessary expenditure on an over-powerful turbine. By 1910 the situation had changed and a number of authors had published clear descriptions of the factors affecting turbine design. Sothern's third edition of *The Marine Steam Turbine* (Sothern 1909) deals comprehensively with practicalities while Professor Ewing's third edition of *The Steam-Engine and Other Heat-Engines* (Ewing 1911) sets out the basic thermodynamic theory. Ewing introduced two new charts of steam properties which had been recently published by Mollier and which were to be of great assistance to turbine designers (Mollier 1904). These were plots of enthalpy versus entropy, and enthalpy versus pressure. When Gerald Stoney gave his Cantor lectures to the Royal Society of Arts, or contibuted to the *Proceedings of the IMecE*, he showed that, left to himself, he was perfectly capable of setting out the principles governing turbine design in language which is comprehensible to a reader with a basic technical understanding (Stoney 1909a, 1914).

When planning a Parsons turbine *ab initio* the designer has to strike a balance between several factors[2]. Fundamental to the calculation are the steam conditions at the inlet and the degree of vacuum at the exhaust. These determine the 'heat drop' or enthalpy, ΔH, which is ideally available in each pound of steam for conversion to useful work. It is analagous to the 'head' available to a water turbine. Depending on the desired power output, the steam flow rate is fixed by this. The corresponding steam velocity c, and steam momentum, is dependent on the number of steps N over which the enthalpy drop is spread. The number of steps N also determines the fractional pressure drop across each set of blades, which can be written as $\mathrm{d}p/p$. This ratio affects the extent of undesirable leakage around the tips of the blades. In the first turbine the ratio was 5% , but generally it was reduced to something like half this. Since the force which turns the shaft depends on the successful harnessing of the momentum of the steam, that is on the reduction of its velocity, the geometry of the passages between the blades becomes critical. For a given mean blade velocity u, there are suitable angles for the inlet and outlet passages of the blades if the flow is to be free of shocks as is evident in figure 80. Finally the appropriate height for the blades is settled by whatever mean blade diameter, blade shape and volume flow rate have been chosen. The actual value of the blade velocity depends on the speed of the shaft, R (rpm), and on the mean diameter of the blades, D, thus $u = \pi DR$.

In his prototype machine, and in each machine after this, Parsons made a practical stab at appropriate values for the various adjustable parameters. Remember that he had only his own experience to guide him at this stage, because no one else was designing reaction turbines. He gave only the slightest hints of his reasoning to his colleagues. He eventually produced

[2] The reader may find appendix 2 helpful in following this discussion.

a formula for them to use. Gerald Stoney described how it was launched (Stoney 1938).

> *Many a puzzle we used to have over bits of paper he left after a conference over a design. I well remember the day when he sprang what is now universally known as Parsons Coefficient, K, upon us, with no more explanation than that it had a value*
>
> $$K = NR^2D^2 10^{-9} \tag{15.1}$$
>
> *where N is the number of rows on the cylinder or spindle, R is the number of revolutions per minute, and D is the mean diameter of the blading in inches, a summation being made for the various stages of the turbine. At that time all we knew was that K was a figure related to both the turbine efficiency, and to the ratio of blade speed (u), to steam speed (c), and it took a good deal of cogitation before we discovered that this ratio u/c, was equal to*
>
> $$\sqrt{\{K/(1.3\Delta H)\}} \tag{15.2}$$
>
> *for a reaction turbine, where ΔH represents the heat drop. I have often wondered what connection there was in Parsons' mind between efficiency, velocity ratio, and heat drop. He never told us, but accepted the above formula when it was suggested to him.*

In general the more advanced the steam conditions, that is the greater the value of ΔH, the larger K should be. Also the closer (u/c) approaches to unity, the higher the theoretical efficiency of the design. In practice this meant aiming for larger values of K. But it must be said that Parsons' coefficient was a rather obscure quantity, by no means constant, and chiefly useful for comparing similar machines.

Stoney does not date the episode when Parsons disclosed his coefficient K to his colleagues. Some years later, in 1925, Parsons responded to a request from Sir Alfred Ewing, by asking Stanley S Cook of the PMSTCo. to check out when knowledge of it first became public (Cook 1925). He replied that a confidential paper had been sent to several licensees in the period 1904 to1905. It was evidently at least 12 pages long, and contained four references to K. He found that the first public mention was made shortly after this, on 24 October 1905 during a discussion of a paper read by Speakman who was an employee of the British Westinghouse Company at that time, but had previously worked for Parsons Foreign Patents Co (Speakman 1906).

Among the different classes of turbine that had now emerged, each had some advantages and some drawbacks. Consider first the magnitude of the 'heat drop', Δh, that can be utilized in a single stage. In the case of Parsons' turbine, with an optimum design, it can be related to the blade speed, u, by $\Delta h = u^2$. If we consider the de Laval machine in a similar fashion we find that $\Delta h = 2u^2$. The same is true for each stage of the Rateau turbine. If we consider the Curtis, velocity compounded, stage with just two rows of moving blades, then $\Delta h = 8u^2$. The total heat drop in a machine, ΔH, is related to the stage drop, Δh, by $\Delta H = N\Delta h$, where N is the number of stages. Clearly the machine which requires the smallest number of stages is a Curtis turbine. Next in order is the Rateau turbine, and finally the Parsons turbine requires the most stages. In consequence the Curtis machine can be much shorter than a Parsons turbine. This can affect capital costs, as well as the critical speed of the shaft. On the other hand, in the Curtis design the velocity of the steam relative to the blade surface is very much higher than in a Parsons machine, so that friction losses are higher. Also the basic diagram efficiency is lower than for a Parsons turbine. In machines of the impulse family there is no pressure drop across the moving blades, and so there is no tendency for steam to leak past the tips of the moving blades without doing useful work, as happens in Parsons' design. Perhaps because Parsons' turbines had a large number of rows of moving blades, there was a tendency to fix them on the surface of a drum as in figure 68. Impulse turbines by contrast tended to be fastened at the periphery of a disc, located on a relatively slender shaft as in figure 94. As a result there was a significant friction or 'windage' loss incurred by discs rotating at high velocity between the fixed diaphragms. In practice no one design had a decisive advantage over the others, and history often decided which one would be favoured by a given firm.

15.4 Patents and licenses

As the legal appeal to extend the protection of the 1884 patents was being assessed, attention was drawn to the appreciable costs of keeping patents active not just in Britain but in foreign countries. For Parsons, his patents were important for financing further development work, and he needed to maximize the income from licenses and royalty payments. When the PMSTCo. was established as a totally separate entity from C A Parsons and Co., the latter sold the rights to use certain patents for marine purposes only, to the former. Taking this financial separation a stage further, in August 1899 C A Parsons and Co. joined with the Marine Steam Turbine Co. Ltd and PMSTCo. Ltd to create the Parsons Foreign Patents Co. Ltd, 'PFPCo'. It was based in London and the shareholders were Lord Rosse, C J Leyland,

C A Parsons, A A Campbell Swinton, W H Lynch, A Baker and F Baker.

S F Prest, who had separated from Parsons after helping him with the Westinghouse Licence, later renewed his connection with him. Indeed he became responsible for looking after licensees. He was a man of tact and continued to work with Parsons until Parsons died. The fact that he was based in London and not Newcastle may have been an advantage in managing his personal relationship with Parsons. An early letter conveys something of this (Prest 1903a),

2 Suffolk Lane, Cannon Street
London E.C.
June 18th 1903

The Hon C A Parsons

My dear Parsons,

I wrote Pirrie early yesterday to ask if he could give me a 5 min^s interview but have no reply. I conclude he thought it was not necessary, having regard to the letter he wrote to you on Tuesday, which your clerk read to me over the telephone this morning.

Please let me have a copy of this letter. It shows he has the matter in hand, and I think it is hopeful.

Biles being brought in makes no difference to me, if I can be of any assistance to you. I was only anxious to know if you thought it would be necessary for me to go also, as of course I have many business details to put in order in case such a visit may be necessary, and it was not any use continuing with these if I was not going.

With regard to remuneration, in addition to my expenses, I consider £5 a day as reasonable, and I think that having regard to the work I have already done, this remuneration should be increased in case of a satisfactory result being arrived at.

How does this appear to you?

I was in no way offended by anything you said about PFPC^o I always like a man to say what he thinks—it saves time, and I always do the same myself.

Nor had your decision regarding the agreements anything to do with the matter.

When you asked me to be your alternate you said you did so as you believed I should not be influenced by the 'presiding genius'. You told me on Tuesday quite plainly that you found that I had fallen under that influence. Therefore my qualification was gone, and I was bound to retire.

There is no need to say anything more on this matter.

Yours sincerely

S F Prest

W J Pirrie was the head of Harland and Wolff, the successful Belfast shipyard which had still not adopted turbines for any of their vessels. Professor Biles of Glasgow University was the naval architect who had been responsible for the design of the Midland Railway's steamers. From the reference to remuneration it would appear that Prest was not a full time employee of Parsons at this juncture. Clearly there had been a rather sharp exchange about the way Prest had handled the board meeting in Parsons' absence, but matters had evidently been put right. Not all of Parsons' colleagues had the skill of Prest in calming relations when tempers flared.

One of the first Continental firms to make an agreement with the PFPCo. was Swiss. The copious availability of hydro-power had alerted the Swiss to the potential importance of electricity supply, and in 1886 Lucerne became the first city anywhere which possessed its own municipal power station. The success of the Elberfeld machines caught the attention of Brown Boveri et Cie, located at Baden some miles to the north of Lucerne. Only a couple of years younger than C A Parsons, this firm had been registered in October 1891 to manufacture electrical motors and generators (Evans 1966). Charles Brown (1827–1905), who was the father of one of the founders, was an Englishman from Uxbridge. He had come to Winterthur in Switzerland to apply his engineering skills at the Locomotive Works and with Sulzer Brothers. He subsequently worked in Zurich and Oerlikon, then in Newcastle and in Pozzuoli in Italy before returning to Baden. He sent his two sons to be educated at the Technical College which had recently been established in Winterthur. When Charles returned to England in 1885, his elder son, C E L Brown (1863–1924), took over the management of the electrical engineering department of the Oerlikon Engineering works at the tender age of 22. He was actively involved on behalf of Oerlikon in 1891, in a demonstration of high voltage, three phase, power transmission. An overhead line, 175 km long, was used to connect a hydro-powered alternator located at Lauffen to an exhibition being held at Frankfurt am Main, where it was used to supply incandescent lamps and a three phase motor of 100 HP. His partner Walter Boveri (1865–1924) was the son of a doctor, and had studied at the Royal School of Engineering at Nuremberg, after which he also joined Oerlikon in 1885.

The two young men became friends and decided to establish their own firm in 1891. Walter Boveri's wealthy father-in-law put up a large part of the necessary capital. The Baden town council had just been persuaded to build a power station as a means of attracting new industry. Brown Boveri were successful in their bid to supply the generators for the water turbines. In 1893 C E L Brown had his proposals for generators for the power station at Frankurt am Main accepted, despite the fact that it meant giving business to a foreigner. The decision by the City Council was swung by a *'very favourable report by the City Engineer Lindley'*. The pair were well matched. Boveri had a good understanding of business, and indeed he developed a separate organization which concentrated on funding the construction of power plants. These were subsequently established as independent entities and sold on. Brown, on the other hand, had little interest in such matters, and shared some of his father's views. For example he was not interested in the savings which batch production made possible. His father

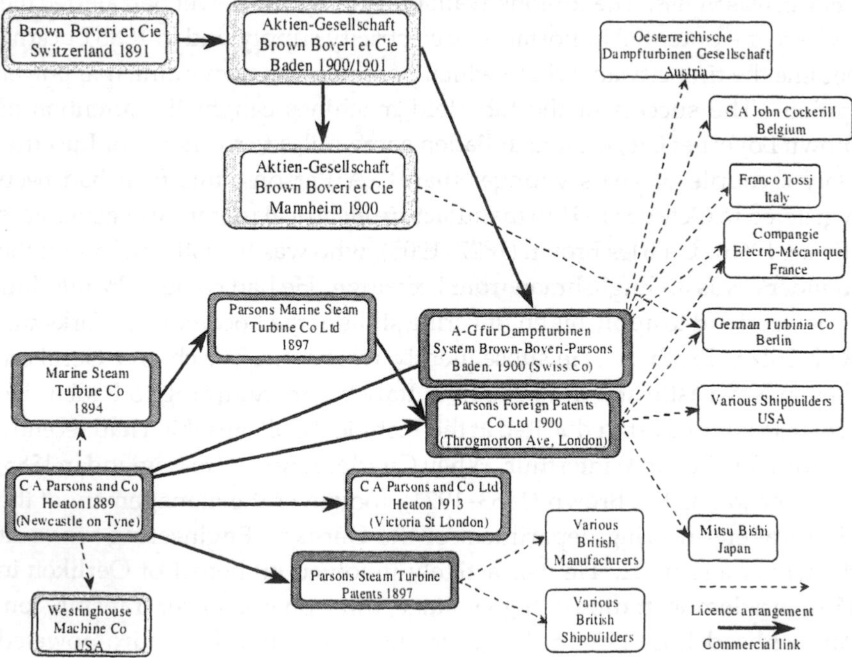

Figure 96. Schematic diagram intended to convey an impression of the relationships between Parsons' companies, and between them and their licensees.

on a visit to the works *'pronounced that this sort of* (batch) *work must be very boring; one should always try to design something new'*. In this at least father and son had much in common with Charles Parsons.

Although in the early years Brown Boveri supplied generators for most of the water powered stations built in Switzerland, there seemed to be little demand for steam power. The home market was small and of necessity most of their business would involve exports, indeed in order to facilitate business with German customers a subsidiary company was established at Mannheim in the Rhineland. However in 1901 Parsons embarked on a joint venture with Brown Boveri. He held a majority stake in the Aktiengesellschaft für Dampfturbinen System Brown Boveri–Parsons which held the exclusive rights from PFPCo. to manufacture steam turbines for land use in Switzerland, Germany, France and Russia. The Mannheim company also joined Parsons in Turbinia Deutsche Parsons Marine AG in Berlin to build turbines for marine use.

Brown Boveri built their first 250 kW turbo-alternator in 1901. By 1904 they delivered a 5 MW single cylinder turbine to the Rheinische-Westfälisches Elektrizitätswerk, and this was followed in 1907 by a 7.5 MW machine for Buenos Aires. The first marine turbine was built at Baden in 1904 with an output of 4.7 MW at 860 rpm. A start on the manufacture of turbo-blowers and compressors was made in 1906, but in collaboration with Rateau rather than Parsons. In 1908 an oil powered servo system was substituted for the linkage mechanism used by Parsons for turbine governors. Differences were emerging in the designs of the two companies, notably the adoption of a Curtis stage for the inlet stage. A somewhat similar development had happened with the continental licensees of General Electric when they abandoned the vertical configuration of the early Curtis turbines in favour of a more conventional horizontal shaft.

After some years Parsons became concerned at what he considered a lack of progress by the Swiss company, and he wrote to Brown (Parsons C A 1911b),

C A Parsons & Co
Heaton Works
Newcastle on Tyne
1st September 1911

Dear Mr Brown,

I have been looking further into the situation of the Swiss Company and of turbine work generally in other Countries, and I find it impossible to understand the tremendous loss in your turbine manufacture during the

last year and the poor results of the preceding year. I am further unable to account for the very great expenditure on turbine plants after they have been erected; it is difficult to understand how so great and so many mistakes have been made in design and manufacture to account for so large a sum. Then I am unable to understand the purport of the great number of patents which have been taken out during the last three or four years, the value of which in protecting your manufacture I must confess, after the most careful study, I am unable to appreciate as having any adequate relation to the heavy expenditure incurred. The position is of course, serious, otherwise I should not taken upon myself to write; in doing so I have refrained from any allusion to the financial side which your letter led me to understand is in the hands of Mr Boveri.

I remain, yours sincerely,
Charles A Parsons

After a meeting with Parsons at his residence in Kirkwhelpington, Prest and Barker of PFPCo. were sent to Paris to meet Boveri and Sydney Brown to obtain their assent to an agreement that effectively ended the joint venture, but without an outward split (Prest 1911). From this distance it is not possible to know exactly what led to this outcome, but the Brown Boveri Company history gives some clues (Evans 1966).

From 1909 on, in spite of a satisfactory state of business, difficulties had to be overcome which were due to the nature of the works themselves. The success hitherto achieved was due to a combination of ingenious design ideas and skilled workmanship of craftsmen in the factory. Now after the first sudden growth, the need for a more rigid organisation became evident. The accent had to be transferred from technically interesting, but commercially unprofitable, 'one-off jobs' to batch production with a resultant improvement in the yield. The organisation of the factory also had to be tightened up and designs standardised, so that a better footing could be obtained for costing.

15.5 Licenses for 'land' use in the United Kingdom

Licensing of companies to build steam turbines for land use, whether in the United States or on mainland Europe, was one thing but it was a different matter with firms in the United Kingdom. The Heaton Works could in principle build any machines required for the home market. Moreover competitors actively sought to get around Parsons' patents and deprive him of any financial benefit which might accrue from all his effort in pioneering the turbine concept. Records survive of dealings with the Willans and Robinson Company at Rugby, and these shed light on this period. As has been mentioned, the Willans central-valve steam engine had achieved a very strong position as the preferred drive for electric generators. Thus the two new stations which were built by the Liverpool Corporation in 1899 and 1900 were each equipped, on the advice of their consulting engineer Dr John Hopkinson, with 12 such engines (Woodward 1992). The largest was rated at 700 kW. By 1904 the Liverpool stations together could supply 25 MW from 78 such engines. In 1903 when tenders were invited for new plant, it was stated that turbines would also be considered if offered.

Peter Willans had died in an accident in 1892 and his colleague Mark Robinson became chief executive of their company. During the discussion of Parsons' 1888 paper Robinson had expressed himself to be an admirer of the turbine, and he realized that the turbine would inevitably displace steam engines. He had now become involved in an attempt to circumvent Parsons' patents. Hugh Francis Fullagar had joined PMSTCo. in 1897 at the age of 25 after completing the Mechanical Science Tripos at the University of Cambridge under Professor Ewing (Cox R 1992). He was a shop foreman when he was dismissed by Parsons. In 1901 he took out patents relating to turbines. Once he had ceased to work for Parsons this was of course quite legal. In order to benefit from his patents he then negotiated with John Tweedy (1856–1916) of Swan Hunter's Neptune Works, A F Yarrow the shipbuilder and M Heaton of Willans and Robinson to form a syndicate to exploit them. The syndicate contracted Fullagar to act as designer and consultant, but did not itself engage directly in manufacture. His patents were also extended to the United States. Not surprisingly Parsons contested Fullagar's 1901 patents in a vigorous action before the Solicitor General. Both he and Gerald Stoney stated that the subject of the patents was not new, and had been widely known at the Heaton Works, but had not been adopted (Robinson M 1903). While this litigation was in train Mark Robinson wrote to Parsons to ask permission to supply Liverpool with turbines,

Hon. C.A. Parsons
17th April 1903

My dear Sir,

The issue of the Liverpool specification accepting turbines as alternative, by such an old supporter of the Willans engine as Mr Holmes, naturally leads us to wish to be in a position to quote for steam turbines.

Would it be agreeable to you to license Messrs Willans and Robinson to make these turbines for Liverpool?

Mark Robinson

Chairman

Parsons replied with a friendly letter written while on holiday at Caragh Lake in Ireland, but Robinson found the royalty of 3/- per kW excessive. On 24 May Robinson questioned Parsons' contention that the supply of metal strips shaped to a blade profile was covered by Parsons' patent 8698 of 1896. He added that if necessary he would challenge this claim, but of course he had no hard feelings. In due course this brought a sharp rejoinder,

Holeyn Hall

8th October 1903

Dear Mr Robinson,

I must ask you further to dismiss from your mind that I shall not consider your further action as personal, for it will give me and my colleagues a great deal of trouble and take up our time which should be devoted to furthering the introduction of the steam turbine, to which we have devoted most of our lives.

... If it becomes necessary, we shall not hesitate to carry through any law cases that may arise.

Charles A. Parsons

Meanwhile Parsons had been successful, and Fullagar's 1901 patent applications 7184 and 8934 were not granted. Robinson now sought to distance himself from Fullagar and congratulated Parsons on his success. Nonetheless he asked Counsel's opinion on Parsons' patents, and by the end of the year Willans had decided to supply machines under the protection of several of Fullagar's other, more recent patents. In December 1903, the Syndicate licensed Allis Chalmers of Chicago to use them, on payment of £2500 and royalties which were guaranteed to reach £1000 for 1905, £2000 for 1906 and so on. Fullagar continued to take out more patents, a total of five in 1904, but Robinson realized that, whatever Fullagar's talents, a deal with Parsons made sense. On 8 November 1904, he presented report number 10, 'on steam turbine manufacture', to his board. In it he wrote

However his (Parsons) *position with regard to Turbines justifies courteous treatment, and something more. It is true that his response to our early*

efforts to come to a friendly arrangement might justify an opposite view, but he probably believed we were insincere, and that we only offered to recognise his moral, in order to evade his legal claims. (Mr Parsons gave me the idea of a man prone to such judgements).

... It is humiliating to take a license in respect of patents we do not work under.

... We do not want Mr Parsons' nameplate; our own is good enough (see below).

In January 1905, at a meeting at the new headquarters of the Institution of Mechanical Engineers, a deal was negotiated by Prest and Robinson. The syndicate was to be wound up and Fullagar was to be recompensed. His patents were purchased by Charles Parsons. Fullagar was a prolific inventor in his own right. Between 1904 and 1906 he worked with some success with Charles and Norbert Merz on a gas turbine of 25 to 30 BHP for the Hebburn Engineering Development Co. (Cox 1992, Hore 1994). He retained contact with Willans and Robinson and in 1920 his first opposed piston oil engine was installed at Rugby.

The 'License for the manufacture of steam turbines', dated 6 February 1905, authorized Willans and Robinson Ltd to build steam turbines in the United Kingdom only, and not for the purpose of marine propulsion, for a period of ten years using the patents listed. It has 29 clauses and a schedule which lists 21 patents starting with 6735 of 1884. Their turbines could be sold to customers in the United Kingdom and its Colonies and Dependencies (except Canada) as well as in Norway, Sweden, Denmark, Spain, Portugal, Egypt, Japan, Brazil and any other South American country. The fee was £1000 and a royalty of one shilling and sixpence per kilowatt of output if coupled to an electric generator. If a direct mechanical drive was used the royalty was one shilling and one penny per brake horsepower. If the turbines were used to power electrical generators manufactured outside the United Kingdom, an additional royalty of one shilling and six pence was payable.

The licensee was to keep records of all machines built, and allow access to the licensor to inspect the manufacture of turbines. Each machine was to carry a name plate which included the name of both the licensee and the licensor. The licensor was required to communicate any improvements or additions relating to the inventions, which were made by him, to the licensee, who in turn was bound to share with the licensor and the PMSTCo., but not with their licensees, any similar improvements made by him. The licensee promised not to challenge or dispute the validity of any of the patents during the duration of the agreement. Nor would he '*assist or (so far as they have power to prevent the same) permit or suffer any person, whether in their employment or not, to apply for or obtain the grant of Letters Patent... without communicating such improvements to the licensor for his own and the PMSTCo's use...*'.

It is clear that the agreement was a carefully thought out legal document and was far more than a permission to use patents. As Robinson found when he sought Counsel's legal opinion, details of the agreement were binding for ten years even though some of the patents might well have expired before the end of this period. Once the agreement was in place considerable correspondence was exchanged between E J Fox in Rugby and W M Johnston at Heaton. The domestic market was depressed and there was concern about one licensee undercutting another and even maligning them. On 31 October 1906 Fox wrote to Johnston,

> *... I consider this is essentially a case where the turbine industry has suffered due to certain licensees of the Parsons Turbine thinking it was in their interest to cry 'stinking fish' at what their competitors were doing.*

On 7 April 1908, Parsons wrote to Robinson to draw attention to the fact that when Stoney and Law read their paper on high speed electrical machines to the Institution of Electrical Engineers (IEE), Fox had contributed to the discussion but lauded the virtues of '*Continental electrical machines.*' A request was made for permission to export to Russia, and Robinson told his Board that '*Mr Prest is ready to use his good offices with Messrs Brown Boveri with regard to importing into Russia in any particular case by special arrangement with them*'. Some of the agreements were very restrictive. One which covered the use of the vacuum augmenter insisted that it could only be used with a Parsons type turbine[3]. Not realizing this fully, in 1908 M Peache sought a special exemption for the case of an installation involving a Zoelly turbine. Parsons' reply was a firm no! Peache wrote an account of a meeting with Prest,

> *Vacuum Augmenter etc*
>
> *I called yesterday upon Mr Prest. He has talked over the vacuum augmenter with Mr Parsons. ... He suggested a calling of a conference of Mr Parsons and all his licensees on the supply of blades and the like—a sign I suppose that the old jealousies are passing.*

In September 1907, with the help of E G Izod, their chief engineer, Fox had prepared a memorandum for Mark Robinson,

[3] The device described as a vacuum augmenter is described in chapter 17.

In accordance with your instructions I beg to set out below the reasons that prompted me to say that Willans and Robinson's interests are prejudiced by any working agreement with Parsons and Co.

To quote one item only,

Blower License

An exclusive license from Parsons for his blower—presumably for a restricted territory—would probably turn out to be of no greater value to Willans and Robinson than an exclusive right for some form of beam engine.

Prest's skills were not always sufficient to smooth relations. Reporting on a meeting in 1911 with Prest about the terms of a revised turbine license, Peache wrote

... Prest ramped on the subject, treated it as looking a gift horse in the mouth... Prest said 'it is no good you running around trying to make a fuss with the other licensees. That won't do you any good'... Prest raved again 'I am not going to bother Mr Parsons further. At my request he has made a great concession'...

15.6 The inventor and society

Parsons' own views on the subject of patents and the inventor were set out in his presidential address to the Engineering Section of the British Association which met at Cambridge in 1904 (Parsons C A 1904c). Rather in the style of his father, he began with a historical overview, noting that in the past there had been hostility to scientific research, discovery and mechanical invention.

The great inventions of the last century in science and the arts... have proved without question that the inventors in the past have, in the widest sense, been among the greatest benefactors of the human race. Yet the lot of the inventor until recent years has been exceptionally trying, and even in our own time I scarcely think that anyone would venture to describe it as a happy one.

He noted that there had been a great increase in the number of research laboratories and of persons trained to scientific methods as a result of money provided by public bodies and private individuals.

> *Perhaps one of the most important results to engineers has been the direct and indirect influence of the more general application of scientific methods to engineering.* Continuing he said, *To return to our subject... let us dismiss from our minds the very common conception which is given in dictionaries and encyclopaedias that invention is a happy thought occurring to an inventive mind... Generally, what is usually called an invention is the work of many individuals, each one adding something to the work of his predecessors, each one suggesting something to overcome some difficulty, trying many things, testing them when possible, rejecting the failures, retaining the best, and by a process of gradual selection arriving at the most perfect method of accomplishing the end in view... Then after the invention, which we will suppose is the successful attempt to unravel some secret of Nature, or some mechanical or other problem, there follows in many cases the perfecting of the invention for general use, the realisation of the advance or its introduction commercially; this after-work often involves as great difficulties and requires for its accomplishment as great a measure of skill as the invention itself, of which it may be considered in many cases as forming a part... If the invention, as is often the case, competes with or is intended to supersede some older method, there is a struggle for existence between the two... The new invention, like a young sapling in a dense forest, struggles to grow up to maturity, but the dense shade of the older and higher trees robs it of the necessary light. If it could only once grow as tall as the rest all would be easy, it would then get its fair share of light and sunshine. Thus it often occurs in the history of inventions that the surroundings are not favourable when the first attack is made, and that subsequently it is repeated by different persons, and finally under different circumstances it may eventually succeed.*

He used the invention of the internal combustion engine to illustrate the role of patents,

> *Surely no fundamental patent could have been granted under the existing patent laws for so obvious an application of known forces. Consequently, patent protection was sought in comparative details, details in some cases essential to success which were evolved or invented in the process of working out the invention... But in answer to the question whether such* (patent) *protection* (of details) *was commensurate with the benefits received by the community at large, there can... be only one reply. Generally those who did most got nothing... and in very few cases can the return be said to be adequate.*

Excellent examples of the importance of '*comparative details*' are to be found in Parsons' approach to achieving the very high speeds essential for small turbo machinery. The two forms of 'damped' bearings, using washers and later using concentric tubes, tamed the inevitable unbalanced forces until the technology of balancing could be perfected, after which they became obsolete. Also his unique use of piano wire to bind conductors on the surface of generator armatures was a vital feature of construction in the initial stages because it was inherently symmetrical and unlikely to generate unbalanced centrifugal forces. This too had a brief, but crucial role during the development period.

He turned to the need for financial backing,

> *To be successful* (the inventor) *must be something of an enthusiast; and usually he is a poor man, or a man of moderate means, and dependent on others for financial support. Generally the problem to be attacked involves a considerable expenditure of money; some problems require great expenditure before any return can thereby accrue, even under the most favourable circumstances... In nearly all cases the inventor has to co-operate with capital: the capitalist may be a sleeping partner, or the capital may be held by a firm or syndicate, the inventor being in such cases a partner—a junior partner—or a member of the staff. The combination may be successful and lasting, but unfortunately the best inventors are bad men of business. The elements of the combination are often unstable and the disturbing forces many and active; especially is this so when the problem to be attacked is one of difficulty, necessitating... expenditure... many times greater than that foreshadowed at the commencement of the undertaking.*

In later years Parsons would give an extensive historical account of the work of previous inventors who had sought to make a steam turbine, but this present awareness was a consequence of the legal research needed in the battle to extend his own patents. He personally did little historical research before starting his own work.

> *In some cases the cost of a thorough search* (of previous patents) *is very great indeed; sometimes it is greater than the cost of a trial attack on the problem.*

But his experience since 1884 had taught him the importance of searching previous patents and he remarked

> *what a waste of time, expense, and disappointment would be avoided if we in England helped the patentee to find out easily what had been done previously,*

on the lines adopted by the United States and German Patent Offices, who advise the patentee, after receipt of his provisional specification of the chief anticipatory patents, dead or alive!

Some tasks are beyond the possibilities for a single inventor, and no patents could be expected to yield an overall financial return. He instanced the need for *'a thorough investigation of the problem of aerial navigation'* and also dealt with a topic to which he had given much thought, no doubt sparked by his interest in diamonds. It was the exploration of the lower depths of the earth. The deepest bore-hole at the time was one mile long. He gave estimates for the costs, time required and the likely rock temperatures at various depths of shaft based on experience of mining in the Rand in South Africa. A hole 12 miles deep might cost £5 million and take 85 years to complete. It was expected that the rock temperature would reach 272 °F. Such a venture has yet to be undertaken.

He could not understand why patent protection lasted only 14 to 16 years while copyright, and the rights of public utilities, extended to 42 years. He cited the enormous benefits that had been conferred on the community at large by steel makers like the inventors of the Bessemer and the Siemens–Martin processes, or by the inventors of electric telegraphs and telephones, and compared them with the relatively small financial rewards for the inventors themselves. For improvement in the inventor's situation he looked to changes in the law, but he was unable to specify how exactly this should be achieved.

15.7 Works management

Traditionally, if a graduate engineer were to find employment, in English industries at least, he would find himself directed towards research and development activity. That was indeed Parsons' role in his partnership with Kitson. When he moved to Clarke Chapman he also undertook some managerial responsibilities. But when he established his own firm at Heaton, Parsons shouldered the entire financial and managerial responsibility, while continuing to press forward several demanding technical programmes. In this respect his firm differed from the limited liability company which had been founded at almost the same time by C E L Brown and W Boveri, at Baden in 1891 (Evans 1966). Later, when Brown and Boveri decided to manufacture steam turbines in 1900, they needed to increase the available capital. By converting to a joint stock company the capital was raised from 615 000 to 12 500 000 Swiss francs. This step involved their enterprise with other financial institutions and gave it some characteristics which Parsons' firm, with its ready access to capital, lacked. Thus it was not until 1913 that there was any formal sharing of authority with a board of directors at Heaton.

Parsons, as far as his education in managerial matters was concerned, was obviously heavily influenced by what he had seen around him during his apprenticeship in Armstrong's Elswick Works. When he came to recruit foremen and managers, he chose, by and large, men who had also served apprenticeships. They too would have imbibed the traditional attitudes which had evolved among employees of the enormous industrial complex which had grown up around the ship building industry on the North East coast. By 1901 even the Heaton Works had a half dozen or so of its own premium apprentices, whose parents had each agreed to pay 100 guineas per year for three years. One such was F G H Bedford, the son of a serving Admiral. In the course of time he succeeded Parsons as Chairman of the company board, so reverting to a traditional practice by appointing a non-engineer to run an engineering enterprise (Bedford). But it must be said that for technical posts Parsons sought out educationally well qualified individuals, just as his father had done when picking assistants to work with the giant telescope at Birr. Both G G Stoney and A H Law were engineering graduates of Trinity College Dublin, while S S Cook was, like Parsons, a mathematics graduate of St John's College Cambridge. Stoney and Cook were to be elected Fellows of the Royal Society, as was A A Campbell Swinton, who, though not an employee, was a member of the board of the PMSTCo.

Even at the present time, when the full power of information technology can be called on, the efficient management of a factory in which sophisticated and expensive machinery is manufactured is a very demanding task. Industrial units like those established by Parsons, Westinghouse and Brown Boveri were a relatively new phenomenon. Pioneers like F W Taylor in the United States were beginning to make systematic studies of the problems that had to be faced in managing them. But, generally speaking, among those who had access to good education in Britain there was a disdain for direct involvement in industry. Notable exceptions to this were the businesses established by Quakers. For long excluded by conscientious scruples from seeking employment in the established church, the judiciary or the armed forces, many members of the Society of Friends did involve themselves in manufacturing in both England and Ireland, and their success showed the potential for industry to gain from well educated executives.

Papers on management topics were presented in Newcastle at meetings of the North East Coast Institution of Engineers and Shipbuilders, which had been founded in 1884. Reading the contributions made from the floor at such meetings, it is clear that many managers were aware of the pitfalls which they faced, and they understood very well what they needed to achieve (Westgarth 1899, Borrowman 1900). Competition was fierce, and the margin between profit and loss was small. But, in general, owners were reluctant to commit the human resources necessary to help managers achieve their objectives. Costs in manufacturing mount up in two ways. Financial outlay to cover the provision of materials is incurred from the moment of purchase until the finished product is paid for by the customer. Obviously any waste is unacceptable, but delays, which tie up capital uselessly, also add to costs by increasing interest payments. Secondly the workforce must be paid. In seeking to keep control of labour costs different approaches can be taken to fixing wages. One is to pay for the number of hours taken to complete a job. Clearly the outcome here is likely to be strongly affected by the climate of human relations between shop floor and management, by whether or not the men are putting their backs into the job. The alternative may be to predetermine a price per 'piece'. For this to be effective, some reliable and objective method is needed to estimate a fair rate for a given task, using techniques like time and motion study. But these were only just beginning to be evolved in American industries (Taylor 1912). Where a task is repetitive, it is not too difficult to assemble reliable data, but most of the work connected with steam turbines varied from one machine to the next. This was worsened by Parsons' insistence that each succeeding machine should be an improvement on its predecessor. Of course when bonus payments were offered to speed the progress of a job, there was another problem: the real risk of poor workmanship. This made further demands on the alertness of the supervisor. If managers were not to operate in the dark they needed timely notice of costs accrued to each job as it progressed. Too often, the data when they became available, were of little more than historical interest.

The critical tool needed for managing a manufacturing operation was a set of drawings with their associated lists of materials required. It was essential that these should be free from errors, and should be available in good time to allow material and bought-in components to be assembled. In fact the drawing office, and the departments responsible for estimates and for the purchasing of materials accounted for most of the remaining staff, and were regarded as costly overheads to be kept to a minimum The foremen of the various shops in the factory then faced the next to impossible task of planning the optimum flow of work, in synchronism with other, related, departments, and without wasting the time of their own labour force.

Sir Alfred Ewing recounts a story which touches on these difficulties, and sheds some light on what it meant to have Parsons as a boss (Ewing 1931). Normally, repetitive jobs, like fixing turbine blading in place, were paid for at a certain rate for given quantity of output. On one occasion a dispute had arisen about what would constitute a fair piecework rate for this job, and the problem came to Parsons' attention. This was of course long before the era of work study. In any case, during the lunch break Parsons took one of his staff with him into the workshop, and closed the door. He took off his coat and set to work. When the men returned after their break they were forced to admit that their ideas about a fair day's work would have to be revised. There was little room for further argument.

With this gesture Parsons demonstrated in a telling way his own skill as a craftsman, and he also showed that he too had beneffited from his time spent in the shops as an apprentice. This sort of ability won him a very strong affection from the shop floor workers, but it is reminiscent of the episode when the crew of the *Viper* had been in dispute with him over rates of pay. The normal autocratic style of management of that period was, no doubt, actually reinforced by the fact that Parsons had inherited the natural habits of a land owner who was all powerful in his dealings with his tenants. What was unusual was his readiness to submit his decisions to a practical test. To some extent the harmful effects of such behaviour were mitigated by another characteristic which he shared with his father, *'the noble Vulcan'* of Birr. It was an ability to get on easily with the worker on the shop floor. His nephew, during his apprenticeship at Elswick, found himself, like his uncle before him, working as mate to

> *old Jack Bonner, who as he did 20 years earlier, loved to munch raw onions from a red handkerchief at frequent intervals throughout the day. We both liked the old man, and that particular smell in later years often brought back pleasant memories of him, and of our days at Elswick'* (Parsons G L 1939).

The foreman was faced with a difficult job: he was faced by a labour force which for many years had been exploited by employers. Not surprisingly, human nature being what it is, the workers sought to minimize the effort required to secure their wages at the end of the week. With large numbers of men to be supervised it was not surprising if confrontation was a common posture.

Parsons' own work rate was extraordinary. Quite apart from

Figure 97. Drawing of Charles A Parsons at the age of 55 by Kathleen Streatfeild (1909), © Birr Scientific and Heritage Foundation by courtesy of Lord Rosse.

involvement in design and development, he ran two quite distinct manufacturing works, one at Heaton, the other at Wallsend. He was continually in negotiation with the Admiralty and other customers, as well as with numerous licensees and business partners. As in the case of so many successful men, he was blessed with a robust constitution. Towards the end of the century, at 45 years of age, he was still powerfully built and capable of considerable physical exertion. Around this time his nephew Geoffry recalled accompanying his uncle and C W Bigge, a friend from Hawthorn Leslie, on a bicycle ride (Parsons G L 1939). The unfortunate Bigge was soon left behind, and even the youthful Geoffry had a hard struggle to keep pace with his uncle. Indeed on one occasion he reckoned the latter had reached a speed of 30 mph on an ordinary bike.

In managing his companies, Parsons did involve members of the extended family. His brother Laurence of course was chairman of PMSTCo.; Clere, who was perhaps the best qualified of all to participate, was a forceful character but he played no direct part in the business. Parsons' own children Algernon, called Tommy in the family, and Rachel, and their cousins Geoffry, who was a son of Laurence, and Arthur, who was a son of Clere, were all given responsibilities of one kind or another in due course. In some ways this brings to mind the highly successful Siemens brothers, who collaborated in the running of ventures in Germany, England and Russia. But it must be said that the involvement of members of the Parsons family was in no way comparable. Laurence, as Chairman of PMSTCo., was largely a figure head, and Charles kept most important decisions very much in his own hands.

Each week a works committee met at Heaton to consider matters requiring attention and a minute book covering the period from mid-1903 to mid-1904 has survived (TWASa). Some excerpts help to convey how Heaton was managed.

> 11th *August 1903 G.L.Parsons (chair), Stoney* (chief engineer), *Williams* (works manager), *Dakyns* (company secretary), *Taylor, Carnegie* (junior engineer), *Charlton*
> *re Progress No 836A stator and rotor*
> *re No 924 pole tips slack,* $1^1/_8$" *drill used instead of* $1^1/_{16}$" *Mr Hood's attention to be drawn to this but committee will pass it this time*
> *re 945, 927*
> *re gear for grinding up turbines in shop Promised this day week*
> *re 900 Further run this AM, but further alterations needed*
> *re 912 To be dispatched without end plate*
> *re circle cutter Bring forward to next meeting*
> *re 904-5 Test house doing preliminary trials*
> *Design of Brass Foundry design and specification submitted to committee, prices being got*
> *Carville Insulation tests Merz*(consulting engineer) *has seen insul^n— perfectly satisfied*
> *New Rates agreed*
> *T Bruce Turner 35/- to 37/-*
> *F Brown Fitter 24/- to 26/-*
> *W Armitage Blacksmith 37/- to 39/-*
> *New Shop Roof Iron Work to be painted*
> *re Test House to be painted black, as new shop, Williams and Howe to arrange*
> *re Stable Mr Williams asked to clear out stable*
> *Lagging on heater It was reported that the lagging on the heater supplied by messrs Par.Mar.St. Turbine Co was an exceedingly bad job.*
> *Re state of affairs in armature shop Hedley Thompson called in. Blackpool armature full of brass turnings, Carville machine samples full of solder. Two*

men interviewed for armature foreman—W J Muir to be started Monday
<u>*re Premium Robson*</u> *mentioned that one of the premiums (Robson) would go to College next Oct for 9 months*

W M Johnston 18/8/03

W M Johnston, as chairman of the succeeding meeting, signed the minutes. He was the manager in charge of correspondence, the preparation of estimates, ordering materials and general commercial work (Bedford). This represented a fairly typical agenda. Progress of work on individual machines was checked, items relating to equipment and buildings were dealt with (the workmanship of the sister company was not spared criticism!). Quality of work was kept under review, and in the case of the armature shop rapid and drastic action was taken. Wage rates were individually reviewed and adjusted. The next meeting was smaller, and one item revealed signs of friction between members of the management.

18th Aug 1903 G L Parsons, R Williams W M Johnston (chair)

<u>*Progress on 836A*</u> *In regard to insulation of rotor coils, Mr Stoney gave separate report, but other members of sub-committee reported that they had not been consulted in accordance with the resolution.*

Bedford recounted details of an episode in which he was involved during his apprenticeship, and which sheds some light on this sort of squabble.

Gerald Stoney, the Engineering Manager, after one of his visits to our Licensees in Switzerland, Brown Boveri & Co. brought over a machine for insulating core plate by rolling paper through a trough of paste on to the plate. Hitherto this work had been done by hand. Alex Law (another premium apprentice) *was given the job of getting this machine to work, and I was appointed as his assistant. The machine gave endless trouble as the paper kept on wrapping itself around the rollers where it was torn to shreds. We tried many schemes for overcoming the defect but without much success. Meantime the programme for coreplate was falling behind and, as soon as John Barker, the General Manager, discovered this, he instructed Law and myself to work through the night in an effort to make up for lost time. When Stoney heard this he forbade us to do as Barker had ordered. He considered that he alone was responsible for the running of the machine. Barker and Stoney had a fierce argument in the shops in full view of all the workpeople nearby, but Barker, being the Manager, had his way.*

Barker was a remarkable young man. He finished a four year pupillage with the North Eastern Railway in 1888, and was active in the Institution of Civil Engineers locally, in which capacity he met Parsons (Smith 1989, Barker 1939). He visited Elvaston Hall and played with Parsons' children. From 1892 he was put in charge by Parsons of the Cambridge Electric Supply Company, after which, in 1900 he came to Heaton as General Manager. Further on, Bedford wrote of Barker

> *When business necessitated a journey to London, Barker, before catching his train, would arrive at the Works wearing a top hat and frock coat, and parade around the shops.*

A meeting on 8 September 1903 carried an item,

> *Re lateness of machines It was pointed out that 23 of the 35 machines on the list are now late, and that during the last months only 500 kWs per month had been sent out, instead of the 2,000 kW necessary to meet financial requirements.*

On 29 September a sub-committee was appointed, consisting of G L Parsons, G Stoney, R Williams and A Q Carnegie,

> *to look into the matter of encouraging apprentices to gain a sound knowledge of technical science. It was pointed out that schemes for this purpose were being adopted at other engineering concerns.*

Stoney later wrote that in regard to education he had

> *always taken a keen interest in promoting it... the best way to get good work out of a man is to explain to him fully the work he has to do and in fact coach him and assist him in every way possible. ... I inaugurated a system of rewards to apprentices who not only did well in the shops but also at evening classes* (Stoney 1917).

The working day was nine and a half hours long, and the week totalled 53 hours. Starting at 6 a.m., there was a half hour break at 8 a.m. for breakfast which was eaten in the workplace. At noon there was a one hour break which was sufficient to allow those who wished to return home for lunch. At 1 p.m. work resumed until 5 p.m. On Saturdays there was normally no afternoon work. Time keeping was the one aspect of performance which was capable of being easily monitored and it received close attention. Everyone in the office had to stamp their card on entering and on leaving (Bedford). Even Parsons' own attendance was recorded.

When Bedford finished his time as an apprentice, he was invited to spend a week-end at Holeyn Hall. He found both Parsons and his wife *'exceedingly kind'* but *'though both, perhaps in different ways, were a little difficult, they could not have been better friends'*. In 1906 after a spell of illness Bedford was sent to Australia where he met customers and his father, who was Governor of Western Australia. When he returned in the following year he was given further responsibility for contact with the firm's offices around Britain. In 1909 he joined the 'staff', a group defined by the privilege of eating in the 'small dining room', and which included the chief engineer, the manager and one or two others. Parsons, when at Heaton, used to preside at table, carving the joint. Bedford wrote ' (Parsons) *had around him in the early days a curious staff, or so it seemed to me. They were, I am sure, excellent in their own line, but they did not appear to me to work as a team'*. He described a row between Bob Howe, the chief testing engineer, *'who was said to have started life as a pitman'* and Gerald Stoney. *'One can recall a few amusing incidents showing the lack of team work, such as the occasion when an argument arose between Stoney and Howe in Howe's office. Stoney was furious, and, as he was leaving the office, he caught sight of Howe's bowler hat hanging on a peg, took it down, threw it on the floor and stamped on it.'* What is perhaps truly remarkable is the fact that men like Parsons and Stoney, products of a privileged university education, were able to work so effectively with colleagues who for the most part had received modest schooling and enjoyed but indifferent apprenticeships.

Such intemperate behaviour must have been given some degree of tolerance by virtue of Charles Parsons' own behaviour. A former apprentice wrote (Turnbull 1939)

> *Few people who met Parsons casually thought of the explosive energy behind his introspective appearance, but it could come out when something happened. The first time I encountered this was when the accident happened to the blades of the axial flow turbine for the Turbinia. The short blades of ordinary turbines were skimmed up in the lathe, but the Turbinia L.P. rotor had blades of unprecedented length, and when the turner tried to machine the tips, the blades were bent right over. Dick Williams (then Works Manager) stopped the lathe instantly, and when Parsons came to look at the mess, the man disappeared in a hurry.*

It appears that the unfortunate turner was sacked on the spot. On another occasion (Rayleigh 1934)

> *a workman whom he* (Parsons) *personally wanted early in the morning was not forthcoming. Parsons became more and more impatient, and when the*

man ultimately arrived half an hour late he said 'You're fired' and would not hear a word in mitigation or excuse. However a member of the engineering staff insisted on stating the man's case, which was that he had been sitting up all night with a sick wife. Parsons was filled with remorse, decreed that the man's wages were to be increased five shillings a week, and personally called at his home with presents of grapes and other delicacies suitable to an invalid.

Given that such stories appeared after Parsons' death, with in one case, an emollient ending, one must assume that such outbursts were not all that rare, and that there was not always a superior who was brave enough to intercede.

It was not just employees who felt his wrath. When George Westinghouse concluded his agreement with Parsons, it was for a period of five years, and as has been noted, progress was slow initially. Parsons became convinced that Westinghouse was using the agreement to 'bottle up' the patents in the United States. After the General Electric Company had been formed by the merging of Thomson-Houston and the Edison General Electric Company in 1892, Westinghouse Electrical and Manufacturing Company was its nearest rival in size. Both companies sought to penetrate the British market. In 1899 British Westinghouse was established with a capital of £1500 000. One of the first members of its board was the Hon. R Clere Parsons, and Lord Kelvin was retained as a technical advisor. Clere Parsons resigned in 1901 over a dispute about the size of a commission payment. The board adopted rather grandiose plans for the manufacturing plant that was built in record time at Trafford Park in Manchester during 1901 and 1902. The year after it was opened, an order was taken from the Metropolitan and District Railway for eight 5.5 MW turbo-alternators for their Lots Road power station. This created a serious disagreement between George Westinghouse and Charles Parsons because Westinghouse had no license agreement for manufacture in Britain. When George Westinghouse was in London, he had a very acrimonious meeting with Parsons (Hodgkinson 1939). The manager of the London office, Charlie Martin, reported that he feared not only for the personal safety of the participants, but for the office furniture. Both adversaries were intent on going to court, when S F Prest, who had negotiated the original agreement between the two men, got to hear of this. He decided to call on Westinghouse himself (Prest 1931).

In conversation with Mr Westinghouse, Mr Prest pointed out the absurdity of the situation in which two Engineers who had both spent much money,

> *time and labour in developing turbines were now about to throw away the money which they had made on a law suit which was bound to be exceedingly costly. Mr Prest asked Mr Westinghouse if nothing could be done which would make an amicable agreement and which would save money on both sides. Mr Westinghouse asked him if he had come to see him with authority to negotiate, and when he heard that he had not obtained such authority he refused to discuss the matter any further, and said that if Mr Prest would obtain authority from Sir Charles to negotiate he could come back again to see him. Mr Prest thereupon telegraphed Sir Charles Parsons at Newcastle and they arranged to meet the next day at York. In the gardens behind the Railway Hotel at York they walked up and down for nearly three hours and after a long time Sir Charles refused to offer to come to an arrangement with Mr Westinghouse and determined to commence the law suit, and if necessary to spend every penny to defeat the rogue. In the end he agreed he should go to see Mr Westinghouse to negotiate a new license, which he did with success. This was not the end of the trouble, for the Westinghouse turbines were not only very badly designed, but also very badly manufactured and the result was a complete fiasco. Not only was every turbine rejected and the order to replace the turbines given to Heaton Works, but in addition the Westinghouse Co. were compelled to pay compensation amounting to £100,000 on account of the inefficiency of their turbines compared with the Parsons machines and the loss of time. The Parsons turbines at Lots Road coupled to the existing Westinghouse generators were very successful and efficient, but the failure of the Westinghouse turbines was such that the Westinghouse Co. realised that they would never again be able to make any money out of Parsons turbines, and that was the end of their dealings with the Heaton Works. They took up the Rateau turbines instead and have continued with this type ever since.*

It was said that one reason for the design failure was the insistence of Westinghouse himself, against the advice of his designers, that the turbine blades should be lengthened. Ultimately Westinghouse was able to have the legal decision reversed by the House of Lords in 1912 (Jones and Marriott 1970). In 1907 the American parent company ran into temporary financial difficulties and was put into receivership. In 1910, in a manner reminiscent of Edison's fate, George Westinghouse was replaced as president of the company by a financier.

Chapter 16

Marine Applications

16.1 Merchant marine

As well as wooing the Admiralty for business, Parsons had also been seeking to interest commercial ship-owners in the advantages of the marine turbine. In his presidential address to the Institution of Junior Engineers in 1899 for example, he commented that *'as regards cross-channel boats, the turbine system presents advantages in speed, absence of vibration, and, owing to the smaller diameter of the propellers, reduced draught of water'* (S.P.2). At a time when *Turbinia* was still the sole turbine driven vessel afloat, Parsons' thoughts were far ahead, on merchant vessels, even on giant ocean liners. When the *Viper* was undergoing her trials, three partners in the Scottish firm of Denny Brothers, James Denny, John Ward and Henry Brock, were present, and their report which was backed by Archibald Denny, was highly favourable. But as Parsons wrote (Parsons C A 1906).

> *About this time very great difficulty was experienced in endeavouring to induce railway companies and owners of mercantile vessels to build a turbine boat. Each company appeared anxious that someone else should make the experiment.*

The earlier notion of replacing the reciprocating engines in the *Violet* with turbines was dropped. Instead, in 1901, the PMSTCo. took a one third interest in Turbine Steamers Ltd, which was formed by Messrs Denny and Captain John Williamson to build a new vessel, the *King Edward*. It was intended that she should ply the Clyde estuary in competition with paddle steamers, vessels whose design had been optimized over a period of many years. She displaced 650 tons and measured 250 ft by 30 ft beam and 6 ft draught. Her machinery was arranged with three propeller shafts, the

Figure 98. *King Edward*, the world's first turbine powered merchant steamer, built by the Scottish shipbuilders Denny Brothers for service on the river Clyde in 1901 (Richardson 1911).

centre one having a high pressure turbine and this fed two low pressure turbines connected in parallel and coupled to the outer shafts. The latter also carried astern turbines. The total power was 3500 HP. She could carry 1994 passengers at $18\frac{1}{2}$ knots. The great exhibition at Glasgow was staged during her first season, and she proved so successful that she was joined by a sister ship the *Queen Alexandra* in the following year. At first, officers unaccustomed to the new form of propulsion had difficulties with manoeuvring, but as they gained experience this ceased to be a problem. The two vessels proved to be significantly faster and more economic of fuel than their competitors. In part at least this was because they were designed to operate continuously at the designed maximum speed, and did not encounter the conflict faced by naval vessels which needed high peak speeds, but yet spent much of their time cruising at a more modest pace.

The success of the Clyde ships was sufficiently encouraging for the South Eastern Railway Company to order from Denny the *Queen*, a mail steamer for the Dover–Calais route. Because this was an international service great interest was shown in its performance, and comparisons were made with three paddle steamers plying the same route. It was

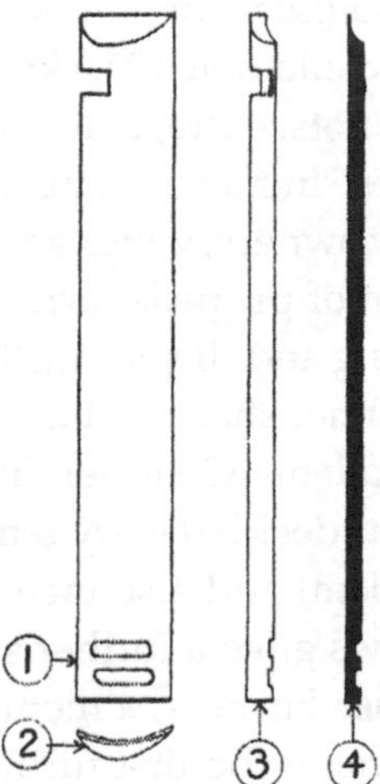

Figure 99. Brass blade with thinned tip, (1) side view showing the root, (2) cross section, (3) front view and (4) vertical section (Sothern 1909).

found that she used less coal, required fewer engine room staff and used less lubricating oil per passenger carried on a double crossing than her competitors. Her service speed of 21 knots was faster than their speed of 17.5 to 18.5 knots, and with her turbine machinery she was less affected by rough weather than the paddle steamers (Parsons C A and Walker 1906).

In 1903 Denny Brothers supplied the London Brighton and South East Coast Railway with the *Brighton* for the Newhaven to Dieppe crossing. Her turbine gave 6500 IHP to three shafts at 21.4 knots. The machinery was rapidly growing in physical size. The shaft speeds had been reduced to a little over 500 rpm. In consequence the blade mean diameter ranged from 41 in at the inlet stages, to 54 in at the exhaust. In order to reduce the amount of steam which leaked past the blade tips without doing useful work, the clearances had to be kept as small as 0.03 in. This clearance was easily lost, given the large size of drums and the temperature gradients that could be created under service conditions. As a result the blading was stripped in an accident, which led to an important development. It was decided that in future the tips of the blades should be thinned appreciably during the manufacturing process. Then, should they foul, they could wear away easily, and so avoid the excessive local heating that would cause greater fouling. The modification was covered by patent 22127 of 1905 (Richardson 1911).

Yarrow built the *Tarantula* for Colonel Mac Calmont in 1902. Displacing only 150 tons, it could reach 25.4 knots with turbines of 2000 HP. For use at speeds up to 15 knots, extra, cruising turbines were fitted ahead of the low pressure turbines. In fact a number of such steam yachts were built at this time for wealthy owners, which gave an opportunity for further developments in the design of propeller arrangements. The *Emerald* was completed in 1903, 198 ft long and displacing 900 tons. The turbines drove three shafts. Although the inner shaft carried only one propeller, the outer, wing shafts had two, in tandem. Considerable vibration emanated from these wing propellers. It was decided to try removing one screw from each shaft. This cured the problem, and also increased the speed. Increasing the surface area of the screws gave a further speed increase. The *Emerald* was built for Sir Christopher Furness, a member of parliament and ship building magnate. She became the first turbine powered vessel to cross the Atlantic, and it naturally aroused much interest in America. The high pressure turbine was later replaced by a reciprocating engine. In 1903 the Summer meeting of the Institution of Naval Architects was held in Ireland. Laurence, Lord Rosse, welcomed the visitors to Dublin. Charles, now a member of the Council of the Institution, reviewed progress in the application of turbines to the propulsion of vessels in a paper which included Ewing's report on *Turbinia* (Parsons C A 1903). He showed a graph from recent tests on *Turbinia* which gave the steam consumption at different speeds for both the original arrangement with three screws per shaft, and also for the case with only a single screw on each shaft. At certain speeds the latter arrangement showed a gain of up to 23% in propulsive HP for a given steam consumption.

In 1904 the Midland Railway decided to build four vessels for use on the Irish Sea (Richardson 1911). They were designed by Professor Biles of Glasgow University. One was built by William Denny and Bros, and another by John Brown and Co. Both were equipped with steam turbines, while the other two were given quadruple expansion engines. The ships were 330 ft long and all displaced around 2200 tons. The weights of engines, boilers and shafting were all less in the turbine powered vessels. The Denny brothers continued to secure orders for turbine ships. The *Loongana* was purchased by the Union Steamship Company of New Zealand. It made the ocean journey to the Antipodes in $30\frac{1}{2}$ days of steaming, eight days of which was done non-stop. In 1904 the British India Steamship Company purchased four ships to service the route between the Persian Gulf and India. The turbines for two of the latter were constructed by Denny Brothers, under a licence, at their own plant at Dumbarton. This firm thus became the first to be licensed to manufacture marine steam turbines in the United Kingdom. Another Scottish firm, the Fairchild Shipbuilding and Engineering Company of Govan was also licensed to build marine turbines and in that year they launched a yacht of 978 tons which was equipped with their own turbines. By 1905 25 United Kingdom firms, and 13 other firms worldwide, had taken licenses from the PMSTCo. (Anon 1905b).

The success of these ships of the merchant marine encouraged the Allan line to build the first ever turbine powered vessels for trans-Atlantic service to Canada. In 1904 the *Virginian* was ordered from Alexander Stephen's yard on the Clyde. She was 530 ft long and had 13 000 tons displacement. Her turbine had one high pressure section which supplied two, parallel, low pressure casings. It developed an estimated total of 12 000 HP for three shafts. This was double the output of each of *Viper*'s two turbines. With a service speed of 19 knots, her propeller speed was dropped further, to 290 rpm. She became the first of many turbine driven, ocean going, passenger ships. Later, after being re-engined, she had an unusually long service life of 50 years (Jung 1982). Her sister ship, the *Victorian*, was built in Belfast by Workman, Clark and Company. The Cunard line also became interested in the potential of the turbine. For many years it had held the 'Blue Riband' for the fastest Atlantic crossing, but in 1897 the Riband was claimed by the *Kaiser Wilhelm der Grosse* of the Norddeutscher Lloyd line. As a result, in the following year this line doubled its share of the massive flow of migrants to the United States. Between 1900 and 1906 the trophy was held by the *Deutschland* which could make 23 to 24 knots with its giant reciprocating engines of 32 000 HP driving four propellers. Moreover International Mercantile Marine, guided by the American financier J P Morgan, had recently joined the White Star Line to two German lines (Jung 1982). The Government was now very concerned that British ships should regain their dominant position on the Atlantic route. As far back as 1901 Cunard had sought proposals for large liners from ship builders, from Vickers Sons and Maxim at Barrow in Furness, and from Swan Hunter and Wigham Richardson on Tyneside. But Cunard could not finance building out of its own resources. In 1903 it asked the government for a subsidy, and parliament agreed to a 20 year loan of £2600 000 at low interest, as well as an annual subsidy. For this Cunard agreed to build two ships capable of a speed of $24\frac{1}{2}$ knots, which could be used for mail service in peacetime, and as cruisers in the event of war. At the time, there were two sister ships with 30 000 tons displacement being built for Cunard by the Clydeside firm of John Brown and Company. In 1904, after their design was already well advanced, Cunard decided to modify one of them, the *Carmania*, by fitting turbines of 21 000 HP to power three shafts. Their turbines were built by John Brown under license. During trials, Carmania proved slightly the faster of the pair, but her coal consumption was relatively high (Clarke 1997). Parsons wrote to Laurence (C.L.(40)),

Holeyn Hall
Wylam on Tyne
23 March 1905

My dear Laurence,

The Carmania has beaten the Caronia's performance by about 15% as a conservative estimate, or turbines in a large liner are at least 15% more economical in coal than the best quadruple expansion engines. The builders and the Cunard company kept things to themselves for reasons of their own, but we shall sooner or later get the exact figures so far as they go, no measuring tanks being fitted to either vessel.

The Carmania did 1 knot faster, and the firing was reported easier than on the Caronia. 1 knot means about 17% , so the figure I stated seems well within the mark. She sails for America on 2 December.

We got 20 duck at C. Strakes before lunch last week and we got 119 pheasants at Hasig 1st day (?) which is not bad considering the trespassing and poaching that goes on.

Your affectionate brother
Charles

At the same time Lord Inverclyde, who was chairman of Cunard, appointed a commission to examine all the information available on steam turbine performance. It included J Bain of Cunard, Vice-Admiral H J Oram, J T Milton of Lloyds Registry, H J Brock of Dennys, T Bell of John Brown, Sir W H White now with Swan Hunter and A Laing of the Wallsend Slipway and Engineering Company. Exceptionally, special tests were carried out on models at the navy's experimental station at Haslar, and the results were made available to the commission. The performance of those turbine driven vessels already in service was reviewed. In addition comparative tests were carried out at the Neptune Bank power station of the Newcastle Electricity Supply Company. This was the only place where a turbine, rated at 1500 kW, and a reciprocating engine, of 800 kW, could be compared. At full load and even at three-quarters load, the steam turbine was significantly more efficient. Moreover it was reported at the

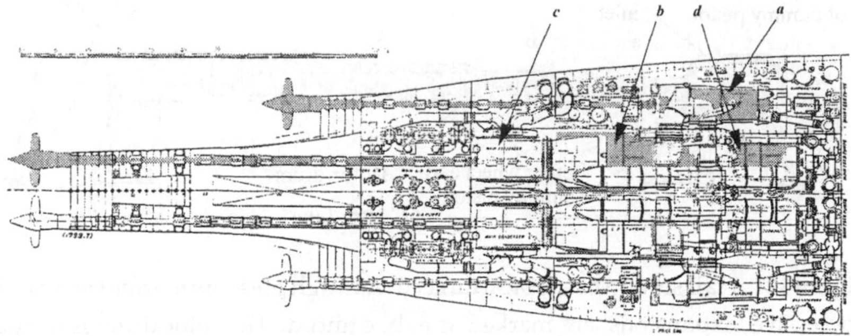

Figure 100. Plan view of the engine room of the *Mauretania* showing the twin turbine sets, each having a condenser *c*, as well as their low pressure forward *b*, and astern *d*, turbines on the two inner shafts. The two outer shafts were driven by the high pressure turbines *a* (Richardson 1911).

1904 British Association meeting in Cambridge that the turbine had run for 7500 hours with no more than 52 hours of outage for inspection. Besides, reciprocating engines had clearly reached a limit in size for practical operation. After considering all these facts it was decided to recommend the giant step of employing two turbines of a size never attempted before. Because the estimate of total power required by the proposed ship was so large, 70 000 HP, it was thought necessary to provide four shafts driven by two turbines. The high-pressure cylinders were to drive the outer screws, and the low-pressure cylinders the inner ones. Astern turbines were placed forward of the low pressure cylinders and on the same shafts, as can be seen in figure 100. The contract for the building of the *Lusitania* was awarded to John Brown, and for the *Mauretania* to Swan Hunter and Wigham Richardson of Wallsend on Tyne. The *Mauretania*, by a small margin, would become the largest and fastest ship afloat. The ships carried some 3000 passengers and crew, and displaced some 36 000 tons each.

In the decade since *Turbinia* had been launched, the power output of marine turbines had increased 16-fold, but in order to accommodate the needs of propeller design it had been necessary to reduce their speed tenfold, from over 2000 rpm to 190 rpm. To create an efficient combination that met such conditions required great ingenuity. In order to minimize the steam velocity, a very large number of stages was employed, each with a very modest pressure drop across it. Take as an example a turbine supplied with steam at an absolute pressure of 215 psi, and exhausting to

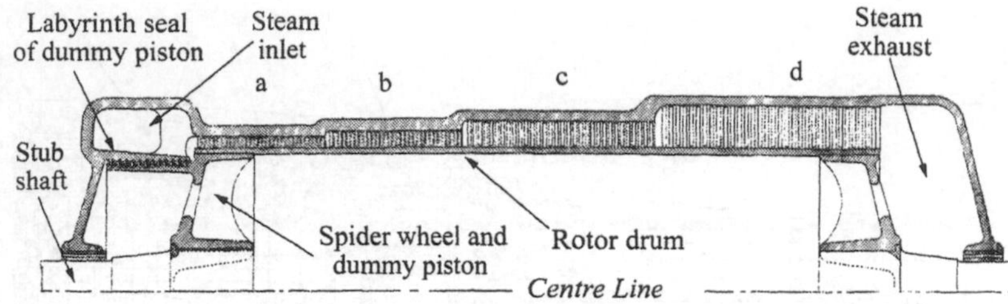

Figure 101. Half section of a typical direct drive high pressure marine steam turbine. Successive 'expansions' are marked as a, b, c and d. The rotor drum is supported on stub shafts by 'spider' wheels at either end. (Sothern 1909).

an absolute condenser pressure of 1.5 psi. The high pressure cylinder might be arranged with four 'expansions' each of twelve stages (a stage consists of a row of fixed blades and a row of moving blades), and the low pressure cylinder might have eight expansions, each of six stages (Sothern 1909). In total this gives 96 stages, or, combining the rotating blades and those fixed to the housing, 192 rows of blades. The pressure drop across any one row would be barely 2.6% of the pressure at that point. But even with such large numbers of blade rows, because of the very low shaft speeds, drums with a large diameter were needed on which to mount blades. Figure 101 shows the sort of relative proportions that resulted. For a given steam flow, increasing the mean blade diameters in order to achieve a reasonable blade speed also meant that for a given steam throughput, the blades must be made shorter. As a result, blade heights were only a small fraction of the drum diameter. Nevertheless the exhaust stages still required blades which were 22 in long. In order to give blading greater stiffness, it was laced with wires that ran circumferentially, and which were silver soldered at points one third, two thirds and at the extremity of the blades. In fact the need for such a measure had first emerged in a turbine supplied for Elberfeld (Anon 1903). In that case, at the suggestion of C E L Brown, lacing was applied to the mid-point of blades which were 10 in long, in order to cure problems with blade vibrations. The total number of blades in one turbine could exceed 100 000. Given their relatively light construction it was not easy to maintain the sort of precision required to keep down leakage across the blade tips and avoid loosening in service Two measures were adopted to improve workmanship. The blades and spacers were pre-assembled in groups of up to 120 or so. Also *'each operation, cutting, drilling, lacing, and caulking, trueing, binding and soldering were each done by separate workmen'*

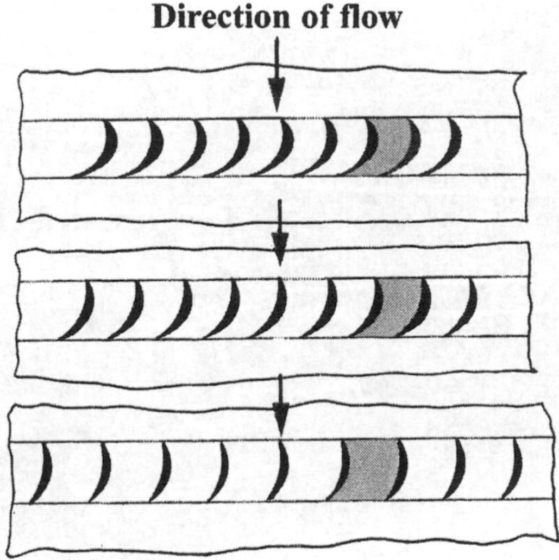

Figure 102. Plan view of three successive rows of moving blades in an expansion. From the darkened areas which represent spaces between blades, it can be seen that the passages for the flow of steam have been progressively increased in width using two different strategies (Sothern 1909).

(Anon 1907). In all of this the staff of the turbine manufacturers at the Wallsend Slipway and Engineering Company collaborated closely with the designers at the Turbinia Works.

Like the very first turbo-generator, the rotors were composite constructions, but now of giant proportions. The cast steel rotor drums were carried on spoked wheels, or spiders, seen in section in figure 101, which had been shrunk on to two stub shafts. The outer casings were made of cast iron, in segments, and were heat treated before being machined to avoid distortion in service. Despite the great size of the rotors, clearances at the tips of the blades in the region of 0.05 to 0.10 in were sought. Great care was needed in both manufacture and assembly in order to achieve this. Static balancing of the completed rotor could be carried out by supporting the shafts on two carefully levelled flat surfaces, though this was not always done.

The design of a turbine's blading involved more than Parsons' *K* constant. As the pressure falls in a turbine, and the steam volume increases, the cross sectional area for flow must be increased proportionately. Ideally blade heights should be increased continuously, but at that time there

Figure 103. One of the low pressure turbines of the *Mauretania* with its upper casing partially raised. The propeller shaft connects to the left hand side. At the right hand the multiple collars can be seen on the shaft for the thrust bearing. The two men give an impression of scale (Richardson 1911).

was difficulty in obtaining cast iron casings with a suitably conical shape. If each successive stage had the same dimensions, as in case of the 'expansions' of figure 101, the steam velocity would rise more and more as the pressure fell. This change in the ratio of steam velocity to the unvarying blade velocity could be accepted up to a point, but other measures were evolved to control steam velocities, even when blade height remained fixed. Two strategies for increasing the area available for flow under these circumstances are shown in figure 102. The first is to flatten the shape of the blades and orient them to lie closer and closer to the axial direction. In this way the cross sectional area available for steam flow increases. The second method is to increase the pitch of the blades, that is to arrange for the spacers which separate the blades to become progressively wider[1]. Parsons expended a great deal of effort in perfecting blade shapes and spacer sizes for specific designs, being guided for the most part by the analysis of experimental trials. In this work Stanley Cook played an important role (Gibb 1952). Professor Ewing had spotted Cook's ability when he was a student at Cambridge, and it was through him that he

[1] See appendix 2.

became apprenticed at Heaton in 1898. There he carried out experimental work on turbine blades under Stoney. After a brief absence working for Renold, the chain manufacturers in Manchester, he returned to Newcastle to work at PMSTCo. in November 1903, where his mathematical skills were harnessed to perfect blade designs for the large, slow speed marine turbines.

At 26 MW each, the output of the turbines that were built for the *Mauretania* and the *Lusitania* far exceeded that of any land machine. Physically they were so big that they have been compared in scale with the units employed in modern day nuclear plants (Jung 1982). The following table illustrates the validity of this observation.

	High pressure rotor	Low pressure rotor
rotor weight (tons)	86	120
rotor diameter (in)	96	140
blade heights (in)	2.75–12.375	8.25–22.0
shaft OD, ID (in)	36, 21	52, 36
number of rows of blades	120	60

The task of manufacturing the drum for the low-pressure turbine rotors of the *Mauretania* demanded specialist skills and was given to Sir W G Armstrong and Co. Ltd in Manchester. The special steel chosen was melted in a Siemens–Whitworth furnace and poured into a mould of 6 ft diameter, where it was subjected to an intense hydraulic pressure while still molten. After it had cooled, it was machined and bored out, and then forged until its diameter had grown to twice its original size. It was heated to release strains and was machined to its final dimensions. Because of its great size it had to be taken by road and then by ship to the Tyne. Large castings were obtained from Paisley in Scotland. In the case of the *Lusitania*, the forgings could be made at John Brown's own Atlas works in Sheffield. From this it is clear that it would have been quite unrealistic for Parsons to have attempted to handle the manufacture of all the turbines built under his patents, and a policy of licensing was essential. Work on the hull of the *Mauretania* began in August 1904. Despite the fact that they only had experience in building reciprocating engines up to this time, the Wallsend Slipway and Engineering Company was given the sub-contract to build the turbines for the *Mauretania* and they began work in June 1905. As well as turbines, 23 boilers were built which could be stoked from either end and

Figure 104. The *Mauretania* in 1907 with the *Turbinia* lying alongside, at the Swan Hunter and Wigham Richardson's yard. When *Mauretania* moved off to begin her service, the *Turbinia* was unable to accompany her due to a fault in her air pump. Courtesy of the Science Museum in London.

weighed 110 tons apiece. The ship was launched on 20 September 1906. The cost of the hull alone was £992 736, and by the time the propelling plant was finished in May 1907 its cost had reached £834 930 (Warren 1987, Clarke 1997).

The sea trials of the 'Scottish built' *Lusitania* were carried out in July 1907 and a speed of 22.8 knots was achieved with an estimated 76 000 HP. The 'English built' *Mauretania* reached 26.03 knots on the measured mile. On her maiden voyage she suffered some damage from a storm and a delay caused by fog, as well as vibration from the propellers. But these problems were soon attended to, and once again Cunard regained the mythical Blue Riband, which *Mauretania* retained until 1929. Her feats over all these years helped to keep Parsons' reputation as a gifted engineer and inventor before the public, in a way that *Turbinia's* exploits could not have done. Measured coal consumption was 1.46 lb $HP^{-1}h^{-1}$, or less than 900 tons per day at full speed. This was comparable to the performance of liners with slower, reciprocating engines. Since all coal handling was manual, 200 stokers and 120 coal trimmers were needed to satisfy the 23 boilers day and night. When oil burning was adopted in 1921, the boiler room crew was reduced to 90. Eventually she surrendered her Blue Riband to the 51 650 ton *Bremen* in 1929 and was broken up in 1935, after 28 years service and having travelled 2100 000 miles (Jung 1982).

In September 1906 Parsons claimed that there were '*31 turbine vessels in service for commercial purposes, representing about 105,000 gross tonnage and 235,000 IHP*'. He used figure 105 to illustrate graphically the progress that had been made in little more than a decade. Ten years after *Turbinia's* display at Spithead, Parsons stood supreme in the field of marine steam turbines. During that decade the maximum power of the turbine engines fitted to a single vessel had doubled in magnitude every two years. Despite all the difficulties in matching the turbine to the screw propeller, he had overwhelmed the reciprocating steam engine for fast passenger service. It was like the triumph of the jet engine in civil aviation in more recent times, but in this case it was essentially the work of one individual in particular. Competitors were now quick to join in, but they walked along a trail that had already been blazed for them. In 1908 the John Brown shipyard on the river Clyde took a license to manufacture Curtis turbines. Stephen Pigott came from the United States, and the Brown–Curtis turbine soon became a serious competitor. It had some advantage in that far fewer stages were required than in the Parsons designs, so that space requirements and weight were reduced.

Since Parsons' marine turbines were manufactured for the most part by the PMSTCo., progress can be followed in the printed annual reports and balance sheets (PAR38/1-11). In the year 1903, the shareholding comprised 800 shares of £100 and 1600 shares of £80 (partly paid up) worth in total £208 000. Patent rights were valued at £110 000, and works, plant machinery etc at £73 208. After four years the Company had accumulated a loss of over £10 000. The first dividend of 5% was not paid until year seven, in 1904. By 1911, that is after 14 years, dividends and occasional bonuses totalled £179 016, which averages about 6% per annum on the total share issue. Nearly all of this could be attributed to income from license fees and royalties which were split between the 'Marine Steam Turbine Co.' and the PMSTCo. The latter received £29 901 as its share of a gross fee income of £58 393, as well as £143 526 as its share of gross royalties of £273 488. The conclusion is that the manufacturing activity contributed almost nothing to profits, though of course the patent activity could not have continued without the existence of manufacturing facilities.

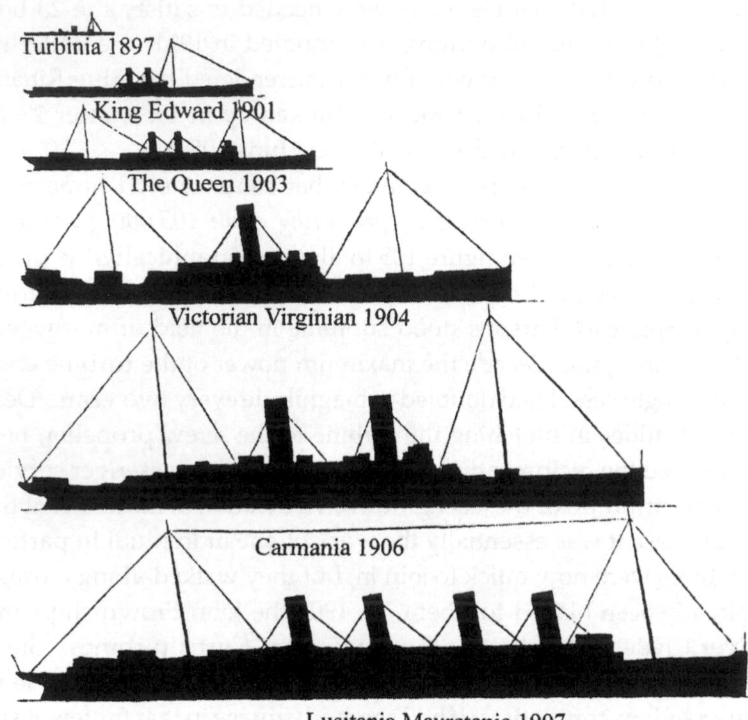

Figure 105. Diagram showing the various advances during a period of ten years in the development of the application of the steam turbine to propulsion in the merchant marine (Parsons C A and Walker 1906).

The annual reports also quote a global figure for the total HP of turbines built by or ordered from Parsons or their licensees world-wide from the date of the first order. In 1904 the figure was 0.34 million HP. By 1906 it had more than doubled to 0.87 million HP. Again by 1908 it was more than doubled to 2 million HP, and by 1910 it reached 4.5 million HP. For seven years the growth rate was 50% p.a. In 1911 the output of turbines for military, i.e. naval, purposes compared with that for civil shipping was in the ratio of 60:40. This is a measure of the central importance for the success of Parsons' plans of a positive engagement with the British Naval establishment, symbolized by the *Turbinia* display.

During the period covered by these annual reports, 1900 to 1911, there were changes in the board of directors. In 1905 Sir William H White joined the board, followed in 1907 by R J Walker, now joint managing director with Parsons. On 29 August 1908 Laurence Parsons died after some years of poor health. If we are to judge by the correspondence between the two brothers, Charles must have felt the loss deeply. Charles followed him as chairman. The following year Norman Cookson, who was the father in law of Clere's son, died and his place was taken by S F Prest. In 1911 Laurence's son Geoffry L Parsons became a director.

16.2 The Royal Navy's requirements

The caution which senior staff in the Royal Navy had shown in regard to fuel economy and manoeuvrability of the *Viper* did not mean that they doubted the value of turbines for use in warships. And yet when the *Viper* and *Cobra* sank, the marine steam turbine had already been a reality for seven years. The turbine powered steamer, *King Edward,* had just completed its trials, but Parsons realized the capital importance of having a full-scale demonstration naval vessel and he lost no time in laying down a replacement destroyer. Once more Christopher Leyland made available sufficient finance from his own resources. The ship was later named the *Velox* and was built by Hawthorn Leslie and Co. who launched it in 1902. The dimensions and general configuration were the same as for the *Viper,* with four propeller shafts driven by two turbines. The high-pressure sections of each turbine powered one of the outer shafts, while the casings of the low-pressure sections which powered the two inner shafts, also housed an astern turbine. The power required to propel a vessel increases very sharply with speed, as can be seen in figure 75 for example. This means that at modest speeds of around 12 or 13 knots the turbines are called on to develop only a fraction of their full power. For a turbine, the steam required per HP, as can be seen in figure 76, is much greater when it is operating at part load. To deal with this problem two small and economical engines were provided. These were triple expansion reciprocating engines rated at 150 IHP, and were only to be used when travelling at modest speeds. When required, they could be connected by a claw coupling to the two inner shafts, and the exhaust steam from them was then taken to the condensers through the two main turbines. This was in fact the first implementation of the idea for combining a reciprocating engine with a low pressure turbine, which had been described in patent 367 of 1894.

It was also decided that a second destroyer, *HMS Eden,* that had been ordered by the Admiralty as one of a batch, should be fitted with Parsons' turbines. She was launched by Hawthorn Leslie in 1903. Her cost of £87 400 may be compared with that of a sister vessel with triple expansion engines, built by the same company at less than £77 000 (Clarke 1997). Power was supplied to three shafts, the high pressure turbine being connected to the central propeller shaft, while its exhaust was fed in parallel to two low pressure turbines, one for each of the outer shafts, as in the *King Edward*. Instead of using reciprocating engines at low speeds, a 'cruising' turbine was fitted. This was designed to have a small maximum power output, and so when in use it could operate at a relatively high efficiency. It had a high pressure cylinder feeding into an intermediate pressure cylinder, which then discharged directly to the main turbines. At the slowest speed, steam was led to the cruising, high pressure, cylinder. Then as speed built up, steam was fed directly to the cruising intermediate pressure turbine, and eventually the cruising turbines were completely bypassed and the main turbines alone were active. The performance of a number of very similar vessels, but fitted with reciprocating engines, was compared with the two turbine powered ships in a series of four hour sea trials. At 11 knots *Velox* used only 58% of the steam required by the conventional destroyers, while *Eden* used 8% more. But at 13 knots *Velox* used 34% more steam and *Eden* used 17% more. At 18 knots *Eden* used 10% less steam than the conventional vessels. In a run at 25 knots, *Eden* achieved 3.39 nautical miles per ton of coal burned, which at least equalled the performance of the conventional vessels. These results showed that there were ways of improving the efficiency of turbine installations at cruising speeds. Eventually the small reciprocating engines in the *Eden* were replaced by cruising turbines, partly because of vibration. Cruising turbines became a common feature in future installations.

Some idea of the rapid growth of the naval construction which was supervised while Sir William White was Director of Naval Construction, can be gained from the monetary value of new construction carried out in succeeding four year periods. From 1887 to 1891, £17 700 000 was spent under the supervision of a staff of 25, from 1892 to 1896 £22 900 000 was spent with 37 staff and from 1897 to 1901 £35 600 000 with 58 staff (Manning 1923). A close watch was kept on developments in the navies of foreign powers, and it was decided to build a number of cruisers. These were fast ships with relatively big guns but not as heavily armoured as battleships. After much consideration, it was decided in 1902 to fit turbines in one of a batch of four 'third class' cruisers that were being built at the time as a counter to developments taking place in the German navy. They were to be of 3000 tons displacement and intended to achieve a speed of $21\frac{3}{4}$ knots. Their engines were rated at 10 000 IHP, and drove two propeller shafts each carrying a single screw. The best speed which the conventional sister ships could maintain was 22.3 knots, while the turbine powered *Amethyst* easily managed 23.6 knots. Comparative trials were carried out between the four vessels. Since the ships all had the same hull design, such tests gave a good overall assessment of the advantages of one system of propulsion over the other, and it was a procedure that would be followed by other navies. It appeared that at 12 knots *Amethyst* had only 70% of the range of its sister ship the *Topaz*, but at 20 knots *Amethyst* used 30% less fuel. This meant that with its bunkers carrying 750 tons of coal, its radius of action was 50% greater than cruisers with reciprocating engines. The *Amethyst* had three shafts with a single screw on each. The propulsive efficiency and the values of slip at different speeds were all considered satisfactory (Richardson 1911).

It was fortunate that Parsons had been able to maintain the pace of turbine development despite the many set-backs, because pressure for change was increasing at the Admiralty. After a spell as commander in chief on the North America and West Indies Station, and later in the Mediterranean, Sir John Fisher returned to Whitehall in 1902, on the invitation of the Earl of Selborne who was the First Lord of the Admiralty. As Third Sea Lord, Fisher's aim was to reform the existing provisions for the education of naval officers, and to combat the harmful effects of class prejudice. By advocating 'common entry' for all branches he sought to compensate for the '*comparative lack of education, polish and family background of the engineer, a differentiation accentuated on shipboard by differences in status*' (Marder 1952). By the time he was appointed First Sea Lord at the Admiralty in October 1904, Fisher had already given thought to many other radical reforms that he intended to set in train. In part these concerned the elimination of waste and useless expenditure on obsolete vessels and dockyard facilities. Although a great deal of money had been spent on new naval construction in the last decade, there had been very rapid advances in technology concerning armour plate, gun designs and propulsion methods. For the most part these developments had been driven by private individuals, capitalists like Armstrong and Parsons in England, and Krupp in Germany. In the light of his recent experience abroad, Fisher had reached the conclusion that in a future conflict it would be Germany and not France that the navy would have to face, and he had his own ideas of how future ships should be designed to deal with this eventuality.

In December 1904 Fisher succeeded in establishing the Committee on Designs under his own chairmanship. Made up largely of his own selection of naval experts, as well as civilian members like Lord Kelvin and Sir John Thornycroft, it formalized proposals for battleships, for armoured cruisers and for torpedo craft (Kemp 1960). Two key features emerged, the decision to build 'all big gun' capital ships, and the enthusiastic adoption of steam turbines. Both were the inevitable consequence of developments that had been pursued by private industrial entrepreneurs, namely Armstrong in the case of gun design, and Parsons in the case of steam turbines. The Committee's report noted that the Director of Naval Construction and the Engineer in Chief of the Navy, Sir John Durston, both urged the use of turbines on the grounds of their simplicity and the potential to reduce weight. But some doubts remained about the speed with which a vessel could be stopped and moved astern. It was decided to await the results of comparative tests on the destroyer *Eden* and her sister ship the *Waveney*, which was equipped with conventional engines. On 17 January 1905 the Hon. Charles Parsons attended a meeting of the committee at its request. He gave the results of the trials just completed. From a speed of 15 knots ahead, *Eden* took 50 seconds to come to a halt, while *Waveney* required only $37\frac{1}{2}$ seconds. Getting under way from standstill it was found that *Eden* was sluggish compared to *Waveney*. This somewhat disappointing result was attributed in part to the use of two screws in tandem, on each propeller shaft of the *Eden*. It was a practice that had been discontinued on all subsequent turbine driven vessels. The tests were to be repeated with single screws on the *Eden*. Also similar tests were to be made on the *Amethyst*. Parsons explained that the astern turbine had only one stage of expansion, but this could easily be changed so that it could achieve 75% of the ahead power. This was already in hand with steamers built for cross Channel service *'so that these steamers could come close up to the jetty before slowing down'*. In the discussions it was pointed out that *Amethyst* could achieve a much tighter turning circle than its sister ship by putting the inner screw into reverse during the turn, which partly made up for other weaknesses in manoeuvring. Parsons was asked how long it would take to turn out turbine machinery of such large power as would be suitable for battleships. He was able to say that *'several large firms had now laid down complete plants for manufacturing turbine machinery, and that any of these should be able to turn out machines of the power required in about 12 months'*.

The committee met the next day and considered the results of a second trial of the *Eden* and *Waveney* which had been made in better weather conditions, and which confirmed the first conclusions. There was much discussion among committee members about propeller designs and actual sea going experience, but in the end *'It was therefore decided to recommend the adoption of turbine propulsion in the new ships, unless the receipt of the further information which the Committee hoped to obtain gave reason to alter the decision'*. An invitation was also sent to the Hon. Charles Parsons to travel down to attend a meeting on 25 January 1905. The main subject of discussion was the use of turbines for the new battleships and cruisers being planned. Parsons accepted that the Committee wished to achieve equal stopping performance with turbines, but eventually a compromise was reached. In view of the many attractions of turbines, such as the possibility of reducing the engine room complement, it was decided to accept some extra weight penalty to allow the fitting of more powerful astern turbines than current practice employed.

Some correspondence was appended to the committee's report. Lord Kelvin's letter dealt with the effect on the magnetic compasses of orienting the heavy guns. Parsons wrote on 19 January explaining that there was no problem in providing a 50% increase in the power of astern turbines for cruisers and battleships, as was being done with the latest cross-channel boats currently being built. There was also a note *'in Explanation of the Figures quoted by Hon. Charles Parsons'*. This referred to values of the coefficient K.

> *It really represents the gradualness of expansion of steam in the turbine. Within limits the higher the function K is, the greater the efficiency, but the efficiency is not directly proportional to K; it is obtained from experiments, and from a curve giving the relation between the efficiency and this function K.*

A further written submission from Parsons was dated 27 January 1905. A breakdown was given of proposed weights for *'the propelling machinery only, in the engine room and aft of the engine room'*. For a 22 knot 30 000 IHP battleship, 819 tons, for a 22 knot 23 000 IHP Battleship, 612 tons, and for a 25 knot 38 000 IHP cruiser, 1039 tons. In each case the turbines accounted for half these figures. He explained that, unlike reciprocating engines, turbines starting from standstill create an accelerating torque that is initially 50% greater than at full speed. He also dealt with the optimum number of propeller shafts, and their effect on the steering capacity of the rudder. What is clear from these minutes is the fact that Parsons spoke not just as a turbine designer, but as a knowledgeable seaman.

Perhaps the most significant outcome of the committee's discussions was the proposal to build the new class of 17 900 ton battleships, the *Dreadnoughts*. These revolutionary vessels were to have a speed of 21 knots, and would be fitted with turbines. They were also armed with no fewer than ten heavy guns, all of the same 12 in calibre. The first vessel was laid down in October 1905 and was completed with extraordinary speed by December 1906. The emergence of this first battleship to be equipped with turbine engines had an immediate effect on the plans for shipbuilding for the German navy, which had now to be radically altered (Padfield 1973). But it was a policy that had critics, among them Sir William White who had recently been forced to retire from ill health in 1902.

A biographer of Lord Fisher, himself a seaman, described with perhaps some slight poetic licence the change which turbines brought to those operating warships (Bacon 1929).

> *When steaming at full speed in a man-of-war fitted with reciprocating engines, the engine room was always a glorified snipe-marsh: water lay on the floor-plates and was splashed about everywhere; the officers often clad in oilskins to avoid being wetted to the skin. The water was necessary to keep the bearings cool. Further the noise was deafening; so much so that telephones were useless and even voice-pipes were of doubtful value. In the Dreadnought when steaming at full speed, it was only possible to tell that the engines were working, and not stopped, by looking at certain gauges. The whole engine room was clean and dry as if the ship were lying at anchor, and not the faintest hum could be heard.* He continued, *Fisher knew that no Fleet at sea in the world could be relied on to steam for eight hours at full speed without one or more ships breaking down.*

In 1905 Prince Louis of Battenberg brought six cruisers with reciprocating engines from New York to Gibraltar. Three of them completed the journey at $18\frac{1}{2}$ knots, and their engines required minor repairs. A year later the *Dreadnought*, though a much larger ship, travelled to Trinidad and back at $17\frac{1}{2}$ knots with absolutely no engine defects.

The first batch of 12 turbine propelled destroyers built between 1905 and 1908 were larger than the *Eden*, being of up to 1100 tons displacement and carrying engines of 21 000 to 27 000 HP (Jung 1982). Despite having oil fired boilers they were relatively inefficient. Fisher sought to push performance still further and a super destroyer of 2170 tons, the *Swift*, was built in 1907. Much larger than *Turbinia*, yet with eight turbines delivering 35 000 HP to four shafts, she was actually faster, reaching 35.3 knots. But she was not a success and after Admiral Lord Fisher resigned in 1910 the idea was dropped for several decades.

Starting in 1908 Brown–Curtis turbines were adopted for a number of vessels of all classes. By the time war broke out in 1914, Brown–Curtis turbines had been fitted to 23 destroyers while Parsons' were used in 108

(Jung 1982). The *Dreadnought* was no sooner completed than larger and more powerful versions were built. Some like the 26 700 ton *Lion* (1909) had Parsons turbines of 70 000 HP and reached 27 knots, but others like the *Tiger* (1911) and the *Renown* (1914) carried Brown–Curtis machinery. The *Renown* displaced 26 500 tons and had engines of 126 000 HP.

16.3 Foreign navies

There was one important personage in particular who was impressed by the display of *Turbinia* at Spithead; he was the brother of Kaiser Wilhelm II. It was therefore to be expected that the German navy should be one of the first to consider adopting steam turbines for the new ships being built in preparation for the looming conflict although at first there was much opposition to this from the advocates of reciprocating engines. The first vessel built in 1902 was the destroyer S 125 (Richardson 1911). It was the only one of six similar vessels to be fitted with turbines built by the PMSTCo. at Wallsend. She was of similar size to the *Eden*, displaced 413 tons, was 210 ft long and had almost identical engines. In comparative trials she proved her worth. The following year one of several cruisers, the *Lübeck*, was fitted with Parsons' turbines. Displacing 3150 tons, she was much like the *Amethyst*, except that she had four propeller shafts instead of three. The turbines for the *Lübeck* and succeeding destroyers were built at the Mannheim works of Brown Boveri and Co. for the German Turbinia Company of Berlin. Again comparative trials were carried out to assess turbines versus reciprocating engines. Around 1910 a number of larger, armoured cruisers of 19 000 tons and capable of 27 knots were fitted with Parsons turbines. They could deliver 70 000 HP and were built by Blohm and Voss of Hamburg. Among these was the *Von der Tann* which was to see action in 1916 at the battle of Jutland. In 1914 the battle cruiser *Derfinger*, displacing 31 000 tons, was fitted with Parsons designed turbines rated at 72 000 HP. In ships for the German merchant marine, Curtis turbines were the first to be used. In 1905 AEG, an associate of GE of America, supplied turbines for a passenger boat, the *Kaiser*.

In France progress was somewhat slower. A torpedo boat of 140 tons displacement was purchased from Yarrow. It had a reciprocating engine as well as a steam turbine of Rateau's design which was built by Oerlikon in Switzerland. Trials were carried out in October 1903 and the results were reported to a meeting of the Institution of Naval Architects (Rateau 1904). As an aside it may be mentioned that, during the discussion of this paper in March of 1904, Parsons' colleague Campbell Swinton apologized for

Parsons' absence, explaining that he was seriously ill, and had travelled abroad to recuperate. Parsons was now in his 50th year and the last 20 years of his life had been devoted to the development of the steam turbine. Evidently the many years of gruelling effort had taken their toll of his health. In the following year, when he delivered his paper to the tenth *International Navigation Congress* in Milan, he was able to quote a figure of 115 000 IHP for the total output of turbines manufactured by the PMSTCo. so far, with a further 37 000 IHP under construction and, including those manufactured by licensees, a grand total of 352 000 IHP (Parsons C A 1905a).

In 1902 the PMSTCo. supplied turbines for a torpedo boat, number 293, for the French navy (Richardson 1911). At 95 tons and 130 ft in length, she was little larger than *Turbinia.* But it was not until 1909 that the first destroyer, the *Chasseur*, was completed. Her turbines were built by the Compangie Électro-Mécanique at Le Bourget. During trials, turbine propelled ships of this class proved to be significantly faster than their counterparts with reciprocating engines. In 1906 it was decided that six battleships of the *Danton* class, displacing 18 374 tons and having a length of 475 ft, should be fitted with turbines. Some of these were built by the Forges et Chantiers Company at their La Seyne works, and the remainder at Saint-Nazaire. From 1908 no more war vessels with reciprocating engines were ordered by the French navy. Actually the first marine steam turbine built in France was constructed by CEM for the *Charles Roux*, a passenger ship which was to ply the Marseilles to Algiers route. The *France*, a trans-Atlantic passenger vessel of 25 460 tons, was built by the Saint-Nazaire Company.

When Prest originally negotiated rights to use the Parsons patents in the United States with Westinghouse, marine applications were specifically excluded. However he later travelled to the States to seek out companies who might be interested in making and selling Parsons turbines for marine purposes(Prest 1903b). The following letter from Prest deals with this (Prest 1903a).

S.F.Prest
2 Suffolk Lane
Cannon Street
London E.C.
30th Sept 1903

The Hon. C.A. Parsons
Messrs Parsons Steam Turbine Co.

Turbinia Works
Wallsend on Tyne

My dear Parsons,

Cramp's enquiry

Gen. Williams brought Mr Rt. W. Parsons, the second Vice-President of the Florida East Coast Ry, to see me last Saturday morning, and it was decided they should take a trip on the 'Queen' on Sunday, and call here again on Monday as they intended to cross in the 'Brighton' to-day on their way to Paris. I have received the enclosed letters from them, from which you will see they have been unable to call here before leaving England. Please return these letters to me.

At the interview Mr Parsons said they intended to order two new boats before the end of the year to run from New Orleans to the West Indies, at not less than 20 knots speed, and that the boats would probably be built at Cramps, and that Mr Cramp had recommended to use turbines, and consequently he was anxious to learn all he could about them. He is not an Engineer, and consequently could not go into details nor was it decided what size the boats will be, so I gave him particulars of those boats you had fitted with turbines, and also told him that negotiations were under way for the sale of Marine Patents in America, but if he wished you to build these particular Turbines, no doubt it could be arranged.

I conclude (that) *nothing more can be done in the matter at the present time.*

Yours faithfully,

S.F.Prest

P.S. I await receipt of Marine license as finally settled before seeing Mr Pirrie.

Cramp's shipyard was at Philadelphia. William J Pirrie headed the great Irish shipyards of Harland and Wolff in Belfast. Prest was negotiating terms under which Harland and Wolff might take a license from Parsons Steam Turbine Patents Co., but the deal seems to have fallen through (Prest 1903c).

The first tentative marine application in America was made in 1903 when a Curtis turbine of 1600 IHP, manufactured by GE, was fitted to a steam yacht. In the end it was the Eastern Steam Ship Co. which fitted the first Parsons turbines to the *Governor Cobb* in 1905 (Richardson 1911). This vessel was built by W and A Fletcher for service on the east coast of America between Boston and St John's. Displacing 2900 tons and 290 ft long, she achieved 20.5 knots with an estimated 4000 IHP. This was followed by the

Yale and the *Harvard*, ordered by the Metropolitan Steam Ship Co. from the same firm, also for service on the east coast. Displacing 4600 tons they were larger vessels and their turbines developed 10000 IHP to achieve a speed of $22\frac{1}{2}$ knots during trials in 1907. Also in 1905 W Cramp and Sons began construction of a 3800 ton vessel, *Old Colony*. The turbine was manufactured at the Quintard Iron Works.

In 1905 the United States Navy decided to try out the steam turbine, and it chose the same path as that followed by the European navies. Three identical light cruisers were built displacing 3750 tons each. The *Chester* was fitted with Parsons' turbines, the *Salem* with Curtis's turbines and the *Birmingham* with triple expansion reciprocating engines. American service requirements emphasized efficiency at cruising speeds to a greater degree than European fleets. The *Chester* was fitted with four shafts and a total of six turbines, and the *Salem* had two shafts. The turbine ships had a maximum power of 16 000 HP, and during the 1907 trials the *Chester* was marginally the faster. At her full speed of 26 knots she was also the most economical, but at cruising speeds the *Birmingham*'s steam consumption was only 86% of the *Chester's*. The *Salem* also suffered some reliability problems, which was not surprising given GE's lack of experience with marine applications (Jung 1982). In 1907 four battleships of 20 000 tons were built. Once more the three engine options were tried, but with Parsons' turbines in two of the group, the *Utah* and the *Florida*. The ships with Parsons' turbines were significantly faster, but the Bureau of Ships decided in 1911 not to build more turbine powered battleships based on considerations of fuel economy. When the *Smith* class of destroyers was begun in 1907, Parsons' turbines were used in five. They were built by the Bath Iron Works, W Cramp and Sons and the New York Shipbuilding Co. In the following year 15 of the *Paulding* class carried Parsons' turbines, three used Curtis's, and one vessel had a Zoelly turbine.

Many other Navies adopted turbine propulsion around this time, among them the Japanese. When the country was opened up under the Meiji dynasty in 1868, the Government espoused a very determined policy of adopting western science and technology. Many young men of noble families travelled to Europe, some legally, others without authority, to acquire a third level education. Some, like Parsons himself, studied at universities such as Cambridge, and some worked as apprentices in shipbuilding and other firms, in Scotland and the north east coast of England. That they spent their time well can be seen from a paper that was presented in 1909. It was by a member of the North East Coast Institution of Engineers and Shipbuilders, Yasuzo Wadagaki (Wadagaki 1909). He described how '*with the permission of Admiral Mukyama, a series of experiments were undertaken some time ago at the Imperial Dockyard at Sasebo with the object of ascertaining the effects which the addition of a propeller casing... would have on the performance of a screw propeller*'. His work is reminiscent of Clere Parsons' paper to the IMechE in 1879, but the data were gathered with much greater precision. The Japanese government also pursued a second approach. Invitations were extended to engineers and scientists in Europe and America to teach in Japan. Under the supervision of the Scotsman Henry Dyer and Y Yamao, a School of Engineering was established in Tokyo with a course of studies extending over six years (Latorre and Hongo 1989). Out of the original contingent of ex-patriates, one stayed on until his death in 1908. He was Professor C D West, an engineering graduate of Trinity College Dublin, who taught mechanical engineering and naval architecture.

In the early years the Japanese navy looked especially to the Royal Navy for guidance, indeed a British naval officer had been present at the battle of Tsushima in 1905 when Admiral Togo destroyed the Russian fleet. Togo had led a high level delegation to visit Lord Armstrong's Elswick shipyard. In 1904 the firm of Mitsubishi obtained a license from Parsons to manufacture turbines for marine and land use throughout all of Asia. In that year an engineer and a craftsman were sent to Parsons at Newcastle for training (Mitsubishi 1991). In 1908 the PMSTCo. supplied turbines for four merchant vessels, two of which were built by Denny and two, the *Tenyo Maru* and the *Chiyo Maru*, were built by Mitsubishi at their Nagasaki shipyard. The turbines followed the general design of the *Dreadnoughts*, with an output of 19 000 HP to three shafts. The *Tenyo Maru* displaced 13 454 tons, was 570 ft in length and achieved 20.6 knots. It was the first turbine powered Pacific Ocean liner (Anon 1908). The first marine turbines to be built in Japan to Parsons' design for the merchant marine were rated at 20 000 HP and were fitted to the *Shunyoo Maru* of the Toyo Shipping Co. in 1911. Before this, in 1908, the first 500 kW turbo-generator was built under license by Mitsubishi for use in its own factory. Considering that it was only in 1903 that marine turbines were built anywhere outside England, this was a remarkably rapid adoption of an advanced technology.

The decision to adopt turbines was made by the Imperial Navy in 1905 (Matsumoto 1990). The Superintendent of Shipbuilding, Mr T Fuji, who agreed with the policy of a '*gradual adoption of the Parsons turbine*', was sent to England to examine the matter. While there, he wrote to ask the advice of A T Bowles, who was the President of the American Fore River Shipbuilding Company which supplied Curtis turbines. When Fuji came to write his report he presented data first for a ship of 4100 tons, and then for a ship of 14 500 tons displacement when these were fitted with either Curtis, or as an alternative, with Parsons turbines. Comparisons were made under 14 headings, and the Parsons turbines appeared to require greater engine room space and to cost significantly more. At that time the cruiser *Salem* had just been ordered, but there was still no vessel in existence of 14 500 tons with a Curtis turbine. Fuji warned of the danger of neglecting rival turbine designs just because of the fame of the Parsons turbine. In fact the navy heeded his warning and decided to order Curtis turbines from the Fore River Shipyard in 1906. A license was purchased for $10 000 000, plus royalties calculated at between 50 and 75 cents per HP. However the first turbine driven vessel supplied to the Imperial Navy was a gun boat, the 23 knot *Mogami*, built by Mitsubishi with turbines supplied by the PMSTCo. The vessel was ordered in 1905, and delivered in 1908. In 1910 the destroyer *Umikaze* was supplied with turbines of 27 000 HP and in 1912 a light cruiser *Yamakaze* with turbines of 22 500 HP was built by Mitsubishi using Parsons' designs. It seems that Bowles' eloquence had badly misled Fuji in his judgement of the two competing systems because eventually, in 1911, the navy approached the PMSTCo. for its own license. The price agreed was £3000 plus royalties at 2 shillings per HP. These figures are significantly less than those charged by the Americans. There were two main shipbuilding firms at that time, the Mitsubishi shipyard, founded in 1884, and the Kawasaki Shipbuilding Co., established in 1896. The Japanese navy decided to back both design options and instructions were given to the former that they should use Parsons technology, while the latter were to use Curtis designs. Kawasaki's first order was for a battleship of 25 000 HP, and it was placed in 1910.

Chapter 17

Technical Developments

17.1 Condensers and combination schemes

The established engine building firms in England and abroad seem to have taken to the manufacture of large, slow, direct drive marine turbines with little difficulty. While the Turbinia Works at Wallsend and its licensees were moving forward rapidly, Heaton was also progressing. The old Jumbo machine, that had been used for experiments to study the Hero's reaction turbine and the radial flow configurations, had been transformed into a test rig. With it changes in blade designs, bearings and other features could be tried out. Once Parsons had succeeded in building a turbine that could exhaust to a vacuum in 1891, attention was focused on the design and operation of condensers. To understand why this should be such a central concern in the case of steam turbines, it is necessary to be aware of a characteristic of steam. In the case of a water turbine, the energy available to do mechanical work increases in direct proportion to the pressure or head available. But in the case of steam, larger and larger amounts of energy can be accessed as the pressure is lowered by equal steps. This can be seen in figure 106. The pressure scale is logarithmic, and it is assumed that we are dealing with a supply of saturated steam with an absolute pressure of 200 lb in^{-2}. The lowest pressure plotted corresponds to 0.25 lb in^{-2}, $\frac{1}{60}$ of an atmosphere or 0.5 in on the barometer. By expanding down to this pressure from the ambient, atmospheric, pressure one can obtain as much energy as can be harnessed in the expansion from the initial pressure down to ambient. Provided an effective condenser is available, a steam turbine can easily handle the huge increase of volume which such low exhaust pressures entail. The reciprocating engine (or pump) on the other hand is limited to a modest pressure ratio in each stage. If this ratio is increased,

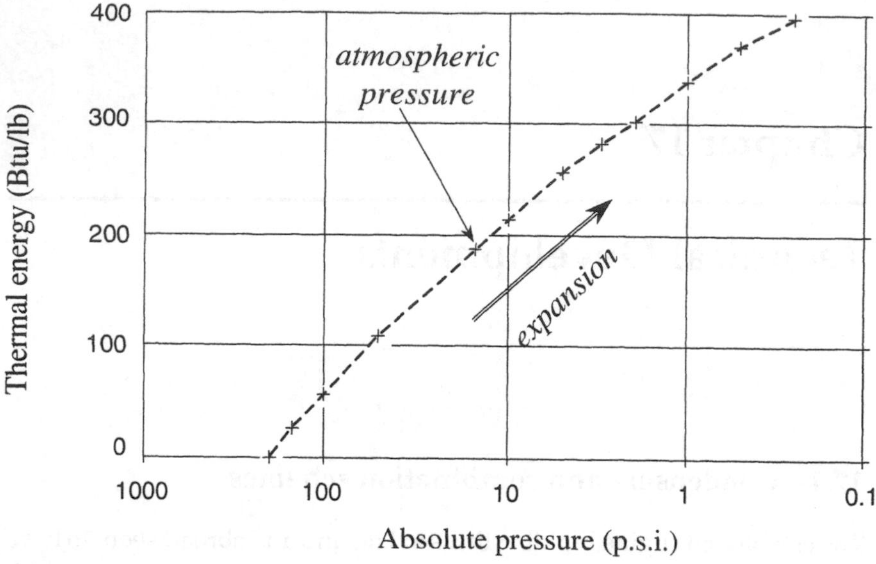

Figure 106. Plot of the ideal thermal energy, in British thermal units, which is available in steam when it is expanded from a pressure of 200 psi. Pressure is in pounds per square inch.

there is a sharp drop in the volume flow rate that can be handled.

As was mentioned in chapter 10, it is important to avoid the build-up of air in the condenser. The invention of the labyrinth type of steam packed gland seals solved the problem of making an air-tight seal where rotating shafts emerge through casings. But as machines grew in size care was also needed to avoid leaks in pipes and casings. Even the compound reciprocating pumps such as Parsons had designed to extract the air from the condenser have difficulty in maintaining the lowest pressures, so he devised a way of augmenting their action. So-called ejector pumps that employed a high velocity jet of water were already in use with some condensers to entrain air and carry it away with the water, and quite low air pressures had been achieved in this way. Parsons used such a device, but substituted a jet of steam for the water. As can be seen in figure 107, the mixture of steam and air from the ejector was led to a separate, augmenter, condenser. From there the reciprocating pump could extract the resulting air and water, and compress them to atmospheric pressure. The difference between the water level in the drain from the condenser and at the inlet to the air pump is a measure of the extra degree of suction which has been achieved by the steam jet. Small as this may appear to be, figure 106 shows

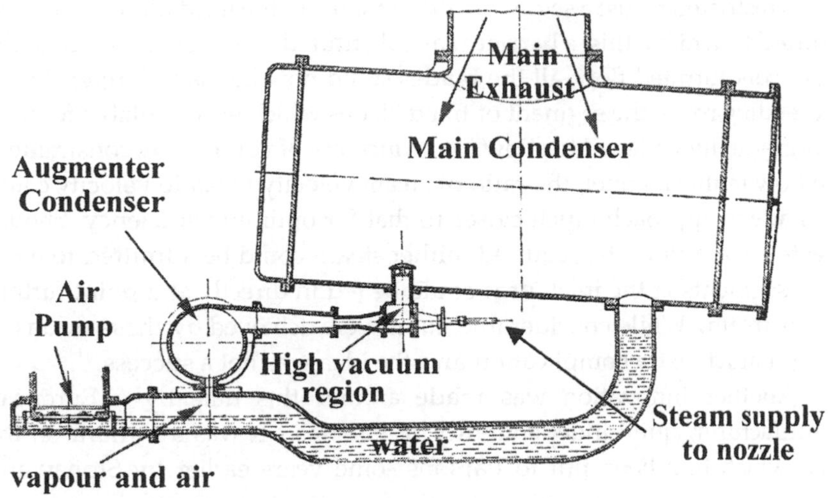

Figure 107. The steam jet of the vacuum augmenter draws air and vapour into the augmenter condenser, from the main condenser space. The condensed vapour joins the main flow of water draining from the condenser in the pipe at the right hand side. Both air and water are extracted by the reciprocating pump at the left hand side. This figure does not show the water filled tubes which condense the steam (Sothern 1909).

that there is a significant benefit. This was an important improvement and he patented it in 1902 (Parsons C A 1902). It has since been widely adopted. At a later date, two such jets were arranged in series to enhance the effect. Other measures were tried at the time to improve the operation of condensers. Most of this work was carried out by Gerald Stoney, Stanley Cook and Robert Howe (Richardson 1911). As the scale of operations grew larger, a considerable part of the development work was now being carried out by staff like these. Tests were also carried out by Professor Weighton at Armstrong College in Newcastle. These confirmed the superiority of the turbine over the reciprocating engine at the highest vacuum levels.

There is a problem in reducing steam flow when running at reduced power, as for example during 'cruising' with marine turbines. If the steam flow is restricted by throttling the pressure, some of the potential in the steam to do useful work is lost. This was tackled by Parsons with his 'partial admission' design (Richardson 1911). This was an attempt to apply the principle shown in figure 94(b) to a reaction turbine. Instead of admitting steam all around the circumference of the first row of blades, it was restricted to just a segment. Where some of the fixed blades had to be omitted to achieve this, a barrier was substituted. As a consequence of the more concentrated flow, all the blades could now be made longer. In the succeeding rows, the segment of fixed blades which was available for flow was gradually widened to allow for an increase of volume. By constraining the flow in this manner, the ratio of steam velocity to blade velocity could be made to approach much closer to that for optimum efficiency. Should greater steam flows be required, either steam could be admitted to more such segments at the inlet, or it could be fed in directly at a point further downstream. While considerable gains were achieved by these means, it caused much extra complication and the idea was not a success.

Another innovation was made around this time by a European manufacturer, quite independently of Parsons. It was a variant on the idea which had been put to Parsons some years earlier by Stoney and Hodgkinson. In 1904 Sulzer Brothers of Winterthur fitted a velocity compounded, Curtis, inlet stage in front of several conventional Parsons reaction stages (Stodola 1927). With two or even three rows of moving blades on the Curtis wheel, a very large drop of pressure could be accomodated, with of course a correspondingly large temperature drop. At that time steam temperatures had begun to reach a point at which the materials used for reaction turbine blades were being damaged. The steel blades of the Curtis stage on the other hand could easily withstand such conditions which was a welcome bonus. Also partial admission in a Curtis wheel can be readily achieved by changing the number of inlet nozzles in use, as in figure 94(b), and this reduced the need for wasteful throttling of the steam supply. Although Brown Boveri and Westinghouse were quick to adopt this idea for land use, Parsons was unwilling to follow their example. His natural conservatism seemed to increase as he grew older, and he had something of a dislike of having to depend on foreign innovations. To quote Lord Rayleigh (Rayleigh 1934),

> *He had a strong dislike of adopting any contrivance from foreign engineering design. His staff were sometimes compelled to adopt unsatisfactory alternative expedients in order to fall in with this sentiment. He would be told that a foreign firm had met the difficulty in such a way. 'Very well then' he would say 'let's do something different.'*

In 1910 Parsons was forced to employ the system in a 5 MW machine built for the London Electricity Supply Corporation at Deptford. After the order had been received it was found necessary to alter the design and to shorten its length so that it would fit into the space available (Parsons R H).

However this machine suffered blading failures as well as other problems which took several years to put right. One suspects that if Parsons had espoused the concept himself, he would have solved the difficulties much more expeditiously.

Despite this, the idea was quickly adopted by PMSTCo. for marine turbines where the competition from builders who used the Brown–Curtis design was growing (Jung 1982). Of the 20 'I' class destroyers purchased by the Royal Navy in 1909/10 only one was fitted with Brown–Curtis turbines, but the proportion grew quickly in succeeding years. In an attempt to compete with this development Parsons fitted a disc at the input end of his high-pressure marine turbines which was equipped as a velocity compounded, Curtis turbine. The first warships to be given such turbines were three destroyers of the 1910/11 'I' class and three cruisers of 5250 tons for the 1910 programme. Although the ideal, diagram, efficiency of the Curtis stage at full power is less than in a reaction stage, the possibility of partial admission yields a gain at part loads, and this is especially important for naval vessels.

The original decision to develop turbines for marine use which were directly coupled to the propeller shaft had proved to be a sound one, if perhaps rather conservative. Marine turbines had been built successfully in the largest sizes, and there had been no failures of the kind that could so easily have discouraged the already cautious ship owners from adopting the new development. Parsons was more successful than de Laval in convincing backers to risk their capital on marine applications. However the chief area of success so far had been in high-speed vessels. The majority of merchant ships steamed at more modest speeds. In order to open up applications with them, Parsons turned to the idea that he had patented in 1894, even before any marine turbine existed. He combined two reciprocating steam engines with a single low pressure steam turbine (Richardson 1911). The pressure of the exhaust steam from the engines that drove the outer propeller shafts was close to atmospheric, and was used to supply the turbine that was coupled to the central shaft, and which exhausted to a vacuum. Thus each component operated under conditions that best suited it. Reasonable blade heights could be used despite the large diameter and slow speed of the turbine, because the steam pressure was so low. Moreover the two reciprocating engines could, if required, be made to operate in reverse so that the need for separate astern turbines was avoided.

Although proposals for such 'combination' installations had been prepared as far back as 1902, once more it fell to Denny Brothers, the

Scottish shipbuilders of Dunbarton, to build the first such ship, the *Otaki*, in 1908. At 9900 tons displacement and having a length of 464 ft, she was designed to steam at 14.6 knots. Her engines developed much the same power, around 5500 HP, as her sister ships, but comparisons showed that she used some 12% less fuel. Later that year the Belfast shipyard of Harland and Wolff applied this approach to a much larger and faster vessel built for the Atlantic route. The liners *Laurentic* and *Megantic* of 20 000 tons displacement were ordered by the White Star line. The former was equipped with triple expansion engines and a turbine, while the latter had only quadruple expansion engines. The success of this experiment led White Star to specify the combination system for both the *Olympic* and the *Titanic*, which each had a displacement of 52 000 tons and an overall length of 882 ft (Anon 1970). The turbines were built by John Brown on the Clyde, but the engines were built in Belfast. Their combined power of 46 000 HP was designed to give a speed of 21 knots. The triple expansion engines were quite enormous, and each developed 15 000 HP or more than 10 MW. Their size can be gauged by the fact that the diameter of the high pressure cylinder was 54 in, and that of the two low pressure cylinders was 97 in. The piston stroke was 75 in. On the 14 April 1912, the ill fated *Titanic* sank on her maiden voyage to New York. She struck an iceberg, and there was a heavy loss of life, which included that of her designer Thomas Andrews.

17.2 Reduction gears for large powers

Perhaps the most important initiative taken by Parsons at this time was his decision to match the speed of marine turbines to that required for propellers by using gearing. In 1908 it had seemed that the combination system was the answer to economy in vessels of lower speed, and yet by 1909 Parsons was ready to apply a large reduction gear to a marine turbine and to demonstrate a full scale test. Already in America George Westinghouse had set his engineers to work on the same task (Jung 1982). When he retired as chief technologist of the US Navy in 1903, Admiral George Melville (1841–1912) began to work with Westinghouse, and John H MacAlpine (1859–1927), a Scotsman who had studied at Glasgow University, was also employed to work on the project. In 1906 a report of a comparison between a 2500 HP reduction gear and a 300 HP de Laval turbine gear was written but not published However in 1909 the results of successful factory tests on a gear rated at 6000 HP were made public.

Gears function by arranging for the surfaces of the teeth to roll over one another. In spur gears the point of contact is a very narrow line on the face of the tooth which runs parallel to the axis of the gear. As the wheels rotate, this line moves up from the base of the tooth on one gear, and down the face of the corresponding gear. If the gears are to function smoothly and without noise it is most important that the tooth geometry be as near perfect as possible. It also helps if more than one pair of teeth is in mesh at the same time, that is if the teeth are made 'fine pitched'. This requires that the tooth be made smaller relative to the circumference of the gear. Cutting the teeth on a spiral is also beneficial. If a pair of spiral cut gears is viewed end on, the point of contact moves progressively up the flank of the tooth as the observer moves in an axial direction. It is as if there were an infinite number of very thin gear wheels stacked behind one another, each at a slightly different point in their meshing cycle. This has the disadvantage that it creates a force which, because of the helix angle, is in an axial direction. The difficulty is solved by using a back to back, 'herring-bone' arrangement of helices, visible in figure 108. De Laval was familiar with all of this, though his largest units could only handle a few hundred horsepower.

The results of Parsons' first trials of gearing which had been made in 1896 had not been encouraging, but since then considerable progress had been made in the technology of gear cutting, especially in the United States and in European countries. The plant that de Laval had established at Trenton in New Jersey in 1901 had contributed significantly to this. Parsons had instructed that '*exhaustive experiments* (be) *undertaken to determine what tooth-speeds could be employed, and what power could be transmitted with safety and durability. The carrying out of these experiments and the compiling of the requisite data fell to the lot of Cook*' (Gibb 1952). In 1909 the PMSTCo. began meticulous preparations for a large scale trial of gearing and purchased the *Vespasian*, a vessel of 4350 tons displacement and 275 ft length for £4825 (Parsons C A 1910). Her triple expansion engines were 20 years old, and so that no criticism could be made of their efficiency afterwards, they were completely overhauled. During the summer she was sent on a return voyage to Malta, fully laden. Runs were also made on the Whitley mile off Tynemouth. A model on a scale of 32:1 was made and tested to provide estimates of resistance. The ship was then returned to dock where the reciprocating engine was removed and the waiting turbines and gearing were fitted in their place. A new condenser with a vacuum augmenter was also installed. In all other respects there was no change. Manufacture of the gears was entrusted to the Power Plant Company of West Drayton. They had been designed to transmit 1100 HP with a gear wheel made with a rim of forged steel which was shrunk on to a cast iron hub. The 398 teeth

Figure 108. The reduction gearing set installed in the engine room of the turbine ship *Vespasian* (Richardson 1911).

had an effective diameter of 8 ft $3\frac{1}{2}$ in. The smaller pinion had 20 teeth on a 5 in diameter pitch circle. It was made of a mild chrome nickel steel to give the teeth adequate toughness and strength. The gear had a rather large reduction ratio of 19.9:1, which reduced the turbine speed of 1432 rpm to give a propeller speed of 72 rpm

A further run on the Whitely mile showed that for all speeds above about 7.7 knots the geared turbine required less steam from the boiler than the original engine. At the upper end of the speed range it was estimated that, for the same rate of fuel consumption, the turbine was 0.8 knots faster. Walker of the PMSTCo. who was present at all the trials, analysed the data with the help of Cook and Stoney. One year later Parsons reported his experiences to the Institute of Naval Architects (Parsons C A and Walker 1911). A series of regular trips had been arranged for *Vespasian*, carrying coal from Tyneside to Rotterdam and Antwerp, and returning in water ballast. In all 20 000 miles had been covered. Careful measurements had been made over a three week period and were found to agree well with the trial tests. Altogether a very full set of data had been amassed as a basis for a truly scientific comparison. It had been suggested that when the propeller of a turbine driven vessel broke the water surface in heavy seas and lost its load, it would tend to race. But it was observed that the huge inertia of rotation of the turbine ensured that no such problem was encountered. With the governors disengaged, the speed variation still had not exceeded 16%. The pinion being smaller was more heavily stressed than the main gear wheel, and replacements were made using a variety of materials. It had been found that a chrome nickel steel with a strength of 55 tons in^{-2} and a carbon steel of 40–45 tons in^{-2} both fared worse than the original steel chosen which had a strength of 37 tons in^{-2}. One of the original pinions was removed after 18 000 miles and was displayed for inspection.

Once more Parsons had set the pace. The ability of gear sets to handle the output of large turbines was firmly established, but he needed experience to uncover what problems would emerge in service. In 1912 two Channel steamers were fitted with geared turbines developing 5000 HP (Parsons C A 1913). A $10\frac{1}{2}$ knot cargo vessel similarly equipped showed a 15% reduction in coal consumption compared to a sister vessel with reciprocating engines. In 1911 the Royal Navy gave Parsons an opportunity to try out gearing for destroyers in the 'I' Class vessels being built. Only the high pressure and cruising turbines were geared; the low pressure turbine was a direct drive. Since the volume flow in the former would be small, there was most to be gained by using turbines which had a small diameter but a high shaft speed. The *Leonidas*, built in 1913, was one of the first destroyers to be fitted with all geared turbines (two of 8.25 MW), and this allowed the propeller speed to be cut to less than half, or 380 rpm (Jung 1982). It was joined in the same year by the cruiser *Calliope* of 4120 tons. However lack of adequate gear cutting capacity meant that the majority of the vessels built during wartime had direct drive turbines. It was not until 1917 that the first battle cruiser, the *Courageous*, was built with geared drives for its four shafts (66 MW). The service record of geared marine turbines during the war was excellent. Roughly 1% of the 600 gear sets experienced failure (Parsons C A 1923).

In 1910 Parsons purchased two gear cutting machines for the PMSTCo. so that he could acquaint himself with the practical difficulties of gear manufacture. They were made by W Muir and Co. of Manchester and cut teeth by using hobs. At the same time two further machines were installed at Heaton to make gears suitable for driving generators, rolling

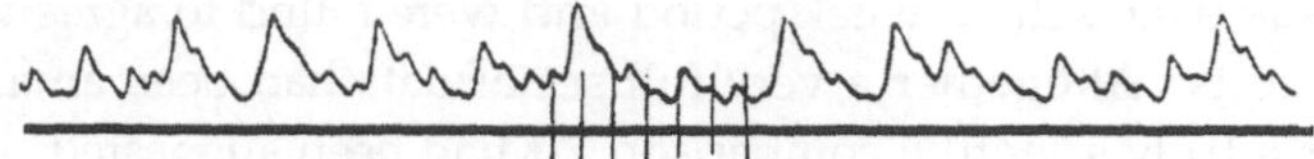

Figure 109. An oscillogram recording of gear noise (Parsons C A 1913).

Figure 110. Spiral hob for cutting the teeth of helical gears (Leonard 1911).

mills etc. Early experience of marine gears quickly revealed a propensity to emit a rather shrill noise. Parsons set out to investigate the problem. Since there were as yet no instruments for quantifying noise, there were different assessments of just how serious the problem was. So he mounted a microphone above a set of gears to pick up the sound emitted by them. In the absence of valve or transistor amplifiers, the microphone was connected to an electro-mechanical oscilloscope whose output was photographed (figure 109). From this it became clear that one particular frequency predominated. This was soon related to a feature of the gear cutting machine itself. It was 160 times the speed of rotation of the gear wheel.

To cut a helical tooth profile, the gear blank is rotated slowly about a vertical axis while the rotating, and angled hobbing tool (figure 110) moves slowly in a vertical direction. By taking power from a common drive these two movements can be synchronized accurately. The table carrying the blank had a 160 tooth wheel which was driven by a worm. Now the exact pitch of any gear, that is the spacing of the teeth, is determined by the gearing in the machine used to manufacture it. Consequently it is no more precise than the pitch of the threads in the gear cutting machine. Any errors in the one are reflected in the other. The art of the toolmaker has been to reduce such errors from one generation of machine to the next. When careful measurements were made, Parsons found variations in the gear being cut of up to 0.004 in, and these were co-periodic with the 160 worm wheel teeth. To reduce these errors, he mounted a second table on top of the main work table (figure 111). The drive to the lower table was reduced in speed by 1% , and the upper table was given added speed relative to the lower one of 1%. Overall the speed ratio remained the same. The effect of giving the work piece a creep of 1% relative to the main table was to spread the errors originating in the main worm drive (c) over one whole revolution of the workpiece. This also resulted in a reduction of the peak value of the error to approximately one-fifth. Moreover the surface profile of those errors which remained was in the form of a pointed cusp, which tends to wear flat quite quickly. This was covered by patent 29380 of 1912.

In the discussion after reading his paper, Parsons said (Parsons C A 1913),

> *I should like to say that in this work I have been much assisted by Mr Cook, Mr Carnegie and Mr Walker, and my thanks are due to them for their highly technical assistance... Last week a gear cut with the new method was tested for a mill-drive in Scotland, and the difference was very marked; there was a great reduction of noise compared with a duplicate gear cut by the usual method.*

Cook in his contribution stated explicitly that it was Parsons who proposed the idea of 'creep' for the hobbing machine. Alfred Quentin Carnegie (1876–1954) had been born and educated in Newcastle and had been employed by Parsons for most of his working life.

During the previous decade Parsons had assembled a talented team around him. The works at Heaton had been provided with both laboratories and experimental shops (Richardson 1911). Full electrical instrumentation was available including a Duddell electromechanical oscillograph as well as a cinematography camera for analysing transient phenomena. A number of boilers which provided steam at pressures up to 350 psi were capable of superheating of up to 100 °F so that turbo-generators could be assembled and tested before despatch. To provide electrical loads for generators under test, electrodes were immersed in water to create resistors. The pond used for this purpose had occasionally been used for towing experiments on models of ships. A special house was built for testing rotating components at full speed when checking for

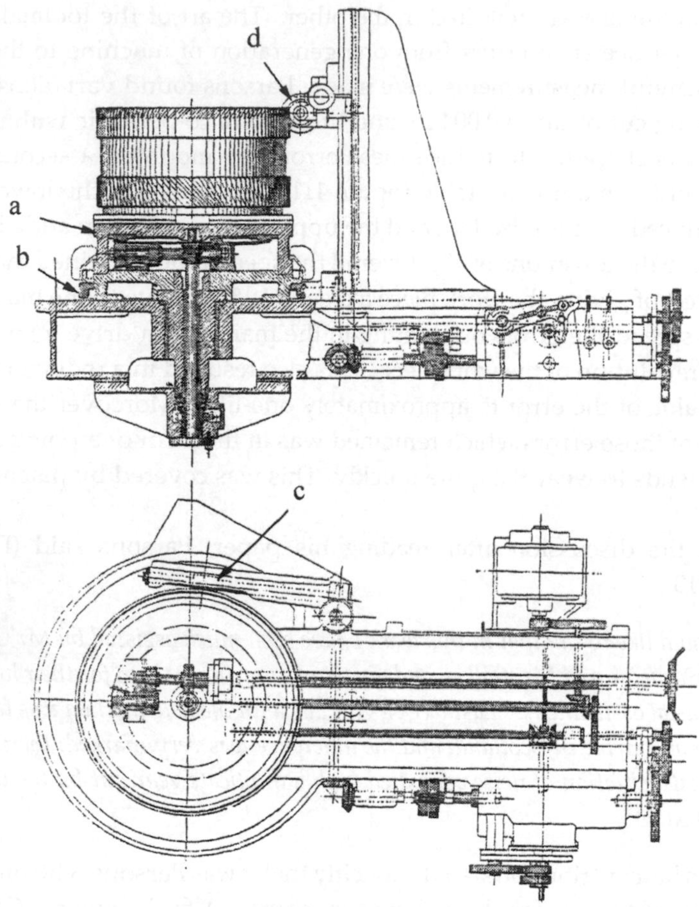

Figure 111. A gear hobbing machine fitted with a creep attachment. The gear blank is mounted on a special upper table a, and the lower table is driven by the worm c, which meshes with the wheel shown in section at b. The hob at d has partly cut the teeth on the top of the workpiece (Parsons C A 1913).

balance. It was constructed of reinforced concrete as a safety precaution in case a component should fail and disintegrate.

Though much work that was relevant to marine turbines was done at Heaton, the Wallsend works also had special experimental facilities. These included rigs for testing and evaluation of blades for marine use and for the design of oil coolers. A tank 100 ft long by 12 ft deep was built for towing

Figure 112. The large scale water tunnel at the Wallsend Works of Parsons Marine Steam Turbine Co. built for testing model propeller screws for cavitation, if necessary under reduced ambient pressure (S.P.(2)).

tests, though in general ship owners were reluctant to pay for model tests, despite the generous pay back from lowered fuel bills achieved by better designs (Clarke 1997). Parsons retained his active interest in propeller design and in the phenomenon of cavitation, and he built a large water tunnel in 1912 (Burrill 1951). This had a circuit of 66 ft in length and the limbs were 36 in diameter (see figure 112). Model propellers up to 12 in diameter could be tested with water speeds up to 14.3 ft s^{-1}. A searchlight and a revolving mirror allowed photographs to be taken with exposures of no more than 0.0002 seconds. Torque and thrust could be accurately measured, while the pressure of the air above the surface could be reduced by an air pump. Large numbers of data were accumulated though none of them was published during Parsons' lifetime.

In the discussions after Parsons' papers on gearing, participants argued in favour of other alternative ways of reducing the output speed of turbines. One very persistent advocate was Dr H Föttinger who had developed a form of hydraulic transmission for the Vulcan shipyard at Stettin. He claimed efficiencies of 86% and even 90%. However Parsons was able to confirm that his own measurements had shown gearing to be 98.5% efficient. In 1910 Westinghouse secured an order from the US Navy for geared turbines for the *Neptune*, a special purpose collier of 20 000 tons. The turbines developed 7200 HP and had a Curtis front end with Parsons reaction stages. Admiral Dyson stated in 1917 that with the exception of the gears which were an unqualified success, the installation was a failure and had to be replaced (Jung 1982). In 1912 Westinghouse applied the de Laval, Trenton, technology and began to cut their own gears.

Other speakers advocated 'electric gearing'. By using the turbine to drive a generator, the turbine design could be optimized. The propellers could then be driven by electric motors at any desired speed, whether forward or in reverse. On the basis of efficiency, Parsons rejected the idea for all but auxiliary vessels, and he added (Parsons C A 1910).

> *in electrical installations of large power on board ship there is a very great danger of short circuits, and if a short circuit of several thousand horse-power took place in an engine room everybody might be killed. I have been in electrical work for twenty five years, and I know the difficulties of avoiding short circuits where there is much salt present in the air. When they occur in large palatial engine-rooms on land the smoke and copper fumes are suffocating, and if anyone happens to be close to the short circuit it is a serious matter for him. It is like a gun going off with a blinding flash; a searchlight is about 12 H.P.; try to imagine one of 5,000 to 10,000 H.P., with a hail of molten copper flying about!*

Knowing how much experimentation Parsons had carried out at first hand, it is clear that such words were a record of fact and not poetic fancy. Despite this, the USN ordered the sister ship of the *Neptune* to be fitted with electric drives (Jung 1982). It was very successful and after the First World War she was refitted as the aircraft carrier *Langley*. The USN subsequently built a number of capital ships with turbo-electric propulsion that gave good service. A later assessment by Somes of GE was that electric propulsion machinery '*had no economic value today* (1959) *in the United States. The electric drive might be quieter and more manoeuvrable with the possibility of higher speeds astern*... (but) *it cost approximately twice as much and weighed twice as much as comparable steam gear machinery*' (Clarke 1997).

Despite the setback with the *Neptune*, Westinghouse persevered with geared drives. As the power to be handled increased it was necessary to lengthen the smaller, pinion gear. J M MacAlpine's 'floating frame' patented in 1905 was used to provide the necessary support for such slender components. During the First World War more than 20 destroyers with two, cross compound, geared turbines (two each of 9 MW) were supplied by Westinghouse to the USN. Altogether over 250 destroyers were built as part of a huge war-time construction programme. Roughly equal numbers of impulse (Curtis) and reaction (Parsons) type turbines were built, two-thirds of these in the engine works of licensees in either federal or privately owned shipyards (Jung 1987a). Westinghouse and General Electric secured around one-third of the orders. A large proportion of these had cross compound gearing. In Britain the battle cruisers for the 1914 programme of construction were the first to have all their turbines geared. Of the *Courageous* class, having 18 600 tons displacement, two had Parsons' and one had Brown–Curtis turbines. The maximum power transmitted through a single pinion gear was 15 500 HP (Parsons C A 1919). Despite their gearing, these vessels had the same propeller speeds as their direct drive predecessors, 340 rpm. The largest such vessel was the *Hood* at 41 300 tons, and it was fitted with Brown–Curtis turbines delivering 144 000 HP. When it completed its trials after the war, it achieved 32 knots, which was close to the top speed of *Turbinia* (Jung 1982).

Practical considerations dictate that large gears should have a ratio somewhere in the range between 5:1 and 10:1. This still leaves relatively high propeller speeds. In 1915 turbines with a two stage reduction gear were supplied by General Electric for a merchant vessel, the *Pacific*, of 6000 tons, built in San Francisco (Dorey 1943). In Britain the first such vessel was the 9770 ton *Somerset* built in 1918. The turbine of 4500 SHP ran at 2970 rpm, and via two reductions of 7.1:1 and 4.92:1, drove the propeller at 85 rpm. Up to 1922—3 1000 000 SHP of double reduction gears had been fitted in Britain, and by 1921 twice that amount in the USA, where gear manufacturers gradually overcame the problems with wear and breakage which were encountered. However during 1922 8% of the 207 British vessels with double reduction gears experienced fractured gear teeth (Parsons C A 1923). Parsons investigated the large number of failures, concentrating on the metallurgy of the components. Although in the case of the *Vespasian* he had shown that ductile alloys were to be preferred, manufacturers had gradually adopted alloys and heat treatments which yielded much harder, and more vulnerable, components. His paper vindicated his original conclusions, but, from this time on, the competition from internal combustion engines rapidly diminished the market for geared merchant vessels.

17.3 Developments in electricity generating plant

The earliest public electricity supply companies employed direct current. In the early stages this had an advantage in that it allowed batteries to be used to even out demand but the area that could be supplied was limited by the voltage drop in the supply lines, and once demand had developed sufficiently alternating current offered an attractive alternative. By transforming the current to a high voltage, wide areas and therefore large numbers of consumers could be supplied from one large power plant. Such plants could be sited where fuel could be delivered economically, where supplies of cooling water were available and where their environmental impact could be minimized. In 1905 the London Power Bills committee of the Houses of Parliament met to consider proposals to build several large, riverside stations to take the place of 61 local stations. Experts, among them Merz, the Hon. C A Parsons, Mr C E L Brown and S Z de Ferranti, gave evidence that large savings in capital and running costs would result. Stations grew larger in the decades that followed and the area within which they were interconnected became more extensive, which brought additional savings. Nowhere was this trend more evident than in the North East.

It is perhaps worth reviewing the broad trends in the first 30 or so years of development of land based turbo-generators, from 1884[1]. We know that Parsons, from very early on, had relied on his designers to implement his ideas. Thus, not long after he joined Clarke Chapman, Gerald Stoney was making the drawings for a turbine of tandem design, with separate high pressure and low pressure cylinders, although this was not built until the argument over patents was settled some years later. Once the PMSTCo. had been established there were two design teams at work, collaborating closely, but driven by the requirements of different products. As the years passed by, it is not always possible to pinpoint Parsons' personal contributions, but he evidently remained close to the designers. His mark is perhaps most clearly evident in cases where some relatively small order, often for an industrial client, was used as a vehicle for trying out some innovation.

There are two influences that operate even to the present day[2]. The first is a demand for single units of ever increasing power, and the second is a need to increase the thermodynamic efficiency by raising steam

[1] For an excellent account of the way the design of reaction turbines evolved during Parsons' lifetime the reader is referred to Bolter (1994).

[2] See also appendix 2.

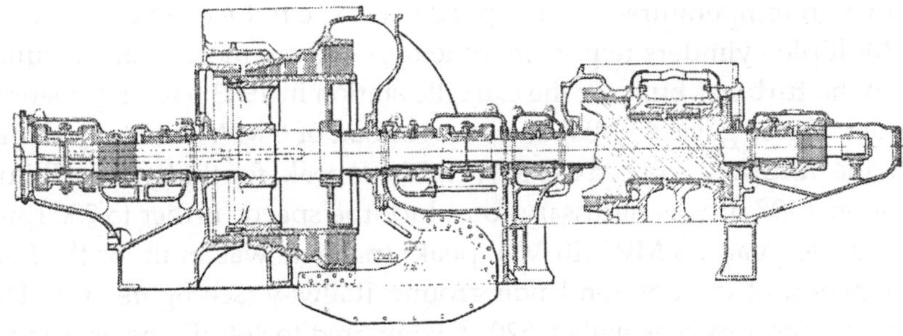

Figure 113. A 6 MW turbine, for Lots Road (1908), running at 1000 rpm, and using steam at a temperature of 520 °F. A single forging is used for the high pressure rotor on the right hand, and a spider supports the drum for the low pressure cylinder on the left hand side. A duct which leads steam from the high pressure exhaust to the inlet of the low pressure cylinder (at the extreme left) is not shown (Parsons R H).

temperatures. In responding to these, the designer must choose a suitable turbine speed. High speeds cause mechanical difficulties. For example components like blade fixings are loaded by centrifugal forces and these increase as the square of speed. High speeds can also cause shafts of both turbines and generators to whirl. But on the other hand the output of a turbine or alternator, for a given weight, increases with speed. At first the chief concern was to reduce shaft speeds from the extraordinary level of 18 000 rpm which had been necessary for the tiny, 6 kW, prototype. By 1889 the speed of the largest machine, a turbo-alternator for NESCO with an output of 75 kW, had been brought down to 4800 rpm. It proved possible to run the first condensing machine, which was rated at 100 kW, at the same speed in spite of the much larger volume of steam at very low pressures, because it was designed for radial flow. The largest machine of this type was *Turbinia*'s first engine, which was rated at 1500 kW. In that case the requirements of screw propulsion had necessitated a further speed reduction to 2000 rpm. After the axial flow patents were recovered in 1894, the first turbo-alternator that was built for Manchester Square had an output of 350 kW at 3000 rpm in a single cylinder. When a new turbine was built for *Turbinia* in 1896, it had much the same speed and power as the radial design, but now there were three cylinders because three independent output shafts were required. This is described as a cross compound arrangement. The idea of building a turbine for land use with two cylinders joined in tandem was first tried out in 1900 for driving the Elberfeld alternator where it helped to minimize the impact of problems due to high temperatures, but its speed had to be reduced to 1500 rpm.

Multiple cylinders require more length and are more costly to build, and so the turbines built for the Carville station in 1903 had only a single cylinder. These gave 3.5 MW, and an overload of 6 MW at 1200 rpm. But when the output was raised to 5 MW (7.5 MW peak) for the Sydney Ultimo station in 1908, it was necessary to reduce the speed further to 750 rpm. In the same year a 6 MW (10 MW peak) machine was built for the Lots Road station of the London Underground Railway (see figure 113). The steam temperature was higher, 520 °F compared to 466 °F, and once more two cylinders in tandem were used at 1000 rpm. The suppliers of forgings continued to make progress, and it was now possible to make the rotor of the high pressure turbine as a single piece, but the low pressure rotor continued to follow current marine practice with a light drum supported on a spider which was shrunk onto the shaft.

Bankfoot was one of the NESCO stations that used 'waste heat' from industries like steel works to generate steam. In 1910 a relatively small machine was built for it which pioneered several new features, some of which were later adopted as standard practice. The two cylinders of this turbine were coupled to two, separate, 2 MW alternators which ran at 2400 rpm, rather in the style of the multiple propeller shafts in the *Turbinia*. The steam inlet temperature was 538 °F, and to avoid the problems encountered when cast iron is used at temperatures above 450 °F, cast steel was used for the high pressure cylinder. Henceforward this became the standard practice. Also in order to increase the area available for steam flow at the exhaust without risking excessive stresses from long blades, a 'balanced flow' arrangement was used. This doubled the area available for the steam flow, with one half flowing in each direction, as in the prototype of 1884.

In 1912 Samuel Insull approached Parsons with a request for a machine of 25 MW output to be used in the Fisk Street station of the Commonwealth Edison Co. in Chicago[3]. The alternator was required to generate a current of 25 Hz. It was in Fisk Street that the first GE Curtis 5 MW turbo-generator had been installed, and by that date it contained several 20 MW machines of the same type. The challenge could not be turned down and Parsons guaranteed a steam consumption that would be no more than 11.25 lb per kWh, with a penalty/bonus for every 0.2 lb per kWh by which the measured value diverged from this (Parsons R H). The steam temperature

[3] Samuel Insull had left London in 1881 at the age of 21 to become Edison's private secretary. After the combination of Thomson–Houston and the Edison General Electric Co in 1892 to form General Electric, Insull left to run the Chicago Edison Company (Josephson 1961).

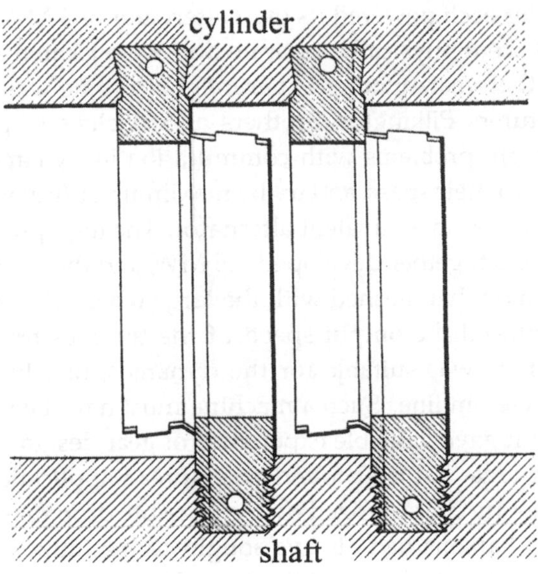

Figure 114. Details of turbine blading fitted with shrouding and suited to end-tightening of the shaft. The clearance around the blades could be reduced by moving the shaft towards the left by adjusting the thrust bearing (Parsons R H).

was 588 °F and it was decided to use a two cylinder tandem arrangement. The shaft speed was kept down to 750 rpm, and, as a further step to handle the large volume of exhaust, the low pressure cylinder had a balanced flow arrangement exhausting to the condenser mounted below. The rotor of the low pressure turbine resembled the Lots Road machine. The last row of blades were 19 in long and spanned 10 ft overall, but the low shaft speed kept the centrifugal forces down to well below one-tenth of those experienced in the original prototype turbine. The slow speed made the whole machine massive. The alternator rotor weighed 49 tons. From end to end the machine stretched 76 ft. Tests were carried out in June 1914. The steam consumption was found to be 10.45 lb per kW h which corresponds to a thermal efficiency of 25.7% , well within guarantee and significantly better than that of the large GE Curtis machines which had recently been installed (Somerscales 1991). It remained in service for 20 years and more, and its service record earned it the tag of 'Old Reliability'.

In 1912 two other novelties were constructed. A cross compound arrangement was used for machines supplied to the Ensenada station of the River Plate Electricity Co.; the high pressure cylinder and the low pressure cylinder each powered its own alternator of 3 MW. The turbines had shafts that were physically separate and were located alongside one another. Also that year, another experimental turbine was made for the glass manufacturers Pilkington Brothers at St Helens to provide direct current. There are problems with commutation in dynamos, and these set upper limits to their speed and to the maximum output which are both much lower than for an equivalent alternator. The high pressure and low pressure cylinders together developed 4.5 MW, and the rotor of each was coupled to a pinion that meshed with the large wheel of a reduction gear. The gearing reduced the output speed of the turbines from 2400 down to 370 rpm, which was suitable for the dynamos, of which there were three, all connected in line. Such a machine must have been uneconomic to produce, but it gave valuable experience of gear design. It served as a test for another idea that later found widespread application. Of course reaction turbines are characterized by the pressure drop that occurs across the blades, and Parsons had always sought to minimize the fraction of the steam flow that leaked around the tips of the blades by maintaining a small radial clearance. In service this could easily be increased if the tips were accidentally worn away. An alternative, which is illustrated in figure 114, was introduced for the first time in the St Helens' machine. A thin metal shroud was brazed to the blade tips at the high-pressure stages of the turbine where the blade heights are small, and then a fine adjustment of the shaft relative to the casing was arranged. Such shaft adjustments had been used first in the radial flow machines, and can be seen in figure 52. The technique was called end-tightening. A significant reduction in steam consumption was achieved and the technique was later adopted for other machines, though it was not taken up by licensees (Harris 1984).

In 1915 an 8.5 MW, 1500 rpm machine was supplied for Neepsend in Sheffield; the height of each successive row of blades was increased progressively. Except for the last four stages, the shaft diameter was also increased. In the last stages, the increased steam volume was accommodated as in marine turbines, by altering the exit angle and spacing of the blades so raising the area available for flow from 0.35 to 0.83 of the exit annulus area. These measures gave both the rotor and the casing a 'conical' shape. This became quite a usual feature of all turbines, but Parsons was the first to adopt it. Makers of impulse machines were slower to appreciate its advantages. In 1917 the Neepsend turbine was the first Parsons machine to be fitted with a governor having an oil operated relay rather than mechanical linkages, an idea pioneered by Brown Boveri.

In 1916 a 3 MW machine for the Priestman Power Co. at Blaydon

Burn contained several unusual features. The rotor of the high pressure cylinder ran at 5000 rpm and that of the low pressure cylinder at 4200 rpm. Each was coupled to a pinion that meshed with a gear wheel coupled to an alternator running at 800 rpm. Some of the steam leaving the high pressure cylinder was taken to heat the boiler feed water. The rest was 'reheated' from 280 to 400 °F in a heat exchanger using live steam from the boiler, before entering the low pressure cylinder. Some steam was also taken from the low pressure cylinder at a point where its temperature was around 150 °F, and was used to raise the temperature of the water leaving the condenser. This 'regenerative' heating raised the temperature of the boiler feed water to 234 °F. The two measures of reheat, and of regenerative feed heating, greatly improve overall thermal efficiency and have become a standard feature of modern steam turbine plants, but they only really come into their own when used with large units and advanced steam conditions.

In 1913 an order was placed for an 18 MW turbine for Lots Road, with an overload capacity of 22.5 MW. A speed of 1000 rpm was chosen and the balanced flow exhaust discharged into two interconnected condensers. Steadily the capability of forgemasters had been improving, so the high pressure rotor and its shafts could now be made as one piece from a single forging. The low pressure blading was carried on several substantial forged wheels that were shrunk onto the shaft. Between 1916 and 1920 a number of 11 MW machines were delivered to the Carville station. These followed the layout of the Lots Road turbine, but ran much faster, at 2400 rpm. By now single forgings were available that could be used for the low pressure rotor. They could be machined to take the low pressure blades and were better able to handle the centrifugal stresses. Steam temperatures of up to 700 °F had to be accomodated and so the high pressure blades were made first of manganese copper and then of mild steel. Even with balanced flow, the blade heights of the last stages were made too small. Testing showed that the resulting high velocity in the exhaust steam gave a 6% loss, or twice what would normally be expected.

The alternator was required to deliver 14.7 MV A at 2400 rpm and the rotor was relatively long and slender. Indeed it was feared that it would tend to whirl and so tests were carried out at 36 000 rpm on a 1/12 scale forging, but it was found that there was no cause for alarm. The actual rotor when balanced was run up to 3000 rpm entirely free of vibration. A tale has been told that sheds some light on how Parsons' reasoning processes functioned on such a problem (Dowson 1937),

Sir Charles' reasoning about the critical speed of a rotating shaft was an even

better (example of a) *generalisation based upon induction. It was found by experiment that the product (angular velocity at the critical speed)*^{2} × *(static deflection) was nearly constant for shafts of similar construction i.e. scale models. Simple theory showed that static deflection of a shaft was proportional to the square of the diameter. Sir Charles deduced that the product (angular velocity)* × *(diameter) was constant, or in other words 'that similar shafts whirl at the same surface speed'.*

As the years have passed Parsons' achievements in turbine design are what have caught public attention, and yet his contributions to the design of electrical generators were also of major significance[4]. It is true that the basic principles of the dynamo had been established by 1884, but such machines operated at speeds between 100 and 1000 rpm, and their electrical efficiency was often quite low. Parsons raised the speed of electrical machinery to 18 000 rpm. and his turbines routinely powered generators at 5000 rpm. This was possible because he had found a practical way with his bearing designs of suppressing vibration in high speed shafts. His use of binding wire to hold conductors in place on the surface of the rotating armature was also crucial. As well as giving the necessary strength to withstand very great centrifugal forces, it offered a practical method of avoiding out of balance masses that excite vibration, and this was important at a time when balancing procedures were still very primitive. Parsons was well aware of his great achievement. It was said that what pleased him most about the 1884 machine was not the turbine, but the dynamo.

Writing in 1922 Parsons threw some light on the problems of designing the early machines when reliable electrical measuring instruments were scarce (Parsons C A 1922),

A simple method of ascertaining the losses that exist in an armature was used a good deal on the early plants. The machine was run up to full speed and voltage on the throttle, and when steady working conditions are established, loads of small amperage are thrown on and off and the steam pressure recorded. From the data thus obtained the relation between the pressure and load in watts can be determined. By separately exciting the machine the power absorbed with and also without excitation is determined. The conductors can then be removed and replaced by wood, and the binding wire replaced to its original smoothness and same windage loss. The eddy-current loss of the conductors can then be calculated.

[4] Both Richardson's and R H Parsons' books are devoted primarily to Parsons' steam *turbine*.

By such means he could estimate not just eddy current and hysteresis loss, but the brush friction and the loss occurring in the steel piano wire as distinct from that in the armature core and shaft. Even in 1891 when tests were made on the Cambridge condensing turbine, there was sufficient doubt about the ability to measure the alternator power output accurately that a specially built direct current armature was made to allow the tests to be cross checked.

Parsons' first alternators were just a simplification of his dynamo, with the complication of the commutator replaced by slip rings as can be seen in figure 46. For machines intended for a single phase supply such as the two 75 kW turbo-alternators made by Clarke Chapman for NESCO's Forth Banks station in 1889, which were the first in the world, this was quite adequate. But as the output of alternators increased, it became more difficult to dissipate the heat generated by the flow of armature currents and by the rapid fluctuations of the magnetic field. The alternator for Manchester Square (1894) had a simple centrifugal fan mounted at one end of the rotor that drew cooling air through axial ducts in the rotor. In the Elberfeld machine, air was drawn in at either end of the rotor and driven through radial ducts in the rotor under the influence of centrifugal force.

Already in 1891 C E L Brown had shown the potential for supplying a three phase motor over a long distance high voltage line. In 1900 Parsons built the world's first three phase turbo-alternator for Lord Masham, the proprietor of the Ackton Hall Colliery in Yorkshire, who was the first to use a three-phase supply for a coal cutting machine. The generator had an output of 150 kW at 350 V and running at 2520 rpm. The output of the first turbo-alternator for public supply which was installed in the Neptune Bank station of NESCO in 1902 was much larger, 1.5 MW at 1200 rpm. Because alternating currents can have their voltage changed up or down by transformers, a higher voltage, of 6600 V, was used. However when three phase windings are involved, conductors with large voltage differences between them must overlap. The result was frequent electrical breakdown of the insulation between conductors which were subject to high centrifugal forces. The alternator was redesigned and the now usual arrangement was adopted, with the field winding mounted on the rotor, and the three phase, armature windings were located on the stator. The field coils were wound on 'salient', that is protruding, pole pieces. Such a configuration was first used in 1887 by F A Haselwander (Neidhöfer 1992). Alternators which are coupled to water turbines run at quite modest speeds, and for them salient poles work very well, but there are major difficulties in building such machines to withstand high speeds. A development that overcame this difficulty, and proved to be of great long term significance, was the 'barrel' rotor invented by Charles E Brown in Switzerland in 1898. With this design the rotor was made of a plain cylindrical steel forging, and axial slots were cut in its surface to accommodate the coils of the field magnet[5].

However the changeover to a rotating field magnet, and the adoption of the Brown barrel rotor created further difficulties in cooling the now stationary armature. Radial air ducts were placed in the stator windings, but it was not until 1915 that forced ventilation was again introduced when fans were fitted at each end of the rotor of the Lots Road 18 MW alternator. Around the same time a more radical solution was used for the Carville 14.7 MV A alternators. The casing surrounding the armature and its stator windings was divided up into a series of cells and an externally mounted fan was used to blow cooling air through one group of cells radially inwards through ducts in the armature. This passed along the rotor surface and exited radially outwards through similar ducts in the neighbouring cell. In order to cool the rotor Parsons returned to the idea, first used in his prototype, of circulating liquid through a hole bored in the rotor, but this time he used water in place of oil, and the water was led back along the rotor in pipes beneath the conductors. In 1919 a further step was taken and the Blaydon Burn machine was modified so that the entire path of the cooling air was made into a closed circuit. This overcame difficulties caused by moisture and dust getting into the windings. It was made possible because the water cooled heat exchanger which extracted heat had been made much more compact by attaching fins to the water tubes in the style of motor car radiators. Such finned tube heat exchangers were pioneered by the French Westinghouse Company some years before in a 4 MW machine.

[5] The rotor of the early fixed field dynamo had to be 'laminated' to prevent currents being generated as the iron armature swept through the magnetic field, heating it and wasting energy. In a rotating field generator the magnetic field moved with the iron rotor, and so no currents were induced and a solid rotor could be safely used.

Chapter 18

Heaton Works Comes of Age

18.1 Charles Parsons and Gerald Stoney

Public recognition came to Parsons in several waves during his lifetime. Undoubtedly it was the Spithead Review that initially put his name in the public arena in a dramatic fashion. In 1900 Dublin University conferred an honorary ScD on him. The first formal recognition from the State came in 1904 when he was made a Companion of the Order of the Bath. This recognized his contribution to both the Royal Navy and the Merchant Marine. At the turn of the century he and his colleagues presented accounts of their achievements to meetings of many learned bodies, and they too honoured him. He had been elected to Fellowship of the Royal Society in 1898, and in November 1902 he was notified that he had been awarded its Rumford medal *'for your application of the Steam Turbine to industrial purposes, and for its recent extension to navigation'*. In 1907 he joined the Council and the following year became its vice-President. Abroad, in France in 1904 La Société des Ingénieurs Civile, and in Germany the Vereins Deutscher Ingenieure made him an honorary member. The VDI awarded their prestigious Grashof medal to two recipients in the same year, Parsons and de Laval, both pioneers of the steam turbine. Parsons received his medal at the Cambridge meeting of the British Association[1]. In 1905 he was the joint recipient, with Gerald Stoney, of the Watt Gold Medal from the Institution of Civil Engineers. These were but the first of the many honours that came his way in later years. Charles Parsons had not followed the example of his father and grandfather in regard to the education of his children; quite simply the demands on his time made such a course

[1] Since the first medals had been awarded in 1894, Charles Parsons in 1905 and George Westinghouse in 1913 were the only recipients who were not from Continental Europe.

impossible. At about this time his eldest child Rachel had just finished her engineering studies at the University of Cambridge. Her brother, who had been sent away to school at Eton, was commissioned in 1906 having completed his studies at the Royal Military Academy at Woolwich.

During the early years at Heaton, Parsons had relied heavily on Gerald Stoney for help in developing his inventions. On 24 July 1893 an agreement was made between them (Bedford).

> *C.A. Parsons and Co. agree to employ and George Gerald Stoney agrees to serve the said employers for a period of five years from the date hereof in the capacity of an electrical engineer, the duties of which shall comprise the management of the* (searchlight) *mirror and testing departments, the carrying out of experiments and such other duties as in the nature of his office shall from time to time be found requisite.*

His knowledge of silvering mirrors that he gained as a youth now found an application. Eventually the mirror department, which supplied the needs for military searchlights, became one of the largest in the world. Since all turbo-generators were assembled and tested in the works before despatch, his other duty brought him close to the heart of design activity. In 1894 he married the daughter of a local farmer, Isabella Lowes. He recalled being given what would now be called a share option (Dowson 1942).

> *Turbinia was started in 1894 and I well remember Mr Parsons telling me that he was forming a company to try the experiment of applying the turbine to marine propulsion. I had not much in those days but scratched up £200 and put it in. I have had some eight or ten times more out of it . . . Mr Parsons gave me £200 more making my total £400.*

In 1895, now thirty two years of age, he was named chief designer of the steam turbine department and chief electrical designer for high speed dynamos and alternators. As Gerald Stoney pointed out, Parsons' earliest dynamos were very well designed given the current state of knowledge about electrical machinery (Stoney 1938). However his comprehension of the principles ruling electrical machines was not quite equal to his grasp of mechanical matters as will be seen. To avoid sparking, the brushes of a dynamo are arranged to contact the commutator segments at a point in their rotation at which there is virtually zero voltage across the corresponding coil. As the dynamo is loaded, the current in the armature (rotor) sets up an additional magnetic field. This combines with the main magnetic field and so alters the correct position for the brushes. As long as dynamos

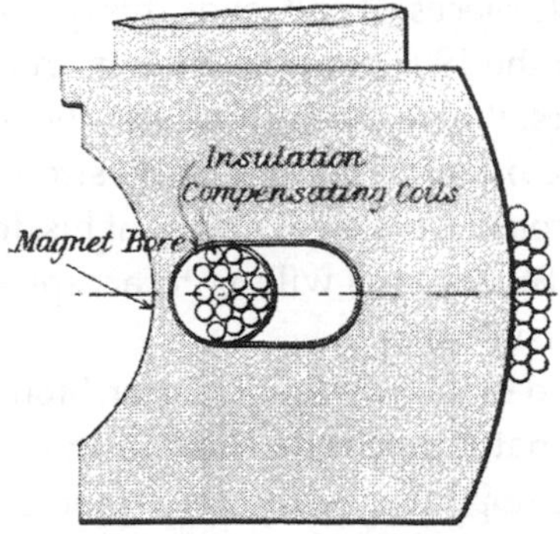

Figure 115. Compensating winding located in the pole piece of a dynamo (1885) (Richardson 1911).

supplied only lighting systems, load changes were slow and the brushes could be moved by an attendant. By contrast, traction systems supplying locomotives or trams could experience very abrupt changes and some means of automatic adjustment was needed for them. In 1885 Parsons had tried using an extra winding which carried the armature current, and which counteracted these changes in the magnetic field (figure 115) (Richardson 1911). However there were insufficient turns in the winding and it was not successful. Later a partial solution was developed with the help of Stoney, who told Dowson (Dowson 1942)

> *The first attempt to overcome this* (need) *was to shift them* (the brushes) *automatically as the load altered. It was effected by a steam cylinder controlled by a spring and connected to the inlet branch to the turbine blading (Parsons–Stoney patent)* (Parsons C A and Stoney 1900). *It worked all right, but there was a time lag and furious sparking if a big load was suddenly thrown on. I wanted to try interpoles, which I had seen on the continent, but Mr Parsons would not have them, and I designed the compensating winding in the pole faces. This worked very well, but owing to the large* (magnetic) *leakage flux, to be effective it required to have at least 2.3 times the armature ampere turns, and the quantity was enormous.*

Stoney had found that if the compensating winding had almost two and a half as many turns as the armature itself, then the response to load changes could be entirely satisfactory (Stoney and Law 1908). Such machines were bulkier and more expensive, and in the long run the solution to the problem was to be found in winding compensating coils on interpoles at locations between the main pole pieces. If one reads this quotation together with the printed discussion of the 1908 paper by Stoney and Law to the Institution of Electrical Engineers, where the authors stoutly defend this system, two things become very evident. The first is the strong inhibiting effect that Parsons' own views could have on the work of his designers, and secondly the degree to which Stoney was willing to accept such constraints while hiding them from outside eyes.

As was mentioned in the previous chapter, Stoney had run into trouble with the 1.5 MW alternator that he had just designed for the Neptune Bank power station, and he hoped to use the new rotor design patented by C E L Brown as a remedy. As chief electrical designer, Stoney made a number of visits to Baden after Brown Boveri had become licensees, and in 1901 he brought back drawings of the barrel rotor. Licences issued by Parsons had, as an integral part of the agreement, a provision that innovations which were introduced by the licensee could be availed of freely by Parsons without charge. Years later Stoney confided that (Dowson 1942)

> *Mr Parsons would not have 'any damned German invention'. I was at my wits' end what to do, and designed the abominable salient pole motor with slide-on pole tips, which was afterwards largely used. At Mr Charles Merz' suggestion a modified barrel rotor was designed... This rotor was a failure owing to the poles being too narrow. As a result Mr Parsons had to adopt the 'damned German invention'.*

This was written after Parsons' death, but in public Stoney valiantly defended design features of electrical machines that he had employed against his better judgement, even when these came under quite strong attack (Stoney and Law 1908). The Neptune Bank power station of the Newcastle upon Tyne Electric Supply Company was just up river from the Wallsend works and in 1903 it broke new ground as the first example of a statutory authority in the United Kingdom supplying three phase power at 5.5 kV, though in this it followed the path already blazed by Westinghouse in the United States.

Alex Law recalled an occasion when Parsons cut across his designers. It related to the use of cast iron in turbines (Dowson 1935).

> *About the time the 942 machine was being built—it was experimental both on the turbine end and the alternator end—the phenomenon of the growth of cast iron began to make itself known. At first I think it was regarded merely as an increase of volume without anyone realising that this increase in volume would necessarily cause cracking and breakage. Since the end-tightened machine depends very largely on the relative positions of the rotor*

and stator blading, someone pointed out to Sir Charles that if the cast iron case were to grow it would upset the end tightening, and he replied that in that case the rotor should also be made of cast iron. I remember the cast iron rotor in the shops quite well, and I think it was actually built and perhaps grooved, but by that time the bad effects of the growth of cast iron had been realized, and no-one ever seriously contemplated finishing off the rotor or running it. I know it lay about for some time, because I think people rather feared to ask Sir Charles for instructions to break it up.

This would be in 1903 when Law had not long finished his apprenticeship. Clearly communication between Parsons and his workforce was not always as easy as Bedford seems to suggest in his recollections.

Stoney's openness to new ideas was evident in the choice of new materials for the plates that were used to construct the magnetic paths in alternators (Dowson 1942). He told Dowson

About 1903 Sir William Barrett tested a lot of alloys (made by Mr Robert Hadfield) for conductivity and permeability. He read a paper to the I.E.E. about this investigation. I saw it and saw that two, a silicon and an aluminium alloy, had important properties (low hysteresis and low conductivity). I wrote to Barrett a letter which he put into the discussion on the paper, and Hadfield asked me to go to Sheffield, which I did. I did not see him there but one of his men, who said that the stuff could not be made, could not be worked, could not be rolled etc. They would do nothing. I then saw Jenkins of Sankeys, who were our suppliers of armature and transformer plates. They would do nothing. I did not want to go to the Germans but a man from the Bismarkhütte of Berlin called and I told him about it. Three months after they quoted for the stuff, and we bought it from them for a long time. Then Sankeys took it up and made it under the name of Stalloy.

On 27 July 1904 Charles Parsons gave evidence to an inquiry into manufacturing and trade under the heading of Electrical Engineering (Parsons C A 1904b). In this he was able to offer comparisons between his own works and that of a 'German' Swiss manufacturer that made very similar machinery. Drawing no doubt on information gathered by Stoney, he stated that overall the cost of production at Newcastle upon Tyne was 3% less on average. Materials in Germany and Switzerland cost 17% more, but labour cost 10% less. He gave figures for wage rates in pence per hour in the two countries for each activity.

The higher costs of selling on the Continent he put down to the larger geographical spread of their market, and to '*more showy offices*'. Bedford later remarked that Parsons

Switzerland (d h^{-1})	England (d h^{-1})	Activity
8	11.2	Borers
5	10.6	Coppersmiths
4.75	8.6	Toolmakers
4.75	7.2	Blacksmiths
4.5	8.6	Fitters
4.5	10	Bladers
4	4.5	Labourers
5.75	10	Turners
5.85	8.6	Pattern-makers
5.25	8.27	Joiners
5.25	9.1	Planers
4.75	8.8	Winders
4.75	8	Painters
4	5.1	Casting cleaners

shunned all forms of publicity. He would not allow the name of the Firm to be painted on the various gates into the Works... the only indication of what went on behind the wooden boundary fence being a small brass plate at the main entrance to the old office block.

Parsons' evidence went on,

> *man for man the English workman produces about 60% more per hour than the foreign workman when engaged in any kind of large work, but this is not so in the case of small work.* He said that *it is my opinion that the external finish of the Swiss manufacture is better than the English, but the essential parts in the machines made in the two countries is equally good.*

He thought that in Switzerland they work 30% longer hours than in England. The chief raw materials that had to be purchased abroad at that time were weldless steel cylinders needed for marine turbine rotors, and iron plates (laminations) for electrical machines, both of which were supplied from Germany. In Parsons' opinion mechanical engineering had not suffered serious disadvantages from lack of technical training. The staff at Newcastle was smaller but better paid and got through more work. '*We have a certain number of Cambridge and Dublin graduates and young men from technical colleges in the higher positions, but workmen and foremen are chiefly recruited from youths of the artizan class, and attend night classes for a term or two*'.

In December 1904 another agreement was made with Stoney and signed by Parsons (Dowson 1942). It read

Confirming what I have mentioned to you verbally, I beg to say that I propose to open a special account with you in our books, and to credit it with five thousand pounds which, during your service with the firm will bear interest at $4\frac{1}{2}$% per annum. This interest you will be at liberty to draw half yearly or yearly. If you continue in the service of the firm for ten years from this time, the capital will be yours and the interest will cease, and you may draw the capital, or by arrangement with the firm, continue it.

This very generous gesture clearly aimed to secure Stoney's' services in the longer term, and it makes one wonder what factors could have made Parsons anxious. Bedford recalled that Parsons had little awareness of salary levels of his staff, and he drew no salary as such himself (Bedford)

Sir Charles, as I have said, was generous, but it did not occur to him that the remuneration of his staff was perhaps inadequate. I remember Carnegie, one of the ablest engineers, found it difficult to provide for his family on his poor salary. So he drew a graph showing the advances he had received over the years and summoned up enough courage to show this to Sir Charles. Until this moment Sir Charles had never realized the low scale of remuneration, nor the long intervals between advances that his staff were receiving. Carnegie immediately received a substantial advance and many others of the staff also. Nevertheless as an employer Sir Charles was in many ways advanced in his ideas. In the early days he supported and encouraged a Foremen's and Workmen's Benevolent Fund, as well as a Staff Pension Fund. He also provided a scheme under which any employee, no matter what his rank, could deposit with the Company any small savings he might have. He looked upon these contributions as showing that the employee had an interest and faith in the Company. He kept a little receipt book in his desk and when approached by an employee, would write out and sign the receipt himself and hand it to him. The deposit carried an interest at $4\frac{1}{2}$% per annum. If it came to his ears that an employee was suffering from some complaint that appeared to be more than could be attended to by his usual doctor, it was not uncommon for him to put his hand in his pocket and tell the sufferer to consult a specialist. He gave encouragement to his employees and was always ready to discuss any problem or new idea that anyone, be it workman, foreman or member of his Staff, might bring to his notice.

18.2 Other turbo-machinery, fans, compressors and pumps

Although Parsons' attention was taken up during these years with pushing marine applications and such serious matters as the giant turbine installations in the *Mauretania* and *Lusitania*, as well as the battleship *Dreadnought*, this did not mean that other possible applications were neglected. The 1884 patent had stated that his invention could serve equally well as a motor, or as a pump or compressor. Indeed the prototype turbine used both a screw propeller as a pump for the lubricating oil and a centrifugal compressor as part of the governor mechanism. So from the beginning he used fluid flow machinery both as motor and pump.

The first application of turbine powered air moving equipment was a fan installed aboard the steam yacht *Speedy* in 1894. The fan was shaped like a screw propeller, and was used to provide forced draught for the boiler fires. The first land application was in a coal mine which had a need for large volumes of air to be moved against relatively small pressure heads to achieve ventilation. In 1896 the Clara Vale Pit at nearby Wylam was supplied with a four bladed screw propeller fan, driven by a steam turbine at 2000 rpm. The fan, measuring 5 ft from tip to tip, shifted 120 000 cubic ft min^{-1}, but the suction pressure was only $2\frac{1}{2}$ in water gauge. Several improvements were required. The very high velocity of the blade tips of such fans caused them to emit a high pitched whine that could be heard over long distances. The problem was partially solved when a crude honeycomb structure was fitted at the outlet from the fan to absorb this sound. Then in 1898 the original fan blades were replaced by a casting of manganese bronze that had eight blades formed integrally. In 1902 a cast steel disc was substituted, with 40 individual cast blades attached. These blades were subject to conditions which were not unlike the last stages of a steam turbine. They were tapered towards the tip, while their angle of attack was reduced from the hub out to the tips as in screw propeller designs. Finally a cone was placed in front of the disc and a diffuser was fitted after the fan discharge. The latter helped to recover some of the energy in the high velocity flow and improved efficiency. In 1903 somewhat similar but smaller diameter axial flow fans, 'turbo-exhausters', were supplied to extract hot gases from blast furnaces producing pig iron. Three fans were used in series, with flow straighteners between them. Their steady action and freedom from clogging were a big improvement on reciprocating compressors, and of course in this situation noise was not a problem.

At the same time turbine driven radial flow compressors and pumps were built. In 1894, a small Gwynne centrifugal pump, designed to run at 1200 rpm, had been driven by a condensing steam turbine, actually the prototype which had been tested by Professor Ewing in 1891 (C.L.(17), Stoney 1938). A number of different impeller designs were tried at that time, the high speeds of turbine drives being especially well suited to enabling centrifugal pumps to reach high delivery pressures. In order to avoid cavitation at the inlet to the impeller at such high speeds, propeller-like blades were fitted on the shaft at the entrance passage. The casing of the Gwynne pump was eventually burst when a speed of 4800 rpm was reached. The first commercial pumps were marketed in 1897 (Parsons R H). Meanwhile Rateau had pioneered centrifugal air compressors. The first such machine to be sold by Parsons was built in 1909 for the Hazelrigg Colliery near Newcastle. Running at 5000 rpm it delivered 4000 cubic feet of air at a pressure of 20 psi to a reciprocating compressor which raised the pressure to 80 psi. The design was reminiscent of the first design of the Jumbo radial flow machine and had seven stages. The diaphragms that separated successive stages contained passages, but instead of carrying a flow of steam to heat the fluid, they carried water to cool the gas. In this way the work required to pressurize a given mass of gas was reduced. Two similar machines were manufactured in 1909 to compress gas in an Australian gas works.

Parsons came closer to making a compressor along the lines of his original axial flow turbine in a design which he patented, number 3060 of 1901. From figure 116 it can be seen that, like the turbine, the rotating blades of this machine are fixed to the surface of a drum that is carried on bearings at either end. The fixed blades, as in a turbine, are carried on the inside of the casing. Since flow is in one direction only, a dummy or balance piston is required to counter the axial force that is created as the air flows over the blades. The shape and relative orientation of blades in a compressor is different from a turbine, and is shown schematically in figure 117 (Dowson 1939). A compressor functions by speeding up the fluid flow in the moving blades. It is then slowed down in the diverging fixed blade passages, with a resulting rise in pressure. Whereas in a turbine the acceleration of flow is relatively insensitive to the exact shape of the blade passages, this is not so for the deceleration that takes place in a compressor. Because this problem was not fully understood, the efficiency achieved in the early compressors was relatively low. Roughly two-thirds of the power supplied to the compressor was degraded to heat. Despite this, such machines driven by steam turbines had advantages over the competing reciprocating compressors, for their delivery was free from pulsations, their size and

Figure 116. The first commercial axial flow compressor, supplied to Cooksons (1901). Running at 4000 rpm it delivered 3000 ft^3 min^{-1} against a pressure of 1.75 p.s.i. A dummy piston is placed at the right hand end (Dowson 1939).

weight were considerably smaller and their maintenance costs were much reduced. A total of 13 turbo-compressors were delivered in the period up to 1908. Of these one-half were exported. In general they delivered large volumes at relatively small pressure rises, and the largest delivered 50 000 ft^3 min^{-1} of air at a pressure of 10 lb in^{-2}. The main application was for the provision of air for blast furnaces, though such a turbo-compressor was supplied in 1904 for Glasgow Corporation's Gas Works (Parsons R H). Driven by a steam turbine at 8000 rpm, it delivered 13 500 ft^3 of gas at a pressure of 24 in water gauge. Earlier experience with marine propeller designs clearly influenced these developments. Efficiency was gradually improved so that losses were reduced to 40%. However in the meantime, Rateau had been developing centrifugal compressors which had the same attractions as axial flow machines but they were more efficient, with losses of only 25%. At a much later date developments in the understanding of fluid mechanics, mostly achieved by the aeronautical industry, opened the way for highly efficient axial flow compressors, but Parsons never pursued the matter further.

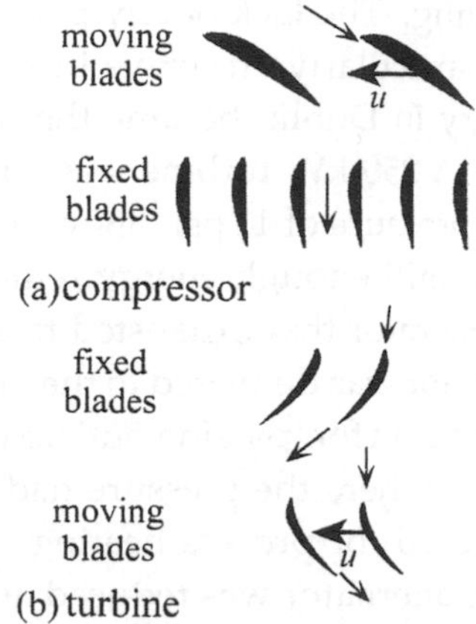

Figure 117. (a) Diagram to illustrate the direction of the flow of fluid in Parsons' first axial flow compressor. (b) The equivalent diagram for steam flow in a Parsons turbine. The velocity of the rotating blades is represented by u. In later machines the fixed blades in compressors were also inclined at an angle to the shaft axis.

At the same time as speed reduction gearing was being developed for marine applications, it was also being used for industrial drives. The first such application was for a rolling mill at the Calderbank Iron and Steel Works. The steam turbine developed 750 HP at 2000 rpm, and drove the three high mill through a double reduction gearing, at 70 rpm. The peak power required by the rolls reached 4 to 5000 HP, but this was supplied from the kinetic energy stored in a large flywheel. Despite the heavy shocks experienced as the iron plates were fed in to the rolls no difficulty was found with the gearing. Gearing opened up other possibilities that vividly demonstrated Parsons' addiction to welcoming a new challenge regardless of the likely profitability of the venture overall. In 1912 an existing plant, an Indian jute mill, was converted to turbine drive by using step down gearing (Parsons R H). Originally powered by two reciprocating engines that distributed their power via rope drives, it was rearranged so that one engine exhausted into the turbine, which exhausted to a condenser. The second engine was removed altogether. The engine and turbine worked quite happily coupled together mechanically, and significant fuel savings were achieved.

Turbines could utilize steam that had been exhausted from a reciprocating engine. Turbines could equally well be used to extract power from a high pressure steam supply before passing it on to some process like heating vats for brewing. The lack of any possibility of oil or grease reaching the steam was particularly attractive in such circumstances. In 1900 the Guinness brewery in Dublin became the first to try this in a so called 'pass-out turbine'. A 250 kW turbine was supplied with steam at 140 psi. The exhaust, at a pressure of 11 psi. above atmospheric, was used to heat the brewing vats. Still enough energy remained in the steam to power a 250 kW turbo-generator that exhausted to a condenser. Later, in 1912, a 1 MW turbo-alternator was delivered to the Edinburgh Ropery and Sailcloth Co. This had provision for 'passing out' up to 5000 lb of steam h^{-1} from a point in the turbine where the pressure had fallen to an absolute pressure of 35 psi, to be used for process heating. Of course when this occurred the output of the alternator was reduced automatically.

18.3 Staff tensions

From 1901 onwards a great deal of Parsons' time had gone into developing his vacuum augmenter system as well, in attempts to perfect his axial flow compressor. The pressure of work was such that he '*relied on Stoney to relieve the burden on his shoulders by conducting the Heaton Works on the managerial as well as on the technical side* (Dowson 1942). At the same time personal rivalries began to emerge between individual staff members. Bedford's recollections shed light on this aspect (Bedford). The company secretary was H G Dakyns and he had a minor set-to with Stoney. After a trip to Switzerland for discussions with Brown Boveri, Stoney submitted his claim for expenses, which included the cost of bringing his wife with him. Dakyns refused to pay her expenses and Stoney appealed to Parsons. He was unsuccessful. Later Stoney submitted an expenses claim which was incorrectly totalled. Dakyns paid up without checking the details, whereupon Stoney reported him to Parsons. In 1910 the unfortunate Dakyns ran into more trouble. After a disagreement with Parsons he was sent to the Far East on business, and Bedford was put in charge of the Secretarial and Accounting offices. When he returned one year later, he found Bedford reluctant to relinquish his newly acquired power,

> *I mentioned to Sir Charles as diplomatically as I could, that as I had acted as Secretary during Dakyns' absence, without as I believed any complaints, I felt it was unfair that I should revert to my former position as Assistant.*

Bedford was made Secretary while Dakyns was moved sideways and soon resigned.

Parsons was a wealthy man. He was the proprietor of the Heaton Works, and had a stake in both the PMSTCo. and the Foreign Patents Co. His style of living was certainly not ostentatious, but one pastime that had given him much pleasure since childhood was fishing and shooting. He rented a grouse moor at Sidwood upstream and north of the Tyne. When an estate at Kirkwhelpington came on the market around 1904, he used some of his personal wealth to purchase it. The residence was called Ray after the nearby Ray Fell. The lands were extensive, covering 4000 acres, and it was largely moorland. Within it lay Sweethope Loch, a stretch of water covering several acres. It was equipped with a boat house, and Parsons delighted in fishing for trout there. A second stretch of water lay 200 or 300 ft higher up than the main lake. It must have brought back memories of his childhood spent on the ornamental lake at Birr Castle. Stoney recalled that *'he had an uncanny power of luring the trout to his fly. I used to row the boat, and he often wanted me to try my hand, but I am no fisherman. Afterwards in the smoke room many and various things were discussed'* (Stoney 1939). When the fishing was spoiled by the growing population of pike, he drained the lake and grew wheat in its place for a year. He then re-flooded it and it was restocked with fish. Katharine Parsons too grew to love the place and set to work on creating a garden, at the same time making the large house comfortable for entertaining guests. She also devoted herself to the running of the farm (Houstoun 1934).

Kirkwhelpington lies some 30 or so miles north-west of Newcastle. There was a halt nearby on the local railway line which provided a connection with Newcastle but no doubt it was the improvements in motor cars that made such a location practical. Quite a lot of entertaining was done at Ray, and in the years before the war Parsons enjoyed the company of his nephews, Geoffry and Arthur, both engineering graduates (Parsons A D C 1939). The latter described shooting parties in which he had participated. On one occasion Parsons took the boat out to finish off what he took to be wounded birds flapping around and unable to leave the water. Later when he asked the game keeper for one of them to demonstrate a point he was making about wings and flight, it was pointed out to him that the birds were immature 'flappers' that had never ever flown. The response of the frequently irascible employer was to sit down and laugh at his own foolishness. He was oblivious to rain and damp. After another shoot the guests returned to Ray for tea having been soaked walking through the wet heather. They went off to change. Parsons didn't bother, until he noticed that only his trousers were still wet. He had the idea of undoing the buckle at his knee and pulling down the trousers over his wet socks to his ankle.

He was quite oblivious of the comic appearance of trouser legs with a foot or so of dry cloth sandwiched between two soaking wet patches. Once more when this was pointed out to him he just laughed.

Arthur Parsons recounted another episode. He remarked to his uncle that Lady X was to visit that evening. Nothing had been said to Sir Charles about this and as he didn't know any such person, the reference books were scoured. Arthur was convinced that she was a personable young woman that he had met. In the event she turned out to be plump and elderly. After dinner the mystery of her identity was still unsolved, until it was realized that she must be the Dowager Lady X. Parsons was piqued and after dinner he would not join the guests. As an excuse he went off to the electric generators, and dimmed the lights a number of times then returned to the smoking room feeling pleased with himself. Arthur explained to the guests that his uncle had to attend to a fault on the engine. However he was quickly located by his wife, whereupon he quite happily rejoined the party. Having worked off his annoyance he played the part of a delightful host for the remainder of the evening.

Charles Parsons was largely indifferent to the food he ate. On one occasion he was staying at the house of a friend in Northumberland and arrived down late for breakfast. Finding the dishes empty he proceeded cheerfully to eat the scraps that had been collected on a plate for the dogs, apparently enjoying them so much that he returned for more (Bedford 1939). He had his own sense of humour. He reckoned that the price of certain brands of wine appeared to be in direct ratio to the length of their description and to make his point he wrote to his wine buff friend (Bruce 1939)

> *My dear—,*
>
> *I have obtained a bottle of Halltgartner-Hendelberg-Riesling-Auslese-Prince-Lowenstein-Wertheim-Rosenberg's Estate and should be delighted if you can come and taste it at lunch on Friday at about 1 p.m.*
>
> *Hoping you can come,*
>
> *Yours very sincerely,*

Honours continued to come his way. In 1908 he was made vice-President of the Royal Society and in March 1910 he was appointed Sheriff of the County of Northumberland, to '*take custody and Charge of the said County during His Majesty's pleasure*'. In the following year he was notified that he was to receive a knighthood, Knight Commander of the Order of the Bath, KCB, on the occasion of the coronation of George V. He was awarded the Albert Medal of the Royal Society of Arts, and delivered papers to the Institute of Marine Engineers, the Institution of Naval Architects and to the Royal Institution in quick succession. He was even proposed as a candidate for a parliamentary by-election at Cambridge, but he subsequently withdrew his name. His two manufacturing plants were each pouring out a stream of revolutionary machines. Through his patent company he dealt with a great variety of licensees all over the world. Research and development went on apace, but now increasingly others attended to details. S S Cook, who had worked for a time as Stoney's assistant, was dealing with blading design and with marine turbines at PMSTCo., while A Q Carnegie was dedicated to the development and application of gearing. Stoney struggled with the design of ever larger generators as well as turbines. In a letter written on Company headed notepaper, he returned to an idea which he had mooted a decade earlier (Stoney 1909b),

C.A.Parsons & Co.
Heaton Works
Newcastle upon Tyne
22nd May 1909

Dear Mr Parsons,

In further reference to my letter of the 18th instant I have been talking the matter of turbines over with Mr Cook and Mr Carnegie, and it seems to us that provided there is no patent in the combination of a Curtis high pressure and a Parsons' low pressure, a test could be easily made by modifying the partial flow 500 kW turbine which we have built, and it would be well worth doing, as, judging from the way things are going both here and on the Continent, there is a strong demand for a turbine with some form of partial flow in the high pressure and a Parsons' low pressure; and since our partial flow does not seem to be giving as good results as were expected, it looks as if a Curtis high pressure might be the solution to the problem... Experiments with the partial flow could be continued on the marine turbine...

Yours very truly,

Gerald Stoney

It is clear from this letter that Parsons was now somewhat distant from the day to day business of turbine design and manufacture. It would appear that Parsons approved Stoney's suggestion at this time because the printed version of the Rede Lecture which he gave at Cambridge in 1911 contains

Figure 118. (a) Experimental 'disc and drum' turbine rated at 500 kW, 3000 rpm, on the test bed in 1909. At the right hand can be seen the large diameter disc for the Curtis stage which feeds the succeeding reaction stages. Problems were encountered in achieving a satisfactory arrangement for the dummy piston (Parsons R H).

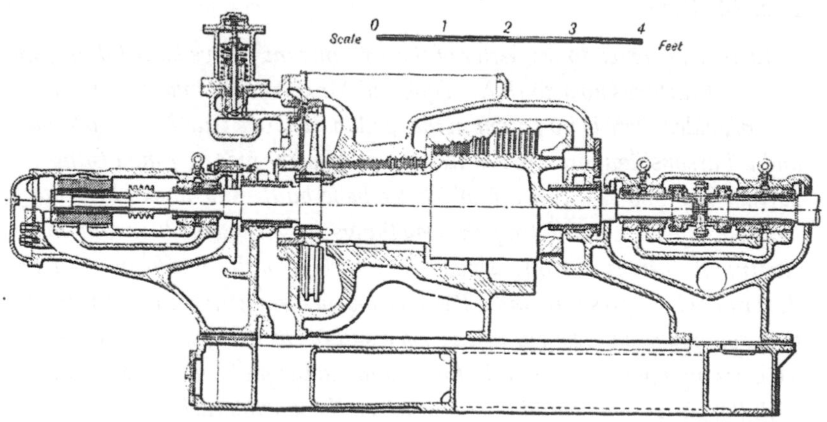

Figure 118. (b) Sectioned drawing of the turbine viewed from the opposite side to (a). Steam at 180 psig and 580 °F was supplied to the Curtis stage with two rows. This was followed by 30 reaction stages in seven 'expansions' (Parsons C A 1911a).

a drawing of such a machine (figure 118). However the Company history of Brown Boveri states (Evans 1966)

The constructional development by Brown Boveri of Parsons' turbine, especially the incorporation of an impulse wheel, led to the termination of the agreement in 1911...

Certainly in later years Parsons, for whatever reason, continued to block the adoption of this approach for turbo-generators even if commercial pressures forced him to use it for marine applications.

A typical assignment for Stoney was the testing of a new, compact single collar, thrust bearing that had been patented by the Australian engineer Michell (Stoney 1913). Parsons' approach up to this time had been to absorb any thrust in an axial direction by cutting many shoulders on the shaft, and these were meshed with matching collars fixed to the housing as can be seen in figure 68 or figure 103. In such a design it is difficult to get lubricant to the points where it is most needed, and it is also almost impossible to ensure that each collar takes its correct share of the load. The designs were bulky and absorbed much energy overcoming friction. In 1905 Michell had applied Osborne Reynolds' mathematical analysis of the cylindrical, journal bearing to the case of two flat surfaces separated by an oil film, one of which was able to tilt as it moved relative to the other. In Michell's design, instead of the thrust of the shoulder on the shaft being supported by a collar, it was taken by a number of roughly kidney shaped pads. These pads were supported at one point only and could tilt. Michell had shown that, under load, the pads would tilt and that oil would be driven with great force between the pad and shoulder, dragged in by the rotating shoulder. He had found it impossible to interest manufacturers in his invention until the advent of geared drives eventually created a demand. Geared drives break the link between turbine and propeller which makes it impossible any longer to balance some of the thrust on the propeller shaft with the thrust in the turbine. Stoney's tests verified the claim that this design could support enormous pressures without any need for a pump to pressurize the oil. His report to Parsons dated 12 May 1912 stated that (Dowson 1942)

bearing metal thrust blocks carried 500 lbs. to 1,100 lbs. per square inch before they seized. White metal thrust blocks were also tried and one carried 1,900 lbs. per square inch before it began to smoke, and on easing to 1,400 lbs. per square inch ran all right, in fact the blocks were hardly marked. Another one carried 2,100 lbs. per square inch with no signs of seizing or wear. In my opinion therefore it is safe to calculate on about 500 lbs. per square inch for white metal thrust blocks under such conditions.

The high pressures that could be used made the bearing very compact. Moreover the total waste of energy by friction was drastically reduced so that a saving of up to 2% was possible in the overall efficiency of a turbine drive. This work on tilting pad bearings enabled Parsons to take out his own patent 8266 in 1912. Michell had calculated the optimum position at which to support the pad, and this was specified to be some distance behind the geometric centre. Parsons verified that if the support were located at the centre, the pad worked quite as well, and he could also claim that it functioned identically no matter in which direction the surfaces moved. No doubt it was Stoney's tests that suggested such an idea to him. He had no compunction in patenting his variant, so effectively purloining Michell's idea. Certainly it was financially valuable to him to have the protection of such a patent.

Robert Dowson, who had joined the company in 1911, had come to know Stoney well. According to Dowson, during the period 1904 to 1907 especially, pressure of work caused Parsons *'to rely on Stoney to relieve the burden on his shoulders by conducting Heaton Works on the managerial as well as the technical side'*. In 1910 Stoney's contributions were recognized by his appointment as technical manager for the whole Heaton Works, in effect as second in command to Parsons himself. At 47 years of age he had risen to the highest position. The following year he was elected a Fellow of the Royal Society. And yet difficulties developed, and on 30 June 1912, *'in a moment of extreme vexation'*, he left the company. Given the very close personal and technical relationship that had developed between Parsons and Stoney over a period of 24 years, this must have been a devastating blow to both men. Both had the reputation for a fiery Irish temper, but the cause of the break lay deeper. Dowson gave his view of the matter (Dowson 1942).

> *This most regrettable turn of events may be attributed to the following causes. Ever since the foundation of Heaton Works in 1889 Parsons looked upon his works as an experimental workshop. When the time came (from 1895 onwards) that licences to manufacture steam turbines were granted to other companies in England and abroad, he regarded his works as a place where experiments could be carried on continuously so that improvements could be made and passed on to the licensees. As regards orders for Heaton Works, his only desire seemed to be to secure enough work to keep his workshops busy and sufficient profit to finance the development and experimental work. He never tried to create a monopoly or corner the market.*
>
> *On the other hand it became clear to the management that if Heaton Works*

Figure 119. Painting of George Gerald Stoney by T B Garvie, courtesy of Mr G A Browbank.

were to continue, it was essential that they should no longer be regarded as an experimental shop, but should be organized on business lines, free from interference with the standardised turbine plant which had to be available for tenders for new work. J.H.Barker, at one time general manager at Heaton Works, wrote apropos of Parsons' wilfulness: 'had he had his way he would never have made two machines alike, each succeeding one would have been improved.'

Parsons' way of settling these controversies was simply to challenge: 'who is paymaster here?' What was required, however was the reorganisation of the staff and the defining of departmental provinces, because besides Stoney there were other energetic and able men in the works, who felt themselves thwarted and were not content merely to do what they were told. Stoney's organising power proved to be not so great as his ability as an engineer, but had he held on for a few more years all might have been well, for ultimately the matter settled itself because its urgency brooked no further delay. Unfortunately his temperament did not allow him to compose these differences on administrative matters or to carry on in spite of them.

The break was not total. Stoney continued to receive a retainer from Parsons, and took up work as a consultant, advising the engine builders Richardsons Westgarth on turbine matters. Their works was located at Hartlepool, a short distance down the coast. In 1903 they had made an agreement with Brown–Boveri to build steam turbines and were currently licensees of Heaton. This suited Stoney whose loyalty to his former employer made him reluctant to work for a competitor.

18.4 Factory management

Parsons' only experience of works management had been at Armstrongs and then at Clarke Chapman. Stoney had never worked anywhere but with Parsons. So it was fortuitous that someone should now join the firm who had experience of one of the large American companies. These had been evolving new approaches to managing the technologically advanced products of the 'second industrial revolution'. Arthur, A D C, Parsons (1881–1955) was the second son of Clere, and in 1904, after graduating from Trinity College Cambridge, he had gone to Pittsburg to serve an apprenticeship with Westinghouse. After a spell with British Westinghouse he came to Heaton in 1913. That he had profited from his time with Westinghouse is evident from the paper that he read in 1916 to the North East Coast Institution on Works Organization (Parsons A D C 1916). He makes a number of points all of which would win modern approval. He emphasized the need for

quickness of manufacture which insures getting the maximum production out of the plant, thus lowering the overhead charges and cheapening the production.

He advocated the use of graphs, for example to represent the '*fullness of the factory*'. He showed a printed form that specified dates when drawings, materials etc were due, and also recorded when they arrived. It was filled out for the case of a 3000 kW turbo-alternator. He emphasized that '*some method must be adopted whereby all material can be kept trace of in a systematic way'*. This was especially true of small sub-components, absence of which could hold up a job completely. He recalled that when he first arrived at Heaton, if a client's inspector decided to make a visit it was necessary to send out *'the test house gang'* to assemble all the components for the client's order in one place so that the state of progress could be verified (Parsons N C 1990). He had ideas for card indexes, and for graphs showing shop loading. He thought that by circulating these charts through the works it would rouse interest in other departments and set up a certain amount of friendly rivalry between them. But he was also aware that

> *care must be taken to guard against the tendency of building up a system of forms only... and of the whole system developing into so much waste paper, and the work being completed before the system designed to assist... can come into use.*

No doubt it was to these fresh ideas that Dowson was pointing when he said that Stoney's problems were soon to be overcome. It is clear that the friction between staff members was not simply the result of quirks of personality, but reflected real organizational difficulties.

Parsons, unlike many of his contemporaries such as George Westinghouse or Brown and Boveri, had been able to finance the Heaton Works as a private company (Bedford).

> *When he required money for extensions or for other purposes he would go to his Solicitor, Mr. T.W. Thompson, who acted for many of the wealthy families of Northumberland, and there appeared to be little difficulty in obtaining such funds as he required by way of loans. Any profits that might be made, (and there were not always profits) were ploughed back into the business. Sir Charles took no salary from the Company.*

Such a policy obviously gave the enterprise one great advantage which was denied to companies that were owned by shareholders, it did not have to fear sudden switches of commercial priority. On 13 June 1913, *'for personal and family reasons'*, Parsons revisited his plan of December 1897 and formally established the business as a Private Limited Company. Shares were to be valued at £100 each. He held 3000 shares himself and gave 110 shares to his wife Lady Parsons and 14 to his son. Senior staff, Johnston and Law were given 15 each and Howe got 12, while Prest, Martin and Carnegie got 10 each and Bedford received four shares. The names of 13 others who held small numbers of shares are recorded in the dividend book (TWASb). The directors were

Sir Charles Parsons (Chairman)
W M Johnston }
A H Law } (Joint Managing Directors)
A G Parsons
C P Martin (London Manager)
S F Prest (Licensee Manager)
F G H Bedford (Secretary).

After the first year's operation a dividend of 5% less tax was paid, while half of Sir Charles' own portion was divided among the other shareholders. At first, board meetings were held in Newcastle at the offices of the company's solicitors, Dees and Thompson. Later they alternated between the works and the London office at 56 Victoria St.

When the business was transferred to the Private Limited Company its capital value was stated to be £421 207 (TWASc). The factory workforce in July 1914 totalled 1077 and comprised 15 women, 202 apprentices and 860 men. In the offices there were a further 181 persons, six women, 18 apprentices and 157 men. Out of 1258 employees 86% were shopfloor workers (TWASd). Compare this with the American General Electric Company newly established in 1892, that employed almost 10 000 workers and was much more heavily capitalized at $50 million (Carlson 1991). One of the first matters to occupy the management of the new Company was a decision to purchase the Close Works across the river in Gateshead, from Scott and Mountain who were in liquidation. One attraction was that it contained a relatively large foundry. By January 1913 a machine shop was up and running. A small cupola furnace from Heaton was installed and within 7 weeks a total of 145 tons of iron had been cast. There were 80 on the pay roll including a small office staff. Algernon G Parsons, Charles' son, was put in charge. Whatever the case about controlling costs at Heaton, the Close Works was to prove more difficult. At this time a designs committee was established at Heaton which would overview all enquiries. It comprised Sir Charles, A H Law, R Howe, A Q Carnegie and W M Johnston.

18.5 Outbreak of war

Parsons had not long entered his 60th year when war broke out between the European powers. As Lord Fisher had envisaged, Germany with the Austro-Hungarian empire was the enemy, and France and Britain were joined together as Allies (Kemp 1960). International links with manufacturers who were now in enemy territory ceased abruptly. In the first months of the war there was apprehension along the North East Coast where industry was very vulnerable to an attack from the sea, until eventually the German Fleet was neutralized (Stoney 1914). In Newcastle the first real impact was felt as men were called to the colours. Charles Parsons' son, Algernon, was one of the first to go. He was recalled for active service as a lieutenant in the Royal Horse and Royal Field Artillery with the British Expeditionary Force in France on 13 September 1914 (Anon 1918). Alex Law, who was in his mid-30s when war broke out, was a member of the Territorial Army and so was called for service in France. Bedford, who was a year or two younger than Law, came from a family with strong ties to the services, and he volunteered for the army. Having signed on, he then informed Sir Charles Parsons. Bedford recalled (Bedford)

> *He was quite indignant with me, but went so far as to say that, with all my naval background, he understood my feelings. He said in a rather pained voice, 'do you want to sit in the trenches and be killed?'. He was determined that I should not go, and having obtained my exemption, he put me in charge of the management of the Mirror Department in order to ease my conscience, there being an urgent need of searchlights and other reflectors for the Services... The only other source of supply was from a small department at Clarke Chapman's, which had been kept turning over since the days when Sir Charles was a partner.*

Quite soon after the outbreak of the war, the demand for ammunition became acute. A Ministry of Munitions had been established to mobilize industrial output. In August 1915 the mayor of Newcastle, who was the President of a North East Coast Armaments Committee, was asked by the ministry to appoint a small board of management, preferably under Sir Charles Parsons, to represent the Tyne and Wear (PAR 31). Membership of the Board was left to the local Engineer's Employers Association. The ministry drew up a form of agreement that would limit the profits that could be made. In September a contract for 15 000 high explosive 4.5 howitzer shells was issued. Parsons had decided to make use of the machines at the Close Works for producing artillery shells, but this was not a success as output never exceeded two or three hundred per week. Eventually, when an offer was made by H I Brackenbury of Armstrong Whitworth and Co. to purchase the works for the same price as had been paid for it originally, it was accepted.

The exodus of senior staff was paralleled by the departure of craftsmen and labourers from the factories. The army was still a volunteer force, and in the early years huge efforts were made to obtain recruits. Charles Parsons was one of '*the chief promoters in the movement for raising a Tyneside Irish Brigade*... (of the Northumberland Fusiliers)... *as honorary colonel he frequently paraded with it*' (Anon 1931a). Gerald Stoney was too old for active service, but he gave his services as joint secretary of the Tyneside Irish Battalions, of which four were raised locally (Dowson 1942). This was not surprising since Newcastle was one of those English cities that had experienced the greatest immigration of Irish families in the 19th century.

An agreement was made between the engineering employers and the trades unions which allowed shop floor vacancies to be filled by recruiting women to take the place of men. There was considerable resistance to such a sharp break with tradition. The trade unions secured a promise from the Government that after the war there would be a return to the *status quo ante*. When Algernon Parsons was called up, his sister Rachel was appointed as a director of the Company in his place, with a responsibility for supervising the growing female workforce. This does not appear to have been altogether successful and she left to work in the Training Department of the Ministry of Munitions. Later she became involved with a separate engineering company, Atalanta Engineering Company Ltd, that was set up by Lady Parsons and which was run by women (Anon 1934).

18.6 Harnessing science for the conduct of war

Parsons' most significant personal contribution to the war effort had been made during the years before the war, when he succeeded in equipping the ships of the Royal Navy with a revolutionary source of power. Unfortunately the Royal Navy was not the only beneficiary. In 1916 every significant ship on both sides at the battle of Jutland was equipped with steam turbines (Nicholas 1995). On the British side Brown–Curtis and Parsons turbines were present in roughly equal numbers. Of course the factories at Heaton and Wallsend were kept busy meeting the wartime demand for turbines both for marine use and for the generation of the electric power that was needed by industry. Heaton was also the chief national supplier of searchlight mirrors.

But a considerable added burden was placed on Parsons' shoulders in his capacity as scientist and engineer. In the spring of 1915 the war was not going well for the Allies and concern was voiced in the *Times* letters columns. On 11 June the famous writer H G Wells wrote,

> *modern war is essentially a struggle of gear and invention... Since this War*

> *began the German methods of fighting have changed again and again, and each novelty has more or less saved the lives of their men and unexpectedly destroyed ours.*

On 16 June Professor J A Fleming of University College also wrote

> *It is not enough to make vague suggestions as to the detection of submarines or the destruction of Zeppelins. Rough ideas have to be hacked into shape, reduced to practice and tested on a large scale. All this means organisation, expenditure, assistants and definite practical experiments. It seems to demand a special government department which will enlist in its service trained and experienced inventors for definite experiments. This war will be won in the laboratories and workshops almost as much as in the field.*

Fleming wrote on 19 June to Sir W Crookes asking

> *how can the powers of scientific research and invention within the British Empire be best organized, not only for victory in war but for consolidation and maintenance of* (her) *position when peace returns*... (Fleming 1915).

On 13 May the education minister J A Pease had announced that the government were to set up an Advisory Council for Industrial Research (Parsons C A 1915a). When Parsons read of this he was concerned that no mention had been made of a role for the Royal Society and he immediately asked for an interview with Pease. He also wrote to several Fellows of the Royal Society to seek their support for '*a proposed scheme in respect of scientific research and the application of science in industry*' (Parsons C A 1915b). This document recalled the role that the Society had played in the past, and the fact that it was as a result of its initiative that the National Physical Laboratory had been established in 1899, and that the Royal Society still had ultimate control of it. Comparative figures were given for annual state grants in different countries; for the National Physical Laboratory it was £7000, for the Bureau of Standards in America £140 000 and for the Reichsanstalt und Material-prüfungsamt in Germany, £70 000. The priorities were seen to be the need (a) to bring about co-ordination in the realms of science, (b) to direct scientific research and (c) (to ensure) the proper appreciation of science and its application to industry. To achieve this the Society ought to be given an annual grant of £100 000 to be dispersed at its sole discretion. Interestingly there is no overt reference to military requirements.

But on 10 June Arthur Henderson at the Board of Education wrote to Sir William Ramsay to say he could not support the scheme drafted by Sir

Charles Parsons (Henderson 1915). He also pointed out that on 1 March the Royal Society had actually sent a memorial to the Prime Minister which had stated that, *inter alia,*

> *it is outside the province and beyond the resources of the Society as a purely scientific organisation to undertake the administration and control of any scheme formulated with the immediate object of developing trade and manufacture.*

This initiative is reminiscent of the efforts made by William Parsons to secure extra funding for Babbage, and of his attempts to get the Admiralty to carry out tests on armour plate for ironclads, and it proved to be just as ineffectual. Nonetheless he pinpointed accurately a real lack in British industry's relations with the State. Not surprisingly perhaps Parsons soon found himself involved in a variety of Committees (PAR4/1).

Chairman, Board of Trade Committee on the Electrical Trades after the War	1915
Advisory Committee for Science and Industrial Research under Privy Council	1915–20
Advisory Council of War Office on artificial limbs for soldiers	1915–17
Committee to report on the efficiency of the Royal Air Force Factory	1915
Board of Trade committee to ensure adequate and economic supply of electric power for all classes of consumers in the United Kingdom	1916
Committee to report on the efficiency of the Royal Flying Corps	1916
Lord Balfour of Burleigh's committee on commercial and industrial policy	1916
Central Committee on Tanks	1916–17
Advisory Council for Scientific and Industrial Research	1916–21
Fuel Research Board under the Privy Council	1917
Central Board of Invention and Research (Lord Fisher)	1915–18
Chairman of Admiralty committee on erosion and corrosion of screw propellers	1916
Provisional Advisory Council—Board of Trade	1920–22

Parsons wrote to Bedford to tell him of his appointment to the one dealing with artificial limbs (Bedford 1939),

9 Hans Mansions
S W
10 July 1916

Telephone
2475 Kensington

My dear Bedford,

Many thanks for your letter which I was very pleased to receive. The committees have been very tiring. The Billing one we expect finishes this week. The Electrical Trades one will go on for months but may be reconstructed or added to because the politicians don't know their minds and have been quarrelling. The Education one will trudge on slowly indefinitely unless they have ? which is unlikely.

I was all day on Saturday at Roehampton (a convalescent home). *Dugald Clerk and I have been put on as mechanical experts by Moulton ostensibly to decide on the best limbs for soldiers amputations, but as far as I can yet see our job is to hunt out jobbery and commissions*

I hope to get North on Friday or Saturday

Charles A Parsons

His interest in aviation led to his appointment to the committees involving the Royal Flying Corps and the Royal Aircraft Factory. In France his son, who was promoted captain in 1915, was appointed Assistant Equipment Officer with the Royal Flying Corps. In this capacity he contributed to several mechanical developments before being transferred to the Royal Artillery once more in November 1916. The Advisory Council for Scientific and Industrial Research became the Department of Scientific and Industrial Research in 1916. When Professor B Hopkinson resigned Parsons took his place, and it was said that (Heath 1931)

His acceptance of the office was a proof of his willingness to subordinate his personal views to the common good, for it was no secret that he thought the national organisation of scientific research would have been better placed under the aegis of the Royal Society than under the State. Yet he was as regular and active a member of the Council at its long and anxious deliberations as any of that brilliant group of seven distinguished men of science.

Parsons was essentially a shy man, and his interventions in discussions at the Council were generally brief, and often by way of critical question than dogmatic statement... The longest speech I ever heard him make was a remarkable analysis of the conditions necessary for the successful application of new inventions or processes in industry... He emphasised the comparative cheapness in time and money of small-scale laboratory research... As a great industrialist himself, he would have liked to see, though he recognized its inherent difficulty, direct assistance given to selected firms in working new ideas on a manufacturing scale. Failing that, he was a whole-hearted supporter of the policy of establishing co-operative research associations in the industries...

Lord Fisher of course knew Parsons well from the days of the Committee on Designs, and in 1912 he had called on Parsons to give evidence to the Royal Commission on Fuel and Engines. Since the time when his brother Lord Rosse had gazed at the 'flare' burning day and night in Pennsylvania, more sources of oil had been found, transportation had been improved and storage facilities established, so that oil was becoming a practical alternative to coal as a fuel. Fisher was enthusiastic about the benefits that oil could confer on the navy as an alternative to coal. It required no laborious manhandling, and diesel engines were at that time more fuel efficient than steam plant. In the hands of continental firms like Sulzer Brothers, diesel engines were already starting to challenge steam for certain applications. Introducing Parsons to members of the Commission Fisher's words displayed a warmth of feeling,

if anybody has reason to be grateful to you, I have personally, in regard to the past history of the Admiralty in reference to the turbine and the Dreadnought... I well remember... Sir Charles coming and sitting on the sofa in my room, as Controller of the Navy, when we developed this new design to him... It looks a sort of ungrateful thing... to ask him to commit Hari-kari and ask him to advise us upon the introduction of the oil engine.

In response Parsons detailed the present difficulties in building very large diesel engines and added '*I do not think the internal combustion turbine will ever come in. The internal combustion turbine is an impossibility*'. This prophecy was uncharacteristic, but his experiments with rocket torpedoes, and some tests made since then, had convinced him that suitable materials were not available. Essentially his judgement was sound in so far as it extended to the forseeable future. Several other pioneers including Fullagar, Franz Stolze in Berlin, A Elling in Norway and R Armengaud in France had all constructed gas turbines in the period 1900 to 1905, but none was successful (Friedrich 1986). Indeed not until 50 years after this did gas turbines secure a significant place in marine propulsion. Parsons' dedication to the development of the steam turbine certainly did not blind him to the advantages of alternative power sources. In his presidential address to the Institute of Marine Engineers he had said (Parsons C A 1905b).

> *If in the future some form of gas producer using ordinary coal is successfully applied for use on board ship (which seems probable) and if the mechanical details for reversing can be satisfactorily arranged (as also seems probable) then the steam engine or steam turbine will have to give place to the gas engine. But changes of this character take considerable time to accomplish*

Lord Fisher was returned to the Admiralty in 1914 as first sea Lord at the invitation of Winston Churchill, but he resigned the following year having disagreed with the latter over the Dardanelles campaign. In June 1915 he was invited to chair a Board of Inventions and Research, the BIR (Hackmann 1984). It was to be independent of navy control and its other members were Sir J J Thomson, Sir Charles A Parsons and Dr George Beilby. In addition there was a consulting panel of scientists and engineers that included Sir W Crookes, Sir Oliver Lodge, Sir Ernest Rutherford, Professor W H Bragg and R J Strutt (later 4th Baron Rayleigh). Their work was shared by six sub-committees. Professor J J Thomson, writing after Parsons' death, recalled of that time (Thomson 1931)

> *He was a very agreeable as well as a most efficient colleague... I have rarely been on a Board where there was less friction and where proceedings were more harmonious. He had the engineering instinct more fully developed than anyone I ever met... Besides being by far the greatest and most original engineer this country has had since the time of Watt, he was one of the kindest and most steadfast friends.*

Some idea of Parsons' contributions can be gained from occasional recorded details. He was for instance involved in the development of a novel kind of creeping mine that would be programmed to move with the tide (Dumas 1915, Edwards 1915). He was involved in anti-submarine measures, and provided components for early Asdic experiments. He was also one of a party of scientists and navy personnel that witnessed an attempt to use trained sea lions to detect a submarine in the Solent (Hackmann 1984). However his most fruitful work was done as chairman of a sub-committee set up in 1915 to investigate the causes of corrosion or erosion of propellers (Parsons C A and Cook 1919). The other members were S W Barnaby, Professor H C H Carpenter, S S Cook, J H Gibson, and Engineering Vice-Admiral Sir George Goodwin.

Corrosion of propellers was not a new problem, but it had become very much worse ever since the advent of high speed vessels like the new fast cruisers and destroyers and the giant trans-Atlantic liners. It had been found that the *Cobra*'s propellers very quickly showed deep erosion, and their metal bosses became dented. Within only three months the giant propellers of the *Mauretania* were so damaged that they had to be scrapped. Their surfaces were pitted over an area of 3 to 4 square feet to a depth of $2\frac{1}{2}$ in in places. The surface of the *Lusitania*'s propellers was honeycombed like a sponge in patches. Some who had examined the problem attributed the phenomenon to chemical corrosion, others to mechanical erosion. Initially opinion among committee members was divided. Professor Carpenter of the Imperial College of Science and Technology at South Kensington carried out a careful metallurgical examination of the damaged blade from a fast cruiser. He could find no evidence of inadequacy in the material. He next tried immersing specimens of the metal in salt water while they were subjected to different mechanical stresses. Some were stressed below the elastic limit, others stressed beyond the yield point, more were subjected to repeated loading at a frequency of 1 Hz. Jets of water under a pressure of 1,500 psi were directed at the surface of the metal at different angles. The results for all these tests were negative.

From the beginning Parsons had a feeling that the explanation was to be found in the phenomenon of cavitation which he had studied personally for many years. At the Oxford meeting of the British Association in 1894, Osborne Reynolds had given a neat display of cavitation in water flowing in an apparatus shaped from glass (Allen 1970). A narrow neck was formed at one point in a length of tube. When the velocity of flow at this point reached 54 ft s^{-1} 'a distinct sharp hiss' was heard, which he attributed to the collapse of vapour bubbles that had formed at the neck. In the Turbinia Works at Wallsend the experiment was repeated on a much larger scale. Centrifugal pumps were used to force water through nozzles shaped with a converging–diverging cross section. These were assembled from two suitably shaped opposing walls of metal, that had been sandwiched between two sheets of plate glass. The water had a velocity of 90 ft s^{-1} at the throat, sufficient, according to Bernouilli's law, to cause cavitation to set in. Along the axis of this nozzle rectangular rods of different metals were placed to examine the effect of this cavitation on their surfaces. It was found that the etching produced was not the same as on propeller blades. It was little affected by stressing of the metal or by dissolving oxygen or carbon dioxide in the water. One last device was tried, a water siren. This, like an air siren, had two discs that carried a number of holes drilled on a circle of 2 in diameter. One disc was fixed and the other was rotated at 1500 rpm, while water at pressure was forced through them. The abrupt starting and stopping of the water flow in the device caused cavities to form. Discs of different materials showed different rates of erosion, from very rapid in the case of brass to imperceptible in the case of specially hardened Novo steel. Whether erosion was caused or not depended to some extent on the shape of the duct carrying away the water from the device. Since the cavities, once created could persist for a time, their collapse, and any resulting damage, could occur somewhere other than at the point at which they had been generated.

At this juncture Parsons produced an apparatus that was a beautiful example of his talent for hitting on a key demonstration of the physical process that concerned him, and that would give an unequivocal answer, and yet was quite simple (figure 120). It consisted of a brass cone A, 18 in long and 2.5 in internal diameter at its base. At D there were apertures, and at B an end piece that continued the conical surface up to a hole that was 0.015 in in diameter. A long handle allowed the device to be immersed in a tank of water. When it had filled up, it was plunged rapidly downwards until it was stopped suddenly by striking a rubber mat. The momentum of the water, as it flowed out at D, created a cavity at the tip of the cone, which gave an audible metallic hammering sound when the cavity collapsed. To estimate the pressure associated with this collapse, the tip of the cone was closed off by a metal plate. Commercial brass 0.0035 in thick was punctured in one operation. This corresponded to a pressure of 15 tons in^{-2}, or over 2200 atmospheres. Thicker plates were dented by repeated actions, but to prevent accumulations of air from spoiling the effect, the nut C had to be slackened so that air could be vented through the holes H beforehand. The momentum effect was greatly increased by extending the cone with a cylinder. This contained a close fitting heavy lead plunger suspended from a coiled spring. With this arrangement a brass plate 0.03 inches thick could be punctured in one operation. Parsons suspected that cavitation was the cause of the erosion experienced in service. Now he had convincingly demonstrated a mechanism that could account for the connection between the two. One speaker after the reading of Parsons' and Cook's paper said

> *I well recall the experience I had on a full-power trial of a fast cruiser. I accompanied the hull overseer on a tour of inspection… of the fastenings of the* (propeller) *shaft brackets. The noise down there was deafening. I can only liken it to the sound of a million pneumatic riveters at work outside… it is now evident to me that the noise was due to the water hammers tracing their repoussé work on the surface of the propeller blades.*

Stanley Cook made a mathematical analysis of the pressures that can be generated if an essentially empty sphere within a fluid suddenly collapses onto a hard spherical surface. He found that the end pressure is

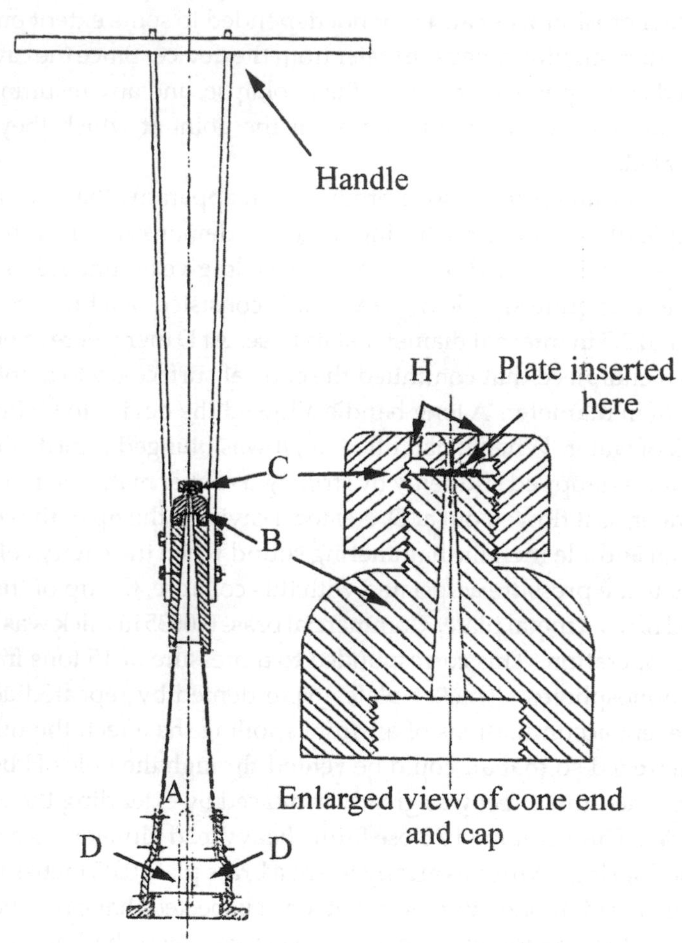

Figure 120. Apparatus for demonstrating the mechanical damage which can be caused by cavitation (see text for details) (Parsons C A and Cook 1919).

independent of the initial cavity size, and depends only on the ratio of the initial to the final radii, R_i/R_f. If R_i/R_f is 10:1, the estimated pressure reaches 24 tons in^{-2}, but if it is 100:1, then it can reach 765 tons in^{-2}. From the earlier high pressure tests that he had carried out, he had values for the compressibility of water at ambient and at very high pressures that showed that actual pressures should not depart dramatically from these estimates. This analysis was verified by J W Strutt, Lord Rayleigh. Some years later Parsons obtained a patent for a method of obtaining high pressures by causing a sphere to collapse under the action of an explosive charge, chapter 20.

Co-operation between the BIR and the navy's own technical establishments was poor. Partly this was due to the low esteem in which naval engineers and lay scientists were held by naval officers; partly it was due to personality clashes of which Lord Fisher was a fertile source as the following memorandum illustrates.

> *... we are doomed to exasperation and failure by not being able to overcome the pigheadedness of Departmental Idiots. We had to deal with three First Lords, all cordial and appreciative, but they were all equally powerless because none of them would kick anyone out... Never has the Admiralty Executive wholeheartedly supported the scientific and thoroughly practical proposals of BIR research.*

In 1918 the scientific work of the BIR was transferred to a Department of Experiments and Research at the Admiralty under C H Merz as director, leaving only the board and its consulting panel in existence.

Chapter 19

Aftermath of War

19.1 The war ends

On 30 November 1917, shortly after being promoted Major, Algernon was wounded in action near Demicourt, close to Cambrai. Alex Law was also wounded in his eye. On 18 December 1917 Law was transferred to the Territorial Reserve and posted to serve the Ministry of Munitions (TWASe). He was then brought back to Heaton as Managing Director when both W M Johnston and R Howe retired (Bedford). In 1918 Charles' nephew, William Edward, who had succeeded as fifth Earl of Rosse, was also invalided home suffering from wounds. In March of that year Algernon, just 31 years old, returned to command his battery in France. On 26 April, six months before the signing of the Armistice, he lost his life in an action close to Ypres. His mother wrote (BirrR/5)

Ray
Kirkwhelpington
May 6th (1918)

Dear Nannie,

I know you will be grieved to hear that Major Algy was 'killed in action' on April 26th. The wire came on Sunday—It is a terrible blow to us, though of course we knew that it might come any day.

I am glad to look back on the time he spent with us so lately when he was wounded—he would not have a home job. Sir Charles happened to be here lucky for me. We both feel it terribly.

Yours faithfully

Katharine Parsons

It can be imagined what a terrible blow this was to his parents, to have lost their only son. For several years father and son had worked together, and Charles must have had high hopes for the future, but all this was to be dashed in the last months of a brutal and senseless conflict. After the war broke out, Charles Parsons decided to live at Ray near Kirkwhelpington, and he had made Holeyn Hall available to the Army so that it could be used as a convalescent home for soldiers. Silent movies were provided to entertain them. But when the war was over Parsons did not return to Holeyn Hall. No doubt it carried too many memories of his dead son. The last of a litany of death and injury was the news that Lord Rosse had died suddenly at Birr Castle on 10 June.

The end of the war brought many problems. The most immediate centred on the women who had filled the vacancies left by the men who had served in the armed forces. The table below shows that while the total workforce at Heaton remained remarkably constant during the War, the number of males did fall by 16% , which was more than compensated by a 12-fold increase in the number of the female workers (TWASe),

	Females	Apprentices	Men	Total
Shops				
July 1914	15	202	860	1077
April 1918	185	187	725	1097
Offices				
July 1914	6	18	157	181
April 1918	72	10	119	201
	Females		Males	
July 1914	21		1237	1258
April 1918	257		1041	1298

19.2 Parsons women in the public eye

In the year after the war ended Katharine, now Lady Parsons, read a paper to the North East Coast Institution of Engineers and Shipbuilders in which she revealed her thoughts about the role of women in industry (Parsons K 1919). She had a strong interest in the subject. At the turn of the century she remarked, girls and women found employment in *'the motor and cycle trades'* following *'the introduction of the new automatic machines'*. But during the war *'there is no doubt that many women developed great mechanical skill and a real love of their work'*. She described the still ongoing Atlanta venture where women were *'cutting out the crankshafts* (for engines) *from the solid forging, using the drilling and slotting machines, and they were making all the tools, jigs and fixtures required for the work of the factory'*. She went on to give examples of women's wartime contributions in research and development. They had carried out metallurgical trials of alloys, tested models of ships and seaplanes in an experimental tank and checked electrical instruments for the Admiralty. She pointed out that women *'had the advantage of receiving a short intensive training in the technical schools set up by the Ministry of Munitions and in private firms. These technical schools proved that . . . an intelligent girl can learn almost any mechanical process in a few weeks, a prolonged apprenticeship not being necessary'*. This all changed after the war, and she thought that the country would be the poorer for wasting the £30 000 000 that had been spent on the training of some $1\frac{1}{2}$ million women. She criticized the Labour Party which sought the right of women to sit in the House of Lords, and to practice at the Bar as solicitors, but *'will not grant to women industrial equality by permitting them to work on the same jobs as men. or to enter the higher ranks of those industries which the war has shown they are perfectly fitted to work.'* She concluded

> *It has been a strange perversion of women's sphere—to make them work at producing the implements of war and destruction, and to deny them the privilege of fashioning the munitions of peace . . . women are merely told to go back to doing what they were doing before, regardless of the fact that, like men, they have now a higher standard of life, and that they also wish to have their economic independence, and freedom to make their way without any artificial restrictions.*

Katharine Parsons was far ahead of her time in her thinking about the place of women in society, which was not altogether surprising since both her husband and her father in law had encouraged the young girls of the family every bit as much as the boys to use the machines in their workshops.

After the war Rachel did not return to Newcastle but developed her interest in politics. In March 1923 she became a member of London County Council representing Finsbury, and sat on the Electricity and Highways Committee. That year at a meeting held in Wigan in Lancashire she was adopted as the Conservative candidate to contest the seat for Ince at the parliamentary elections (Anon 1923a). Despite her party allegiance she propounded ideas that would not have disgraced a socialist. *'I am in favour of social legislation to improve the condition of the people. I am in favour of housing improvements... I am also in favour of keeping the prices of food down'.* She favoured tariffs to protect home industries. *'Your trade unionists... are the greatest exponents of protection that I have ever heard of. I have had a good deal of experience of trades unions, because I have worked with them during the war in my father's engineering works. Sometimes I agreed with them, and sometimes, I must confess I did not.'*

When the votes were counted, the Rt Hon. S Walsh (Soc.) had 17 365 and Miss R Parsons (U) received 6262 votes. This appears to have been the only serious brush with national politics which any member of the Parsons family had.

19.3 Changes at Heaton; the market for turbo-generators develops

The Board documents for 19 November 1918 make no reference to the Armistice, but do carry a report on the work of the Rate Fixing Department that had been functioning since March. The aims were to relieve foremen of detail work related to rate fixing and to readjust prices up or down with agreement by the men... *'It is not an object with the Department to cut prices merely because good money is being earned. The aim is to ensure that a maximum of work is charged to a job, and a minimum to "charges"'.* The payroll was scrutinized to identify excessive items and the management was alerted to abnormal expenditure other than labour. The idle time on each big machine or group of small machines was studied with the foreman and the reason identified under one of eight heads. The hand of A D C Parsons is visible in all this. A report of a sub-committee of Company and Workmen's representatives held on 1 November considered three matters, a Home Office suggestion of a joint committee to encourage a 'safety' attitude, a strong plea to allow smoking on the day shift, which was refused, and arrangements whereby men 'on leave from the Front' were allowed to visit friends at the works.

In 1918, in a letter to senior staff, Parsons formalized his practise of giving access to shares in the Heaton company (Bedford).

> *With a view to affording a more direct personal interest in the Company to those holding important positions, I have decided to set aside a certain number of my shares in the Company upon the terms of the annexed Statement.*
>
> *The shares are fully paid up shares of £100 each, and the number to be set aside is, in your case ......*

Stated generally, the arrangement is that the shares shall be credited with the dividends received in respect of them, and that for each £100 of accumulated dividends a share shall become yours, and so on until the whole of the shares set aside for you have been accounted for. It shall be open to you to accelerate your ownership by payments out of your own money.

There are provisions to meet the case of death or cessation of employment which are, I think clearly set out in the Statement.

I hope this arrangement will be beneficial both to you and the Company.

Yours faithfully,

Charles A. Parsons

The statement provided that, should the employee die or leave his employment, Parsons could repurchase the shares at their par value. A total of 364 shares were set aside for this scheme, with further additions made later.

Early in 1919 the first works canteen was built where men could have their midday meal in some comfort (Anon 1949). The reduction in demand for searchlight mirrors made it possible to reallocate some of the extra space that had been provided during the war for use by a tool room, a workshop for experimental projects and for laboratories. These moves made space available for an apprentice training school where youths could be taught the use of machinery before entering the works. Demand for turbines was such that an extra bay was added to the erection shops in 1921. Much of the experimental work since 1900 had been done in a temporary building. It was nick-named the 'Block House' after similar structures which had been used during the Boer War. In 1923 this was removed and work was started on a three-storey building to house the Mechanical Research Laboratory. The manufacture of electrical power transformers had been part of the range of products of the Heaton Works originally but had been discontinued in 1900. In 1924 it was decided to resume this activity and two more shops were built to accommodate coil winding, core plate stamping and assembly.

The conclusion of the war meant that the demand for marine turbines for fighting vessels abruptly diminished. Fortunately the demand for electricity continued to be buoyant and the Heaton Works received plenty of business. This led to another confrontation between Parsons and a senior colleague, Alex Law, who was the managing director at Heaton. Parsons decided that some of the extra work should be redirected to the Turbinia Works (Bedford). Law felt that Heaton could cope with the increase, and no doubt resented the inevitable complication if the same products were manufactured in two locations. As on past occasions, the difference of opinion led to a rift between the two. In October 1919 Law left to join the English Electric Company which had been newly formed by the amalgamation of a number of firms: among them were Willans and Robinson who were Parsons' licensees, the Siemens Dynamo Co. and an AEG subsidiary (Jones and Marriott 1970). Law was replaced as managing director by H I Brackenbury.

Turbines and generators continued to increase in size and improve in performance even during the War. In 1918 an order was received for a 15 MW machine for the power station at Dunston. Like Carville it supplied a 40 Hz network, and the highest speed of 2400 rpm was chosen. This required that the low pressure exhaust cylinder should have a double flow arrangement[1]. In the following year a 12.5 MW, 3000 rpm machine of similar layout was supplied to the Mersey Power Co. When the output was raised to 18.75 MW for a South Wales Electric Power Co. turbine, it was possible to use a single cylinder design but only by restricting speed to 1500 rpm. In 1922 the first of many machines was built for the Barking station in London. A pair of 20 MW 3000 rpm two cylinder tandem machines with double flow low pressure cylinders were built. Additional, 40 MW, machines were ordered, and while these were also designed for 3000 rpm, a cross compound arrangement had to be used. The size of alternators was still limited and at 20 MW this made them the largest alternators yet built for 3000 rpm. The steam supply was at 700 °F and 350 p.s.i. After expansion in the high pressure cylinder of one turbine, it was reheated to 700 °F before being led to the intermediate pressure cylinder on the other turbine. Both turbines carried identical low pressure, double flow cylinders, and between them they shared the steam that was exhausted from the intermediate pressure cylinder. The re-heating and cross compounding required special safety features to be installed in the governors. Steam could be bled from the turbines at several points for regenerative feed water heating. Tests showed that the adoption of reheat gave an energy saving of 7.5% if feed heating was not used, and 5.5% if it was used.

In 1922 Samuel Insul placed another order with Parsons, this time for a 50 MW set for the Crawford Avenue station of Commonwealth Edison Co. of Chicago. It was said that the order was placed without a price having been fixed, such was Parsons' reputation for fair dealing

[1] See appendix 2.

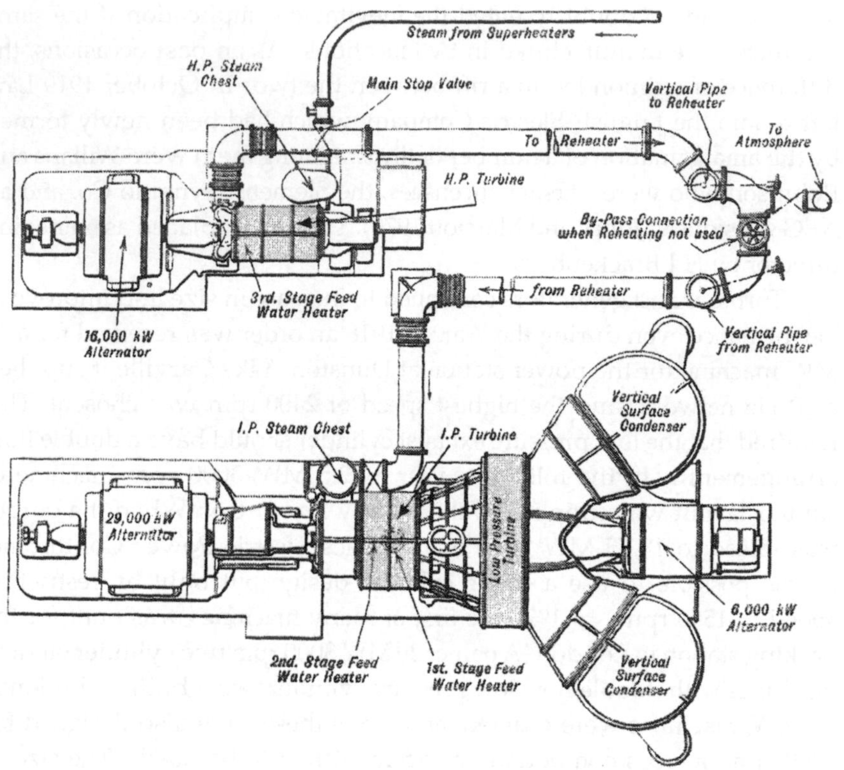

Figure 121. Plan view of the 50 MW cross compound turbine for Crawford Avenue power station, Chicago (1922). Although the intermediate pressure and the low pressure turbines were placed in line, their shafts were physically separated and ran at 1800 and 750 rpm respectively (Parsons R H).

(Bedford). The performance that Parsons offered to achieve was 10 265 Btu per kW h, which corresponds to an energy efficiency 'from steam to electricity' of 33.3%. Steam conditions were 750 °F and 550 psig. In order to deal with the massive volume of low pressure steam a cross compound arrangement was used (see figure 121). The high pressure and the intermediate pressure turbines powered alternators of 16 MW and 29 MW respectively at 1800 rpm. The low pressure turbine ran at 720 rpm and was coupled to a 6 MW alternator. The IP and LP turbines were arranged in line but their shafts were of course not joined. Considerations of space required the HP turbine to be located to one side. The LP turbine contained five stages only and developed quite a small output. However

Figure 122. View of the intermediate pressure and low pressure turbines, with the vertical condenser, for the 50 MW set while being assembled at the Heaton Works prior to despatch to Crawford Avenue station (Parsons R H).

it had to cope with a steam flow of 72 000 $ft^3\ s^{-1}$, which corresponds to an average velocity of 465 mph. At the last stage the steam had become tenuous in the extreme, its density being less than 5% of atmospheric air. The last row of blades had a mean diameter of 13 ft 4 in and were 42 in long. They were each rolled from a separate billet of mild steel. Each was weighed and balanced carefully. The speed of their blade tips was 635 ft s^{-1}. The number of blades in the whole machine totalled 34 370; consequently the force exerted on individual blades by the steam flow was modest and ranged from 3.75 lb at the inlet stage to 7 lbs at the exhaust stage (Dowson 1935). Both re-heating and feed heating were employed, and when the acceptance tests were carried out the promised performance was exceeded by 2%, with a figure of 10 038 Btu kW h^{-1}, which corresponds to a thermal efficiency of 34%. Parsons was now but one of several technically advanced, and often much larger, manufacturers. He kept pace with them, but he no longer had a significant lead. For example, at the same power station in Chicago GE was putting into service one of their impulse turbines, a 60 MW cross compound, 1800/1200 rpm design.

It was around this time, on 20 November 1923, that a catastrophic accident destroyed a 20 MW turbine being commissioned in Shanghai for the Municipal Council. Parsons learned of the accident as he was about to attend a dinner of the British Electrical and Allied Manufacturers Association. A meeting of the directors that night decided that they would accept full responsibility for repairing the damage (Appleyard 1933). Their cablegram to Shanghai read

> *Deepest regrets at serious accident and loss of life. Sending immediately two chief experts to investigate. Keep all parts for evidence. Will replace turbine and recondition whole of our plant entirely at our cost.*
>
> TURBO

At the Board meeting held on 16 January 1924 a report on the accident was discussed. Fourteen people had been present while a Chinese fitter was running up the machine. Donaldson, who was the Parsons' erector, was using a tachometer at the time of the accident and the speed was 1400 rpm. The overspeed governor had not tripped. There had been no report of serious vibration before the accident. The alternator was not coupled to the turbine. When the turbine disintegrated the fitter and Donaldson were killed instantly. One piece of the rotor weighing 5 or 6 tons was found 300 ft away buried 10 ft in the ground. Because the people in the shops had reported problems with the steel discs that had been supplied recently, the first thoughts were that a disc of a low pressure stage had failed. By 5 February a cable was sent to Ranft, the representative in the Far East stating '*agree cause of accident defective shaft forging*'. On 11 March an estimate for the total monetary cost of claims for the incident was £26 121. Later in the year £10 000 was added to the fund used for making grants or pensions to deserving employees.

A meticulous inquiry was conducted and it was concluded that the ingot from which the turbine shaft had been forged contained a 'clinck' (Liversidge and Fearn 1967). Examination of the rotor showed that it was '*full of internal flaws and in such a state of strain that, when it was cut up for examination, cracks developed spontaneously*' (Stoney 1938). When an ingot is cast there is a tendency for internal cracks to form as it cools. This can be worsened if the cooling is uneven, or if during heat treatment large gradients in temperature are created. Sometimes, but not always, such cracks may be welded together during forging. As a result of the investigation a quality assurance programme was introduced. '*Parsons took up the matter with his usual energy, and insisted that large forgings should never be allowed to cool below a good red heat from the time of pouring the ingot until all forging operations were complete. Any re-heating necessary during the process had to be carried out in accordance with strict rules*'. Furthermore, before the forging was machined, a hole was bored through its centre from end to end. This allowed the internal surface to be examined visually at high magnification and in strong illumination for flaws, using a borescope. As a final precaution, machined forgings were subjected to several minutes of overspeed in a special armoured chamber, though it should be said that the Shanghai rotor had survived overspeeding without any problem.

This episode serves to highlight the success that Parsons had, for the most part, enjoyed as he continually increased the demands placed on materials and components. He was by no means alone in encountering such calamities. For example, between 1917 and 1920, when gearing became available for marine turbines, and higher shaft speeds came into use, a new phenomenon of disc and blade flutter (vibration) emerged in impulse turbines (Jung 1982). Discs can be made to vibrate like a bell, and blades like a tuning fork. Impulse turbines that employ partial admission subject individual blades, and so discs also, to abrupt pulses of thrust as they pass through the jet of the inlet nozzle to a region where no steam is being admitted. If the frequency of these pulses coincides with the natural frequencies of the components, their vibratory motion can build to a point where fatigue damage is done, and ultimately a fracture results. In America General Electric were able to bring their problem under control in 1924 when their English born engineer, Wilfred Campbell, published his analysis of the phenomenon of disc flutter (vibration) (Campbell 1924). In 1923 General Electric had 4400 machines larger than 5 MW in service, of which they tested 3600, changed 500 and tuned 200. The turbines supplied by Brown–Curtis to the Royal Navy also suffered similar difficulties, but they had less success in finding a remedy for them. Combined with difficulties over gear manufacture this eventually caused the Royal Navy to exclude Brown–Curtis from new construction after 1930 (Jung 1987a).

When looking back at the post-war period it is useful to set Parsons' experience in some sort of perspective. The war had been good to the shareholders of the neighbouring Armstrong–Whitworth partnership. The share capital reached £9 000 000 in 1918, and an average of £798 569 per year, equal to $12\frac{1}{2}$% tax paid, had been paid to shareholders. The year 1919 was full of optimism on the North East Coast as men returned from the war to end all wars, ready to replace the ships and other material that had been destroyed or damaged during the last four years. There was something like a frenzy among ship owners who scrambled for the vessels that they anticipated would be needed for a trade boom. Giants like Armstrong–Whitworth planned to replace the manufacture of gun turrets and suchlike with railway locomotives, motor cars and consumer goods

(Scott 1962). But in the spring of 1920 business slumped abruptly. High wages and foreign competition caused industrial unrest. Also German war reparations meant that quantities of surplus merchant shipping clogged the market. The shipbuilding industry which was the mainstay of the region was badly hit. In the three years between 1920 and 1923 the annual tonnage of vessels built on the North East Coast fell from close to one million tons to barely a quarter of this (Clarke 1997). The plans for peacetime manufacture at Armstrong–Whitworth failed to be profitable and the Bank of England had to provide heavy overdrafts. Eventually a merger was arranged with another northern giant, Vickers, in October 1927.

Chapter 20

Other Inventive Endeavours

20.1 An acoustic amplifier

Inventors may well earn a reputation for their success with one specific product, and yet their activity cannot be neatly pigeon-holed. Westinghouse was renowned for his patent railway brake and Edison for his phonograph, but both men had hundreds of patents to their name covering a great variety of fields, many of which were by no means financially rewarding. Long after Orville Wright had ceased to play a central role in aviation he continued to apply his inventiveness to making improvements in the plumbing and other features of his home (Crouch 1989). Given the enormous satisfaction that he derived from his inventions, it is therefore not surprising that Charles Parsons should have followed some unconventional enterprises. One such was his auxetophone.

At the turn of the century the efforts of Edison and others had made it possible for the first time for speech and musical performances by renowned artists to be reproduced in any home (Chew 1967). The one drawback was that the sound level was low, and no means of amplification was available. Sometime around 1901 Parsons began to consider the way in which the mechanical motion of a gramophone needle was turned into sound waves (Carnegie 1934). Essentially it had been connected to a diaphragm that was made to push on an air column and so launch an acoustic wave. Parsons decided that instead he would make the movement of the needle control a flow of air directly by means of a valve. On a much larger scale he had used his 'gust governor' to control the flow of steam on a similar principle. He learned that Edison in 1877 had proposed to drive a diaphragm with an air relay, but neither Edison, nor successive inventors, had been successful in producing a practical device. This of course did

Figure 123. Parsons' air valve reproducer, auxetophone, coupled to a cello (1904). (S.P.2)

not deter Parsons from pursuing what seemed to him a perfectly valid principle. He carried out his experiments in the little workshop that he had built in his home at Holeyn Hall, and where he had made his classic studies of cavitation in propellers. Starting after his evening meal he would work on his 'hobby', often late into the night. For him this was a pleasant relaxation after the stress of the day's work. He tried a variety of valve types, double-beat valves, and slide valves with several openings before settling on a 'comb-like' structure (Parsons C A 1921b).

As a result of some preliminary measurements of the rate of flow of air through very narrow slits, he found that the flow rate reduces much more rapidly than does the area of the slit. He made calculations of the forces involved in moving components in response to a needle vibrating

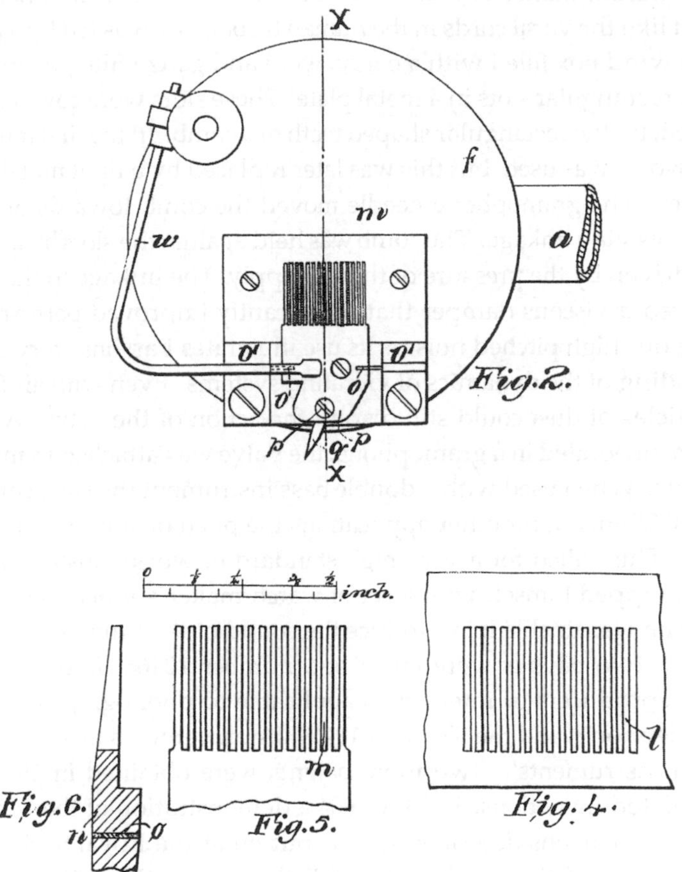

Figure 124. Details of the components of the comb style valve for the auxetophone. 'Fig. 2' is a side view of the comb valve with a gramophone needle in place at *p*, and showing the connection to the acoustic horn at *a*. 'Fig. 4' is the fixed grating and 'Fig. 5' is the moving comb (which is held in place by the arm *w* in 'Fig. 2'). 'Fig. 6' is a section of the comb teeth. (PAR2/2/6.)

at several hundred times per second. For example at 500 Hz an amplitude of only 3/1000 of an inch required an acceleration which was 77 times greater than that due to gravity. He computed that with a pressure of 3 psi, and a sinusoidal vibration at a frequency of 500 Hz, the quantity of air that will pass through a fine slit is 3 to 5 in^3 for every square inch of slit opening. But having established his design on a firm basis, it was a different matter to realize it. The comblike structure functioned somewhat like the vocal cords in the human throat. Air was fed by a pump through a wind box filled with a cotton wool and gauze filter, to an array of narrow rectangular slots in a metal plate. These slots were covered, and overlapped, by the rectangular shaped teeth of a comb. In the first tentative trials boxwood was used, but this was later replaced by a light metal alloy, magnalium. The gramophone needle moved the comb towards or away from the slots via a linkage. The comb was held against the slots by a piston that was driven by the pressure of the air supply. The linkage to the comb incorporated a viscous damper that significantly improved performance by cutting out high pitched noises. Its use illustrates Parsons' very secure understanding of the dynamics of vibrating systems. Even with air filters, small particles of dust could still disrupt the action of the valve. As well as being incorporated in a gramophone, the valve was attached to musical instruments. When used with a double bass instrument the slots could be spaced at 0.25 in, but for other applications the pitch of slots was as small as 0.02 in. This called for a very high standard of workmanship, and so Parsons equipped himself with a set of watch-maker's tools. Despite all his efforts he was unable to reproduce the sound of the human voice.

In 1903 two patents were taken out, number 10468 'Improvements in sound reproducers or intensifiers applicable to phonographs, gramophones, telephones and the like' and 10469 'Improvements in and relating to musical instruments'. Two more patents were obtained in 1903 and 1904 for related inventions. In May 1904 a demonstration of the *'air valve reproducer'*, as Parsons described it, was put on at the Royal Society. Dr Johnstone Stoney, father of Gerald, and Fellow of the Royal Society suggested a name for it, the auxetophone. He wrote a letter to Parsons on 28 May 1904 with some mathematical analysis of the device. This showed that it tended to enhance the higher harmonics of a sound. In the case of a double bass this was considered to be pleasing. Some time before this, Parsons found that he had been forestalled. Horace L Short had made a demonstration with a similar device from the Eiffel Tower some years earlier. Parsons later recalled that Short was associated with the Edison Company through Colonel Gouroagh, though perhaps this was Colonel Gouraud (Josephson 1961). Between 1902 and 1904, with the help of S F Prest, Parsons negotiated the purchase of the patents from Short for £700 (PAR2/2a). Because Short was very skilled with his hands, Parsons agreed to employ him at the large salary of £400 per year for four years to work on perfecting the instrument. Short stayed on until 1914 when he left to join his brother in the aircraft firm of Short Brothers. Some time after he had acquired Short's patent, Parsons sold the rights both to it and to his own patents for application to phonographs and gramophones world wide to the Gramophone Company for £5000. He reserved the rights to its use with musical instruments for himself, but at least his hobby certainly had not cost him anything!

In its final form the instrument performed very well. It was demonstrated as recently as 1984 playing 78 rpm gramophone records during the celebrations for the Turbine Centenary held at the Works of C A Parsons at Heaton. Short developed other versions that could be fitted to stringed musical instruments like the cello and double bass, with which it showed to best advantage. Large horns which were modelled on brass band instruments and resembled enormous acoustic gramophones were used to launch the sound output. During the 1906 Winter Season of Promenade Concerts in the Queen's Hall Sir Henry Wood used the device to augment the sound of the strings at concerts of the orchestra. Sir Henry wrote appreciatively about their performance, but the musicians were not so keen (Appleyard 1933). In an effort to attract publicity for the invention, Parsons made a deal with van Biene, the 'well known actor and musician', to give cello recitals at a number of venues, starting in 1909. The surviving correspondence shows that Parsons kept a close eye on the financial side of the arrangements, and was in no way a soft touch (PAR2/2/4). Given the weighty matters that occupied his attention at this time, this underlines Parsons' careful attention to money matters. Attempts to adapt the instrument for use with harps or pianos were not successful. The Victor Talking Machine Company in the United States and the Gramophone Company in England made components for auxetophones for some years. The instruments were used in hotels and restaurants.

20.2 Synthetic diamonds

The 'air valve reproducer' was a device that was developed successfully by following thoroughly scientific methods. Parsons spent far greater sums, between £20 000 and £30 000, and much more time on another scientific project that in the end yielded nothing at all, the synthesis of diamond (Stoney 1931a). When his biography was written two years after his death it must have seemed that he had been engaged in the senseless pursuit of a chimera, but since then his objective has been fully realized. In 1955 scientists at the General Electric Company's laboratories in America evolved a practical process for turning graphite into diamond. What is notable is that it has been effected using essentially the same methods as

Parsons pioneered. He used a hydraulic press to compress a charge of graphite, and he heated it by passing electrical currents through it.

Carbon is an element that can exist in several forms, amorphous, as graphite sheets, or as hard crystalline diamond. At ambient conditions, graphite is the stable form. It can exist as soot for instance, or it can be mined in certain locations in varying degrees of purity. Diamond, which is the denser form, is metastable, that is it tends to transform into graphite. However at ambient conditions this is a very slow process. To reach a region where diamond is stable it is necessary to raise the pressure greatly. It was not until 1938 that an accurate estimate could be given of suitable values of pressure and temperature for such a region (Rossini 1938). Moreover if a graphite sample is to be transformed into diamond crystals with any rapidity, not only must the temperature be raised, but a catalyst must be employed. Typically temperatures of around 1500 °C and pressures 50–60 000 atmospheres are used (Davies 1984).

It will be recalled that Parsons first became acquainted with powerful hydraulic presses at Armstrongs, and one of his earliest patents concerned hydraulic machinery. While at Clarke Chapman he became interested in transforming the industrial carbon that was available to him for manufacturing arc lamp electrodes, into a denser form. In his paper to the Royal Society in 1888 he described experiments that he made for this purpose. He used an 80 ton press to produce pressures in his specimens of up to 4600 atmospheres. Because carbon is an electrical conductor, he was able to heat it up by passing currents of up to 300 amperes through it. Current was supplied by a 35 HP turbo-generator. He observed that some 'bort like' powder was produced, but not diamond. Some time after this, a French scientist Henri Moissan (1852–1907), who was to be awarded the Nobel prize for his work isolating fluorine and for exploiting the electric arc furnace, also claimed to have created diamonds. His technique was to dissolve carbon in iron which was heated to white heat in an arc furnace. The sample was then plunged into water. He reasoned that as the solidifying iron shrank, it would raise the pressure of the still molten core high enough to create diamonds.

After he acquired his own works at Heaton and using his own workshop at Holeyn Hall, Parsons turned his attention to the specific objective of creating '*artificial diamonds*'. Two brief reports were published (Parsons C A 1907, Parsons C A and Swinton 1908). In the first he summarized the results of '*about 100 experiments*'. But now his installation was altogether more ambitious (Parsons C A 1918a, b). He had a new hydraulic press which was capable of developing a force of 2000 tons.

His specimen was placed in thick walled cylinders made of gun steel. These could have a bore that was up to 9 in diameter, but whenever the highest pressures were employed it was only 2 or 4 in. The specimen was electrically insulated from the walls of the cylinder and an electric current was fed in through the steel plugs that closed off the cylinder at either end. An electric battery capable of delivering 330 kW at voltages between 4 and 48 volts supplied currents that could reach 50 000 A. The upper limit was set by the steel plugs used to close the bore of the vessel. These melted at high currents despite efforts to cool them by water jackets. The press was placed in a small building specially constructed partly below ground level with walls 2 ft thick and made of reinforced concrete. The roof was made of light galvanized iron, and the door was made of steel 3 in thick. In addition steel rings 2 in thick could be raised by counterweights to surround the test cylinder.

> *These precautions, as experience proved were necessary as several violent explosions occurred which cracked the steel rings and blew off the roof. A charge of iron and carbon when confined and raised to a high temperature may be very violent if suddenly released by the melting of the pole pieces* (end plugs).

In an effort to concentrate the heating effect he tried placing a disc of insulating material like magnesium oxide at the centre of the test volume (figure 125). A central hole of 0.625 in diameter, which was filled with graphite, created a neck between the charge above and below. The test cylinder in this case had a bore of 4 in. Pressure could be raised to 15 000 atmospheres. At such pressures the cylinder became 'permanently strained' and had to be re-bored before being reused. A current of 8000 amps was applied for 7 s. He found the graphite changed into large soft flakes, but there was no sign of crystals. He estimated that four times more energy had been supplied to the material in the neck than would be needed to raise its temperature to 5000 °C. This temperature, we now know, is enough to melt carbon even at the highest pressures, but diamond can only be recovered if the cooling occurs at much higher pressures. Next he poured chilled, liquid carbon dioxide around the central charge of carbon in the test cylinder and applied a pressure of 4500 atmospheres, causing the volume to be reduced by 20%. Current was then supplied to heat the charge. He was able to contain the hot gases successfully even at such a high pressure. When the vessel had cooled and the gas had been allowed to escape it was found to be 95% carbon monoxide, while a quantity of carbon had been deposited as a '*woolly nest*'. In other experiments he substituted

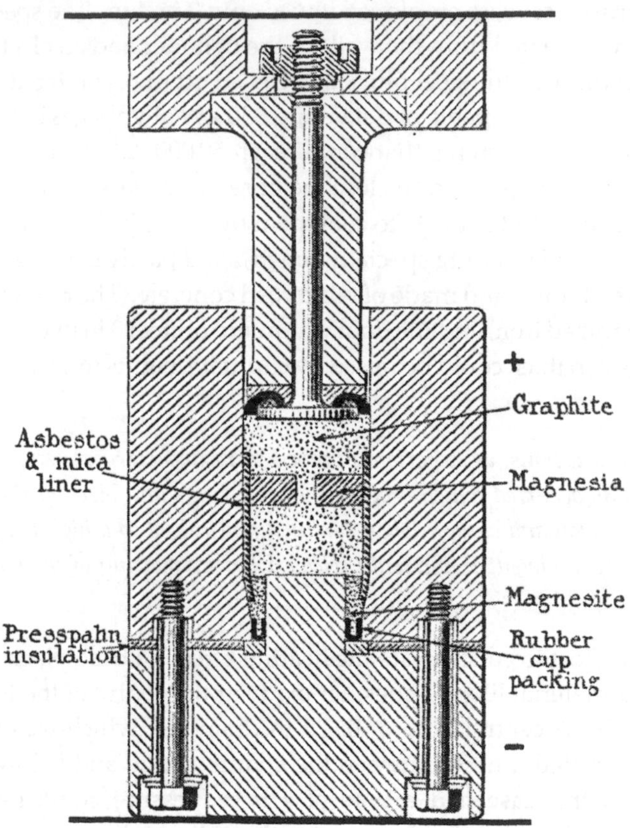

Figure 125. Graphite specimen inside a test cylinder with a 4 in bore, compressed by an upper sealing plug and supplied with electric current + and −. The horizontal lines, top and bottom, represent the jaws of the hydraulic press. The insulating constriction made of magnesia increases the current density locally and hence raises the graphite temperature at this neck (Parsons C A 1918b).

iron rods, or iron tubes filled with iron filings and lamp black, for the carbon core. His purpose was to test Moissan's conjecture that carbon dissolved in molten iron is precipitated as diamond if the iron cools under pressure. He also poured molten iron heavily charged with carbon into a mould, and, before it solidified, applied a pressure of 11 000 atmospheres. He concluded that *'in none of the experiments... has the residue more than a suspicion of black or transparent diamond'*. Also there was no sign of carbon becoming a non-conductor. The very numerous chemical analyses of residues were performed for him by J Trevor Cart.

Two observations may be made about this work. The first is that it employed some of the highest pressures ever reached to that date. For example the work of the great exponent of high pressure physics, E H Amagat, had been restricted to 3 or 4000 atmospheres. In 1912 the American scientist P W Bridgman, though just starting his high pressure studies, published results obtained at 20 000 atmospheres, but his pressure vessels were of much smaller bore (Bridgman 1949). The second point is that the configuration shown in figure 125, where the compressed and electrically heated specimen is surrounded by an insulating barrier, is very like the 'belt and anvil' which is used in the modern production of diamonds (Davies 1984).

The second brief paper on the 'Conversion of diamond into coke in high vacuum by cathode rays' was written with A A Campbell Swinton (Parsons C A and Swinton 1908). It established the temperature above which diamond ceases to be metastable at low pressure. Swinton had been one of the first people in Britain to experiment with the use of the newly discovered x-rays, and he adapted his equipment for this study. A small diamond was placed on an iridium plate in a continuously evacuated glass vessel. An alternating voltage of 5000 up to 12 000 volts was applied to two electrodes. These were located in such a way as to direct the resulting cathode rays, that is beams of electrons, onto the sample. The sample turned red, then white hot and swelled up. An optical pyrometer gave the transition temperature as 1890 °C. The residue looked like coke after it had cooled.

20.3 Some pure science

The high pressure laboratory that Parsons had constructed was a unique facility. In his earlier tests he had observed the contraction in volume caused by the application of pressure. Accurate measurements of this 'compressibility' can shed light on molecular forces existing in the material under study. In 1908, with the assistance of Stanley Cook and of Bob Howe who was in charge of the test department at the time, Parsons set about improving the precision of his measurements. His test vessel was a cylinder 12 in in diameter, with a bore of 4 in. It was made of gun steel having an

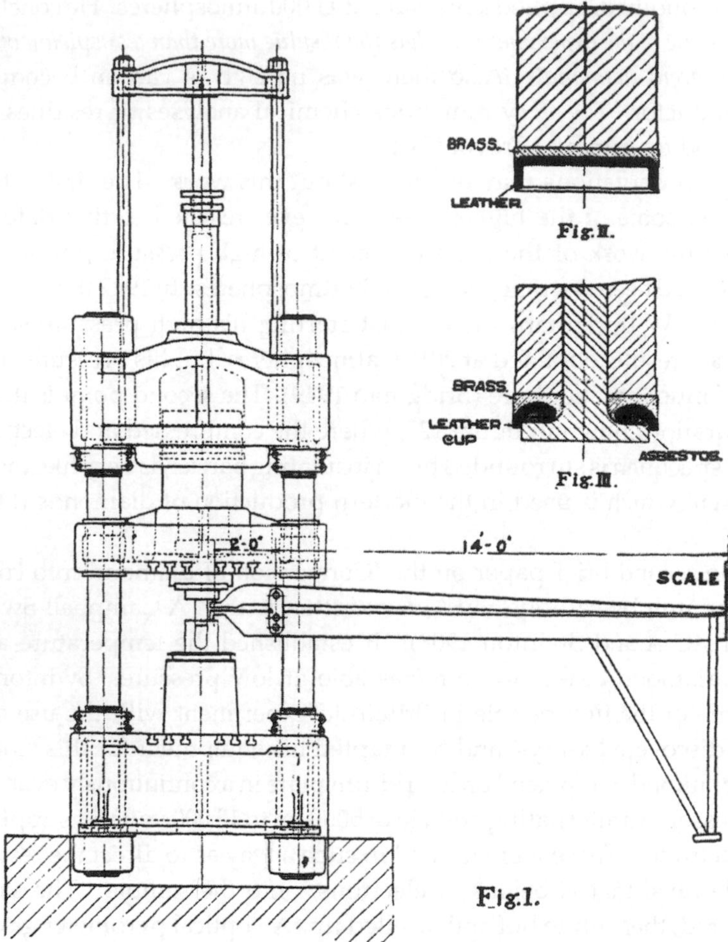

Figure 126. Details of the 2000 ton press showing the levers for measuring the movement of the ram, 'Fig. I'; first piston seal, 'Fig. II', and second seal arrangement using the unsupported area principle, 'Fig. III' (Parsons C A and Cook 1911).

elastic limit of 6200 atmospheres and bored from one end only, so that the test sample could be contained by inserting a single piston, equipped with a suitable seal, at the open end. Arrangements were made to control the sample temperature.

The sample volume was measured by the movement of the piston, as indicated by a lever device that gave a sevenfold magnification (figure 126) (Parsons C A and Cook 1911). For safety's sake readings were taken by a telescope from outside the building. A calibrated Bourdon gauge monitored the pressure of the water supplied to the press. Knowing the diameter of the press ram, the force transmitted to the piston was calculated. When a plot was made of the apparent pressure against the volume of the specimen, it was found that the curve was different depending on whether the piston was moving into or out of the cylinder bore. This was attributed to friction of the seal between the piston and the cylinder. An estimate of the true pressure at a given volume was taken as the average of these two pressures. A procedure was adopted that allowed errors due to the, unknown, deformation of the piston seal etc to be determined. The samples were of relatively enormous volume, 2 litres. In the case of water it was found that the volume was reduced to 90.2% at 3000 atmospheres and 4 °C. This compared with Amagat's value of 89.8% at 0 °C. A similar measurement, made at 35 °C, on ether gave 82.9% compared with Amagat's value of 82.2%. Tests were also made on oils and two samples of graphite.

When a fluid expands without any flow of heat, that is adiabatically, its temperature drops. One way to ensure this is by insulating the specimen to prevent heat transfer. Alternatively the expansion can be carried out so rapidly that there is no time for appreciable amounts of heat to enter the sample. To do this an accumulator was fitted at the water supply to the press which allowed the pressure to be raised or lowered in seconds. Several attempts were made to insert electrical temperature measuring devices in the sample to measure the effect of such sudden changes, but without success. Eventually the piston seal was modified (III in figure 126) so that the vessel could be opened in seconds, instead of taking 15 min. The temperature of the sample could then be taken with a thermometer before it had had time to change significantly. When the pressure of ether at 35 °C and 4400 atmospheres was reduced suddenly to ambient, its temperature fell to −11 °C. Similar measurements were made on water and a paraffin oil. An appendix, written by Cook, analysed the various thermodynamic parameters that had been measured, to check their self-consistency.

20.4 Back to diamonds

This piece of pure science grew from the technology that Parsons had developed in his pursuit of diamond synthesis. While he documented the results for others to avail of, it was only a diversion. His main aim remained unaltered and in the following decade he exercised his ingenuity in a variety of quite novel ways for reaching high temperatures and pressures.

In his Bakerian lecture to the Royal Society in April 1918 he summarized his earlier work and gave accounts of some short duration experiments (Parsons C A 1918b). In one case he used a duck gun with a bore of 0.9 in. A massive steel block was screwed to the muzzle. A quantity of acetylene gas and oxygen was trapped between the block and a light, steel, piston located 36 inches down the barrel. A charge of gunpowder propelled the piston. As the powder burned, the trapped acetylene ignited. Its volume was reduced in a ratio of 288 to 1. It was estimated that the temperature reached between 15 250 °C and 17 700 °C, while the observed deformation of the apparatus was consistent with pressures of 15 000 atmospheres. No diamonds were found. It is now known that diamonds do form in milliseconds but it requires pressures of 130 000 atmospheres and temperatures of 3300 °C to 4300 °C. He tried using a 0.303 rifle with almost twice the normal charge of explosive. It was estimated that it propelled the special bullet at 5000 ft s^{-1}. It was aimed at a steel block with a blind hole of the same diameter as the bullet. A great variety of materials were placed at the bottom of the hole including graphite, bisulphide of carbon and olivine. The pressures in this case were estimated to have reached 300 000 atmospheres. Once more no signs of bulk transformation of carbon were found. Stoney reckoned that a sum of around £20 000 was expended in efforts to make diamonds, and he threw some light on Parsons' motivation by quoting a remark which was made to him, '*We have now made a bit of money and deserve to have some fun*' (Stoney 1938).

Numerous experiments were made at atmospheric pressure, under vacuum, and under 'x-ray vacuum'. Parsons called on all his technical resources to address the task in hand: for the vacuum experiments he deployed steam ejector pumps as auxiliaries. Literally hundreds of experiments were carried out, and even more chemical analyses performed. In his summary, the results of the great majority of experiments are marked '*Nil*'. Some two dozen only are shown as having produced '*crystals*', just one or two '*resembled diamond*'. For once his father's caution deserted him. Parsons believed that he had indeed produced some artificial diamonds. He concluded his paper by thanking Sir Dugald Clerk, Professor Jeans, Stanley Cook, Campbell Swinton and H M Duncan. Duncan worked from January 1911 to August 1914 as assistant to Parsons and as his analyst.

But as time passed Duncan became more and more convinced that Parsons had been mistaken in his belief that the crystals were diamonds (Lonsdale 1962). The problem lay in the difficulty in identifying diamond positively. In 1912 the work of Max von Laue and of W H and his son W L

Bragg created the tool of x-ray diffraction that would solve this problem, but it was some years before it could be used on very small crystals. Meanwhile reliance was placed on diamond's resistance to aggressive solvents like hydrofluoric acid, on its hardness and its optical properties, and on the fact that it burned completely in oxygen to leave only carbon dioxide. The latter test of course destroyed the evidence. No doubt it required some courage on his part, but Duncan persisted in suggesting that Parsons had been mistaken so that, as he recalled, '*he at last decided to test, by burning in oxygen, the characteristic crystals which he had carefully mounted on slides, and it was a bitter blow for him when he found that they would not burn*'. It must have been a shattering experience; nevertheless he consulted the President of the Royal Society on the best way of acknowledging that he had been mistaken. As a result, an article, signed only C.H.D., was printed in *Nature* stating the facts. '*In one of these tests, five very characteristic crystals, photographs of which had been exhibited by Sir Charles Parsons at the Royal Society soirée in 1918, were tested in this way* (by heating to 900 °C in oxygen) *and were found to be unburnt*' (Desch 1928). While this article stated the facts, it did not make it clear that it was written at the instigation of Parsons. Subsequently, when the news of successful diamond synthesis experiments became public in the 1960s, Professor Kathleen Lonsdale, the expert crystallographer, published a further account. This quoted from correspondence that she had had with both 'C.H.D.' and H M Duncan and which put the matter beyond doubt.

It is interesting what this reveals about Parsons' attitude to science. Obviously, like his father, the esteem of other scientists, Fellows of the Royal Society in particular, meant a great deal to him, more perhaps than his fame as an inventor, so much so that it seems to have caused him to draw back from correcting the claims which he had made at the Bakerian lecture in a more direct way. In fact Parsons did allude to the matter in public when he responded to the award of the Kelvin Gold Medal in 1926 (PAR4/1),

> *I worked for many years in trying to make artificial diamonds (Laughter), and I spent a great many thousand pounds upon it. I tried to carry out Moisson's and Cripps' work, and eventually there were rumours from Paris and my chemist, Mr H M Duncan, a most painstaking person and genius, thought he would try most rigorously to find out whether the things which we got, and which Moisson got, were really diamonds or not, and we found they were not.*

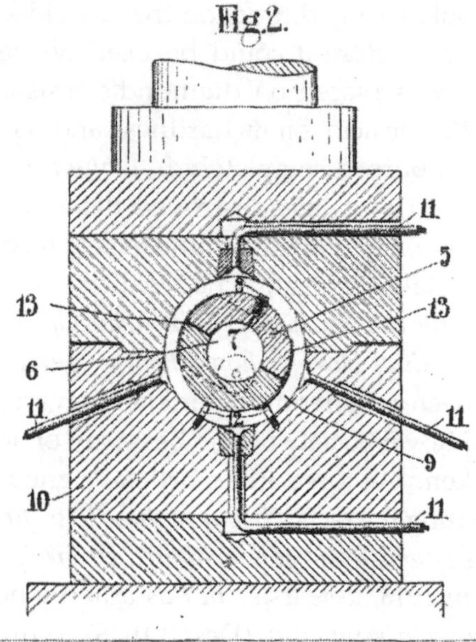

Figure 127. Arrangement for producing very high pressures by implosion of an evacuated sphere, 5. This diagram is from patent application 169 274 of 1921.

20.5 A mechanism for detonating nuclear bombs?

Before this final debacle there was one more intriguing episode. In June 1920 Parsons applied for a patent for a '*Means for producing very high pressures*' (Parsons C A 1921a). This arose from an investigation that Parsons had been carrying out with Stanley S Cook into the erosion of the surface of structures that can be caused by the collapse of small cavities in water (Parsons C A and Cook 1919). It had been shown by Cook and confirmed by Lord Rayleigh that when 'vacuous' cavities collapsed, extremely high pressures are created in the final stage of collapse. Parsons therefore proposed to cause a hollow sphere made of lead, 5 in figure 127, to collapse suddenly by applying an explosive such as TNT to its surface in the space 9, and detonating it everywhere simultaneously by fuses, 11. The object to be compressed was to be located at the centre of the evacuated sphere, 6. He calculated that if a pressure on the surface of the sphere reached 3000 atmospheres, then when the inner cavity had compressed down to 1% of its initial diameter, the pressure there would reach over 100 000 atmospheres. Arrangements were made with Nobel's Explosives to test the idea at their factory at Ardeer (PAR30/4). Writing from there on

29 July 1920 W Rintoul said

> *Since our meeting in town I have been trying to establish a standard method of operation and this has given a lot more trouble than I anticipated. All efforts to explode a half inch layer of T.N.T. on the lead sphere failed because of the difficulty of getting a sufficient initiating charge within the depth of explosive. With cast picric acid we were more successful, provided we used a very strong initiating explosive, but, on the whole, probably the Velox Gelatine we used initially is the most suitable composition for our purposes.*
>
> *The most recent experiments have been carried out with evacuated spheres and the results have been most interesting. In the first shot fired in this way the sphere disappeared altogether, and could not be found, in the second the sphere was recovered, but torn into fragments and showing considerable signs of having been fused in places. The next shot gave us an explanation of what happened with the others, as the sphere was recovered practically intact, but with the appearance of there having been an internal expansion at its centre. I take it that we have here succeeded in developing exceedingly high pressure and temperature at the meeting point of the waves, sufficiently high indeed to cause the vaporisation of some of the lead, with the result that a reverse pressure wave was set up which gave the appearance of the ball having been ruptured by an internal pressure.*

The striking photographs which accompany the letter bear out the writer's assessment. Parsons' reply on 4 August emphasizes the need *'for simultaneous detonation at several places around the sphere.'*

These efforts are interesting because the inventors of the first atomic bombs at Los Alamos used two different methods to bring the mass of fissile core material to a critical size very rapidly. One method was to fire one half sphere at its partner with a gun. The other was to cause an assembly to 'implode' using explosives. The latter idea was identical with Parsons', though it does not appear that anyone at Los Alamos was aware of the fact (Rhodes 1986). The idea was simple, but to implement it required a huge human effort. Novel electronics had to be evolved so that simultaneous detonation could be achieved. Also the need to guarantee that the detonation waves converged uniformly on the centre of the sphere called for a highly sophisticated design for the explosive charges. In short, the technology to achieve this simply did not exist in 1921.

It is only in more recent times that a reliable map has been produced showing those regions of pressure and temperature where carbon exists either as graphite, as diamond or as a liquid (figure 128). The lower hatched area is where the modern manufacturing process operates. This requires

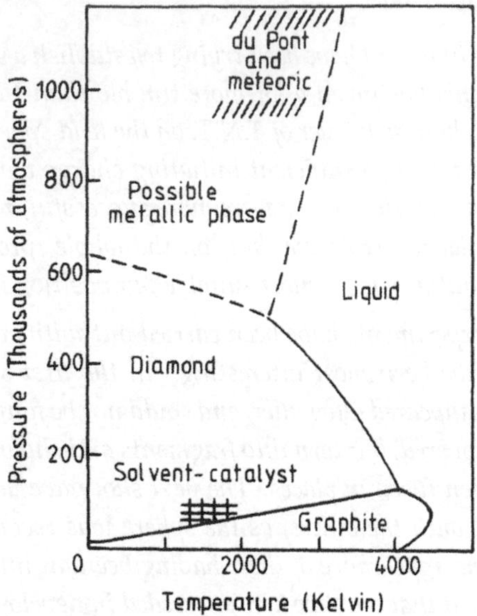

Figure 128. Phase diagram for carbon as established by F P Bundy, H M Strong and R M Wentorf at the General Electric Research Centre. The hatched area at the bottom of the diagram defines the region used for industrial diamond synthesis while that at the top represents a region where very short duration experiments yield results (Davies 1984).

pressures some three times higher than Parsons achieved in his hydraulic press. The upper hatched area represents conditions achieved when explosives have been used successfully to make diamonds. Experimenters in the first half of the 20th century had to work without the guidance of such a map.

In 1919, when Parsons gave his presidential address to the British Association at Bournemouth, he reviewed advances in engineering (Parsons C A 1919). In a section entitled 'Science and the War' he referred back to his father's address in 1858 to the Mechanical Section of the Association in which he had said *'There seems however, something still wanting. Science may yet do more for the Navy and Army if more called upon'*. Charles' review of war-time developments included the work of the Board of Inventions and Research. Trying to convey the enormity of the forces which were by then embodied in weaponry, he reckoned that if all the guns of a single warship like the *Queen Elizabeth* fired simultaneously, they would develop energy at a rate of 13 132 000 horse-power (10 000 MW). Looking to the future he reviewed the existing sources of power, water-power and fossil fuel, and recognized that, *'failing new and unexpected discoveries in science, such as the harnessing of the latent molecular and atomic energy in matter, as foreshadowed by Clerk Maxwell, Kelvin, Rutherford, and others'*, England would not be able to maintain its industrial eminence. He concluded his address very much in the style of his father,

> *The possibility of the uncontrolled use on the part of a nation of the power which Science has placed within its reach is so great a menace to civilisation that the ardent wish of all reasonable people is to possess some radical means of prevention through the establishment of some form of wide and powerful control.* And in a footnote to this he wrote, *For example, it may some day be discovered how to liberate instantaneously the energy in radium, and radium contains $2\frac{1}{2}$ million times the energy of the same weight of T.N.T.*

Just one generation later atomic energy was harnessed, in the first instance 'instantaneously', in a bomb.

Chapter 21

Optical Industries

21.1 Moulded searchlight mirrors

William Parsons presented two legacies to his youngest son. The first was an excellent introduction to the craft of the mechanical engineer. The second was a familiarity with the application of optical science. Both bore fruit early in his career. At the same time that the first turbines were being developed in the stables of Park House, which were part of Clarke Chapman and Parsons premises at Gateshead, work was begun on a novel method of manufacturing curved glass mirrors suited to use in searchlights (Stoney 1933b). Once arc lamps and sources of electric power suitable for use aboard ship became available, searchlights began to have an important place in naval warfare. In order to ensure that their beam of light remained concentrated over long distances, it was necessary that the mirror surface should approximate to a parabola. The German manufacturer Schukert had achieved this by developing a mechanism for generating a parabolic surface, but this was an expensive process. In France the Mangin design retained spherical surfaces that are easier to create. It used a relatively thick glass blank, with a larger radius for the back surface than for the front face. As a result the path for the light that was reflected from the silver coating at the back was made to traverse a greater length of glass at the edge than near the centre. This in large measure corrected the tendency for the spherical reflecting surface to produce a diverging beam. However it did result in heavy mirrors. In the case of a 36 in diameter mirror, the glass was some 4 in thick at the rim but only 1/4 to 3/8 in thick at the centre point.

Parsons had devised a quite novel method that cut the cost of mirrors in half. By heating a sheet of plate glass, he was able to mould it accurately. The mould was made of cast iron, carefully shaped to a parabolic surface.

This was much like the speculum for a telescope, but made of iron instead of bronze. A disc of glass was placed over the mould, and the assembly was heated in a gas fired oven. When the glass had softened and sagged as far as gravity could achieve, a vacuum was created beneath the glass to draw it down into intimate contact with the mould. This produced a light weight and accurately figured mirror at reasonable cost. (The key to success, the use of a vacuum, was kept a closely guarded secret.) Mirrors were then placed on a rotating table and polished by hand using pumice at first, though soon this was replaced by jeweller's rouge. Unusually details of the moulding process were not patented. Parsons did not follow the example of Shuckert who obtained many patents in Britain for his product between 1885 and 1892. A searchlight mirror concentrates only that fraction of the arc's light output that is intercepted by the mirror. Clearly it is advantageous to place the arc as close as possible to the mirror. Appleyard states that it was Parsons who replaced the long focus by a short focus mirror (Appleyard 1933). This was practicable because his mirrors were relatively thin and uniform in thickness, and so they were better able to withstand the heating caused by an arc. Stoney records that by October 1888, Parsons was already supplying mirrors to the Admiralty of 24 in diameter and 17.125 in focal length, as well as of 20 in by 16 in. At first the mirrors were sent to an outside firm, Reed Millican, to be silvered, but after Stoney's arrival sufficient expertise in silvering was available for it to be undertaken in house.

When Parsons moved to his own factory at Heaton he built a wooden shed 40 ft by 20 ft to house a 4 ft diameter gas fired furnace as well as several items of machinery. A polisher was built along the lines of the machine used for the 6 ft mirrors at Birr. Other machines with wooden frames were built for both grinding and polishing. As the demand for mirrors grew, the shed was extended and a brick building was also added. Despite the potential fire hazard further gas fired furnaces were installed in the wooden structures. In 1905 the department was moved to another corner of the site. It will be appreciated that since the moulds were repeatedly heated and cooled, they required continuous attention. Over a period of time, improvements were made to the precision of the templates used in machining the moulds. Compared to mirrors of German manufacture, moulded mirrors at that time lacked accuracy. An attempt was made to remedy this by using a specially built machine. It carried a fast rotating grinding tool 2 in in diameter, which was traversed across the surface while guided by two templates. This did improve matters, but there were difficulties due to the effects of vibration. Meanwhile the moulding process was improved, and a thermocouple was introduced for monitoring temperature, so this machine was discarded. Gradually the narrowness of the beam was improved so that a distant source of light created a beam that converged to a spot 1/8 inch in diameter. This is small compared with the size of the crater of the electric arc that could be six times greater.

Ships which were navigating the Suez Canal had an economic pressure to continue their journey during the hours of darkness. To do this it was necessary for them to illuminate the buoys at night on both sides of the passage. Regulations covering such night time operation were issued by the Suez Canal Company in 1887. At the prompting of Sir J Duglas, Parsons developed another unique product that met this need (C.L.(1)), see chapter 14. This was a mirror that had a double curvature, a parabolic surface in a vertical plane, but which was ellipsoidal or hyperbolic in the plane at right angles to this. It produced a beam that remained narrow vertically, but that fanned out horizontally at an angle of 12°. Previously the practice had been to achieve this effect by using lenses.

Because there was no machine available that could generate such a surface automatically, the mould was formed first as an ellipsoidal surface and then it was finished by hand on either side to a parabolic shape. The mathematical equation that represents such a surface is in fact of fourth order. From the time that Stoney became manager of the Mirror Department in 1898, this equation was used to make a series of templates to guide the process of machining the mould. He went on to design a machine that could grind a surface of double curvature successfully (Anon 1931b). Later it was found that, by dividing the mirror vertically into two halves, it was possible to create two independent beams from a single light source with a dark region directly ahead so that the pilots of oncoming ships were not dazzled (Parsons C A and Stoney 1909). Large numbers of double curvature mirrors were also supplied for military use. Since their beam at first converged to a narrow vertical line before fanning out, they could be projected through narrow slits, and so protected from rifle fire. By 1906 both Parsons and Stoney were using double curvature mirrors on their motor cars, but they could not compete in cost with small Mangin mirrors.

At least 11 patents relating to mirrors were taken out by Parsons in the period 1894 and 1923. A great deal of effort was devoted to protecting the silver layer from damage. The practice followed by foreign manufacturers, of depositing a layer of copper on the outside surface of the silver, was also adopted. To give protection against corrosion a layer of specially formulated paint was applied over this. A wire mesh clamped by a band around the edge of searchlight mirrors helped to hold mirrors together even when they had been fragmented in battle (Parsons, Stoney and Bennett 1907). The lack of an in-house chemist's expertise was remarked on by Stoney. In 1912 as war approached, the demand for mirrors increased sharply and several large new bays were built, 225 ft by 32 ft wide. In wartime the output of mirrors rose to several hundred per month, ranging in size from 12 in to 60 in diameter. This made a significant contribution to the overall turnover of the company. In the last year of the war, the financial report to the board showed that mirrors contributed £98 100 to the total income of £513 014, a sizeable 19%. Orders decreased sharply after the Armistice, and by 1920 only a skeleton staff was employed. Business picked up somewhat from 1923 when talking films created a new demand for silent lamps. Arc lamps can create a loud hissing noise that makes them unsuitable for situations where sound is recorded live. High powered incandescent lamps with glass mirrors proved to be the answer to the problem. All the same the value of sales remained below £5000 a year.

21.2 Instruments and optical quality glass

Other optical equipment than mirrors had been in strong demand during hostilities. The firm of Ross Ltd of Clapham Common near London had for many years produced binoculars, camera lenses as well as other equipment of high quality such as cinema projectors (Campbell R S). Heaton Works supplied the mirrors for the last named items and in January 1921 an opportunity occurred for Charles Parsons to acquire a controlling interest in the firm, and he became its chairman. His nephew A D C Parsons, who had joined Heaton in 1913, and had risen to become the head of the Production Department in 1922, parted company with his uncle in that year to join the Mersey Railway Co., perhaps after disagreements about management, but in 1924 he accepted an invitation to join the board of Ross.

Before the war roughly 60% of the world supplies of optical glass came from Jena in Germany, 30% from France and 10% from Chance Brothers of Birmingham (Manville 1968). The demand during the war became so great that the Government arranged support for the Derby Crown Glass Co. to install machinery and secure manpower to greatly increase its output. But when peace came the firm faced collapse. In July of 1921 Sir Charles Parsons stepped in and acquired all the shares and discharged the firm's debts. He had no prospect of making a profit, but was determined that an important national asset should be preserved. The firm was now known as the Parsons Optical Glass Company. He obviously had his sights set on the manufacture of much larger components than any they had yet made. He lost no time in studying the production methods and making improvements. The glass was made in a pot that was heated in an individual oven. It was a frequent occurrence for pots to fracture, resulting in the loss of their entire charge. He redesigned the furnace so that heating now took place from above, but without allowing flames to play on the surface of the pot; also the exit for the gases to the flue was located beneath the pot. The result was a much more uniform temperature distribution in the furnace and pot breakage practically ceased altogether.

To ensure homogeneity in the glass it has to be stirred, so a study of various methods of stirring was made. In the end he chose a two bladed paddle that rotated about a vertical axis. The axis itself executed an orbital motion. Previously the glass was allowed to cool in the pot, but it tended to shatter into pieces of every shape and then had to be remoulded into blocks or other desired shapes. A new procedure was adopted whereby the pot was cooled somewhat while in the furnace. It was then transferred to a special pre-heated furnace. It was inverted over a heated plate that also carried a mould ring. A hole was made in the bottom of the pot to allow air to enter, and by means of special levers the pot was pulled upwards, while the viscous glass 'flowed out' onto the plate. The yield was increased appreciably and discs as large as 42 in diameter by 7 in thick were regularly made. A study of the materials used in making pots was made with a view to minimizing the tendency of certain components to dissolve in the glass and so degrade its quality.

A further essential precaution to secure good quality is perfect annealing. To achieve this the blanks were placed in a container and covered with sand, and an insulated canopy was lowered over them. Gas or electricity was used to heat the oven. An externally driven fan, with suitable deflectors, ensured that the temperature remained uniform throughout. Lord Rayleigh told how (Rayleigh 1934)

> *Parsons had examined the annealing process, and had determined by the permanent bending of a glass thread what was the temperature at which it became plastic. This investigation pleased him very much by its simplicity and showed the way to a considerable economy of fuel in the annealing process.*

One project was begun, though not completed before Parsons' death. Its aim was to improve the speed and efficiency of the process of cleaning a batch of glass at the end of the melt. It involved lowering a special hood over the pot and creating a vacuum while the process of stirring continued unchecked.

A number of machines were installed for grinding discs or blocks before despatch to customers. These were based on the Heaton design. Experiments were also made on moulding lens blanks to a shape close to the desired finished dimensions with the aim of minimizing the amount of grinding to be carried by lens manufacturers. In all it was reckoned that an expenditure of over £60 000 was involved in improving manufacture of optical glass.

21.3 A tradition of telescope building

Otto Boeddicker continued to work at Birr until 1916 when, being a German citizen, he was allowed to return home. But long before Laurence's death in 1908, the giant telescopes had ceased to be operational at Birr. Neither William Edward, who succeeded as fifth Earl of Rosse, nor Lois Lister-Kaye, who became his wife, had any interest in astronomy. In 1912 Clere and Charles sent one of the mirrors of the 6 ft telescope, mounted on its carriage, together with other items, to the Science Museum in London for safe keeping. The rebellion in Dublin at Easter in 1916 began a new period of unrest in Ireland. In 1919 while Laurence Michael, the sixth Earl, was still a child, violence broke out and gradually intensified. In that year Birr Castle was damaged severely by a fire which destroyed the library. It was started during the night by a fault in the electrical wiring (Anon 1919). Lady Rosse, who slept in the bedroom above the library, was successful in getting everyone to safety with the exception of a French governess who died in the fire. Civil war broke out in Ireland after the Treaty was ratified on 7 January 1922. The barracks at Crinkle outside Parsonstown were burned by anti-Treaty forces on the night of 14 July, and this led to the Castle being once again garrisoned for a time by government troops. When the family returned late in 1925 the Leviathan, like the Castle itself, was in a state of disrepair. Although the structure of the 36 in telescope remained largely intact until 1927, the walkways, platforms and counterweights of the Leviathan had to be removed for safety reasons (Moore 1971). No doubt these events must have greatly distressed Charles Parsons, now in the last decade of his life; nevertheless he had been offered another opportunity to renew his acquaintance with large telescopes.

In 1901 Sir Howard Grubb, the renowned Dublin optical instrument maker, had designed and patented a periscope for use in the first Holland submarines being built for the Royal Navy by Vickers at Barrow-in-

Furness. Without such a device the crew of a submarine were of course at a grave disadvantage. In February 1902 Vickers cabled Isaac Rice of the Electric Boat Co. who owned the Holland patent rights, *'Course can be accurately kept by Sir Howard Grubb's periscope'* (Scott 1962). The steel tubes for periscopes had to be precision made and strong enough not to deflect when the vessel was moving. They were manufactured by Vickers and were then sent to Dublin where Grubb had supplied and fitted the optics for up to 95% of all periscopes made, from his works (Glass 1997). However the navy became alarmed about the hazard of the double crossing of the Irish Sea in wartime and it was also concerned about dangers to production from republican sympathizers in the workforce. As a result Grubb was forced to move his works to England, to a factory at Saint Albans in Hertfordshire, though the actual move was not accomplished until 1919 when of course the war had ended. This disruption, together with the impact of the war on sales of his telescopes, eventually forced Grubb into voluntary liquidation in 1925. Two projects were in hand at the time, a 26.5 in refractor telescope that had been ordered in 1909 for Johannesburg, and a 32 in refractor and a 1 m reflector that were ordered by the Imperial Russian Government in 1912. After Grubb encountered difficulties with Chance Bros, Parsons had agreed to supply glass for the lens of the South African instrument from his newly acquired glass factory. By April 1925 Parsons had purchased *'the goodwill, drawings and sundry machines'* of Grubb's from the liquidator. The Saint Albans works was closed and a new premises was built close to Heaton at Walkergate. The company was henceforth to be known as Sir Howard Grubb, Parsons and Co. Some of the Grubb staff were retained, Cyril Young as general manager, Romney R Grubb, a son of Howard Grubb, the optician John A Armstrong and two others. By the time of Parsons' death the number had grown to 25 and the firm had become a wholly owned subsidiary of C A Parsons and Co. It was planned to build large telescopes, the larger components of which were to be built at Heaton and assembled at Walkergate.

In 1924, while Grubb was teetering on the verge of collapse, R T A Innes travelled from Johannesburg to report on progress. In February 1925 he wrote that (Glass 1997)

> *Soon after I got to England and work on the two disks (for the object glass) progressed, it was discovered that the flint disk supplied through Sir Charles Parsons' firm* [it appears that the Chance flint turned out unsatisfactory] *was not perfect and Sir Howard Grubb actually recommended me to decline it... If I rejected the disk, then we might feel*

Figure 129. The 36 in diameter reflector telescope for the Royal Observatory at Edinburgh, being tested at the Walkergate Works (Manville 1968).

> *sure that Messrs Grubb would go into liquidation forthwith. Sir Charles Parsons in my presence instructed his manager to cast another flint disk forthwith and privately told me that if Grubb's went under 'he would see me through'—I asked him if he would mind putting that in writing but he declined, because, he said that Grubb's liquidator might use such a promise as an asset.*

In the event the '*completed telescope was found to be a very fine one*'.

After the revolution in Russia, the order for the 32 in refractor and dome being built for Pulkova were cancelled. This followed lengthy and fruitless discussions about progress between Sir Charles Parsons, Sir Howard Grubb, Sir Frank Dyson (Astronomer Royal), Romney Grubb and Young on the one hand, and three Russian Professors, Kriloff, Blumback and Mme Foehringer for the Observatory on the other. Among the issues considered was a proposal to increase the diameter of the lens to 41 in. Indeed the Parsons Optical Company had successfully cast four 42 in diameter discs for the object glass, weighing 800 lb each (Sisson 1968). The 1 m silvered glass reflector built for Simeis in the Crimea on the other hand was completed with its dome and delivered in March 1925. It too turned out to be an excellent instrument but it was destroyed in the Second World War. The first large telescope designed and built by the new company was a 36 in reflector for the Royal Observatory in Edinburgh and it was completed during Parsons' lifetime (figure 129) (Manville 1968). In addition, two telescopes were built for Stockholm Observatory, a 40 in reflector delivered in 1930, and twin refractors with 20 in lenses (Anon 1931c). At the time of his death a telescope for Toronto with a 74 in reflector was under construction in the works. It would be the largest built in Europe (Hasselkus 1931). Very large reflector telescopes were being planned at this time including one with a 200 in mirror in the United States. The challenge of returning to the field in which his father had won fame led Parsons to make a series of attempts at constructing glass discs for mirrors of such a size. Like his father he experimented with composite structures. First he tried fusing a stack of discs each between 3/8 and 3/4 in thick. After many failures he did complete a disc of 52 in diameter, only to encounter problems when polishing, caused by imperfections in fusing and unequal thermal expansion in the different layers. He then tried another approach, which involved creating an assembly in which two discs were separated by squares of glass set on edge, the whole being carefully fused and annealed (figure 130). Once again he encountered problems when grinding and polishing. The differences in the routes by which heat generated during these processes, was conducted away made it impossible to achieve the precision of a few millionths of an inch that was required for a good figure. In the end the Pyrex glass that had been developed by the Corning Glass Co. in America, proved to be superior.

He made one last, unsuccessful, experiment harping back to bronze specula. This time a welded steel structure was made with a nickel plated surface. No doubt it was his extensive first hand experience with large mirrors that caused Professor G Hale to consult him about the projected 200 in telescope for Mount Wilson (Anon 1934). After Parsons' death the firm of Sir Howard Grubb, Parsons and Co. continued to prosper as a scientific instrument manufacturer. It eventually met its demise in 1984, shortly after it had delivered the 4.2 m William Herschel altazimuthal telescope.

Figure 130. Composite glass mirror of 3 ft diameter (1930) (S.P.(2)).

During all his involvement with the optical industry, Parsons' objective was always to improve the manufacturing process. He never sought to emulate his father's achievements as an astronomer. When a staff member proposed as the title for a paper to be read at a meeting of the British Association in 1931 'The practical limit to our vision into space', which certainly had a resonance with his father's work, it was rejected with the assertion that 'we are manufacturers not astronomers!' (Dowson 1931).

Chapter 22

Autumn of Life

22.1 Advancing years

Early in 1923, after a distinguished career, Richard Clere Parsons died (Anon 1923b). Although Clere never became involved directly in Charles' projects yet he had always been intensely proud of his brother's achievements. In 1926 Dr John Bell Simpson, the wealthy mine owner who had backed the development of the marine turbine from the start, and who was a close friend, also died. In the summer of 1924 Parsons marked his 70th year by crossing the Atlantic with Lady Parsons in the company of scientists who were travelling to the British Association meeting in Canada. He read a paper to the International Mathematical Congress at Toronto which was unique in actually containing some mathematical symbols such as an integral sign (Parsons C A 1924a). With the physicist Sir Ernest Rutherford, Major General Sir David Bruce of the Lister Institute and Sir John Russell of the Rothamstead Experimental Station, he was awarded an honorary degree by the University of Toronto (Anon 1924). He used the opportunity to visit the Allis Chalmers plant at Milwaukee and the Westinghouse works in Pennsylvania. George Westinghouse had died ten years before, but Frank Hodgkinson, who was by now an engineer of renown in his own right, would have given his mentor a warm welcome. Seventy is the normal age for retirement, but of course Parsons did not retire. Perhaps he should have. At any rate it is a convenient point to take stock of his affairs.

The two firms of C A Parsons and Company based at the Heaton works, and the Parsons Marine Steam Turbine Company located some distance away at the Turbinia works at Wallsend, were quite separate financial entities. Parsons owned most of the equity in the first, but was only a relatively modest shareholder in the second. At Heaton the Managing Director was H I Brackenbury who had come from Armstrong–Whitworth in 1919 to fill the vacancy left by Alex Law. The senior engineer was A Q Carnegie who had joined as an apprentice and had been made a member of the board in 1918. Robert Dowson (1888–1944) had spent summer vacations at Heaton while he was a student at University College London. When he graduated with a degree in engineering in 1911 he came to Heaton and was now responsible for turbine research and development there (Anon 1945). In 1929 he was given responsibility for developing new designs and was freed from routine work. By 1930 he had risen to become head of all mechanical research. He died unexpectedly in 1944 two years after achieving a PhD from London University. His obituarist tells us that in writing to a friend he had posed the question *'what have I done for my fellow men to help the community in general?'*.

The Wallsend factory was headed by the naval architect R J Walker, and the personnel there seem to have been less prone to change; perhaps the physical distance from Heaton helped to avoid personal conflicts between Parsons and his employees. The technical manager was S S Cook, who had played a major part in evolving suitable strategies for the blading of marine turbines, at a time when their power increased at an extraordinary pace in the period leading up to the building of the *Mauretania*. He introduced more modern ideas about the flow and momentum of fluids acting on blading surfaces and he also used the unique water tunnel at Wallsend for testing propellers under conditions of reduced pressure. Though the tunnel had been used to carry out much systematic testing for shipbuilding clients, none of it had been published (Burrill 1951). As a sideline he had also carried out much of Parsons' programme of high pressure research, the equipment for which was of course located at Heaton. At the same time he did much of the preparatory work needed before the adoption of turbine gearing for high powers. In technical matters he had in many respects taken Stoney's place as Parsons' right hand man. He was named with Parsons as the author of 75 patents, as well as sharing in a further 20 with other employees. In 1928 he was elected Fellow of the Royal Society and in 1930 he became a director of the Marine Company.

Two other talented younger men joined the Heaton Works. One was Jessel Rosen. He had worked closely with Parsons on alternative designs of alternator rotors just before the First World War (Rosen 1922). He was also responsible for developing, with Parsons, the novel concentric conductors for high voltage alternator windings (Parsons C A and Rosen 1929). These made it possible to generate currents at a voltage of 36 kV, which made it possible to connect directly to the supply cables without the extra cost of a step up transformer. After a long interval it is an idea which is being pursued once more by a large manufacturer (Leijon 1998). The other recruit was Claude Dixon Gibb (1898–1959), an Australian airman who had graduated from Adelaide University with a degree in engineering after wartime service. In 1924 he had come to Heaton as an apprentice. In the following year his resourcefulness in dealing with a problem which had been encountered during the commissioning of a turbine at the Barking station in London brought him to Parsons' attention. He was immediately put in charge of the Test House, then moved to the Design Offices with a commission to overhaul the organization there. In 1929, he was called to Parsons' office to discuss some incident when, out of the blue, '*Sir Charles said quite suddenly and unexpectedly, "I am making you a Director, come and have a cup of tea"*'. It seems that Parsons acted without reference to other members of the board in appointing him a director and Chief Engineer (Hinton 1959). In later years he achieved distinction for his success in steering the firm through the period after the Second World War, when plans were being laid for the building of some of the largest turbines ever built. He was elected Fellow of the Royal Society in 1946. Some insight to the reasons for his astonishing progress made under the elderly Parsons can perhaps be glimpsed from the comments of his obituarist,

> *one of the qualities most admired by all employees in the factories for which he was responsible was the pleasure which he took in meeting and mixing with the people on the shop floor; he did this as it can only be done by a man who has worked on the shop floor himself.*

And

> *Claude Gibb had an honest and healthy liking for publicity; it arose not merely from personal taste for it but also from a proper evaluation of its business value...*

Gibb was obviously both resourceful and had some of the self-confidence for which Australians are renowned. Besides he was able to handle Parsons and had not been cowed by him. Another consideration in his selection may well have been the need to find a replacement for A Q Carnegie who had been taken ill in the summer of 1928. When eventually he recovered his health, he was posted to the London office. However his ailment returned and he died in 1934 (Bedford).

The total value of sales at Heaton for the year 1923/24 was £714 000 of which 'machines' accounted for 85% (TWASf). This did not change greatly in the next five years. Profits were sufficient to pay a dividend after tax of £32 854, which represents $12\frac{1}{2}$ % on shares worth £332 000. The surplus was transferred to investments, which stood at £414 736 in that year, or were used to fund extensions to the site and buildings. The workforce at Heaton stood at 1473, an increase of one-sixth since 1914. Wages and salaries totalled around £235 000 a year. The average weekly wage for apprentices was £0.725, for women £1.175, and for men the combined average for skilled and unskilled was £3. The salaries totalled £6176 for 274 persons, three-quarters of them male, which averages £4.50 weekly. Business after the war was very competitive. Quite a number of manufacturers now supplied the home market, only some of which were licensed by Parsons. A request from Vickers for a license was refused due to 'excess capacity'. In the six year period 1922–1927 the home market for turbo generators had been shared out as follows (TWASg):

Firm	Output (MW)
BTH	805
Metropolitan Vickers	675
C A Parsons	587
General Electric Co.	437
English Electric	285
Brush	243
Bellis and Morcom	106.7
Richardson Westgarth	76.6
D Adamson	46.8
James Howden	19.7
Others	257.6
Total	3539.4

This places C A Parsons and Co. in a modest third place, taking one-sixth of the total. Attempts were made to manage the home market so as to maintain prices while sharing out work among the firms building turbines. Parsons had an interest in this both as a manufacturer himself and as the recipient of royalties. In 1927 Brackenbury reported to the board that roughly half their orders were 'open price' and half 'closed price', and that the former made a loss of 4% , while the latter gave a profit of 16%. The resulting net profit was 5.6%. There was concern over the action of Metropolitan Vickers who '*took an order for Northampton contrary to the terms of the agreement*'. In November 1927 a meeting of the British Electrical and Allied Manufacturers Association was held to discuss the 'Turbine Price Maintenance Agreement'. To allay fears that foreign imports were depressing prices, the secretary produced figures for the years from 1919 to 1926, which showed that the British signatories had supplied a total of 4081.4 MW, while Swiss firms had delivered only 383.2 MW.

The Japanese firm Mitsu Bishi Goshi had paid £1000 to Parsons in 1904 for the right to build turbines for use on land only (TWASh). In 1907 a 15 year agreement covering turbine powered dynamos and alternators was concluded with a royalty of 1/- (£0.05) per kW. The first machine, built that year, was a 500 kW three phase alternator. The initial agreement covered turbines of all types and in 1921 royalties were actually paid on three Ljungström machines. The agreement expired in 1926. It was not renewed and Zoelly designs were adopted instead. Despite this, a five year license covering pivoted thrust bearings for sale in the Far East was signed in October 1926. The board meeting in January 1927 considered a letter from R H Rooksby in Tokyo (TWASi). He reported that

> *the quality of your product is higher than that of all cheaper makes* but that *they* (Mitsubishi) *found the cost of manufacturing Parsons turbines considerably higher than the cost of manufacturing Impulse turbines such as the Zoelly which they now sell. In addition for land machines they are able to guarantee efficiencies for the Zoelly type equal to those of the Parsons type... According to Mitsubishi the marine turbine has a further disadvantage in that it requires considerably more engine room space than that of machinery of Impulse design.*

By 1928 the workforce had risen to nearly 1800, though the number of women in the shops had been halved since the end of the War.

	Females	Apprentices	Men	Total
Shops	89	249	1138	1476
Offices	68	4	249	321
Totals	157	253	1387	1797

When the factories at Heaton and Wallsend were established they were equipped with the latest machinery, but, a quarter of a century later, much of it was in need of replacement (Bedford). Bedford recalled that

> *sometime after the First World War a party of German engineers from Blohm and Voss, the large shipbuilding and engineering company of Hamburg,*

visited the Parsons Marine Works at Wallsend. As licensees of the Parsons Foreign Patents Co. they had come over to see the latest practice in marine turbine construction.

After their inspection at Wallsend, Sir Charles brought them to Heaton and showed them around the shops. At tea later, one of the delegation said to me, 'I think it is wonderful how you achieve such accuracy with such old fashioned machinery.'

This remark struck me very forcibly and after their departure I prepared a list of the principal machine tools, planes, boring machines, lathes etc, showing the dates when they were installed. They were all pretty old, some of them had been in use for over 20 years. They were evidently doing their job, but no one seemed to have realized the advantages in speed and efficiency that could be gained by the provision of modern machine tools.

On seeing the list, Sir Charles agreed that we should embark on a programme of re-tooling the main machine shop.

With all the talented engineers in Parsons' employ, it is extraordinary that it was left to foreign visitors to nudge the Secretary into action. Once the go-ahead had been given, the potential for gains in productivity were quickly grasped. The 'Works report' to the board dated 9 March 1927 noted that the newly ordered radial drill was guaranteed to cut times to one-third, while the two special purpose milling machines that had been purchased had already cut times by 50 and even 60%. In 1929 an investigation into the poor quality of laminations used for alternators revealed that the machine for punching holes, which dated back to 1903, was in urgent need of replacement. Dowson and Crocker submitted a report with proposals for scrapping the belt drives on old machines and for purchasing new machines at a cost of £20 550. At the same time tests were made that showed that the lighting fittings in the large machine shop had so deteriorated that the lighting levels had dropped by one-half. A 330% increase was planned, with no more than an 80% increase in power.

Bedford had become a member of the board in 1922. In 1925 he was invited with his wife to join Sir Charles and Lady Parsons in a holiday to the West Indies. In the course of the holiday there were good opportunities to discuss future plans. According to Bedford, there was a disagreement between Parsons and Brackenbury in 1926, and Parsons appointed Bedford as joint Managing Director. Not surprisingly Brackenbury resigned at the end of 1927, and so in a strange way the sequence of events in which Dakyns had departed from the company was repeated. In fact Bedford appears to have been one of the few senior executives who remained all his working life at Heaton. Engineers who became involved in management decisions had soon found themselves in the firing line, but he at least survived.

Parsons' delight in tackling new projects without much regard for likely financial reward never deserted him (Parsons R H). One example may be instanced. In 1925 an 817 BHP turbine was built for a pumping station at Wapping. The relatively small turbine ran at 8000 rpm and this was geared down to 1500 rpm to drive a water pump. The turbine had a Curtis stage followed by four Rateau and five Parsons stages! Another challenge was met in 1929 when an order was taken from the neighbouring firm of Reyrolle. They wanted a special purpose alternator which would allow them to subject their high voltage circuit breaker switches to the sort of currents that they must encounter in service when clearing short circuits. The alternator was designed to give a peak output of 1500 MVA. While the enormous currents subjected the conductors to very large mechanical forces, they were of short duration and so heating was not a problem. Rosen's design was a complete success.

The invention of the steam turbine led to intense interest in the physics of expanding steam. Service experience with impulse designs had shown that even steel turbine blades could be eroded rapidly by very high velocity steam, though in general Parsons' reaction stages with their lower steam velocities were free from this problem. In time it was recognized that as steam expands, and its pressure falls the temperature drops below the boiling point and some vapour condenses as droplets of water. If these droplets grow, and if the steam velocity relative to that of the blade surfaces is high, their impact can erode the metal, just as the collapse of bubbles caused by cavitation erodes propellers. Other questions faced designers. If steam expanded very rapidly, did it remain a gas even at temperatures below the boiling point? Parsons thought it did (Parsons C A 1924b). In order to calculate the mass that will flow through a given passage it is necessary to know its density. This will obviously be higher if the flow contains some liquid as droplets. There are great experimental difficulties in measuring the true temperature and pressure in such circumstances which had been brought home to Parsons and Stoney when they were exploring the flow of hot gases for a gas turbine (Richardson 1911). Years before, they arranged for gas from a flame to expand to a low pressure in a divergent nozzle. Platinum wires were placed in the flow at the inlet and outlet of the nozzle, and they were viewed through glass windows. Both wires glowed bright red, so indicating that any cooling that occurs during expansion is almost completely reversed when the velocity of the jet is abruptly destroyed by striking the surface of the wire. In March 1925 it was suggested that the advice of Professor C T R Wilson, who had invented the cloud chamber, should be sought since he was expert on the subject of causing supersaturated vapour to condense (Wilson 1925). Stoney had recently visited Ray and must have discussed the matter because on 15 April he wrote to Parsons pointing out that the measurements of flow rate on recent turbines would require impossibly high efficiencies if steam did remain supersaturated (Stoney 1925)[1]. Nevertheless a rig was built by Dowson to try out Wilson's idea of adding very fine dust to the flow. Writing to Parsons on 11 June, Dowson reported that in fact there was no effect on the mass flow rate, so confirming Stoney's assessment (Dowson 1925).

22.2 Gerald Stoney returns to Heaton

When Gerald Stoney broke with Parsons in 1912 he worked at first as a consulting engineer with Richardson Westgarth at Hartlepool on turbines, and with H T Newbigin on Michell bearings. During the war he was a member of the panel of experts of Lord Fisher's BIR and numerous other committees. In 1917, on the suggestion of Charles Merz, he was chosen for the chair of Mechanical Engineering in the Faculty of Technology at the Victoria University in Manchester. The Institution of Mechanical Engineers set up a committee to study the behaviour of steam nozzles, of which he was a member and he acted as the reporter for the experimental work that was carried out in Manchester. The study related chiefly to the design of impulse turbines, and Stoney reckoned that, for a total expenditure of £8000, fuel savings of £50 000 per year had been achieved (Stoney 1931b). At this time his wife was an invalid and she remained at Newcastle. Her health deteriorated, and increasingly heavy demands were placed on Stoney, travelling to and from Manchester. In 1926 when Parsons became aware of this, he invited Stoney to return to Heaton as Director of Research.

One of Stoney's first tasks was to re-organize the mirror department when the manager E Bennett resigned in April 1927. He was able to report that they had (TWASj)

> *improved the accuracy of the templates, moulds etc and introduced improved methods of bending, grinding, polishing and testing the reflectors which had*

[1] When some steam condenses to water during expansion through a given pressure range, some extra thermal energy is given up which can perform extra mechanical work. If this contribution is overlooked the turbine will seem to have achieved its mechanical output from a smaller thermal input.

resulted in them having now more than double the accuracy, while losses due to breakage and rejection were reduced, as well as the time required for grinding and polishing.

Stoney's report for the board at the end of 1927 gives an overview of the state of research and development (TWASj). The test rig 'Alice', that had been used for experiments on blading, was over 20 years old. It was overhauled and re-launched as 'Alicia'. Studies were being made in collaboration with Stanley Cook and '*from the preliminary results on wing blades in Alicia it was possible to formulate a theory of the flow of steam at the exhaust end of turbines where there is a rapid expansion and large carry over*'. Noting that still, after 40 years, no reliable data were available for the effects of blade tip clearance on efficiency, a systematic study was made that settled the matter. Curtis blading and components of Rateau turbines were being tested. He was conscious of the fact, no doubt influenced by his sojourn in Manchester, that tests to study losses incurred at turbine exhausts could be made on models using air, but only if the correct Reynolds number were used[2]. Tests on blades were also made with air instead of steam. Dowson was working on designs of condensate extraction pumps and steam ejectors that it was hoped to commercialize soon. Air coolers and rotor fans were being studied to improve performance, as well as many component parts such carbon glands, governors and oil pumps. Meanwhile, at Wallsend, Cook was carrying out tests to understand the cause of erosion in the tubes of condensers.

Evidently much of the research and development was being carried out by staff, largely on their own initiative. The arrival of such an experienced person as Stoney was very timely. His past service with the company was invaluable in establishing a corporate research capability that could survive the inevitable departure of Parsons. Parsons' own direct involvement had been much reduced. He did chair a meeting between Cook and Douglas who represented the PMSTCo., and Gardner and Dowson who represented Heaton. The subject for discussion was the variety of cross sections and gauges used by both companies in the manufacture of turbine blades. It was agreed that they would adopt common standards for the future. Stoney reported that '*a special blade with a long entrance, as suggested by Sir Charles Parsons is being tried*'.

[2] Osborne Reynolds had shown that different measurements on the effect of fluids moving relative to solid structures could be compared, even if the structures were of different scales, provided a number, the 'Reynolds' number', which embodies three physical quantities, is kept the same in each case.

The design office continued to advance the power rating of turbo-alternators (Parsons R H). In 1929 two turbines with an output of 30 MW at 1500 rpm were built for Velsen in Northern Holland. These were tandem machines with the low pressure cylinder arranged for double flow. In that year another machine rated at 30 MW, but running at 3000 rpm, was supplied to the Hackney municipal station in London. The benefit of the higher speed can be seen by the fact that, at 46.5 ft overall, the Hackney turbine was 9.5 ft shorter than the Velsen machine.

The most advanced machine built under Parsons' supervision was a 50 MW ordered for the Dunston station of NESCO in 1930 (figures 131(a) and (b). The output was the same as the 1922 Chicago machine, but it was supplied by a single alternator rated at 62.5 MV A, which was driven by a tandem two cylinder, single flow, turbine. Also the pressure and temperature of the steam supply were higher. The Elberfeld machine had set new standards of efficiency when it was built in 1900, and a comparison between the two machines serves to illustrate the great advances made in 30 years (Parsons R H). Neither had an impulse stage at the inlet so that 115 pairs of blade rows were required for the Dunston machine. Particular attention had to be paid to the prevention of cylinder distortion given the high temperatures and long casings. Numerous refinements were incorporated in this machine: for example the blades in the final row were made of hollow construction in order to ease the centrifugal stresses. Also, for the first time damage due to erosion of blade surfaces by droplets of water impacting at high velocity was becoming a problem because of the adoption of more advanced steam inlet conditions, and higher blade velocities. A test rig was built to help in the search for a material that would best resist erosion. A tungsten alloy was identified. As this was expensive, small shields of the material were brazed on locations which had been found to be affected. This work benefited from the earlier studies of cavitation damage. In general the designs of turbo-alternators had been kept up to date, but the initiative now lay more and more with the firm's designers than with Parsons himself. In terms of maximum output American firms already held a clear lead.

22.3 Still tackling new challenges

Parsons himself concentrated his attention on other matters. One example was the problem of failures in condenser tubes. This had become a matter for serious concern, especially to ship owners. Cooling water frequently contains salt and other contaminants, and if these reach the boiler and

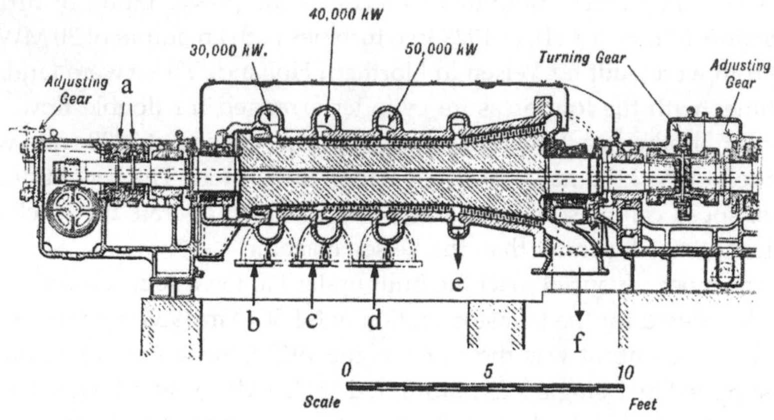

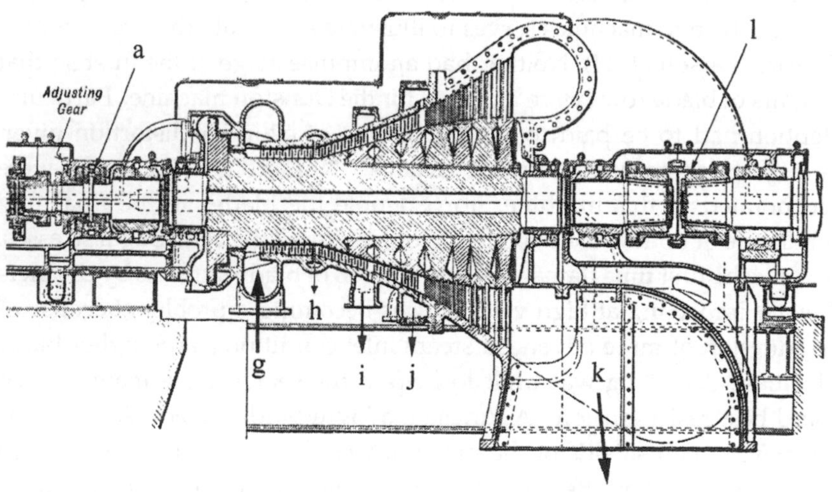

Figure 131. (top) A sectioned view showing the high pressure turbine of the 50 MW machine for the Dunston power station. A two collar, tilting pad, thrust block is visible at a. Steam inlets at b, c and d come into play as the load increases. Steam is returned to the boiler for re-heating at f. (Parsons R H).

(bottom) Section of the Dunston low pressure turbine. Re-heated steam is admitted at g to the low pressure turbine. Steam is bled off at e (upper figure), and at h, i and j, to warm the water supplied to the boiler. The exhaust enters the condenser at k. The claw coupling to the alternator can be seen at l (Parsons R H).

	Elberfeld	Dunston
Output (MW)	1.25	50
Speed (rpm)	1500	1500
Configuration	Tandem, single ended LP	Tandem, single ended LP
Steam pressure (p.s.i.)	130	600
Steam temperature (?F)	480	800
Reheat point	None	Between HP and LP cylinder
Reheat pressure (p.s.i.)		120
Reheat temperature (°F)		800
Feed water heating	None	At 4 points
Exhaust vacuum (in Hg)	28.4	29.0
Heat consumption BThU $kW^{-1} h^{-1}$	22 000	9280
Thermal efficiency (%)	15.5	36.8
Overall length (ft)	51	80.5
Power density* kW ft^{-2}	1.82	31.0

* This relates the power output of the turbo-generator to the floor area required to house it.

superheater tubes because of leaks into the condenser, they can cause deposits to form and can lead to further damage to the boiler. The question to be answered was 'were the condenser tubes damaged by corrosion (chemical attack) or erosion (a mechanical action)?'. In the case of Parsons' earlier study of propellers the evidence had pointed strongly to the second process. In 1927 he presented a paper to the Institution of Naval Architects in which he sought to show that, at least, erosion is also a contributory factor to tube failures. But the problem with condensers in general is too varied and complex for any one explanation (Parsons C A 1927b).

Parsons' attempts to make larger telescope mirrors have already been mentioned, as well as his efforts to improve the process of glass making. Not entirely unrelated to the latter was a project that was sparked off by the catastrophic accident in Shanghai. There it was found that the rotor had failed because of a large internal defect in the ingot. The demands being made by manufacturers on the suppliers of forgings were stretching the suppliers to the limit. At a recent joint meeting of turbine builders with the Forgemasters association, the chairman had said '*Gentlemen you must not go in for larger forgings; you have reached the limit, and must moderate your demands*'. Indeed a committee of the Iron and Steel Institute was established in 1924 to study inclusions and heterogeneity in steel ingots. Its reports in 1926 and 1928 disclosed the presence of such weaknesses in every one of the ingots that were studied. Parsons sought to remedy this situation by changing the way that ingots of steel are poured, thereby improving the uniformity of texture of the metal throughout. The 20 ton trial ingot was made much fatter than normal, having a 70 in diameter and being only 45 in high. The base of the mould was a massive steel structure that acted as a chill, extracting heat from the bottom. The walls of firebrick were thermally insulating and were pre-heated. The cover had provision for mounting burners for keeping the top of the casting hot. In this way he caused the metal to freeze from the bottom upwards, with temperature largely uniform in a horizontal plane, but decreasing from top to bottom. It is very evident that his efforts were strongly influenced by his father's approach when casting the large bronze discs for his telescopes.

The ingot was extensively examined visually, chemically and by means of mechanical tests on specimens cut out of the bulk. In addition samples were provided to other researchers for them to examine. Results were reported in a paper to the Iron and Steel Institute (Parsons C A and Duncan 1929). The segregation of impurities along the central axis had been eliminated and mechanical properties had been improved. The first trial was only partially successful and the mould was modified, after which five more castings were made. There was general agreement among the audience that the research was a valuable contribution to solving a pressing problem, though the economics of the technique seemed to rule it out for regular use. Of course the technique was covered by filing a patent, number 278 033 of 1928. In order to be able to test the method it had been necessary to persuade a steel maker to co-operate in what was a large scale and expensive exercise. Early in 1928 Parsons got S F Prest to write to W Beardmore and Co to secure their co-operation in casting a 20 ton ingot at their Parkhead Works in Glasgow. In May 1929 Parsons' contributions to the science of metallurgy were acknowledged by the award of the Bessemer medal of the Iron and Steel Institute.

In 1923, during a visit to Newcastle, Edward Prince of Wales was shown around the works by Sir Charles Parsons. When it was decided to stage an exhibition of industrial firms on Tyneside in 1929, Parsons was asked to be the president. During the post-war years Parsons' reputation as an engineer and inventor acquired a legendary quality. The most notable honour to be conferred on him, and one which gave him much pleasure, was his appointment to the Order of Merit in the birthday Honours List of King George V in 1927. The Order had been created in 1902 by Edward VII,

Figure 132. Sir Charles A Parsons showing a 500 kW turbine to HRH Edward Prince of Wales during his visit to Heaton in 1923 (Anon 1998).

to recognize eminent service in the armed forces, or distinction achieved in science, art or literature. It was limited to a membership of 24 at any one time. Women, like Florence Nightingale, were occasionally included. Lord Kelvin was one of the first members, but Parsons was the first professional engineer to join the Order. He wrote from Ray to his nephew Laurence in September 1927 (BirrR/5)

> *Very many thanks for your kind congratulations. I knew nothing about their thinking about giving me the O.M. till a letter reached me from Lord Stamfordham about ten days before it was in the papers. From the many letters that reached me from engineers, they seem very pleased that their class for the first time have been considered, and I was surprised to receive a telegram of congratulations from Mussolini on behalf of the Italian Marines.*

Several distinguished people were sounded out beforehand, and the response of Sir Alfred Ewing may be quoted (Bedford 1939),

The Athenœum
Pall Mall
S.W.1
December 1st 1926

Dear —,

I entirely agree with your suggestion—Parsons pre-eminently deserves the Order of Merit; the wonder is he has not received it before.

His development of the steam turbine is incomparably the greatest achievement of any living engineer, the greatest both in its effects and in the genius and courage it has required. Parsons as an inventor has proceeded from the first on scientific lines. He has poured out (and still pours out) a stream of new ideas. But unlike most inventors, he has also had the energy and resolution to reduce them to practise.

It was this combination of qualities that made him first create the steam turbine, then improve it stage by stage, and so gradually apply it on an a constantly increasing scale to the production of power. Parsons has made the modern central station possible, and has revolutionised marine propulsion.

It has been my privilege from time to time to test his turbine—the early types used in central stations and the experimental engines of the first vessel, the 'Turbinia'. This gave me some insight into the working of his amazingly fertile mind, and some knowledge of the steady advance by which difficulties were overcome. His success is now so universally acknowledged and so complete that the early struggles may hardly be realized by people who are only familiar with present-day results.

It has been a splendid progress, compelling at every step one's admiration. And all through it he has kept his modesty. I can imagine no award of the Order of Merit that would be more unquestionably regarded as right, or that would give greater pleasure to those who keep in touch with practical science.

J. A. Ewing

Professor Alfred Ewing was particularly well placed to evaluate Parsons' contribution from a British point of view. J H Bruce was an engineer with Parsons Foreign Patents Company, and he had plenty of contact with engineers throughout Europe (Bruce 1939). He noted that of late when speakers at technical meetings referred to Sir Charles Parsons, they spoke of 'Parsons', as one might speak of Shakespeare, Watt or Lister or other eminent persons.

Frequently, one heard during meetings of his Continental friends and colleagues the question, 'Was sagte der Parsons dazu' or 'Wie denkt der Parsons daruber'—always 'der Parsons'—that retiring genius.

Probably Parsons' last visit to Northern Europe was the one during which, and much against his inclination, he was prominently before the public eye when, as one of the representatives of H.M. Government and President of the Engineering Section, he attended the World Power Conference in Berlin during 1930[3].

Punctilious in the discharge of his self-imposed duties, his retiring disposition was a constant source of embarrassment to those who wished to do him honour.

Few greater privileges can be accorded to distinguished foreigners visiting Berlin than that of signing the Golden Book. The attendant formalities, however, proved too much for the inventor of the steam turbine, and while his signature reposes in the archives in Berlin, there is support for the story that the signature was written on a slip of velum in the privacy of Sir Charles' sitting room and pasted, with equal tranquillity, in the famous volume.

Parsons was very conscious of the thermodynamic reality that the efficiency of a turbine depends on the inlet temperature and as he sought to raise this ever higher, he needed data for the performance of metals under these conditions. Sometime around 1929 he arranged a very neat superheater to raise the temperature of steam to 1000 °F (538 °C) (Gibb 1947). Steam from a boiler was made to flow through the low voltage winding of a transformer which had been made in the form of a tube. A heavy current heated the wall of this tube to the required temperature. The resulting jet of steam was played on surfaces of different materials for long periods and tests satisfied him that erosion was not a problem, and that suitable materials could be found for use in either steam, or gas, turbines.

Although higher temperatures and pressures were being adopted for land turbines, there had been a reluctance to follow this example in marine practice. Even as early as 1897 Parsons had anticipated raising steam pressures to 300 p.s.i. At the First World Power Conference in London in 1924 he again argued the case for higher steam pressures. Eventually he enlisted the support of Turbine Steamers Ltd who had been involved in building the pioneering *King Edward*, and they fitted a new Clyde steamer, the *King George V*, with boilers delivering steam at 550 lb in^{-2} and 750 °F. This was twice the highest pressure yet used for marine purposes. Her

[3] Another celebrity who attended this conference was Albert Einstein.

two geared turbines were designed to give 3500 SHP to her twin shafts. She was launched on 29 July 1926 (Appleyard 1933). Towards the end of the following summer season, on 29 September, there was an explosion on board that killed two of the crew. It was a severe set-back and an intensive enquiry was carried out. It was found that scale had built up in the tubes of the boiler so insulating them from the water within. The tubes overheated and deformed until one of them eventually burst with disastrous consequences. The original intention had been that the boiler should use only distilled water, but difficulties with the distillation equipment led to the practice of using town water to which lime was added to counter acidity. When this error was put right the vessel returned to satisfactory service.

There was little new naval construction after the war but the Admiralty decided to order a destroyer, built by Yarrow, the *Acheron* (Jung 1987a). In Germany and the United States the experience with higher pressures had been good and, partly on Parsons' urging, the *Acheron's* machinery was designed for steam at 750 °F and 550 psi. Each of the geared turbines developed 17 000 SHP (12.7 MW), and the aim was to achieve 36 knots. The high pressure and intermediate pressure turbines had solid forged rotors but the low pressure rotor had a hollow forging with a shrink fit stub axle at one end. Steam was admitted at the centre of the low pressure cylinder and flowed in both directions. *Acheron* was delivered in 1930. Her trials were a disaster. Although steam consumption figures of 7.7 lb $SHP^{-1}h^{-1}$ and a fuel consumption of 0.608 lb oil $SHP^{-1}h^{-1}$ were returned, sporadic and violent vibration was encountered in the low pressure turbine (Cook 1939, Nicholas 1995). It appears that it was caused because temperature changes affected the shrink fit of the low pressure rotor. This set-back had very long term consequences. When rearmament began after 1935, Parsons was dead and the *Acheron* experience caused the Royal Navy to shrink from adopting such radical developments for new vessels. Furthermore Parsons had abandoned his attempts to develop double stage reduction gearing, and single stage gearing necessitated relatively low turbine speeds. American manufacturers, on the other hand, persisted with their efforts, and as a result manufacturers of high speed impulse turbines with double reduction gears were able to displace Parsons' licensees as suppliers to the United States Navy (Jung 1987a). Slow speed turbines together with low steam temperatures made the propulsive machinery in British warships heavier and less efficient than in those of her Allies until well after the Second World War. It is a commentary on the personal influence that Parsons carried with the navy that his passing should so affect policy.

In March 1930, when A Campbell Swinton died, S F Prest took his place on the board of the Parsons Foreign Patents Co. In the same month Gerald Stoney retired at the age of 66. The board minute relating to his last research report for the year to 30 September 1929 stated that '*The Chairman noted inadequate co-operation between the Research Department and design and production. Future research programmes should be approved by the Works Director*'. This would suggest that friction had developed between Stoney and Gibb. Of course Gibb, unlike Carnegie or Dowson, had not known Stoney in the days before he had left Heaton.

22.4 Death of the Hon. Sir Charles Parsons O.M.

In January of 1931, and approaching 77 years of age, Charles Parsons travelled with his wife for a holiday in the West Indies. As so often before, he sought relaxation in travel by sea. The journey to Barbados in the *Duchess of Richmond* was pleasant. He visited Trinidad and then crossed to Venezuela and drove to Caracas. But he became unwell. A heavy smoker all his life, both his lungs and his circulatory system had suffered. Confined to his bunk, he returned to Kingston in Jamaica. In the evening of 11 February he died. Katharine Parsons wrote to Bedford the following day (Bedford)

> *I had a terrible time. Sir Charles became ill six days ago, just one of his usual attacks, high temperature etc. and I thought nothing of it, but after three days it did not go down so I called the ship's doctor. Later Dr. Llewellyn, a fellow passenger, a very well known Brook Street doctor whom Sir Charles knew at the Athenaeum Club and had seen a great deal of, saw him. He was very pessimistic. Sir Charles became delirious at times, but the day before he died, he was considerably better, smoking a pipe and chatting. In the evening he became wandering again with ups and downs; finally a terrible change came over him in the afternoon and one could see that it must be the end.*

His body was brought home to the North East, to the parish church of the little village of Kirkwhelpington. As a mark of honour, the coffin was draped with the White Ensign of the Royal Navy. Katharine, Lady Parsons, arrived, accompanied only by Rachel, and she took her customary seat beneath the memorial to her son. At her request there were neither hymns nor anthems. The church was packed and over 150 wreaths were laid out on the snow covered ground outside. At the same hour, a service was held in

St Nicholas' Cathedral in Newcastle and in Westminster Abbey in London. As well as family and employees, civic and industrial representatives attended. A two minute silence was observed by employees in the various works. Condolences were received from all over the world, among them a message from the Imperial Japanese Navy on the death of *'a brilliant engineer and inventor'*.

An anonymous 'marine engineer' recorded his impressions of the man in his last years (Anon 1931d).

Sir Charles Parsons was one of the most retiring and aloof men it was possible to meet. He was the last person one would imagine to be an engineer or scientist. He rather suggested the scholar and writer, with his keen observant eyes, generally atwinkle, revealing a restrained humour. Yet his personality was that of a man of precision. There was nothing loose about him, nothing slipshod or vague. He dressed well, was never ill groomed, and he carried himself buoyantly, and at 76 was as alert as a man of 40, both in mind and body. He was an aristocrat who had good right to be regarded as the salt of the earth.

Katharine Parsons was to join her husband not long afterwards, on 16 October 1933. Something of her character is revealed in two letters that she wrote a few weeks before her death (Redmayne 1942). One was addressed to her friend the mining expert Sir Richard Redmayne.

Ray Demesne
Kirkwhelpington
Northumberland
10th August 1933

Dear Sir Richard,

Thank you very much for your kind letter. I am a complete cripple from cancer of the spine and, of course it is quite incurable. I am very fortunate in being at home and am able to come downstairs most days and sometimes into the garden and that is the extent of my activities. My friends are all most kind in coming to see me.

We are collecting Charles's papers and hope to make a good book of them; several people are helping Geoffry with the editing; of course a great deal has been cut out and many of the papers are joint papers and many are repetitions, so there is a great amount of editing to do. I had in view the volume Charles collected of his father's papers which is very interesting but unfortunately there are a great many more papers of Charles, and they cover a wide field.

It is so very good of Lord Rayleigh to write a preface and I am sure it will add very considerably to the value of the book.

Thank you very much for your letter.

Sincerely yours,

Katharine Parsons

The second letter was read to the staff at Ray by the Vicar of Kirkwhelpington,

Letter to Staff
August 1933

Dear Friends,

I leave this letter to be opened after my death because none of us would like the strain of an individual 'Farewell'. I have valued your friendship and service very highly. No one could have had more willing service or ungrudging help than I, and I thank you all for contributing to make my life here so easy and happy, especially during the last two years.

Some of you may have thought that I leave everything here—my beautiful place and my many friends—with too light a heart, but I remembered that all these good things are lent us for the short time of our lives to make the best use of, and that we must be ready and willing to answer the call whenever it comes and start the last journey.

Faithfully yours

Katharine Parsons

Charles Parsons had chosen a remarkable woman to share his life, and her strength must have been a great asset to him as he lived out a life of extraordinary fruitfulness.

25 October 1929, the day of the collapse of the New York Stock Exchange, has been dubbed Black Friday. The effects of the Great Depression that followed were still being felt by the industrial nations. For long the fact that the Heaton Works was largely owned by one person had been very beneficial in that it removed uncertainty about future policy, but when Parsons died it faced an acute crisis. His only son was dead and his daughter had not been involved in the business for many years. His estate was valued at £1214 355 gross. In order to pay death duties and carry out the bequests in his will, the company had to be put in the public domain.

Figure 133. Photograph of Katharine Lady Parsons.

Just before he left on his last voyage, Parsons had invited Bedford, his Managing Director, to dinner in his London house. He had appointed Bedford as one of his executors, along with his wife and Vincent Thompson, but he said nothing about that to Bedford, only urging him to carry on as usual while he was away. After Parsons' death Lady Parsons wrote to Bedford to say that he should chair the board. However S F Prest had always served when Parsons had been absent, and he was prevailed on to fill the position. Prest also took the chair of the PMSTCo. But a brief six months later he too died, and Bedford was asked to take charge as acting chairman at Heaton, while Walker became the chairman at the PMSTCo. Richard Clere's son, A D C Parsons, who was manager of Ross Ltd, was invited to join the board of C A Parsons as a member of the family. Business was now so bad that some members of staff and shop floor workers were paid off, while those who remained took cuts in their salaries.

One of the first steps taken by the executors was to renew a claim made previously by the PMSTCo., suing the United States Government for $2687 000 *'for manufacture of turbines by, and for use by the United States for marine propulsion within six years prior to filing of original petition herein dated 9th July 1924, and without license of the owner'* (United States Court of Claims No D503, BirrR/5). On 14 November 1932 the Court of Claims ruled that the plaintiffs were not entitled to recover damages.

Parsons had drawn no salary from C A Parsons and Co., and relatively modest dividends were paid. As a result the company built up cash reserves that were mostly held in government securities. In effect the shares were worth more than twice their nominal value. The company now issued redeemable debentures to the shareholders of an amount equivalent to the reserves. The shareholders cashed these debentures and the company's reserves were depleted accordingly. But now the executors, who held the greater part of the shareholding, had sufficient cash to meet death duties as well as some other obligations. Under the Employees Share Agreement of 1918, if the employee died or left the company, Parsons retained the right, for a period of up to one year after the event, to re-purchase the share. One of the beneficiaries pressed the executors to test the possibility that such shares might be repossessed, with the associated debenture payments (Bedford). But the courts ruled that this was out of order. Bedford now approached Charles Merz for advice about selling Parsons' shares to meet his bequests. He was advised that an approach should be made to Charles' brother Norbert who was a director of A Reyrolle and Co. Ltd at Hebburn-on-Tyne. This progressive firm manufactured switchgear for the Electricity Supply industry, and had been at Hebburn since 1901 (Lightie 1994). Reyrolle purchased 2099 shares at £100 each which gave it a 60% stake in the company. The liaison suited the employees of C A Parsons very well because the two firms were not competitors, and indeed they complemented one another.

There were two firms in which Parsons had only a minority shareholding, Parsons Foreign Patents Company and the PMSTCo., and consequently his death had less immediate impact on their financial situation. But it did have an effect on the development of marine turbines in the longer term. Parsons had framed the legal agreements with his licensees in such a way that they tended to rely heavily on the lead given by Wallsend to maintain development. After his death there was no one of equal calibre to maintain progress. In 1942 it was observed '*that there was a comparative standstill of about 20 years in equipment, methods, experience and application of ideas* (among engine builders in the North East)' (Clarke 1997). In May 1944, with a view to improving designs, the Admiralty encouraged C A Parsons and Co. and its British licensees to set up Pametrada, the Parsons and Marine Engineering Turbine Research and Development Association. This had some success but was eventually disbanded in 1966 (Jung 1987b). In 1957 Richardson Westgarth purchased the PMSTCo. for an issue of their own fully paid up shares valued at £544 700. The plant finally ceased production in 1964 (Clarke 1997).

At the end of the Second World War, following wartime service with the Ministry of Supply for which he received a knighthood, Claude Gibb returned to C A Parsons and Co. and took over as chairman and managing director of the firm in September 1945. He pushed forward vigorously with a programme of re-equipment and research; in that year the first industrial gas turbine, rated at 500 BHP, was commissioned. The award in 1953 by the United Kingdom Atomic Authority of a contract for the turbo-generators for the world's first nuclear power station at Calder Hall marked the start of a significant involvement in the supply of turbines for nuclear powered stations for both the home and export markets. In 1968 the financial interest held by A Reyrolle and Co. Ltd in C A Parsons and Co. Ltd was taken a step further when a holding company, Reyrolle Parsons Ltd, was established. In 1977 Clarke Chapman joined to form Northern Engineering Industries, NEI, and turbines continued to be manufactured by NEI Parsons Ltd. The shake up of British heavy engineering continued apace, and in 1989 NEI merged with Rolls–Royce who built a significant range of gas turbines for industrial use. At this point it is worth drawing attention to the fact that in the century since the world's first steam turbo-generator was built the business at Heaton had remained an indigenous industry. All the other large manufacturers of such equipment had come into existence as offshoots of American firms like Westinghouse, Thomson-Houston or General Electric, or of continental firms like Siemens. This fact is eloquent testimony to the unique contribution which the life's work of Charles Parsons made to British industry. Eventually in 1997, a little over a century after its birth, the Heaton plant was purchased from Rolls–Royce by the Siemens organization, which had also recently acquired the Westinghouse power generation business in the United States. Thus were two of the pioneers of the steam turbine, Westinghouse and Parsons, linked again. Although the giant Siemens organization is older than Parsons, it did not take up turbine manufacture until 1927. As a mark of respect for the inventor, Siemens have decided to retain the name 'C A Parsons Works' for the Heaton plant (Siemens 1997).

Chapter 23

An Assessment

When Charles Parsons died he was honoured as a great engineer and scientist. As his colleagues cast their minds back to remember the character of the man, there was general agreement that his achievements were founded above all on his enormous courage and determination. It was his fellow engineers and inventors who appreciated best how important was this quality. After all Osborne Reynolds had built a steam turbine, but could not contemplate the task of developing it commercially. De Laval had built a marine turbine, but had failed to secure financial backing. Curtis had had to pass on his invention to the engineers of GE to be developed. Parsons' first two projects, his torpedo and his high speed rotary engine, were commercial failures, but that did not deter him from tackling the turbo-generator. When his plan to push its development was thwarted and he lost his patent rights, he hardly hesitated before taking a quite different path. His initial experience with *Turbinia* was equally trying, yet he never seems to have doubted that he would succeed in bringing his project to maturity. The decision to design and construct the giant turbines for the *Mauretania* involved a huge leap in scale, not to mention the enormous expenditure which was put at hazard. It was striking testimony to a wonderful confidence in his own ability. In this respect he emulated his father when the latter decided to go ahead with the construction of the Leviathan. He was like him, too, in being able to recognize when a window of opportunity existed, and in the speed with which he rushed to open it while it was still worthwhile to do so. When William Parsons perfected the giant metallic reflector telescope, it was the only instrument which was capable of making visible the structure of the spiral nebulae. Less than 50 years later advances in photography had removed this monopoly. In a similar fashion there was a relatively brief period, which lasted no more than 50 years, during which the marine steam turbine was the only option for large high speed vessels. Its dominance has since been brought to an end, with the development of large diesel engines and of jet propelled passenger aircraft. Both men made their mark by seizing their opportunity and pursuing it with the greatest vigour.

Charles Parsons was evidently generously blessed with the gifts of a sharp intelligence, technical understanding and imagination and these no doubt would have found expression whatever his circumstances had been. But he also enjoyed other advantages. He was born into the British aristocracy and he inherited significant personal wealth, and, perhaps even more important, financial connections which permitted him to manage his business pretty much on his own terms. It is clear that he had problems working harmoniously with others, and, if he had been no more than an employee, he would most likely have suffered the same fate as other gifted engineers like Fullagar, Stoney and Law. He was fortunate that his grandfather and father were educated and thoughtful men who understood science. All these advantages were synergized by one other crucial factor, his Irishness. It was this that served to distance the family somewhat from other members of an upper class which, in England especially, enjoyed the wealth made possible by industrial activity, yet which considered it beneath themselves to engage in advancing industry. Almost uniquely, Charles' father was as much at home with scientists, engineers and industrialists as with other land owners and parliamentarians. Charles was spared the experience of having to attend one of the public schools of the time which were permeated with a dismissive attitude to those who might dirty their hands in industrial enterprises. In consequence he retained the mind-set passed on to him by his father, and was able to study at Cambridge without becoming diverted from his objective. It is true that there were rare individuals like John William Strutt (1842–1919), third Baron Rayleigh, who was both an aristocrat and a formidable scientist and mathematician, but his son Robert John (1875–1947) who succeeded him as the fourth Baron stayed with a respectable career in science and experienced no temptation to follow a career in industry.

The financial independence which Parsons enjoyed permitted him to embark on long term developments with a freedom which the shareholders of few companies then, or indeed now, would be willing to underwrite. But one must also recognize his skill in protecting his inventions and securing a sufficient financial yield from them to finance continuous improvements, yet without attempting to create any kind of monopoly position. It was a philosophy which men like Yarrow scarcely appreciated. Though Yarrow admired Parsons' inventive skills, he was known to have stated that '*he was a stupid sort of chap in business*'. He was not alone in that estimation. But it is important to remember that Parsons' first objective was not to make a fortune, but to ensure the speediest adoption and development of the steam turbine for use on land and sea. Parsons was a wealthy man when he died, but by comparison with entrepreneurs like Sir Christopher Furness, whose estate was valued at £184 million, his was a modest fortune (Clarke 1997). As he had noted, '*the world at large has received a present of ten-thousandfold greater value* (from their inventions) *than all the money spent and misspent by the small band of past inventors*' (Parsons C A 1904c). Nor was Parsons stupid, for he was successful in protecting his patents which numbered over 300. He heeded the good advice which he received from Marks and Clerk who were his patent agents. The legal agreements which he made with his licensees forbade litigation and in return provided them with generous technical support. Although he had had one or two encounters with the law to do with the 1884 patents, and with the Fullagar affair, he avoided a really damaging contest of the sort encountered by Edison and Westinghouse. He may perhaps have failed to extract the maximum efficiency from his factories, but he did keep track of the considerable expenditure on research and development, even on relatively unimportant projects like the 'auxetophone'.

One contemporary who matched him in terms of his inventive gifts was Thomas Alva Edison (1847–1931). Edison contributed in a major way to telegraphy, telephony, sound recording, electric lighting, storage batteries and movies. And yet during his career he lost several fortunes. He found out that, if they wished, the wealthy could always gain control of an invention whether by imitation, evasion or the attrition of the legal process (Josephson 1961). When Edison–GE joined Thomson–Houston to form General Electric in 1892, Edison himself was excluded from any managerial role. A similar fate later befell George Westinghouse, and in each case the root cause lay in their failure to control finances. Parsons was fortunate that not only did he inherit wealth, but he had the backing of wealthy men like C J Leyland, J B Simpson and N C Cookson. Their loyalty to him over the years is testimony to the confidence that was inspired in them by his ability, his determination and his fair dealing. There was an interesting parallel in the experience of Parsons and Edison during the War. Though Edison averred that '*making things which kill men is against my fiber. I leave that death dealing to my friends the Maxim brothers*', he accepted in 1915 when he was asked by the Secretary of the Navy to chair an advisory board of inventors and scientists, the Navy Consulting Board (Josephson 1961). A fellow board member, the scientist Robert Millikan, echoing the reaction of Professor J J Thomson to Parsons, was somewhat surprised to discover Edison to be *'a much greater man than I expected to find, simple direct, intelligent and unspoiled.'*

With those close to him, scientists and engineers like Lord Rayleigh, Sir Alfred Ewing and Gerald Stoney, Parsons could be a delightful host. Indeed where there was a shared interest, as when he had helped in person to sort out the problems at the Manchester Square station, he could get along just as well with companions 'who got their hands dirty'. During his many sea voyages he would often seek out members of the ship's crew in charge of the engine room, and on occasion he could be found, puffing away at his pipe, sitting in the stern, listening to the noises and discussing with them the source of vibrations (Bruce 1939). When his attention was engaged, he could be very kind to employees or their dependants, and in this he inherited his parent's habits. The trouble was that he was so often preoccupied with his thoughts that he failed to sense the human impact that he made on others. This failing was no doubt a feature of the personality he was born with, but it must have been exaggerated by the unconventional nature of his early upbringing. The problems that flowed from this were compounded by the fact that he was the 'absolute' owner and manager of the Heaton works, not to mention the affliction of possessing a fiery temperament. However one should see this failing as one side of a coin, because the obverse was his ability to direct his attention totally towards the solution of a difficulty, a gift which made him so effective as an inventor. A friend remembered that

> *when he set himself to a task, he would work at it continuously, so he told me, for twenty-four or thirty-six hours or longer at a stretch. This intensity of application and speed must often have left his collaborators and assistants gasping behind him. Yet no man, I believe, did more than he to train his men in the method of his work. Had he been one half as careful of his own health as he was for that of his staff, this grave loss to industry, to science and to Great Britain might have been postponed* (Heath 1931).

His wife Katharine was tough enough to cope with his vagaries but his children may have been affected by a lack of warmth in the emotional atmosphere between their parents. When both children were young and opportunity offered, he encouraged them to become involved in the experiments that he carried out in his workshop at home. But he often worked till the early hours of the morning, and before long the time came for the children to be sent away to boarding school. Unlike his father and grandfather who were landed gentlemen, he could not afford the time to supervise the education of his children with the help of tutors. Certainly in middle age Rachel distanced herself from her parents. His nephews were a different matter and he delighted in entertaining them for holidays at Ray. His wife would watch from the lake shore '*while* (he) *fished, sometimes with a boatload of noisy fidgety little boys, whom he taught to throw a fly with infinite patience and kindness*' (Houstoun 1934). Clere's son Arthur recalled a day's shooting (Parsons A D C 1939),

> *before the drive started we were sitting by his butt and he was gazing down the line of butts with that far-away look he so often had. I thought he was thinking about some deep problem, but he suddenly turned around to me and said 'Arthur do you think they are all enjoying themselves?' I assured him there was no doubt about it. His one concern was that we should have a delightful day.*

One might have expected that because of their shared interest in the behaviour of fluids, in turbines, pumps and bearings, Osborne Reynolds and Charles Parsons would have interacted positively, but the opposite seems to have been the case, and in later years some of the Reynolds family harboured very bitter feelings towards Parsons and his children. In 1965 Reynolds' daughter Margaret, then in her late 70s, wrote to her niece Rachel (Reynolds 1965)

> *... father was no advertiser, he invented things, patented and then forgot them, so that the greatest of his inventions—the steam turbine—passed into other hands. The man who put it on the market, the Hon. Sir Charles Parsons O.M. F.R.S., who never (so far as I know) invented anything, claimed father's invention, after the correct lapse of 10 years as his own and got away with it. Quite correctly father did nothing. I can remember as a small child being at some childish play in the dining room at Manchester, when the door burst open. Father 6 ft, and mother 5 ft 1 in, rushed to the bookcase which completely covered one wall to the ceiling. After a short time they went out much more subdued, and some time later I enquired what all the fuss was about. Mother explained and said Father had written to Sir Charles to ask if he could take the son of one of the college servants in his works—that he was a promising mechanical engineer and had written on the subject of turbines. Sir Charles' answer gave him away. He was not only rude, Mother said, but he would have nobody in his works who thought they knew anything about turbines. Father had his turbine working in the Labs. Parsons was fittingly*

punished. His only son was killed in the first world war, and his daughter became eccentric, took to breeding race horses, but paid her grooms so little that one in a rage a few years ago shot and killed her. Father had said in his patent specifications that the turbine would have great commercial value to anyone who liked to exploit it. Parsons was acquainted with Father and knew what a hard worker he was, but he was too mean not to try and sneak it away for nothing.

It is indeed surprising that Parsons gave Reynolds no recognition for his turbine because, in the numerous presidential addresses that Parsons wrote over the years, he gave generous credit to those other inventors who had contributed to the advance of engineering. Even Fullagar received a mention by name.

Mark Robinson observed that it seemed to be in Parsons' nature to suspect the good faith of those with whom he did business and this no doubt comes close to identifying the cause of his behaviour. Parsons never doubted his ability to overcome technical difficulties, but he was often uncertain about, and indeed misread, the attitude of others towards himself. Yarrow, who crossed swords with Parsons over the affair of Fullagar's Turbine Advisory Syndicate, teamed up with Rateau, and in 1904 he refused Prest's attempts to interest him in a Parsons' license (Appleyard 1933). Speaking in 1931 Yarrow said

Parsons was a very jealous man, and he wrote me a very rude letter, practically saying, 'I don't want to have anything to do with you'. So we took up with another kind.

In 1905 when Parsons was courting Sir William White to have him join the board of PMSTCo., White refused unless Parsons settled his quarrel with Yarrow. Yarrow continued

Then Parsons turned up at the factory one day and said 'You do not seem to be inclined to work with me' and I replied 'You have written me such a rude letter I cannot continue dealings with you.' Parsons said he did not remember the letter... and I showed it to him. Parsons read it and confessed 'I don't know how I came to write such a letter, but it was just at the time the Cobra went down (1901).*' I thereupon forgave him and asked 'why didn't you add to your letter that you were not in a fit state?' We parted friends and remained so ever after.*

Writing to Sir William White on 26 March 1904, Yarrow, among other matters, stated his opinion of Parsons more bluntly (Yarrow 1904),

> *With regard to Fullagar, I cannot go into any private quarrel between him and Parsons. It appears to me that Parsons falls out with a lot of his staff; because some who have occupied high positions in his place have applied to me for employment. No one can blame me for employing Fullagar after Parsons discharged him; but I don't think anybody can blame Fullagar for making use of information he obtained while in Parsons' employ, more particularly because Fullagar thinks that Parsons sent him off without just cause. That Parsons valued Fullagar's services there can be no question, because they took out a patent in their joint names. It is universally recognized that Parsons is a man who is difficult to get along with; therefore when any of his staff come to me and say they could not get on with Parsons, my natural inclination is to believe them and attribute their leaving to Parsons' peculiar temperament.*

Lord Rayleigh wrote that '*Parsons was a singularly modest man, and did not seem at all to realise his own standing in the world. His lack of self-assertion was at times almost comic*' (Rayleigh 1934). In 1922 Parsons travelled to a conference of the International Astronomical Union at Rome. Arriving at his hotel he was piqued not to have received the customary invitations to various receptions. '*He seemed to think that this was a sign that he was not appreciated and was much hurt.*' Instead of protesting or making enquiries, he stayed silent, only to find on his return home that the missing invitations were awaiting him there. The episode makes it plain that, while he was in no way boastful, yet he did value the approbation of others.

Love of money certainly did not motivate Parsons. Bedford told of a time when one of the firm's sales representatives wrote with a request from a manufacturer in 'one of the British Dominions' for a license to manufacture turbines on payment of a royalty (Bedford 1939). Parsons granted the request, but decided that the royalty offered was far too high and cut it to one quarter, much to the dismay of the representative. On another occasion he was urged to take on more overhauls and repairs, where it would be possible to make far better margins, but he wrote to Bedford,

> *Such a suggestion is disgusting to me who have always preached that it was bad practice to take advantage of old customers and friends who have placed confidence in us.* In Bedford's words, *he detested anything that savoured of dishonesty or sharp practice,* but he added, *he often told me that he was naturally suspicious, and sometimes this resulted, unintentionally, in his doing an injustice to people.* Even Bedford had his problems, *on one occasion Sir Charles made a serious accusation against me in relation to his*

personal accounts, and it required the Company's Auditor to explain to Sir Charles that what I had done was quite in order. Sir Charles' form of apology was to come along to my office a few minutes later and ask me to spend the week-end with him.

There seems to have been a special rapport between Fisher and Parsons (Appleyard 1933). One thing they had in common was their high estimation of the importance of the professional engineer to the successful functioning of the Royal Navy. Fisher had fought hard and successfully to integrate such men into the ranks of officers with his scheme for 'common entry', but in 1925 his reforms were in part reversed. From this time engineer officers were to be distinguished from executive officers. Their uniform was to bear distinctive markings and their names would appear separately on the Navy List. The First Lord explained to the House of Commons that their chances of promotion and rates of pay would be in no way affected by the new order; nevertheless its most important consequence was that engineers could not aspire to serve on the Board of Admiralty. Parsons was joined by Sir John Thornycroft, Sir Archibald Denny and Lord Weir, all engineers who had a close acquaintance with the navy, in denouncing the change,

At no time in the history of our fighting Services is it more necessary to avoid differentiation of status between the executive and the technical officers. Anything done today to cause parents to question the future status or dignity of a technical career, in any of the services for their sons, anything that appears to detract from the position and prestige of the engineering profession, can only, in our opinion, react harmfully on the future efficiency of the Services.

The following January a deputation went to the Admiralty to protest, and Parsons' message was that

It is very hard that Engineers of great ability who had gone into the Navy should be debarred from getting beyond a certain line of demarcation.

Although the protest was concerned with the treatment of naval personnel only, it touched on a difficulty which has afflicted the engineering profession in Britain right up to the present time, namely the low esteem in which it is held by the public with a consequent difficulty in attracting its share of the brightest youngsters to its ranks.

Motor cars began to make an appearance in Newcastle at the turn of the century and Parsons' affair with the internal combustion engine sheds light on the man. He did not always use a chauffeur but drove his Daimler himself. He rather enjoyed the challenges that the early models presented to the mechanical engineer. Lord Rayleigh recounted stories, like the time when Parsons explained that his late arrival at a picnic was caused because *we had to undress the parlour maid* (Rayleigh 1934). Being translated this meant that a running repair to the distributor in the engine could be dealt with, provided a piece of spring steel were removed from the parlour maid's stays. Road surfaces were still pretty rough and on one occasion a bent axle was straightened by lighting a fire at the road side and heating it to red heat.

Parsons' driving style was rather idiosyncratic. Around 1911 he collected the future Lord Raleigh from the train at Newcastle and drove him the 20 miles to Ray.

> *Parsons was a severe critic of the drivers of the cars that we passed, and called out his criticism in emphatic language. His voice, however, had not much carrying power, and this circumstance may not have been altogether unfortunate... Sometimes he expressed his protest by blowing his horn, and would continue to blow it for a long time afterwards, much to the surprise of the passers by, who had not seen the earlier phases of the incident.*

Bedford had his own tale to tell (Bedford 1939).

> *Sir Charles took up motoring in the very early days, and I suppose he and the late Campbell Swinton were the first to be fined for speeding. They were both caught in a speed trap in the village of Horsley one Sunday afternoon. It was a rather terrifying job motoring with Sir Charles. One day coming in from Ray, we came across some men mending the road. Half the road was covered with large rough stones and the other half was blocked with a water-cart. Sir Charles blew his horn expecting the men would move the water cart and let him pass, but they signalled him to go over the rough stones. This was too much for Sir Charles. He said to me 'Come on, we'll move the water-cart for them.' So we jumped out and he picked up one of the large stones... and threatened the poor men with all sorts of tortures if they did not move the cart. This however had no effect, so he and I had to do it ourselves whilst the men stood by greatly amused. Sir Charles was scarlet with rage at the time, but he had an amazing way of calming himself down, and before we had gone very far was looking upon the whole incident as a huge joke.*

In later years newspapers carried reports of accidents which involved both Sir Charles and Lady Parsons. In August 1926 Lady Parsons was driving alone along the Great North Road in her two seater Vauxhall, when a tyre burst some 50 miles south of Newcastle (Anon 1926). The car overturned and she was knocked unconscious, but when the 70 year old driver was brought to Ripon Cottage Hospital, it was found that she had suffered no broken bones. A year later the couple were travelling together near Newcastle when they ran into a stray sheep (Anon 1927). The farmer was taken to court and fined 10 shillings.

It has been noted that when it came to explaining his reasoning in relation to some technical problem such as the choice of a suitable profile for the blade of a reaction turbine, or the genesis of his K constant, Parsons often failed miserably. Those who knew him personally remarked that he was frequently embarrassed at being unable to express himself adequately at meetings. *'It was difficult or impossible to persuade him that others, who were more vocal and made their views heard when his were practically a sealed book, did not mean him any injury. But these resentments did not last long and vanished like a puff of smoke if a friendly advance came from the other side'.* Despite this, his collected papers contain numerous surveys of topics such as the development of prime movers, or the difficulties faced by inventors, which are entirely lucid. No doubt he received help in researching historical detail, but they were certainly his own work. Lord Rayleigh remembered that he was shown a draft of Parsons' presidential address to the British Association with a request for his observations (Rayleigh 1934). These were duly made and accepted, but when the address was printed they had all been discarded. For the most part Parsons neither sought out the ideas and opinions of others, nor did he adopt them. Alex Law who was chief electrical designer recalled (Law 1939)

> *I have a vivid recollection of a scene when I tried to get Sir Charles to listen to an explanation of the different functions of compensating windings and a commutating pole... The obvious cure was a small commutating pole, but this was forbidden.*

As he grew older, his views and behaviour did change significantly, as can be seen in his address to the Institute of Physics in 1924 (S.P.2),

> *Within the last three-quarters of a century the fields of science and the arts have so much developed in extent, complexity and refinement that it has become impossible for a single brain to grasp effectively more than a small part of those wide fields with any prospect of making any advance, and it has become necessary to specialise, to collaborate and to delegate work, in order to realise success.*

Parsons held strong views about the importance of a well educated workforce and the general inadequacy of current arrangements, and these views were not confined to technical personnel only. At a prize-giving ceremony at the Rutherford College, in December 1909, he spoke on the subject (Parsons C A 1909),

> *I think it is universally agreed by all who have given careful thought to the subject, that for a commercial community to be successful it must comprise a large percentage of intellectual and well educated men, and also that the intellectually trained must be found (though in different proportions possibly) throughout all classes and grades of the community.*

He went on to commend the courses run by the Commercial Department, commercial English, *'how few people can express themselves accurately in business matters, even when they try'*, commercial geography, book-keeping and commercial law, banking and accounting and modern languages.

> *To the average business man a grounding in these subjects is infinitely more valuable than the most intimate knowledge of the dead languages'.* And, perhaps over-optimistically, *industrial leaders in this district recognise that the rule of thumb methods which formerly prevailed and answered, while this Country enjoyed the industrial monopoly of the World's trade, are now obsolete, and must be replaced by scientific methods, and by thorough grounding and education, of masters, managers, foremen and men.*

At all times Parsons had taken care to recruit men of the highest ability to share in his ventures, even if he had rather cramped their style in the early years. When he was responding to the conferring of the Freedom of the City of Newcastle upon Tyne in 1914, he said (Parsons C A 1914)

> *I consider myself the representative of a class of men who, whether self-taught or trained in schools and colleges, devote their energies in after life to scientific research and to the improvement of the Arts and Manufactures... I would venture to claim brotherhood with this large and increasingly useful class of skilled workers, and to be allowed to attribute some of the honour which you have conferred on me, also to them.*

Stoney, Cook and Gibb were each in turn made Fellows of the Royal Society. Others like Carnegie and Walker, though highly intelligent, had never entered a university yet had absorbed from Parsons himself a scientific approach to technical problems. However it was left to the Americans to apply to the management of human resources in factories that same scientific approach which Parsons had brought to bear so effectively on, for example, the production of high quality optical glass.

The grandfather of Charles Parsons was involved in politics but he was nothing like as active as the Duke of Wellington would have wished him to be. Charles' father William was, if anything, even less enamoured of party politics. Charles himself really had no involvement, though his wife Katharine participated in local Conservative Party activities. If Charles Parsons found it difficult to explain to others his reasoning in matters like turbine design, he seems to have made no effort at all to address his inner, religious beliefs in writing as his grandfather had done. We know the family did attend the church in Kirkwhelpington, but if we are seeking for clues to his attitude to the conundrum of human existence we can only study the way in which he lived out his life. Born to an aristocratic and wealthy family he could of course have opted for the life of a man of leisure, amusing himself with scientific hobbies. Instead he chose, regardless of the personal cost to himself and his immediate family, to apply those precious gifts with which he had been endowed, in the service of mankind everywhere. At all times his actions were guided by a conscience which, according to its lights, aspired to responsible, honest and charitable behaviour.

Appendix 1

(a) Fluid mechanics

When we are dealing with machines like centrifugal pumps, fans, propellers or turbines it is necessary to understand how a moving fluid and a solid structure interact. Two physical principles apply. Isaac Newton (1642–1727) propounded a law that states that if the direction of travel, or the speed of a body, or flow of fluid is changed, a force must be exerted on that body or flow. The second law was stated by the Swiss mathematician Daniel Bernoulli (1699–1782) in 1738. He identified three components of energy which are present in a flow of fluid, and he noted that the totality of these will in certain *ideal* circumstances remain constant. That is, he assumed the following to be true—that friction would be negligible, that the flow at a given point is unchanging with time and that there would be no change in the temperature of the fluid. His relation can be written,

$$gz + \frac{p}{\rho} + \frac{u^2}{2} = \text{const.}$$

Here g is the acceleration due to gravity, z is the height above some reference datum, p is the hydrostatic pressure, ρ is the fluid density and u is the velocity. The first two terms represent potential energy and the third represents kinetic energy. If z does not change and if the fluid is incompressible, that is if ρ does not depend on p, this implies that when u increases then p must fall, and vice versa. For practical purposes water can be considered as incompressible, and for relatively small variations in pressure air and steam can also be taken to be incompressible.

(b) Propellers

The action of a screw propeller was described by Rankine by treating it as a disc through which a fluid flows. As the fluid crosses the plane of

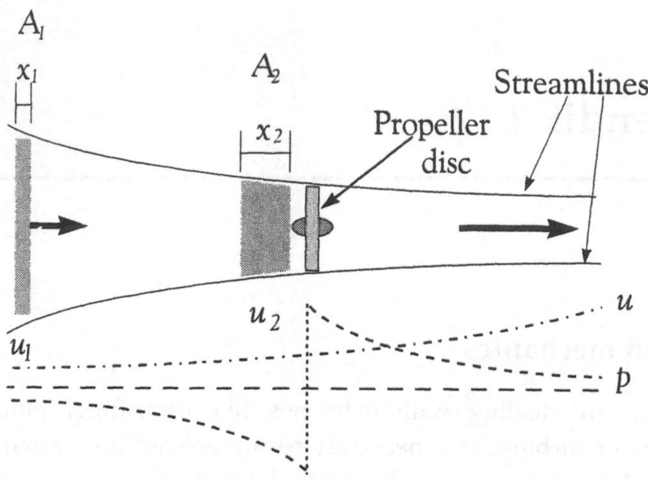

Figure A1. Rankine's model of a propeller. Fluid flows from left to right through the propeller disc, and the flow area A reduces with increasing velocity u. The pressure p falls below ambient, but at the disc it makes a jump and then declines gradually to ambient again.

the disc, its pressure jumps. This jump in pressure when multiplied by the area of the disc gives the 'thrust' experienced by the propeller. To understand the process it is necessary to consider 'streamlines'. If the flow of an ideal fluid is steady and orderly, it will be undisturbed if very thin walls are located in such a way that they lie parallel to the direction of the flow at all points. These walls lie on 'streamlines'. Clearly no fluid crosses a streamline. Referring to figure A1, if an imaginary plane surface A is located in a flow confined between streamlines, and it is oriented at right angles to the velocity of the fluid at this point, the volume of fluid passing in unit time will be the product of the area of the plane and the fluid velocity. This quantity will be the same at any point along the passage formed by the streamlines, $u_1 A_1 = u_2 A_2$. If the fluid moves a distance x in a short time, then $A_1 x_1 = A_2 x_2$.

Consider now a propeller disc located in such a passage with walls which lie on streamlines. At very great distances from a propeller the fluid will be almost stationary; consequently the streamlines must be very widely separated. As the fluid approaches the propeller, it speeds up, and so the streamlines approach each other and the passage narrows. Now

Bernoulli's equation states that as the fluid speeds up its pressure must fall. As it approaches the disc the fluid pressure will be below ambient, but as it emerges on the far side of the propeller disc it will have jumped to a value above ambient. The fluid will continue to speed up until its pressure falls to the ambient pressure once more. Eventually in a real fluid the kinetic energy in this fast moving flow will be turned into heat because of friction with the stationary fluid outside the passage. This waste of energy depends on the total mass involved and on the square of its velocity. If the propeller disc is made larger, the same mass can be handled with a lower velocity. Consequently the energy lost will drop significantly. However the larger area will be associated with greater friction losses at the screw surfaces.

(c) Visualizing propeller action

To visualize the action of a propeller, we can imagine a screw surface which is infinitely thin, that is, razor sharp, and which is immersed in a jelly-like medium. Now consider three cases. In case *one*, imagine that the screw rotates very slowly. The screw will leave behind a phantom screw profile where the jelly has closed up afterwards. The screw moves forward a distance L given by the pitch of the screw P multiplied by the number of rotations n of the screw, $L = nP$, but the medium itself does not move except to allow the passage of the screw surface (see figure A2).

In case *two*, the screw is speeded up. Now it moves forward by a smaller distance, $(1 - S)nP$ because of a slip, S, while at the same time a tube of the medium is squirted backwards by the screw, through an otherwise largely stationary medium, at a rate related to the amount of slip. The slip, which lies between 0 and 1, is defined as

$$S = (nP - L)/(nP).$$

This requires that a quantity of the medium is drawn towards the screw from ahead, and is discharged behind. The momentum and energy imparted to this tube of fluid is provided by the screw which must be supplied with power.

In case *one*, the velocity of the medium relative to the screw surface is parallel to the screw surface. In case *two*, the fact that the medium through which the screw passes is now moving as a tube causes the velocity of the fluid to lie at an angle ϕ to the surface of the screw (see for example figure 85). It is this angle which generates the thrust of the screw. Consider

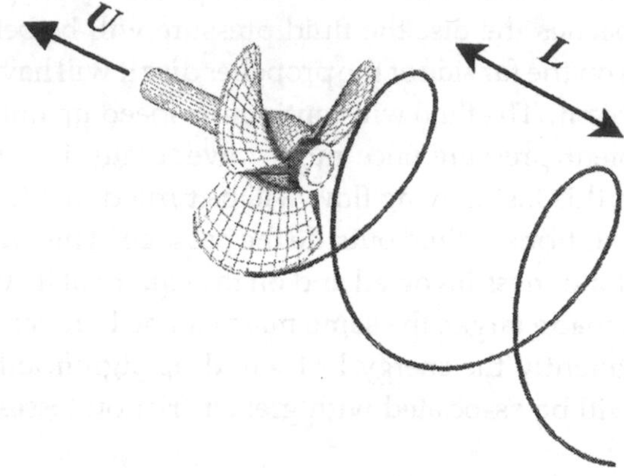

Figure A2. A screw propeller moves through a stationary medium with a velocity U, travelling a distance L for each complete revolution of the screw.

case *three* where more power is fed to the screw, and this angle ϕ becomes excessive. The flow of the fluid can now become increasingly chaotic so that more of the energy which is absorbed by the screw is turned to heat.

The 'propulsive efficiency' of a screw is that proportion of the energy supplied to the screw which is converted to a useful thrust in the direction in which the screw is travelling. There are *two* different features which reduce this efficiency. In case two above, the tube of fluid which is speeded up carries away energy which is wasted. The energy of the fluid in this tube is proportional to the volume flow rate and to the square of its velocity. By making the cross sectional area of the tube larger—using a propeller with a larger diameter—the volume flow rate of fluid can be maintained while reducing its speed. Because the wasted energy is related to the square of speed there is a gain in efficiency. The second feature is the friction experienced between fluid and the screw surface. It is a matter for the designer to trade off the benefit of a larger diameter against a greater friction loss (Walshaw and Johnson 1972).

In case *three* just how much energy is wasted depends on the exact details of the direction and magnitude of the speed of the fluid relative to the screw surface. It is an extremely complex matter which still demands the help of experimentation to handle.

Appendix 2

This account is based on Stoney's Cantor lecture (Stoney 1909a).

(a) The design of Parsons' axial flow turbines—scale and output

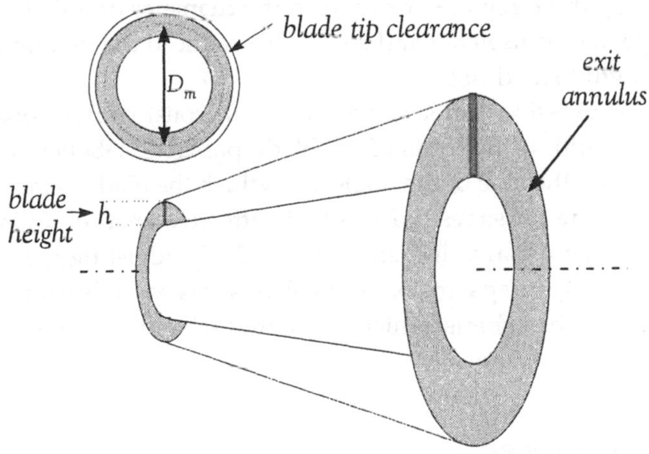

Figure A3. Diagram showing the geometry of an axial flow turbine

The power output of a turbine is roughly proportional to the mass flow rate of steam. The energy Δh, which is released from a unit mass of steam when it expands with a small drop in pressure Δp, is $v\Delta p$, where v is the specific volume of the steam. This appears as kinetic energy and is proportional to c^2 where c is the resulting steam velocity. Now for a restricted range of pressure the quantity $v\Delta p$ can be written as {const $\times$ $\Delta p/p$}, where p is the absolute pressure. The ratio of inlet pressure to exhaust pressure $p_i/p_0 = (1 + \Delta p/p)^N$ where N is the number of pairs of blade rows. In order to limit the fraction of the steam flow which leaks through the clearance space around the blade tips, the value of the fractional pressure drop, $\Delta p/p$, was kept low; values of from 2 to 4% were used.

Ideally the steam velocity c should be close to the blade velocity u, but in practice values of 1.7 to 2.5 times u were employed. Once the ratio u/c was chosen, the inlet and exit angle of the blade were settled, as well as the component of the velocity c_a of steam moving axially down the turbine. If the desired power is known, the mass flow rate of steam, $\dot{m}$, can be estimated. This can be expressed as $\dot{m}/\rho = Ac_a$, where ρ is the density of steam and A is the area of the annulus through which the steam flows. For a blade of height h, $A \approx \text{const}\,\pi D_m h$. At the inlet stages, D_m will be chosen as small as possible to ensure that h is a manageable size.

As the steam expands, its volume may increase 100-fold. The appropriate increase in Ac can be achieved by augmenting D_m, as well as lengthening the blades. Early machines raised D_m in two steps of $\sqrt{2} : 1$, giving three regions over which the diameter doubled. Doubling the diameter also doubled u and so c_a. This reduced the required increase of blade length to 25 times.

In the expression for A above, the term const comes about chiefly because the width of the exits of the blade passages, labelled w in figure A4, depends on the size of the angle α at which the blade is inclined. The inclination of the passages reduces their effective area to $\approx \sin\alpha\,\pi D_m h$. When $\alpha = 20°$, a usual value, $\sin\alpha$ is $\approx 1/3$. By adjusting the exit angle, and by increasing the spacing between blades, it is possible to manage with blade lengths at the exhaust which are no more than 12.5 times longer than at the inlet stage.

(b) Scale and speed

An upper limit to the speed of the shaft is set by the stress caused by the centrifugal force acting on the turbine blades. This force is proportional to the mass and to the square of the mean diameter of the blades and of the shaft speed. Consider a turbine which is scaled in size by a factor of 2. Areas double and volumes, and therefore masses, increase eightfold, but the stresses (force per unit area) can be kept the same if the shaft speed is reduced by one-half. Since both D_m and h have been doubled, the area available for steam flow has increased fourfold. The blade and steam velocity will have remained unchanged, so mass flow and power output rise fourfold. In other words, power varies inversely as the square of speed.

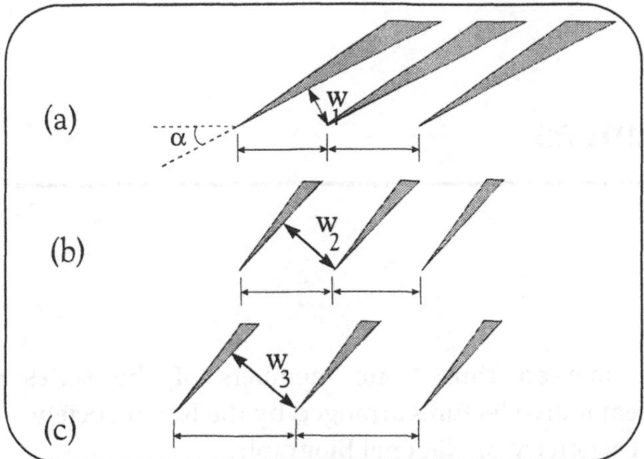

Figure A4. Options for increasing the effective flow area between blades. Looking down on the exit region of a row of blades, cases (a) and (b) have the same number of blades around the circumference, but in case (b) the angle α is increased, and $w_2 > w_1$. In case (c) the angle α is unchanged, but $w_3 > w_2$ and, though there are fewer spaces, the ratio of total blade thickness to space is less.

The same relation applies to the output of alternators, which makes them well matched to turbines. Note however that to increase power fourfold a machine which is eight times heavier is required.

If it were possible to double the speed of a machine without changing its dimensions, and yet still withstand the higher centrifugal forces, then c_a would double, as would the mass flow. Thus the doubled power output would involve only a minimal weight increase. This explains the attraction of raising shaft speed if at all possible.

References

- Papers marked thus * are members of the series of Parsons commemorative lectures arranged by the Royal Society
- DNB, Dictionary of National Biography
- TWAS, Tyne and Wear Archives Service, Newcastle upon Tyne
- PAR, Parsons Archive, Science Museum London
- Birr, Birr Scientific and Heritage Foundation by courtesy of Lord Rosse, County Offaly, Ireland

Airy G B 1849 Lecture delivered by the Astronomer Royal on the large Reflecting Telescopes of the Earl of Rosse and Mr Lassell at the last Nov. meeting *Mon. Not. R. Astron. Soc.* **9** 110–121

Airy G B 1896 *Autobiography* ed W Airy (Cambridge) pp 196–201

Allen J 1970 *The life and work of Osborne Reynolds* in *Osborne Reynolds and Engineering Science Today* ed D M McDowell and J D Jackson (Manchester)

Andrews J H 1985 *Plantation acres* (Belfast)

Anon 1826 *The Picture of Parsonstown in the King's County* (Dublin)

Anon 1843 The largest telescope in the world *Illustrated London News* 9 September

Anon 1844 *The Monster Telescopes erected by the Earl of Rosse* (Parsonstown and London)

Anon 1867a *King's County Chronicle* 4 December

Anon 1867b *Inst. Civil Eng.* **27** 1867–8, 584–7

Anon 1869 *King's County Chronicle* Wednesday 1 September

Anon 1872 Obituary in *Mon. Not. R. Astron. Soc.* **32** 102–9

Anon 1877 Calendar of University of Cambridge

Anon 1878 Calendar of University of Cambridge

Anon 1885a *Official Catalogue International Inventions Exhibition* (London)

Anon 1885b Steam Engines at the Inventions Exhibition *The Engineer* 8 May p 347

Anon 1885c King's County Chronicle July 30

Anon 1885d High speed engines at the Inventions Exhibition *Engineering* **39** 449–50, 459–60

Anon 1887 Exhibition guide TWAS L3666 91

Anon 1891 Notes taken at arbitration hearing between Clarke Chapman and Co and C A Parsons PAR 17/2

Anon 1898 *Reports of Patent, Design and Trade Mark Cases* **15** No 15 349–359 (patent extension case)

Anon 1899 *Engineering* **68** 206

Anon 1900a Annual report and balance Sheet for PMST Co June 1900 PAR 38/1

Anon 1900b *Engineering* **69** 223

Anon 1901a *The Times* 15 July

Anon 1901b *Engineering* **72** 553–5

Anon 1903 *Notes on trial of the Queen on the Clyde 10 June* PAR5

Anon 1905a The Compound Steam Turbine *Engineering* (3 February) 137–8

Anon 1905b *Eighth annual report of PMST Co* PAR38/6

Anon 1907 *Engineering* **84** 639

Anon 1908 *Engineering* **86** 444

Anon 1918 *Engineering* **105** 518

Anon 1919 Fire at Birr Castle *King's County Chronicle* 21 August

Anon 1923 *Wigan Observer* 1 December TWAS reference PAR2402

Anon 1923b Obituary of Hon R C Parsons *Engineering*)(20 February) 150

Anon 1924 *Montreal Daily Star* 13 August

Anon 1926 *Daily Mail* 11 August

Anon 1927 *Evening Chronicle* 24 August

Anon 1931a *Newcastle Journal* TWAS reference PAR2402

Anon 1931b Birr R/4 Typescript *Historical outline of the growth of the mirror department* (Nov 1931)

Anon 1931c 40 ft revolving dome for Stockholm Observatory *Engineering* (20 February) 257

Anon 1931d Press cutting from *Northern Mail* in 'Robert Dowson's Book of Cuttings' TWAS

Anon 1934 Typescript Sir Howard Grubb Parsons and Co, Optical Works PAR4/2

Anon 1936 Obituary of R J Walker CBE *The Shipbuilder* **43** 582

Anon 1945 Obituary of R Dowson *Heaton Works J.* **4** 69

Anon 1949 Heaton Works 1889–1949 *Heaton Works J.* 179–202

Anon 1970 *Ocean Liners of the Past, Olympic and Titanic* reprints from *The Shipbuilder* (London)

Anon 1984 VW research, new flat four *Car Magazine* June 316–7

Anon 1997 Quoted by Clarke J F in *Building Ships on the North East Coast* Newcastle upon Tyne **1** 113

Anon 1998 *Parsons the pioneers of power* (Newcastle upon Tyne) brochure re-issued by Siemens Power Generation Ltd

Appleyard R 1933 *Charles A Parsons—his life and work* (London)

Armstrong W 1888a A vague cry for technical education *Nineteenth Century* **24** 45–52

Armstrong W 1888b Nineteenth Century **24** 653–68

Ashbrook J 1984 Articles in *The Astronomical Scrapbook* ed L J Robinson (Cambridge), J Hershel's large 20 ft telescope, the visual Orion Nebula, and the discovery of spiral structure in galaxies

Atkinson N D 1961 *Sir Laurence Parsons, 2nd Earl of Rosse 1758–1841* PhD thesis, Dublin University

Bacon R H 1929 *Life of Lord Fisher of Kilverstone* (London) **1** 263–64

Baines F E 1895 *Forty Years of the Post Office* (London) **2** p 60 *et seq*

Barker J H 1939 Reminiscences *Heaton Works J.* 311–4

Barnaby K C 1960 *A History of the Institution of Naval Architects* (London)

Bedford F G H *Sixty Years with C A Parsons and Co Ltd 1901–1960* 24 Typescript TWAS reference PA1274

Bedford F G H 1939 Reminiscences *Heaton Works J.* **3** 316

Bennett J A 1990 *Church, State and Astronomy in Ireland—200 years of Armagh Observatory* (Belfast)

Bennett S 1979 *A History of Control Engineering 1800–1930* (London)

Bergh W 1882 On a centrifugal separator for liquids of different specific gravities *Proc. I. Mech. E.* 519–27

Birr B/29 1835 Poster from Lord Oxmantown withdrawing from election

Birr D/20 1822 Correspondence

Birr D/21 Letters from Maria Edgeworth to second Earl

Birr E/22 1827 Letters from William to John Parsons

Birr E/23 Copy book of second Earl labelled *'Recollections'*

Birr F/20 Poems written for Lord Oxmantown to turn into Latin verse

Birr G/60 Letters, accounts valuations of estate of John Wilmer Field

Birr J/19 Letter from Queen Victoria to Lady Rosse

Birr J/25 1865 Laurence Parsons' diary for 1865

Birr M/8/1 1884 Diary of Laurence Parsons

Birr R/5 Collection of letters to and from Charles Parsons and Lady Parsons

Boeddicker O 1883–87 On the changes of the radiation of heat from the Moon during the total eclipse of 1884 *Trans. RDS* **2** ser 2

Bolter J R 1994* Sir Charles Parsons and electrical power generation—a turbine designer's perspective *Proc. I. Mech. E.* **208** 159–76

Borrowman W C 1900 Some considerations affecting the training of young engineers *Trans. NECIES* **16** 113–265

Bourne J 1867 *A treatise on the screw propeller, screw vessels and screw engines* (London)

Bridge J A 1998 Sir William Brooke O'Shaughnessy *Notes Rec. Roy. Soc. (London)* **52** (1) 103–20

Bruce J H 1939 Reminiscences *Heaton Works J.* **3** 300–2

Burrill L C 1951* Sir Charles Parsons and Cavitation *Trans. I. Mar. E.* **63** 149–67

Burnett J E and Morrison-Low A D 1989 *Vulgar and Mechanick* (Dublin)

Campbell W. 1924 *The protection of steam turbine disc wheels from axial vibration* Trans. ASME **46**

Campbell R S *Typescript Account of Derby Crown Glass Works* PAR8

Cardwell D S L 1975 *One of the contributions to electrical engineering in nineteenth century Manchester* IEE meeting on the History of Electrical Engineering

Carlson W E 1991 *Innovation as a social process* (Cambridge)

Carnegie A Q 1934 *The Parsons Auxetophone* Appendix A in SP(2)

Chew V K 1967 *Talking machines (1877–1914)* (London)

Chirnside R C 1979 Sir Joseph Swan and the invention of the electric lamp *Electronics and Power* February 96–100

C L (*n*) Letters from C A Parsons held by St John's College, University of Cambridge; *n* is number placed on the original by G L Parsons

Clark E F and Linfoot J 1983 *George Parker Bidder* (Bedford), also private communication from E F Clark

Clark E K 1938 *Kitsons of Leeds* (London) 147

Clark K 1921 Letter from C A Parsons 24/3/21, courtesy of E F Clark

Clarke J F 1984 *A century of service to Engineering and Shipbuilding* (Newcastle upon Tyne)

Clarke J F 1997 *Building Ships on the North East Coast* (Newcastle upon Tyne) vol 2

Clarke J F 1998 *An almost unknown great man, Charles A Parsons* re-issue of 1984 publication (Newcastle upon Tyne) p 6

Clarke Chapman records DS/CC1/17/1–3 at TWAS

Clarke W *et al* 1889 *Agreement made on dissolution of partnership between Clarke, Chapman, Parsons and Furneaux* TWAS

Cochrane A 1909 *The Early History of Elswick* (Newcastle upon Tyne) TWAS

Conte-Helm M 1989 *Japan and the North East of England* (London)

Cook S S 1925 Letters to Sir C A Parsons 11 November and 14 December PAR 4/1

Cook S S 1939* Sir Charles Parsons and marine propulsion *Proc. I. Mech. E.* 133–55

Cooke T L 1875 *The early history of the town of Birr or Parsonstown* (Dublin) reprinted 1990 (Tullamore) with introduction by M Hogan

Cox R 1992 Hugh Francis Fullagar *The Newcomen Bulletin* **154** 18–20

Cox R C 1990 *Bindon Blood Stoney, biography of a Port Engineer* (Dublin)

Cranny N 1982 *The Upstart Earl—a study of the social and mental world of Richard Boyle first Earl of Cork* (London)

Crichton A T 1989–90 William and R E Froude and the evolution of the ship model experimental tank *Trans. Newcomen Soc.* **61** 33–49

Crotty M 1847 *A narrative of the Reformation at Birr* (London)

Crouch T 1989 *The Bishop's Boys* (London)

Cullen L M 1987 *An economic history of Ireland* (London)

Davies G 1984 *Diamond* (Bristol)

Davison D H 1989 *Impressions of an Irish Countess* (Birr)

de Laval C G P 1889 *Improvement in steam turbines* British patent 7143

Desch C H 1928 The problem of artificial production of diamonds *Nature* **121** 799–800

Dickinson H W 1939 *A Short History of the Steam Engine* (Cambridge)

DNB a **15** 419–21

DNB b **15** 408–9

DNB c **15** 329, 425–7

DNB d (1931–1940) 926–8

DNB e (1922–1940) 839–41

Dorey S F 1943* Sir Charles Parsons and mechanical gearing *Trans. I. Mar. E.* **54** 111–26

Dowson R 1925 Letter 11/6/25 to Hon Sir Charles Parsons, 1 Upper Brook St, London PAR 4/1

Dowson R 1931 PAR 4/2

Dowson R 1935 The development of the steam turbine VII *Heaton Works J.* **2** 102–9

Dowson R 1936 The development of the steam turbine VIII *Heaton Works J.* **2** 170–6

Dowson R 1937 Memo, April 1937 PAR4/2

Dowson R 1939 The Parsons steam turbine and parallel flow compressor *Heaton Works J.* **3** 408–16

Dowson R 1942 George Gerald Stoney *Obituary Not. Fellows of Roy. Soc.* **4** No 11 183–96

Dreyer J L E and Turner H H (1987) *History of the Royal Astronomical Society 1820–1920* (Oxford)

Dreyer J L E 1888 New General Catalogue of Clusters and Nebulae *Mem. R. Astron. Soc.* **44** 1–237

Dumas P 1915 Letter to C A Parsons from Admiralty PAR 46/2

Edwards A J 1915 Letter to C A Parsons, draft patent specification for mine, PAR46/7

Elliott I 1998 Entry on Laurence Parsons for new DNB

Emmerson G S 1977 *John Scott Russell* (London)

Evans K E 1966 (editor) *75 Years Brown Boveri* (Baden)

Ewing J A 1911 *The Steam-Engine and other Heat-engines* (Cambridge)

Ewing J A 1931 Obituary of Sir Charles Parsons *Proc. Roy. Soc. (Lond.)* **A 131** v-xxv

Fairburn A R *Elswick* Typescript reference L3506 TWAS

Fisher J 1918 *How the Great War was carried on* quoted in Hackmann (1984) 37

FitzGerald G F 1888 On a non-sparking dynamo *Sci. Proc. Roy. Dublin Soc.*

Fleming J A 1915 Letter of Prof J A Fleming to Sir William Crookes 19 June (NECIES archive)

Francis J B 1855 *Lowell Hydraulic Experiments* (Boston)

Friedrich R 1986 The origins of today's gas turbine *VGB Kraftwerkstechnik* **66** 380–8

Gibb C D 1947* Parsons—the man and his work *Proc. I. Mech. E.* 213–8

Gibb C D 1952 Stanley Smith Cook 1875–1952 *Biog. Mem. Roy. Soc.* **8** no 21 119–27

Gibb C D 1954 A tribute to the memory of Sir Charles Parsons *Heaton Works J.* **31** 283

Gibbs-Smith C H 1985 *Aviation* (London)

Gibson A H 1948 *Osborne Reynolds* (London)

Glass I S 1997 *Victorian Telescope Makers* (Bristol)

Goodall S V 1942* Sir Charles Parsons and the Royal Navy *TINA* **84** 1–16

Gray E 1975 *The Devil's Device, the story of Robert Whitehead* (London)

Green A 1985 Technical education and state formation in nineteenth century England and France *History of Education* **24** 123–39

Greenwood J N, Miller D R and Suiter J W 1954 Intergranular cavitation in stressed metals *Acta Met.* **2** 250–8

Griffiths A R G 1987 *The Irish Board of Works 1831–1879* (London)

Hackmann W 1984 *Seek and strike* (London)

Hadfield A 1994 The Newcastle upon Tyne Electric Supply Company *Conf. Newcomen Soc.*

Hansson S 1955 *Birger and Fredrik Ljungström* (Stockholm)

Harris F R 1984 *The Parsons Centenary—a hundred years of steam turbines Proc. I. Mech. E.* **198** 1–42

Harry O 1988 The Hon. Mrs Ward (1827–1869) in *Science in Ireland* ed Nudds *et al* pp 187–97

Hasselkus J 1931 Typescript 24 March PAR 4/2

Hawley R 1979 Parsons man of power *Electronics and Power* January 46–50

Heath F 1931 *Nature* **127** 315–6

Hele-Shaw H J S 1898 Investigation of the Surface resistance of water and of streamline motion under certain experimental conditions *TINA* **40**

Henderson A 1915 Letter from Arthur Henderson to Sir W Ramsay 10 June (NECIES archive)

Herschel J F W 1833a Astronomy, in *Cabinet Cyclopoedia* 228

Herschel J F W 1833b Observations of nebulae and clusters of stars *XIX Phil. Trans. Roy. Soc. (London)* 359

Hinton C 1959 Claude Dixon Gibb *Biog. Mem. Roy. Soc. (London)* **5** 87–93

Hockey T 1998 *Galileo's Planet, observing Jupiter before photography* (Bristol)

Hodgkinson F 1898 Letter to C A Parsons PAR35/1

Hodgkinson F 1939 Reminiscences *Heaton Works J.* **3** 293–300

Holbrow C A 1939 Reminiscences *Heaton Works J.* **3** 309

Hopkinson J 1886 Dynamo-electric machinery *Phil. Trans. R. Soc. (London)*

Hore R A 1994 Pioneering work of the Tyneside electricity industry *Conf. Newcomen Soc.*

Hoskin M 1982 The first drawing of a spiral nebula *J. Hist of Astronomy* **13** 97–101

Houstoun M 1934 Obituary of Lady Katherine Parsons *Trans. NECIES* **50** 181–3

Humbolt Per Nar V

Jarrett F E C, Osler A G and Grieve G R 1980 *Turbine and Gearing for Launch No 2* (London)

Jones K G 1968 *Messier's nebulae and star clusters* (London)

Jones R and Marriott O 1970 *Anatomy of a merger* (London)

Josephson M 1961 *Edison* (London)

Jung I 1968 *Gustaf de Laval, the high speeds and the gear* de Laval Memorial Lecture *Roy. Swed. Acad. of Eng. Sciences*

Jung I 1973 *Gustaf de Laval—the flexible shaft and the gas turbine* de Laval Memorial Lecture *Roy. Swed. Acad. of Eng. Sciences*

Jung I 1982 *The marine turbine* 1897–1927, Maritime monograph 50 (London)
Jung I 1987a *The marine turbine* 1928–1980, Maritime monograph 60 (London)
Jung I 1987b *The marine turbine* 1928–1980, Maritime monograph 61 (London)
Kemp P K 1960 (editor) *The Papers of Admiral Sir John Fisher* part 1 (London)
Kennedy A B W 1898 *The Mechanics of Machinery* (London)
Kilgour C M 1844 A letter entitled *Discoveries in the Moon* in A Guide to Life number 17, 11 May (London)
King H C 1980 *The telescope* (London)
Kipling R Letter in the Carpenter Collection, Library of Congress, Washington DC, USA
Kline R R 1992 *Steinmetz 1865–1923* (London)
Langley S P 1889 The temperature of the Moon *Mem. Nat. Acad. Sci.* **4** 195
Laplace P-S 1802 *Mecanique Celeste livre* 3 No 43 *'De la figure des corps celestes'*
Latorre R and Hongo K 1989 Relevant features from the benchmark period of Japanese engineering education (1866–1900) *Int. J. Appl. Eng. Ed.* **5** 341–9
Law A H 1939 Reminiscences *Heaton Works J.* **3** 311
Lecky W E H 1972 *A history of Ireland in the eighteenth century* (Chicago)
Leedham-Green 1998 private communication
Leonard W S 1911 *Machine-shop tools and methods* (New York)
Leijon M. 1998 Powerformer—a radically new rotating machine *ABB Review* **2** 21–6
Leyland C J 1935 Turbinia Jottings *Heaton Works J.* 25–31
Lightie D 1994 The beginnings and early history of Reyrolle *Conf. Newcomen Soc. (Newcastle upon Tyne)*
Liversidge D B and Fearn G A 1967 The origin, detection and identification of defects in steel forgings *Brit. J. Non-Destructive Testing* **9**
Lonsdale K 1962 Further comments on attempts by H Moisson, J B Hannay and Sir Charles Parsons to make diamonds in the laboratory *Nature* **196** 104–6
Mackay R F 1973 *Fisher of Kilverstone* (Oxford)
Mackie G C and Hutchinson K W 1997 Turbinia—state of the art? Then and now RINA *Proc.Conf. on Marine Propulsion* pp 1–19
Macleod R M 1971 Of medals and merit: a reward system in Victorian science 1826–1914 *Notes and records of the Roy. Soc. of London* **26** 91
Macleod R and Collins P (ed) 1981 *The Parliament of Science* (London)
Macpherson H 1926 *Modern Astronomy* (London)
Malcomson A P W(a) 1982 *The pursuit of the heiress—aristocratic marriage in Ireland 1750–1820* (Belfast)
Malcomson A P W (b) 1982 Introduction to calendar of the Rosse papers held at Birr Castle *Public Records Office of Northern Ireland*
Mallet R 1856 *On the physical conditions involved in the construction of artillery* (London)
Manning F 1923 *Life of Sir William White* (London)
Manville G E 1968 Two fathers and two sons *Heaton Works J.* **11**
Marder A J 1952 *Fear God and Dread Nought* (London)

Matsumoto Miwao 1990 The structure of technology transfer in the Japanese shipbuilding industry in the first decade of the 20th century: the Navy connection *12th World Congress of Sociology (Madrid)*

Maxwell J C 1867/68 On Governors *Proc. Roy. Soc.* **16** 270–83

Mayr O 1971 Victorian physicists and speed regulation: an encounter between science and technology *Notes and Records of the Roy. Soc. London* **26**

McDowell R B 1964 The Irish Administration 1801–1914 (London)

McDowell R B and Webb D A 1982 *Trinity College Dublin 1592–1952* (Cambridge)

McDowell R B 1997 *The fate of southern Unionists—Crisis and decline* (Dublin)

McKenna-Lawlor S M P 1988 Astronomy in Ireland, in *Science in Ireland* ed Nudds *et al* pp 85–96

McLaughlin P J 1965 *Nicholas Callan* (London)

Meehan P F 1983 *The members of Parliament for Laois and Offaly* (Portlaoise)

Mitsubishi 1991 Details from the official company history, kindly supplied in 1991

Mollier R. 1904 Neue Diagramme zur technischen Wärmlehre *Z. des VDI* 271

Moore P 1971 *The Astronomy of Birr Castle* (Birr)

Moor D I 1984* Ship model experiment tanks *NECIES* **101** 19–31

Neidhöfer G 1992 *The evolution of the synchronous machine* ABB Review (supplement)

Neilson R M 1912 *The steam turbine* (London)

Nicholas D G 1995 A brief history of the marine steam turbine *Trans. I. Mar. E.* **108** 67–88

Noble A 1906 *Artillery and Explosives* (London)

Nudds *et al* 1988 (editors) *Science In Ireland*

Orange A D 1981 The Beginnings of the British Association, in *The Parliament of Science* ed R Macleod and P Collins

Osler A G and Grieve G R 1978 *Turbinia (Report No2)* (London)

Osler A G and Grieve G R 1980a *Sir Charles Parsons' notebook (Report No 1)* (London)

Osler A G and Grieve G R 1980b *The report by Prof J A Ewing* (London)

Osler A G and Grieve G R 1980c *Turbinia (Report No4)* (London)

Padfield P 1973 *Guns at Sea* (London) 239 *et seq*

Pakenham T 1991 *The Scramble for Africa* (London) p 46, 375

Parl P 1847 Board of Works report **52** 159

Parl P 1852 Report of Select Committee of House of Lords to Enquire into the Operation of the Acts relating to the Drainage of Lands in Ireland *Parliamentary Papers* **26** 1

Parl P 1868 *Report of Select Committee appointed to inquire into the Provisions for giving Instruction in Theoretical and Applied Science to the Industrial Classes* p iii, iv

Parsons A D C 1916 Works Organisation *Trans. NECIES* **33** 34–49

Parsons A D C 1939 Reminiscences *Heaton Works J.* **3** 305–8

Parsons C A 1877 Patent 2344 *Improvements in steam engines* (15 June 1877)

Parsons C A and Cross W 1880 Patent 846 *Communicating fluid pressure to moveable machinery*

Parsons C A 1881a Notebook on *Experiments on Rocket Torpedoes* PAR 9

Parsons C A 1881b Notebook on *Gas engine experiments* PAR/10/1

Parsons C A, Parsons R C and Kitson J H 1882 Patent 4797 *Steam and other fluid pressure engines*

Parsons C A 1884 Letter to J Swan, TWAS

Parsons C A 1887 Patent 5312 Improvements in rotary motors

Parsons C A 1888a The compound steam turbine and its theory, as applied to the working of dynamo-electric machines *Trans. NECIES* **4** 127–49

Parsons C A 1888b *Description of the compound steam turbine and turbo-electric generator Proc. I. Mech. E.* 516

Parsons C A 1888c *Proc. Roy. Soc. (London)* **44** 320–3

Parsons C A 1893 *Turbinia design calculations* Science Museum Library PAR12

Parsons C A 1894 *Improvements in the means of supplying forced draught to boiler furnaces when used in connection with steam turbines* Patent 394A applied for 8 January, accepted 8 February 1895

Parsons C A 1896 *Flying Engines*, a letter to the editor of *Nature* **54** 148–9

Parsons C A 1897a The application of the compound steam turbine to the purpose of marine propulsion *Trans. Inst. Naval Architects* **38** 232–42

Parsons C A 1897b *Turbinia Trials Notebook* PAR 13/1

Parsons C A and Stoney G G 1900 Patent 9203

Parsons C A 1902 *Improvements in condensers working in conjunction with air pumps* Patent 840

Parsons C A 1903 The steam turbine and its application to the propulsion of vessels *TINA* **45** 284–311

Parsons C A Stoney G G and Martin C P 1904a The steam turbine as applied to electrical engineering *Proc. IEE* **133** 794–837

Parsons C A 1904b Typescript of evidence PAR37/13

Parsons C A 1904c British Association Meeting, Cambridge **SP(2)**

Parsons C A and Stoney G G 1905 The steam turbine Minutes of *Proc. ICE* **163** 167–230

Parsons C A 1905a Parsons' Marine Steam Turbine *10th Congress at Milan* (Brussels)

Parsons C A 1905b Residential address *Inst. Mar. Eng.*

Parsons C A and Walker R J 1906 The development of the steam turbine *Trans. I. Mar. E.* 5–24

Parsons C A 1907 Some notes on Carbon at High Temperatures and Pressures *Proc. Roy. Soc. (London)* **79A** 532–5

Parsons C A, Stoney G G and Bennett E 1907 Patent 16599 *Improvements in Reflectors for Searchlights and the like* (1907), and 1908 Patent 17881

Parsons C A and Swinton A A C 1908 The Conversion of Diamond into Coke in High Vacuum by Cathode Rays *Proc. Roy. Soc. (London)* **80A** 184–5

Parsons C A and Stoney G G 1909 *Improvements in and relating to reflectors for projectors and the like* Patent 16550

Parsons C A 1909 Speech on 15/12/09 at Rutherford College Prizegiving PAR5/1

Parsons C A 1910 The application of the marine steam turbine and mechanical gearing to merchant ships *TINA* **52** 168–83. Also Report and *Accounts of PMST Co, 5th July 1912* PAR38/12

Parsons C A 1911a *The Steam Turbine* Reid Lecture (Cambridge)

Parsons C A 1911b Letter to Brown, 1 September PAR 39/3

Parsons C A 1911c Contribution to discussion on paper by R E Froude, The acceleration in front of a propeller *TINA* **53** 158

Parsons C A and Walker 1911 Twelve months' experience with geared turbines in the cargo steamer Vespasian *TINA* **53** 29–36

Parsons C A and Cook S S 1911 Experiments on the compression of liquids at high pressures *Proc. Roy. Soc. (London)* **85A** 332–48

Parsons C A 1913 Mechanical gearing for the propulsion of ships *TINA* **55** pt 1 48–68

Parsons C A 1914 On receiving Freedom of Newcastle upon Tyne *Trans. NECIES* **30** 582–593

Parsons C A 1915a Letter Parsons to L Smith 16/5/15 (NECIES archive)

Parsons C A 1915b *An introduction to a proposed scheme* (NECIES archive)

Parsons C A 1918a The formation of diamond *Proc. Inst. Metals* **20** 5–24

Parsons C A 1918b Bakerian Lecture—Experiments on the Artificial Production of Diamond *Phil. Trans. Roy. Soc. (London)* **220A** 67–107

Parsons C A and Cook S S 1919 Investigations into the causes of corrosion or erosion of propellers *TINA* **61** 232–47

Parsons C A 1919 Presidential Address to Bournemouth meeting of the British Association **SP(2)** 131–50

Parsons C A 1920 Typescript *Some reminiscences of early days of turbine development* extract from *J. Franklin Inst.* **190** PAR 4/1

Parsons C A 1921a *Means for producing very high pressures* Patent 169274, applied for 22 June 1920, accepted 22 September 1921

Parsons C A 1921b Letter to Sir Ambrose Fleming, quoted in Appleyard p 204

Parsons C A 1922 Contribution to Commemorative Meeting *JIEE* **60** 411–3

Parsons C A, Cook S S and Duncan H M 1923 Mechanical Gearing *TINA* **65** 60–90

Parsons C A 1924a The steam turbine *Proc. Int. Math. Congress, Toronto* 465–72

Parsons C A 1924b Steam turbines *First World Power Conference (London)* section D

Parsons C A 1926a *The Scientific Papers of William Parsons* (1926) published by Sir C A Parsons

Parsons C A 1926b Response to presentation of Kelvin Medal at the Institution of Civil Engineers PAR 4/1

Parsons C A 1927a Typescript of reminiscences 6 September 1927 PAR4/1

Parsons C A 1927b Some investigations into the cause of erosion of the tubes of surface condensers *TINA* **69** 1–28

Parsons C A and Duncan H M 1929 A new method for the production of sound steel *J. Iron and Steel Inst.* 255–304

Parsons C A and Rosen J 1929 Direct generation of alternating currents at high voltages *JIEE* **67** 1065

Parsons G L 1939 Reminiscences *Heaton Works J.* **3** 302

Parsons K 1919 Women's work in engineering and shipbuilding during the war *Trans. NECIES* **35** 227–36

Parsons L 1795 *Observations on the Bequest of Henry Flood Esq to Trinity College Dublin, with a Defence of the Ancient History of Ireland* (Dublin)

Parsons L 1834 *An argument to prove the truth of Christian revelation* (London)

Parsons L 1866 Description of an equatorial clock *Monthly Notices of Roy. Astr. Soc.* **26** 265

Parsons L 1868 An account of the observations on the great nebula in Orion made at Birr Castle with the 3-feet and 6-feet telescopes between 1848 and 1867 *III Phil. Trans. Roy. Soc.* p 190

Parsons L 1869 *Proc. Roy. Soc.* **17** 436–43

Parsons L 1870 *Proc. Roy. Soc.* **19** 9–14

Parsons L 1873 On the radiation of heat from the Moon, the law of absorption by our atmosphere, and its variation in amount with her phases *Phil. Trans. Roy. Soc.* 143

Parsons L 1877 *Nature* 20 Sept

Parsons L 1880a Observations of nebulae and clusters of stars made at Birr 1848–1878 *Sci. Trans. Royal Dublin Society* **2** series 2, 1880–1882 with appendix by G Johnstone Stoney

Parsons L 1880b *Nature* **22** 75

Parsons L 1880c On some recent Improvements made in the Mountings of the Telescopes at Birr Castle *VI Phil. Trans. Roy. Soc.* 1880 (SP 153–160)

Parsons L 1896 Report of William Coghlan's death *Monthly Notes of the Royal Astronomical Society* **19** No 240

Parsons L 1968 William Parsons third Earl of Rosse *Hermathena* **107** 5 (Dublin)

Parsons N C Undated letter to Clere Parsons in possession of N C Parsons

Parsons N C 1987* 1884: the rebirth of steam power *Proc. IEE* **134 A** 359–68

Parsons N C 1990 Personal recollection

Parsons R *Reminiscences* (printed for private circulation)

Parsons R C 1877a The theory of centrifugal pumps as supported by experiment *Minutes of Proc. ICE* **47** 267–82

Parsons R C 1877b Provisional patent 2330, *Pumps* (14 June 1877), Patent 2331 *Propelling Vessels* (14 June, sealed 11 December 1877)

Parsons R C 1878a *On the working of Punkahs in India as at present carried out by Coolie labour and the same operation effected by machinery* (London).

Parsons R C and Palliser E 1878b patent 1015 *Working of punkhas by compressed air*, patent 1353 *Working of punkhas by fluid pressure*

Parsons R C 1879 On the loss of power in the screw propeller and the means of improving its efficiency *Proc. I. Mech. E.* 588–609

Parsons R C 1880 Patent 1267 *Water pressure engines*

Parsons R H 1936 *The development of the Parsons steam turbine* (London)

Parsons R H 1940 *The early days of the power station industry* (Cambridge)

Parsons W 1828a Account of a new reflecting telescope *Edin. J. of Sc.* **IX** 25

Parsons W 1828b Account of apparatus for grinding and polishing the specula of reflecting telescopes *Edin J. of Sc.* **IX** 213

Parsons W 1830 Account of a series of experiments on the construction of large reflecting telescopes *Edin J. of Sc.* **II** 136

Parsons W 1840 An account of experiments on the reflecting telescope *XXII Phil. Trans. Roy. Soc.*

Parsons W 1843 Address to the Parsonstown Agricultural Society, quoted by M Hogan in *The Great Famine, Birr and District* 1996, Birr Historical Society

Parsons W 1847 *Letters on the State of Ireland* (London)

Parsons W 1850 Observations on the Nebulae *XXV Phil. Trans. Roy. Soc. (London)*

Parsons W 1851 Plain specula of silver *BA report* 12 (SP(1))

Parsons W 1854a *Monthly Notices of the Roy. Astron. Soc.* **XIV** 199

Parsons W 1854b Presidential address to Royal Society, 30 November 1854 (SP(1))

Parsons W 1859 Presidential address to the Mathematics and Physics division of the BA (SP(1))

Parsons W 1861 On the construction of specula of six-feet diameter; and a selection of observations of nebulae made with them *Phil. Trans. Roy. Soc. (London)* XXVIII 681–745

Perry J 1900 *The steam engine* (London)

Petree J F 1963 *Henry Maudslay* in Engineering Heritage (London)

Playfair L 1888 *Nineteenth Century* **24** 325–33

Pollard S and Robertson P 1979 *The British Shipbuilding Industry 1870–1914* (Harvard)

Preece R 1982 The Durham Students of 1838 *Trans. Architectural and Archaeological Society of Durham and Northumberland* **6** 71–4

Prest G S 1903a Letters to C A Parsons PAR50/22

Prest G S 1903b Letter to C A Parsons PAR35/7

Prest G S 1903c Letter to C A Parsons PAR36/46

Prest G S 1911 Letter to Parsons, 30 September PAR39/3

Prest G S 1931 *Notes prepared for R Dowson* PAR4/2

Rankine W J M 1869a On the outflow of steam *Engineer* **28** 352

Rankine W J M 1869b Engineer **27** 249

Rankine W J M 1881 *Miscellaneous Scientific Papers* (London)

Rateau A 1904 Different applications of steam turbines *Proc. I. Mech. E.* 737–85

Rayleigh 1917 (3rd Baron) On the pressure developed in a liquid during the collapse of a spherical cavity *Phil. Mag.* **34** 94–8

Rayleigh 1934 (4th Baron) *Some personal reminiscences of Sir Charles Parsons* in SP (2) xv- xxviii

Redmayne R A S 1942 *Men, Mines and Memories* (London)

Reynolds O 1886 On the theory of lubrication and its application to M. Beauchamp Tower's experiments *Phil. Trans. Roy. Soc.* **177** 157–234

Reynolds O 1875 Patent 724 *Obtaining Motive Power* 21 August

Reynolds M C 1965 Letter in possession of Prof J D Jackson of University of Manchester

Rhodes R 1986 *The making of the atomic bomb* (London)

Richardson A 1911 *The Evolution of the Parsons Steam Turbine* (London)

Robinson L J 1984 (editor) *Astronomical Scrapbook* (Cambridge)

Robinson M 1903 This collection of letters and reports is held by GEC Alstom at their archive in Rugby

Robinson T R 1845 On Lord Rosse's Telescope *Proc. RIA* **III** 114–33

Robinson T R 1848 On Lord Rosse's Telescope *Proc. RIA* **IV** 119–28

Rosen J. 1922 Some problems in high speed alternators, and their solution *JIEE* **60** 439–58

Rossini F D Jessup R S 1938 *J. Res. Natl. Bur. Stand.* **21** 491

Scaife W G 1984 Parsons' 1884 patents and his prototype turbo-generator *First Parsons Int. Turbine Conf. (Dublin, 1984)* 5–12

Scaife W G 1985 The Parsons Steam Turbine *Scientific American* **252** 132–9

Scaife W G 1988 Charles Parsons' experiments with rocket torpedoes—the precursors of the steam turbine *Trans. Newcomen Soc.* **60** 17–29

Scaife W G 1997 Technical education and the application of Technology in Ireland 1800–1950 In *Science and Society in Ireland 1800–1950* ed P J Bowler and N Whyte (Belfast)

Scott J D 1962 *Vickers* (London)

Siemens Werner von, 1892 *Inventor and Entrepreneur, Recollections of Werner von Siemens* republished (London) 1966

Siemens 1997 *Parsons, the pioneers of power* (Siemens Power Generation Ltd)

Sisson G M 1968 Sir Charles Parsons and astronomy *Trans. NECIES* **85** 27–34

Smith F E 1936* Sir Charles Parsons and Steam *Trans. NECIES* **53** 35

Smith K 1996 *Turbinia* (Newcastle upon Tyne)

Smith R A 1989 *Parsons and the Cambridge Electric Supply Company* (London) 2nd PITC 5–10

Sobel D 1996 *Longitude* (London)

Somerscales E F C 1991 The vertical Curtis steam turbine *Trans. Newcomen Soc.* **63** 1–52

Sothern J W M 1909 *The Marine Steam Turbine* (Glasgow).

SP (1) 1926 *Scientific Papers of the Third Earl of Rosse* ed C A Parsons

SP (2) 1934 *Scientific papers and addresses of the Hon Sir Charles A Parsons* ed Hon G L Parsons (Cambridge)

Speakman E M 1906 The determination of the principal dimensions of the steam turbine with special reference to marine work *Trans. Inst. of Eng. and Shipbuilders in Scotland* **49** 2–40

Stodola A 1927 *Steam and Gas Turbines* reprinted 1945 (New York) vol 1

Stoney G G 1897 letter to Charles Parsons PAR 29/1

Stoney G G and Law A H 1908 High speed electrical machinery *JIEE* **41** 286–329

Stoney G G 1909a Steam Turbines *J. Roy. Soc. Arts* **57** 951–60, 964–77, 983–9

Stoney G G 1909b Letter to Hon C A Parsons CB, 22/5/09 PAR5

Stoney G G 1913 High speed bearings *Trans. NECIES* **30** 218–58

Stoney G G 1914 The effect of vacuum in steam turbines *Proc. I. Mech. E.* 741–59

Stoney G G 1917 Letter dated 28/6/17 to Prof J M Garnett Archive at UMIST

Stoney G G 1925 Letter to Sir Charles Parsons 15/4/25 PAR4/1

Stoney G G 1931 The late Sir Charles Parsons—an Appreciation *Engineering* 47

Stoney G G 1931b Contribution to paper by C D Gibb 'Post War Land Turbine Development' *Proc. I. Mech. E.*

Stoney G G 1933a Some reminiscences *Heaton Works J.* December 135–9

Stoney G G 1933b The development of the Parsons Mirror from 1886 to 1930, typescript PAR8

Stoney G G 1938* Scientific activities of the late Hon Sir Charles A Parsons *JIEE* **82** 248–64

Stoney G G 1939 Reminiscences *Heaton Works J.* **3** 309

Strandh S 1989 *The History of the Machine* (London)

Swade D 1991 *Charles Babbage and his Calculating Engines* (London)

Swinton C A A 1930 *Autobiographical and other writings* (London)

Taylor F W 1912 *Shop Management* (London)

Thomson J J 1931 Wartime work on the Board of Inventions and Research *The Times* 16 February

Thornycroft J I 1883 Efficiency of guide blade propellers *TINA* **24** 42–54

Tubridy M 1998 *Reconstruction of the Rosse Six-foot Telescope* (Birr)

Turnbull C 1939 Reminiscences *Heaton Works J.* 309

TWASa Minute book no 2 Tyne and Wear Archive, Newcastle upon Tyne

TWASb Dividend book for C A Parsons and Co Ltd TWAS

TWASc *Private transfer Journal No1* TWAS

TWASd *Board meeting agenda for 1925* TWAS

TWASe *Board Minutes and Agenda 1918* PA 2402/2

TWASf *Board Minutes and Agenda 1923/4* PA 2402/2

TWASg 1927 *Board meeting agenda for 14/12/27* TWAS 2402/1/2

TWASh *Mitsu Bishi License Book* TWAS 2402/15/16

TWASi *Board meeting agenda for 12/1/27* TWAS 2402/1/2

TWASj 1927 *Board Agenda—Stoney's report on research at Heaton 2/8/26 to 31/12/27*

Vincze S A 1985 101 years of HVAC power transmission *Electronics and Power* 551–2

Wadagaki Yasuzo 1909 On the adaptation of steam turbines for the propulsion of vessels at moderate speeds *Trans. NECIES* **25** 263–74

Walshaw A C and Jobson D A 1972 *Mechanics of Fluids* (London) p 156 *et seq*

Warren M D 1987 *Mauretania* (Wellingborough)

Wayman P A 1986 A visit to Canada in 1884 by Sir Robert Ball *Irish Astron. J.* **17** 185–96

Wayman P A 1987 *Dunsink Observatory 1785–1985* (Dublin)

Wayman P A 1995 Counterpoises of the Birr 6-foot telescope *Irish Astr. J.* 213–5

Westgarth T 1899 On Works Organisation *Trans. NECIES* **15** 87–123

Wilson C T R 1925 Letter to Sir Charles Parsons 10/3/25 PAR 4/1

Woods J L 1984 Rival forms of Steam Turbine *Trans. Newcomen Soc.* **56** 33–8

Woods T 1845 *The monster telescopes erected by the Earl of Rosse, Parsonstown* (Parsonstown)

Woodward G 1992 Electricity in Victorian Liverpool 1851-1901 *Eng. Sci. and Educ. J.* 183–91

Wright O 1988 *How we invented the Airplane* (New York)

Yarrow A F 1904 Copy letter with Report by M Robinson to Directors of Willans and Robinson Ltd, GEC-Alstom Archive, Rugby

Index

on Keynes UK
m Content Group UK Ltd.
W021930071024
27UK00022B/1752

9 780367 400057